VOLUME TWO HUNDRED AND TWENTY FOUR

ADVANCES IN IMAGING AND ELECTRON PHYSICS

Quadrupoles in Electron Lens Design

VOLUME TWO HUNDRED AND TWENTY FOUR

ADVANCES IN IMAGING AND ELECTRON PHYSICS

Quadrupoles in Electron Lens Design

PETER W. HAWKES
Formerly Cavendish Laboratory, Cambridge, England

ELSEVIER

ACADEMIC PRESS
An imprint of Elsevier

Cover photo credit: **Fig. 1** (Above) Electrostatic quadrupoles and octopoles. (Left) Quadrupole possessing both symmetry and antisymmetry. (Right) Octopole possessing both symmetry and antisymmetry. *By Peter W. Hawkes.*
Fig. 2 (Below) Magnetic quadrupoles and octopoles. (Left) Quadrupole possessing both symmetry and antisymmetry. (Right) Octopole possessing both symmetry and antisymmetry. *By Peter W. Hawkes.*

Academic Press is an imprint of Elsevier
125 London Wall, London EC2Y 5AS, United Kingdom
525 B Street, Suite 1650, San Diego, CA 92101, United States
50 Hampshire Street, 5th Floor, Cambridge, MA 02139, United States
The Boulevard, Langford Lane, Kidlington, Oxford OX5 1GB, United Kingdom

ISBN: 978-0-323-98865-0
ISSN: 1076-5670

For information on all Academic Press publications
visit our website at https://www.elsevier.com/books-and-journals

Publisher: Zoe Kruze
Acquisitions Editor: Jason Mitchell
Developmental Editor: Jhon Michael Peñano
Production Project Manager: Vijayaraj Purushothaman
Cover Designer: Greg Harris

Typeset by STRAIVE, India

Contents

Preface

Theoretical electron optics, in its early times, concentrated on the examination and evolution of axially symmetrical systems. I do not mean to imply that systems without this symmetry were completely neglected; after all, the classical work of Störmer is living proof for the very early development of systems without practically any symmetry at all. The more detailed investigation of dipole-type systems started, however, only after the axially symmetrical ones were put on a firm footing and the impetus to examine quadrupole and octopole systems came partially from Scherzer's classic paper (Scherzer, 1947) and partly from the requirements of high energy physics.

The present review of quadrupoles in electron lens design by Dr. Hawkes is oriented essentially toward "low" energy applications, although the equations are relativistically correct. It is my conviction that this excellent review, containing much of the author's own fine work, will become as much of a classic in this limited field as Glaser's treatise is for axially symmetrical systems.

L. Marton
August, 1969
Washington, DC

CHAPTER ONE

Introduction

Quadrupole studies have now reached a point at which we know that corrector units for reducing the spherical aberration of a round lens can be successfully designed and that good quadrupole projector lenses can be built. We know too that quadrupoles can be made achromatic and that their chromatic aberration can be made opposite in sign to that of a round lens. Furthermore, quadrupole multiplets can be built that, especially at high accelerating voltages, rival or surpass the focusing properties of good round objective lenses and have lower spherical and chromatic aberrations. Why then are quadrupole correctors not in common use and why have the potentialities of quadrupoles not been exploited more? There are two main reasons for this disappointing situation. The first is a very excellent one: quadrupoles are difficult to use, the number of variables in any quadrupole system is high, and every experimental study of these lenses has shown how vexatious is the problem of mechanical tolerances. Nevertheless, the rewards to be gained from a successful quadrupole system are very great, and it seems unlikely that they can be achieved by any other method with less difficulty. The second reason, with which we are primarily concerned in this review, is the lack of a convenient collection of data about quadrupoles and of guidance about the various quadrupole multiplets. A number of books, chapters, and review articles have been written about quadrupoles, but a large amount of numerical material has only recently become available, and we have made some attempt to bring all this together in the hope that designers of electron optical systems may at least be able to decide whether they cannot usefully contemplate using quadrupoles and octopoles. The first monograph on quadrupoles and related systems was published by Strashkevich (1959); only electrostatic lenses are considered, and much of the book is devoted to potential problems. In 1961, Septier collected most of the experimental results then available and the basic theory into a long and lucid review article for these *Advances*, and 5 years later, he described more recent work in a review of methods of overcoming spherical aberration

Advances in Imaging and Electron Physics, Volume 224
ISSN 1076-5670
https://doi.org/10.1016/B978-0-323-98865-0.00001-7

(Septier, 1966). In the same year, we published a short monograph[a] on quadrupole optics (Hawkes, 1966a) in an attempt to provide a coherent account of the theory together with some experimental results. Meanwhile, quadrupoles had been discussed from the point of view of strong-focusing lenses by Steffen (1965) and Banford (1966). A Russian monograph on quadrupoles has recently been published (Yavor, 1968). Several textbooks of electron optics devote space to quadrupoles (Bonshtedt & Markovich, 1967; Glaser, 1956a; Grivet, 1965; Kel'man & Yavor, 1963b; Picht, 1963; Septier, 1967; Strashkevich, 1966; Sturrock, 1955). The list of references in QO is reasonably complete to autumn 1966, and in the present review, we have concentrated upon more recent work; for details of reviews and papers published before 1966, therefore, QO should be consulted as we only mention key papers below.

Our object, then, is to provide guidance for designers of electron optical systems who may wish to use quadrupoles, and to recapitulate in detail the solutions that have been proposed of the attendant problems. We begin by defining the relevant parameters and listing formulae for the aberration coefficients that concern us. Both round lenses and quadrupoles are included in these lists, as a few of the formulae have remained unpublished and we feel that the quadrupole analysis is easier to follow when the more familiar round lens equations are given in parallel. In the next sections, we describe the various methods of establishing the potential functions that characterize quadrupole behavior and discuss which types of systems are more suitable for the various roles they may play. The final section contains a large collection of data about individual quadrupoles and quadrupole systems in tabular and graphical form. Several of the tables have been computed for this review, and have not been published elsewhere.

[a] This is referred to in the text as QO.

CHAPTER TWO

Definitions, notation, and methods of analysis

Contents

1. Equations of motion

The behavior of electrons in the electric or magnetic fields of well-aligned electron lenses is completely determined by the magnetic field on the optic axis and by the electrostatic and magnetostatic potential functions. In round magnetic lenses, the trajectories are wholly defined by $B_z(0,0,z)$, which we write $B(z)$, where the z-axis of a Cartesian coordinate system coincides with the optic axis. The paraxial equations of motion in the familiar rotating coordinate system are then

$$\frac{d^2x}{dz^2} + \frac{1}{4}\eta^2\frac{B^2}{V_0}x = 0 \tag{1}$$

with a similar equation for y. It is often convenient to write $\Omega(z) = \eta B(z)/2V_0^{1/2}$ so that Eq. (1) becomes

$$x'' + \Omega^2 x = 0 \tag{2}$$

Advances in Imaging and Electron Physics, Volume 224
ISSN 1076-5670
https://doi.org/10.1016/B978-0-323-98865-0.00002-9

where primes signify differentiation with respect to z. V_0 represents the relativistically modified accelerating voltage,

$$V_0 = \Phi_0(1 + \varepsilon\Phi_0) \tag{3}$$

in which Φ_0 is the real accelerating voltage; η and ε are constants:

$$\begin{aligned} \eta &= (e/2m_0)^{1/2} \simeq 3 \times 10^5 \mathrm{C}^{1/2}\mathrm{kg}^{-1/2} \\ \varepsilon &= e/2m_0c^2 \simeq 10^{-6}\mathrm{V}^{-1} \end{aligned} \tag{4}$$

where $-e$ and m_0 are the charge and rest mass respectively of the electron. When relativistic effects are negligible, we set ε equal to zero.

The potential in an electrostatic system possessing two planes of symmetry may be expanded in the form

$$\begin{aligned} \phi(x, y, z) = \Phi(z) &- \frac{1}{4}\Phi''\left(x^2 + y^2\right) + \frac{1}{64}\Phi^{(\mathrm{iv})}\left(x^2 + y^2\right)^2 - \ldots \\ &+ \frac{1}{4}D(z)\left(x^2 - y^2\right) - \frac{1}{48}D''\left(x^4 - y^4\right) + \ldots \\ &+ D_1(z)\left(x^4 - 6x^2y^2 + y^4\right) - \ldots + \ldots \end{aligned} \tag{5a}$$

or

$$\begin{aligned} \phi(r, \theta, z) &= \Phi(z) - \frac{1}{4}\Phi''r^2 + \frac{1}{64}\Phi^{(\mathrm{iv})}r^4 - \ldots \\ &\quad + \frac{1}{4}\left\{D(z) - \frac{1}{12}D''r^2 + \ldots\right\}r^2\cos 2\theta \\ &\quad + \{D_1(z) - \ldots\}r^4\cos 4\theta + \ldots \end{aligned} \tag{5b}$$

$$\begin{aligned} &\equiv \phi_0\left(r^2, z\right) + \phi_2\left(r^2, z\right)\cos 2\theta + \phi_4\left(r^2, z\right)\cos 4\theta \\ &\quad + \phi_6\left(r^2, z\right)\cos 6\theta + \ldots \end{aligned} \tag{5c}$$

The functions $\Phi(z)$, $D(z)$, $D_1(z)$, ... are each characteristic of particular types of element. In round electrostatic systems, the potential distribution is characterized by $\Phi(z)$ alone; in electrostatic quadrupoles possessing a plane of antisymmetry as well as a plane of symmetry (Fig. 1A), only ϕ_2-, ϕ_6- and $\phi_{(4n+2)}$- terms occur, so that if we halt the series expansion at terms in r^4, only $D(z)$ is required to specify the potential distribution. In symmetrically excited octopoles (Fig. 1B), the potential is characterized by terms in ϕ_4, ϕ_{12} and $\phi_{(8n+4)}$ so that the principal potential function is $D_1(z)$. Electrostatic quadrupoles not possessing planes of antisymmetry (Fig. 1C) require $\Phi(z)$, $D(z)$ and $D_1(z)$ in their potential expansions.

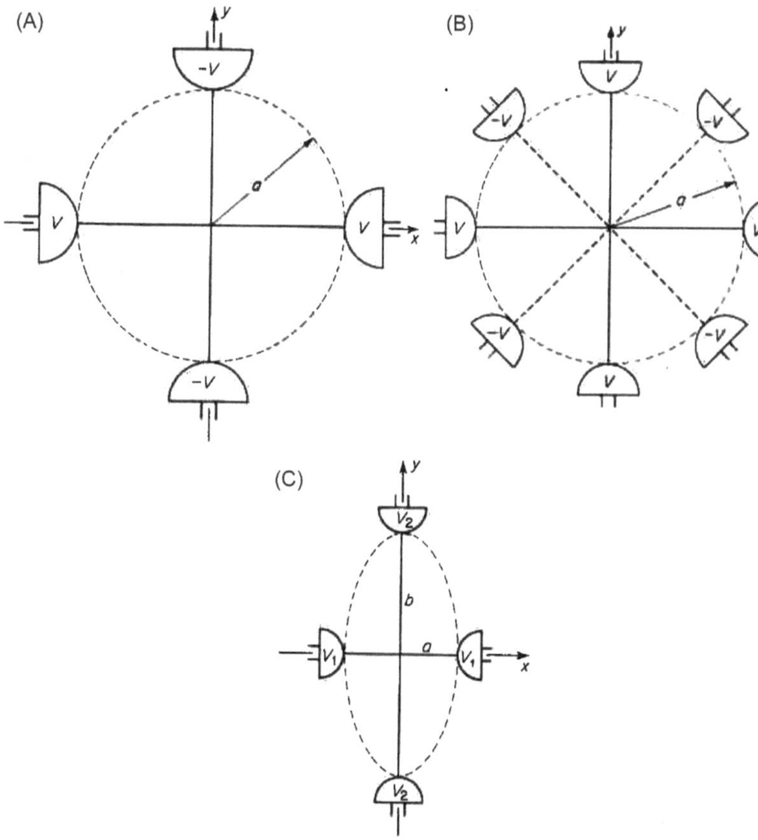

Fig. 1 Electrostatic quadrupoles and octopoles. (A) Quadrupole possessing both symmetry and antisymmetry. (B) Octopole possessing both symmetry and antisymmetry. (C) Quadrupole not possessing antisymmetry.

The magnetostatic potential $\psi(x, y, z)$, $\mathbf{B} = -\operatorname{grad}\psi$, may be expanded similarly; in a system with planes of antisymmetry, we have

$$\psi(x, y, z) = Q(z)xy - \frac{1}{12}Q''xy\left(x^2 + y^2\right) + \ldots + 4Q_1(z)xy\left(x^2 - y^2\right) - \ldots \quad \text{(6a)}$$

or

$$\psi(r, \theta, z) = \left\{\frac{1}{2}Q(z) - \frac{1}{24}Q''r^2 + \ldots\right\}r^2 \sin 2\theta + \{Q_1(z) - \ldots\}r^4 \sin 4\theta \quad \text{(6b)}$$

$$\equiv \psi_2\left(r^2, z\right)\sin 2\theta + \psi_4\left(r^2, z\right)\sin 4\theta + \psi_6\left(r^2, z\right)\sin 6\theta + \ldots \quad \text{(6c)}$$

In a round lens system, we employ not the potential functions $Q(z)$, $Q_1(z)$, … but the axial magnetic field $B(z) \equiv B_z(0, 0, z)$. The function $Q(z)$ characterizes the potential in a symmetric quadrupole (Fig. 2A) while $Q_1(z)$ characterizes that in a symmetric octopole (Fig. 2B). When the magnetic poles

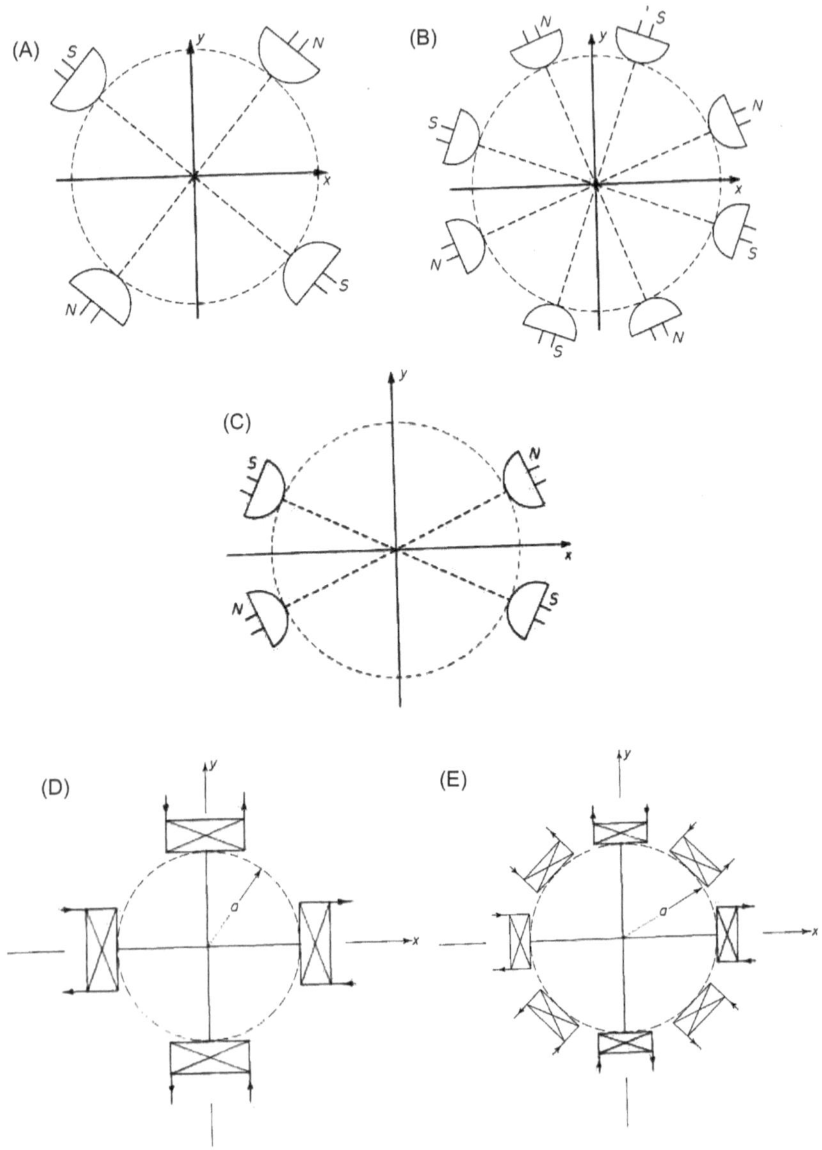

Fig. 2 Magnetic quadrupoles and octopoles. (A) Quadrupole possessing both symmetry and antisymmetry. (B) Octopole possessing both symmetry and antisymmetry. (C) Quadrupole not possessing symmetry. (D) Tangential solenoid quadrupole. (E) Tangential solenoid octopole.

are not inclined to one another at right angles, both $Q(z)$ and $Q_1(z)$ are required (Fig. 2C). Schleich (1960) and more recently, Schleich, Hoffmeister, Koops and Lenz (1968) have discussed a different type of magnetic quadrupole and octopole design, in which the field is produced by solenoids tangent to the lens bore (Fig. 2D and E). These are the exact analogues of the electrostatic elements of Fig. 1, and they allow an octopole component to be introduced with a quadrupole in an equally simple way.

In Figs. 1 and 2, the Cartesian axes *Ox* and *Oy* are shown intersecting the electrostatic electrodes and passing midway between the magnetic poles, since these are the configurations that allow the paraxial equations of motion to be separated. (A more detailed analysis is given in QO.) We shall refer to this as the *standard orientation.*

In the discussion of the parasitic aberrations of quadrupoles, we shall require the terms in the potential expansion that arise when the symmetry is imperfect. The general form of Eqs. (5a)–(5c) and (6a)–(6c) may be written

$$\begin{aligned}\phi(r,\theta,z) &= \phi_0 + \phi_{1c}\cos\theta + \phi_{1s}\sin\theta + \phi_{2c}\cos 2\theta + \phi_{2s}\sin 2\theta \\ &\quad + \phi_{3c}\cos 3\theta + \phi_{3s}\sin 3\theta + \phi_{4c}\cos 4\theta + \phi_{4s}\sin 4\theta + \ldots\end{aligned} \tag{7a}$$

$$\psi(r,\theta,z) = \psi_0 + \sum_{n\geqslant 1}(\psi_{nc}\cos n\theta + \psi_{ns}\sin n\theta). \tag{7b}$$

The paraxial equations of motion of electrons passing through a system of electrostatic round lenses and electrostatic and magnetic quadrupoles in the standard orientation take the following form (octopoles do not affect the primordial properties of optical systems):

$$x'' + \sigma\frac{\Phi'}{2V}x' + \left(\sigma\frac{\Phi''}{4V} - \frac{\sigma D - 4\eta Q V^{1/2}}{4V}\right)x = 0 \tag{8a}$$

$$y'' + \sigma\frac{\Phi'}{2V}y' + \left(\sigma\frac{\Phi''}{4V} + \frac{\sigma D - 4\eta Q V^{1/2}}{4V}\right)y = 0 \tag{8b}$$

where

$$\begin{aligned}\sigma &= 1 + 2\varepsilon\Phi \\ V &= \Phi(1+\varepsilon\Phi).\end{aligned} \tag{9}$$

In the absence of round electrostatic lenses and of any round lens component arising from asymmetrically excited quadrupoles, these equations simplify to

$$\begin{aligned}x'' - \frac{\sigma D - 4\eta Q V_0^{1/2}}{4V_0}x &= 0 \\ y'' + \frac{\sigma D - 4\eta Q V_0^{1/2}}{4V_0}y &= 0.\end{aligned} \tag{10}$$

In most quadrupole systems, electrostatic and magnetic quadrupoles are not used simultaneously—the exception occurs when achromatism is desired—and it is convenient to write

$$\sigma\frac{D(z)}{4V_0} = \beta_E^2 k(z) \tag{11a}$$

$$\frac{\eta Q(z)}{V_0^{1/2}} = \beta_M^2 k(z) \tag{11b}$$

$$\beta_M^2 - \beta_E^2 = \beta^2 \tag{11c}$$

$$n = \beta_E^2/\beta^2 = \beta_M^2/\beta^2 - 1 \tag{11d}$$

in which $k(z)$ is a normalized "shape function" having a maximum value of unity. We find

$$\sigma D(z) = 4V_0\beta^2 nk(z);\ \eta Q(z) = V_0^{1/2}\beta^2(n+1)k(z) \tag{11e}$$

$$\frac{\sigma D - 4\eta Q V_0^{1/2}}{4V_0} = -\beta^2 k(z) \tag{11f}$$

and the equations of motion (10) become

$$\begin{aligned} x'' + \beta^2 k(z)x &= 0 \\ y'' - \beta^2 k(z)y &= 0. \end{aligned} \tag{12}$$

In quadrupole systems, therefore, electrons incident in the x–z plane will follow quite different paths from those incident in the y–z plane; the lens will be convergent in one plane and divergent in the other, like a toric glass lens. It is for this reason that quadrupoles must almost always be grouped in multiplets. We refer to these two privileged planes, the x–z plane and the y–z plane when all quadrupoles are in the standard orientation, as the *principal sections* (*Hauptschnitte* in Scherzer, 1965) and the directions Ox, Oy as the *principal directions*.

In the foregoing discussion, we have separated round magnetic lenses from all the other lenses; the reason for this is that when the magnetic field of a magnetic round lens overlaps the field of a quadrupole, the equations of motion can no longer be separated and the coupled equations that take their place can only be solved analytically in special cases (Schiske, 1966–67). If this mathematical inconvenience were the only drawback, we should have recourse to numerical techniques; in fact, however, the absence of planes of symmetry increases the number and variety of the aberrations so that more corrector elements would be needed to annul them (Hawkes, 1965a,b;

Rose, 1966–67a, 1966–67b). The task of canceling the aberrations of orthogonal quadrupole systems (that is, those in which the equations of motion separate) is sufficiently difficult, without inviting this extra complication. This does not of course prevent us from using quadrupole correctors to reduce the spherical aberration of round magnetic lenses; the only requirement is that the magnetic field of the round lens must have fallen to a negligible value when the electrons enter the corrector field. (See, however, Reichenbach & Rose, 1968–69.)

The paraxial equations of motion provide a complete description of the primordial imagery of round lenses and quadrupoles. In order to study geometrical aberrations, we retain higher order terms in the potential expansion. The primary aberrations of a system with a straight optic axis are third order, and the third-order equations of motion in systems of quadrupoles, octopoles and electrostatic round lenses have the form

$$
\begin{aligned}
x'' &+ \frac{\sigma\Phi'}{2V}x' + \frac{\sigma\Phi'' - \sigma D + 4\eta Q V^{1/2}}{4V}x \\
&= \left\{ -\frac{1+\sigma^2}{32}\left(\frac{\Phi''-D}{V}\right)^2 + \frac{\sigma}{32}\frac{\Phi^{(iv)}}{V} - \frac{\sigma}{24}\frac{D''}{V} - \frac{\sigma}{8}\eta Q \frac{\Phi''-D}{V^{3/2}} \right. \\
&\quad \left. + \frac{1}{12}\eta\frac{Q''}{V^{1/2}} + 2\frac{\sigma D_1}{V} - 4\eta\frac{Q_1}{V^{1/2}} \right\} x^3 \\
&\quad + \left(-\frac{1+\sigma^2}{32}\frac{\Phi''^2 - D^2}{V^2} + \frac{\sigma}{32}\frac{\Phi^{(iv)}}{V} - \frac{\sigma}{8}\eta Q\frac{\Phi''+D}{V^{3/2}} \right. \\
&\quad \left. + \frac{1}{4}\eta\frac{Q''}{V^{1/2}} - 6\frac{\sigma D_1}{V} + 12\eta\frac{Q_1}{V^{1/2}} \right) x y^2 \\
&\quad + \left(\frac{\sigma}{8}\frac{\Phi''' - D'}{V} - \frac{1+\sigma^2}{16}\frac{\Phi''-D}{V^2}\Phi' \right) x^2 x' \\
&\quad + \left(\frac{\sigma}{8}\frac{\Phi''' + D'}{V} - \frac{1+\sigma^2}{16}\frac{\Phi''+D}{V^2}\Phi' \right) y^2 x' \\
&\quad + \eta\frac{Q'}{V^{1/2}}xyy' - \frac{\sigma(\Phi''-D) + 6\eta Q V^{1/2}}{4V}xx'^2 \\
&\quad - \frac{\sigma(\Phi''-D) + 2\eta Q V^{1/2}}{4V}xy'^2 + \eta\frac{Q}{V^{1/2}}yx'y' - \frac{\sigma}{2}\frac{\Phi'}{V}x'\left(x'^2 + y'^2\right)
\end{aligned}
\tag{13a}
$$

$$
\begin{aligned}
&y'' + \frac{\sigma\Phi'}{2V'}y' + \frac{\sigma\Phi'' + \sigma D - 4\eta Q V^{1/2}}{4V}y \\
&= \left\{ -\frac{1+\sigma^2}{32}\left(\frac{\Phi''+D}{V}\right)^2 + \frac{\sigma}{32}\frac{\Phi^{(iv)}}{V} + \frac{\sigma}{24}\frac{D''}{V} + \frac{\sigma}{8}\eta Q\frac{\Phi''+D}{V^{3/2}} \right.\\
&\left. -\frac{1}{12}\eta\frac{Q''}{V^{1/2}} + \frac{2\sigma D_1}{V} - 4\eta\frac{Q_1}{V^{1/2}} \right\} y^3 \\
&+ \left(-\frac{1+\sigma^2}{32}\frac{\Phi''^2 - D^2}{V^2} + \frac{\sigma}{32}\frac{\Phi^{(iv)}}{V} + \frac{\sigma}{8}\eta Q\frac{\Phi''-D}{V^{3/2}} \right.\\
&\left. -\frac{1}{4}\eta\frac{Q''}{V^{1/2}} - \frac{6\sigma D_1}{V} + 12\eta\frac{Q_1}{V^{1/2}} \right) x^2 y \\
&+ \left(\frac{\sigma}{8}\frac{\Phi''' - D'}{V} - \frac{1+\sigma^2}{16}\frac{\Phi''-D}{V^2}\Phi' \right) x^2 y' \\
&+ \left(\frac{\sigma}{8}\frac{\Phi''' + D'}{V} - \frac{1+\sigma^2}{16}\frac{\Phi''+D}{V^2}\Phi' \right) y^2 y' \\
&-\eta\frac{Q'}{V^{1/2}}xyx' - \frac{\sigma(\Phi''+D) - 2\eta Q V^{1/2}}{4V}yx'^2 \\
&-\frac{\sigma(\Phi''+D) - 6\eta Q V^{1/2}}{4V}yy'^2 - \eta\frac{Q}{V^{1/2}}xx'y' - \frac{\sigma}{2}\cdot\frac{\Phi'}{V}y'\left(x'^2 + y'^2\right).
\end{aligned} \tag{13b}
$$

In the absence of any round lens component ($\Phi = \text{const.}$), Eqs. (13a–13b) reduce to

$$
\begin{aligned}
&x'' - \frac{\sigma D - 4\eta Q V^{1/2}}{4V}x \\
&= \left\{ -\frac{1+\sigma^2}{32}\left(\frac{D}{V}\right)^2 - \frac{\sigma}{24}\frac{D''}{V} + \frac{\sigma}{8}\eta\frac{DQ}{V^{3/2}} + \frac{1}{12}\eta\frac{Q''}{V^{1/2}} + \Xi \right\} x^3 \\
&+ \left\{ \frac{1+\sigma^2}{32}\left(\frac{D}{V}\right)^2 - \frac{\sigma}{8}\eta\frac{DQ}{V^{3/2}} + \frac{1}{4}\eta\frac{Q''}{V^{1/2}} - 3\Xi \right\} xy^2 \\
&-\frac{\sigma D'}{8V}\left(x^2 - y^2\right)x' + \eta\frac{Q'}{V^{1/2}}xyy' \\
&+ \frac{\sigma D - 6\eta Q V^{1/2}}{4V}xx'^2 + \frac{\sigma D - 2\eta Q V^{1/2}}{4V}xy'^2 + \eta\frac{Q}{V^{1/2}}yx'y'
\end{aligned} \tag{14a}
$$

$$
\begin{aligned}
&y'' + \frac{\sigma D - 4\eta Q V^{1/2}}{4V} y \\
&= \left\{ -\frac{1+\sigma^2}{32}\left(\frac{D}{V}\right)^2 + \frac{\sigma}{24}\frac{D''}{V} + \frac{\sigma}{8}\eta\frac{DQ}{V^{3/2}} - \frac{1}{12}\eta\frac{Q''}{V^{1/2}} + \Xi \right\} y^3 \\
&\quad + \left\{ \frac{1+\sigma^2}{32}\left(\frac{D}{V}\right)^2 - \frac{\sigma}{8}\eta\frac{DQ}{V^{3/2}} - \frac{1}{4}\eta\frac{Q''}{V^{1/2}} - 3\Xi \right\} x^2 y \\
&\quad - \frac{\sigma D'}{8V}\left(x^2 - y^2\right) y' - \eta\frac{Q'}{V^{1/2}} xyx' \\
&\quad - \frac{\sigma D - 2\eta Q V^{1/2}}{4V} yx'^2 - \frac{\sigma D - 6\eta Q V^{1/2}}{4V} yy'^2 - \eta\frac{Q}{V^{1/2}} xx'y'
\end{aligned}
\tag{14b}
$$

where

$$
\Xi = 2\frac{\sigma D_1 - 2\eta Q_1 V^{1/2}}{V}. \tag{15}
$$

In terms of the notation introduced in Eqs. (11a–11f), these become

$$
\begin{aligned}
x'' + \beta^2 kx = -\beta^2 \Bigg[&\left\{ \frac{\beta^2 k^2}{2\sigma^2} n\left(n - \sigma^2\right) + \frac{1}{12} k''(n-1) \right\} x^3 \\
&- \left\{ \frac{\beta^2 k^2}{2\sigma^2} n\left(n - \sigma^2\right) + \frac{1}{4} k''(n+1) \right\} xy^2 \\
&+ \frac{1}{2} nk'\left(x^2 - y^2\right) x' - (n+1)k'xyy' + \frac{1}{2}(n+3)kxx'^2 \\
&- \frac{1}{2}(n-1)kxy'^2 - (n+1)kyx'y' \Bigg] + \Xi x\left(x^2 - 3y^2\right)
\end{aligned}
\tag{16a}
$$

$$
\begin{aligned}
y'' - \beta^2 ky = \beta^2 \Bigg[&\left\{ \frac{\beta^2 k^2}{2\sigma^2} n\left(\sigma^2 - n\right) + \frac{1}{12} k''(n-1) \right\} y^3 \\
&+ \left\{ \frac{\beta^2 k^2}{2\sigma^2} n\left(n - \sigma^2\right) - \frac{1}{4} k''(n+1) \right\} x^2 y \\
&- \frac{1}{2} nk'\left(x^2 - y^2\right) y' - (n+1)k'xyx' + \frac{1}{2}(n+3)kyy'^2 \\
&- \frac{1}{2}(n-1)kyx'^2 - (n+1)kxx'y' \Bigg] + \Xi y\left(y^2 - 3x^2\right).
\end{aligned}
\tag{16b}
$$

Eqs. (16a) and (16b) reduce to those given explicitly by Ovsyannikova and Yavor (1965) and implicitly by Dymnikov, Fishkova, and Yavor (1964a) in the nonrelativistic approximation, when $\sigma = 1$.

2. Cardinal elements

2.1 Round lenses

The paraxial equations of motion of round electrostatic and magnetic lenses are linear, homogeneous, second-order differential equations, and the equations for $x(z)$ and $y(z)$ are identical. The general solution is therefore a linear combination of any pair of linearly independent solutions. If $z = z_o$ is some real object plane and a general ray intersects this at (x_o, y_o), then we may write

$$x(z) = x_o g(z) + \alpha h(z)$$
$$y(z) = y_o g(z) + \gamma h(z) \tag{17}$$

where $g(z)$ and $h(z)$ are solutions of the paraxial equation satisfying the boundary conditions

$$g(z_o) = 1 \quad h(z_o) = 0$$
$$g'(z_o) = 0 \quad h'(z_o) = 1 \tag{18}$$

and $\alpha = x'(z_o)$, $\gamma = y'(z_o)$.

When the object plane lies within the field of the lens, not all of the lens field contributes to the image formation and hence the "effective lens" that creates the image varies with object position. The cardinal elements will in turn be dependent upon z_o and, if the lens is to be used with the object in more than one position, the behavior of the "osculating" cardinal elements must be established. They are defined in Fig. 3. In practice, the objective lens of an electron microscope (which is the only lens considered here that forms

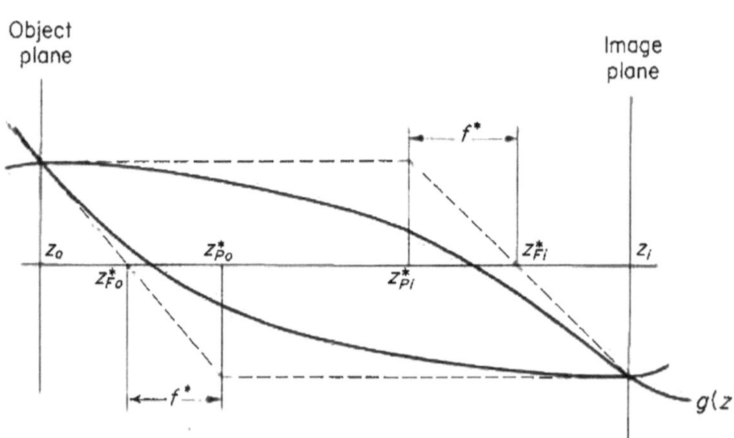

Fig. 3 Osculating cardinal elements.

the image of a real immersed object) is regularly operated at high magnification so that the dependence of the cardinal elements upon z_0 is of little interest. There is, however, a class of fields, Newtonian fields, possessing the property that the cardinal elements are independent of object position. The simplest of these Newtonian fields, Glaser's bell-shaped field, $B(z) = B_0/\{1+(z/d)^2\}$ in the magnetic case, is of considerable importance since it provides a sufficiently faithful representation of the magnetic field in microscope lenses for many purposes and is easily treated analytically.

Of the five or more lenses in an electron microscope, only the objective forms an image of a real object. The intermediate and projector lenses of microscopes and the probe-forming lenses of microanalyzers all form images of virtual objects and their imagery is characterized by the asymptotic cardinal elements. These are illustrated in Fig. 4; they are established in practice by considering the behavior of a pair of solutions of the equations of motion, $G(z)$ and $H(z)$, satisfying the boundary conditions.

$$\begin{aligned} \lim_{z\to-\infty} G(z) &= 1 \\ \lim_{z\to-\infty} H(z) &= z - z_o \end{aligned} \tag{19}$$

where z_0 is now the position of the virtual object. We use the following notation for the asymptotic cardinal elements, and when we need to distinguish them from the real or osculating cardinal elements the latter will be identified by an asterisk:

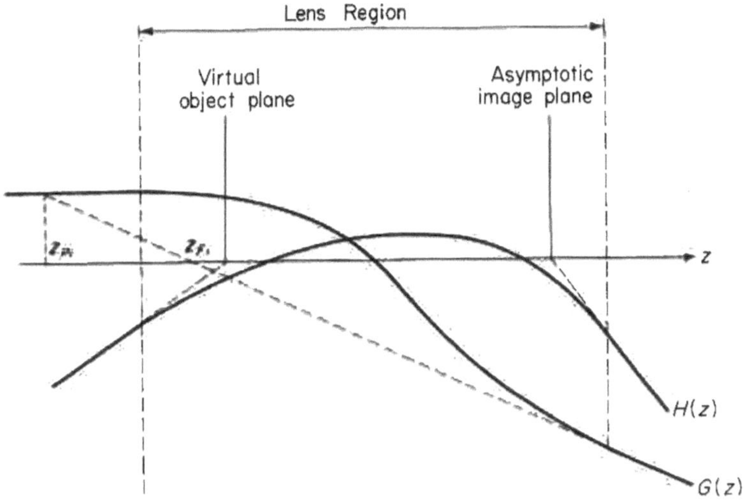

Fig. 4 Construction of asymptotic cardinal elements (round lens).

$$
\begin{aligned}
&\text{Object focus}: \quad z = z_{Fo}\\
&\text{Image focus}: \quad z = z_{Fi}\\
&\text{Object principal point}: \quad z = z_{Po}\\
&\text{Image principal point}: \quad z = z_{Pi}\\
&\text{Focal length } (z_{Po} - z_{Fo} = z_{Fi} - z_{Pi}) : f.
\end{aligned} \tag{20}
$$

The object and image focal lengths are always equal provided the lens has no overall accelerating or decelerating effect; in symmetric lenses, for which $B(-z) = B(z)$, $z_{Fo} = -z_{Fi}$ and $z_{P_o} = -z_{P_i}$.

Part of the complexity of quadrupole analysis stems from the fact that these lenses must always be considered in combination. The numerous parameters can be systematically collected into compact groups with the aid of various matrix notations; we now explain how these different conventions arise and how best to use them, first in the case of round lenses and then in connection with quadrupoles.

The type of transfer matrix most commonly found in the literature relates the position and slope of a ray in some plane ($z = z_1$) to its position and slope in another plane ($z = z_2$). Restricting the discussion to asymptotes, we find

$$
\begin{pmatrix} x_2 \\ x_2' \end{pmatrix} = \begin{pmatrix} A & B \\ C & D \end{pmatrix} \begin{pmatrix} x_1 \\ x_1' \end{pmatrix} \tag{21}
$$

with a similar expression for y, where x_1, $x_1{}'$ and x_2, $x_2{}'$ denote the position and slope of the ray at the planes $z = z_1$ and $z = z_2$, respectively. The matrix elements A, B, C, and D may be written in terms of the cardinal elements of the lens:

$$
\begin{pmatrix} x_2 \\ x_2' \end{pmatrix} = \begin{pmatrix} -\dfrac{z_2 - z_{Fi}}{f} & f\left\{\dfrac{(z_2 - z_{Fi})(z_1 - z_{Fo})}{f^2} + 1\right. \\ -\dfrac{1}{f} & \dfrac{z_1 - z_{Fo}}{f} \end{pmatrix} \begin{pmatrix} x_1 \\ x_1' \end{pmatrix} \tag{22}
$$

Setting $\mathbf{x} = \begin{pmatrix} x \\ x' \end{pmatrix}$ we may write Eq. (22) in the form

$$
\mathbf{x}_2 = \mathbf{T}\mathbf{x}_1 \tag{23}
$$

where

$$\mathbf{T} = \begin{pmatrix} -\dfrac{z_2 - z_{Fi}}{f} & f\left\{\dfrac{(z_2 - z_{Fi})(z_1 - z_{Fo})}{f^2} + 1\right\} \\ -\dfrac{1}{f} & \dfrac{z_1 - z_{Fo}}{f} \end{pmatrix}. \tag{24}$$

From this general matrix, we derive the various forms that lend themselves especially well to specific situations. We mention in passing that if the space between z_1 and z_2 is a field-free region, which is known as a "drift space," the transfer matrix is given by

$$\mathbf{T} = \begin{pmatrix} 1 & D \\ 0 & 1 \end{pmatrix} \tag{25}$$

where $z_2 - z_1 = D$. We note too that det $(\mathbf{T})$ is always unity.

a. *Dušek matrices.* The matrix $\mathbf{T}$ of Eq. (24) is particularly convenient for relating the input and output rays of fields that may be regarded as falling rapidly to zero at a pair of end-planes z_1 and z_2. Electron lens fields fall rather gradually, however, and Dušek (1959) observed that for such fields it is most convenient to relate the position and slope of an incident asymptote in some plane to the position and slope of the outgoing asymptote at the same plane (the mid-plane if the lens is symmetric, for example). Writing $z_1 = z_2 = \bar{z}$, and selecting $z = \bar{z}$ as the origin of coordinates so that $\bar{z} = 0$, we have

$$\mathbf{T} = \begin{pmatrix} \dfrac{z_{Fi}}{f} & \dfrac{z_{Fo}z_{Fi}}{f} + f \\ -\dfrac{1}{f} & -\dfrac{z_{Fo}}{f} \end{pmatrix}. \tag{26}$$

The matrices used by Dušek are slightly different from this, since he writes the incident and emergent asymptotes in the form

$$\begin{aligned} x(z) &= M_-z + N_- \quad \text{incident} \\ x(z) &= M_+z + N_+ \quad \text{emergent} \end{aligned} \tag{27a}$$

and puts

$$\begin{pmatrix} M_+ \\ N_+ \end{pmatrix} = t\begin{pmatrix} M_- \\ N_- \end{pmatrix} \tag{27b}$$

Since $x_1 = N_-$, $x_1{}' = M_-$, $x_2 = N_+$ and $x_2{}' = M_+$,

$$\mathbf{t} = \begin{pmatrix} -\dfrac{z_{Fo}}{f} & -\dfrac{1}{f} \\ \dfrac{z_{Fo}z_{Fi}}{f} + f & \dfrac{z_{Fi}}{f} \end{pmatrix}. \tag{28}$$

A drift space is then characterized by

$$\mathbf{t} = \begin{pmatrix} 1 & 0 \\ D & 1 \end{pmatrix}. \tag{29}$$

b. *Conjugate planes.* If z_2 is conjugate to z_1, and the magnification is $\boldsymbol{M}$, then **T** may be written

$$\mathbf{T} = \begin{pmatrix} M & 0 \\ -\dfrac{1}{f} & \dfrac{1}{M} \end{pmatrix} \tag{30}$$

c. *Regenstreif matrices.* In an attempt to simplify the matrix products that result when quadrupole multiplets are studied, Regenstreif (1966a, 1966b) and Regenstreif (1967a) introduced the matrix relation between the object and image foci: $z_1 = z_{Fo}$ and $z_2 = z_{Fi}$. This yields

$$\mathbf{T} = \begin{pmatrix} 0 & f \\ -\dfrac{1}{f} & 0 \end{pmatrix}. \tag{31}$$

d. *Principal planes.* A special case of the matrix (30) for conjugate planes arises when these are the principal planes. **T** then reduces to

$$\mathbf{T} = \begin{pmatrix} 1 & 0 \\ -\dfrac{1}{f} & 1 \end{pmatrix} \tag{32}$$

and we show below that recurrence formulae for the cardinal elements of multiplets can be written down easily in terms of these transfer matrices (Dymnikov, 1968; Hawkes, 1967a).

2.2 Quadrupole lenses

The equations of motion are now different in the x–z and y–z planes, although both are again linear, homogeneous, second-order differential equations. The imagery will therefore be characterized by two sets of cardinal elements, one set corresponding to each principal section. In quadrupole lens analysis, we rarely need to consider the real or osculating cardinal elements, for even if a quadrupole multiplet replaces the objective lens of a microscope the object is unlikely to be immersed in the field. Furthermore, only for the first lens of the multiplet would the object be real, and quadrupoles have so far been considered more suitable for correctors, projector lenses or probe-forming lenses; in all these latter cases, the asymptotic cardinal elements are required.

The asymptotic cardinal elements of a quadrupole lens, or of a sequence of such lenses, are defined in terms of pairs of linearly independent solutions of the paraxial equations of motion (8 or 10). Let $G_x(z)$, $H_x(z)$ be such a pair of solutions of Eq. (8a) and $G_y(z)$ and $H_y(z)$ be solutions of Eq. (8b), satisfying the boundary conditions

$$\begin{aligned} &\lim_{z\to-\infty} G_x(z) = G_y(z) = 1 \\ &\lim_{z\to-\infty} H_x(z) = z - z_{ox} \\ &\lim_{z\to-\infty} H_y(z) = z - z_{oy}. \end{aligned} \tag{33}$$

The points at which the asymptotes to $G_x(z)$ and $G_y(z)$ as $z \to \infty$ intersect the optic axis will then be the image foci and the points of intersection of the emergent and incident asymptotes will define the image principal planes. The object cardinal elements are obtained by considering rays, $\bar{G}_x(z), \bar{G}_y(z)$, that satisfy the conditions $\lim_{z\to\infty} \bar{G}_x(z) = \bar{G}_y(z) = 1$. The cardinal elements in the two planes are illustrated in Fig. 5.

The definitions of the *H*-rays introduce the idea, peculiar to quadrupoles, of an astigmatic object :z_{ox} denotes the object in the *x*–*z* plane and z_{oy} denotes that in the *y*–*z* plane. If the asymptote (as $z \to \infty$) to $H_x(z)$ intersects the optic axis at z_{ix} and the asymptote to $H_y(z)$ intersects it at z_{iy}, then the pairs (z_{ox}, z_{ix}) and (z_{oy}, z_{iv}) satisfy Newton's lens equation. Introducing the notation

	$x-z plane$	$y-z plane$	
Object focus :	$z=z_{Fo}$	$z=z_{Go}$	
Image focus :	$z=z_{Fi}$	$z=z_{Gi}$	
Object principal point :	$z=z_{Po}$	$z=z_{Qo}$	(34)
Image principal point :	$z=z_{Pi}$	$z=z_{Qi}$	
Focal length $(z_{Po}-z_{Fo}=z_{Fi}-z_{Pi}$ or $z_{Qo}-z_{Go}=z_{Gi}-z_{Qi})$:	f	g	

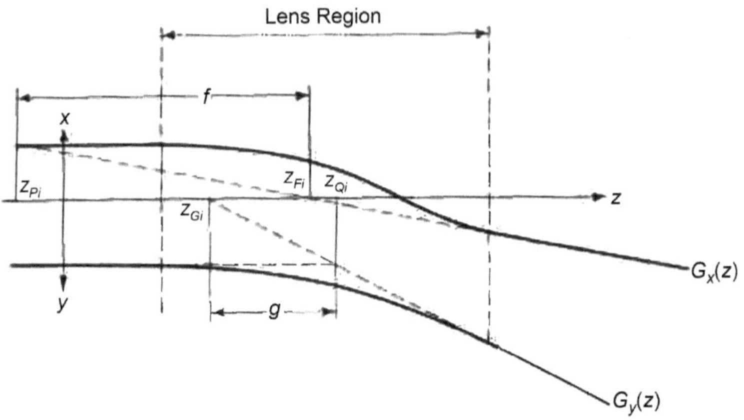

Fig. 5 Construction of asymptotic cardinal elements in the principal sections of a quadrupole lens; notice the standard practice with quadrupoles of plotting the mutually perpendicular principal sections above and below the optic axis.

we find that

$$\begin{aligned}(z_{ox} - z_{Fo})(z_{ix} - z_{Fi}) &= -f^2 \\ (z_{oy} - z_{Go})(z_{iy} - z_{Gi}) &= -g^2.\end{aligned} \tag{35}$$

If the magnification between z_{ox} and z_{ix} is M_x and that between z_{oy} and z_{iy} is M_y then

$$\begin{aligned}M_x &= \frac{f}{z_{ox} - z_{Fo}} = -\frac{z_{ix} - z_{Fi}}{f} \\ M_y &= \frac{g}{z_{oy} - z_{Go}} = -\frac{z_{iy} - z_{Gi}}{g}.\end{aligned} \tag{36}$$

A number of other relations are derived in QO; notice that in QO, the sign convention is such that the image focal length is opposite in sign to the object focal length, since both are defined by z (principal plane) – z (focus).

The matrix notation introduced above [Eqs. (21)–(32)] is extremely well suited to quadrupole analysis. The transfer matrices are now different in the two orthogonal planes and we replace Eq. (23) by

$$\mathbf{x}_2 = \mathbf{T}_x\mathbf{x}_1; \quad \mathbf{y}_3 = \mathbf{T}_y\mathbf{y}_4 \tag{37}$$

where

$$\mathbf{T}_x = \begin{pmatrix} -\dfrac{z_2 - z_{Fi}}{f} & f\left\{\dfrac{(z_2 - z_{Fi})(z_1 - z_{Fo})}{f^2} + 1\right\} \\ -\dfrac{1}{f} & \dfrac{z_1 - z_{Fo}}{f} \end{pmatrix} \tag{38a}$$

$$\mathbf{T}_y = \begin{pmatrix} -\dfrac{z_4 - z_{Gi}}{g} & g\left\{\dfrac{(z_4 - z_{Gi})(z_3 - z_{Go})}{g^2} + 1\right\} \\ -\dfrac{1}{g} & \dfrac{z_3 - z_{Go}}{g} \end{pmatrix}. \tag{38b}$$

a. *Dušek matrices.* These are probably the most convenient for studying multiplets of short quadrupoles when a computer is available. We now have $z_1 = z_2 = z_3 = z_4 = 0$ and

$$\mathbf{T}_x = \begin{pmatrix} \dfrac{z_{Fi}}{f} & \dfrac{z_{Fo}z_{Fi}}{f} + f \\ -\dfrac{1}{f} & -\dfrac{z_{Fo}}{f} \end{pmatrix}$$

$$\mathbf{T}_y = \begin{pmatrix} \dfrac{z_{Gi}}{g} & \dfrac{z_{Go}z_{Gi}}{g} + g \\ -\dfrac{1}{g} & -\dfrac{z_{Go}}{g} \end{pmatrix} \tag{39}$$

or

$$\mathbf{t}_x = \begin{pmatrix} -\dfrac{z_{Fo}}{f} & -\dfrac{1}{f} \\ \dfrac{z_{Fo}z_{Fi}}{f} + f & \dfrac{z_{Fi}}{f} \end{pmatrix}$$

$$\mathbf{t}_y = \begin{pmatrix} -\dfrac{z_{Go}}{g} & -\dfrac{1}{g} \\ \dfrac{z_{Go}z_{Gi}}{g} + g & \dfrac{z_{Gi}}{g} \end{pmatrix} \tag{40}$$

2.2.1 Conjugate planes

If z_2 is conjugate to z_1 and z_4 to z_3, then

$$\mathbf{T}_x = \begin{pmatrix} M_x & 0 \\ -1/f & 1/M_x \end{pmatrix}; \quad \mathbf{T}_y = \begin{pmatrix} M_y & 0 \\ -1/g & 1/M_y \end{pmatrix}. \tag{41}$$

2.2.2 Principal planes

The principal planes form a special case of conjugate planes; setting $z_1 = z_{Po}$, $z_2 = z_{Pi}$, $z_3 = z_{Qo}$ and $z_4 = z_{Qi}$, we have

$$\mathbf{T}_x = \begin{pmatrix} 1 & 0 \\ -1/f & 1 \end{pmatrix}; \quad \mathbf{T}_y = \begin{pmatrix} 1 & 0 \\ -1/g & 1 \end{pmatrix} \tag{42}$$

so that the transfer matrix between an arbitrary pair of planes $z = z_\alpha$ and $z = z_\beta$ separated by $(n-1)$ quadrupoles has the following form (in either the x–z or the y–z plane; the suffix is to be understood):

$$\mathbf{T}(z_\alpha, z_\beta) = \mathbf{D}_n\mathbf{T}_{n-1}\mathbf{D}_{n-1}\mathbf{T}_{n-2}\ldots\mathbf{T}_2\mathbf{D}_2\mathbf{T}_1\mathbf{D}_1 \tag{43}$$

in which

$$\mathbf{D}_i = \begin{pmatrix} 1 & D^{(i)} \\ 0 & 1 \end{pmatrix} \tag{44a}$$

and

$$\begin{aligned} D^{(i)} &= z_{Po}^{(i)} - z_{Pi}^{(i-1)} \qquad 2 \leqslant i \leqslant n-1 \\ D^{(1)} &= z_{Po}^{(1)} - z_\alpha, \qquad D^{(n)} = z_\beta - z_{Pi}^{(n-1)}. \end{aligned} \tag{44b}$$

The elements of $\mathbf{T}_x(z_\alpha, z_\beta)$ can be written as Gaussian brackets, defined as follows:

$$\begin{aligned} &[x] = x, \quad [x_1x_2] = x_1x_2 + 1, \quad [x_1x_2x_3] = x_1x_2x_3 + x_1 + x_3 \\ &[x_1x_2x_3\ldots x_n] = [x_1x_2x_3\ldots x_{n-2}] + [x_1x_2x_3\ldots x_{n-1}]x_n. \end{aligned} \tag{45}$$

If we set

$$\mathbf{T}_x(z_\alpha, z_\beta) = \begin{pmatrix} \alpha_{n-1} & \beta_{n-1} \\ \gamma_{n-1} & \delta_{n-1} \end{pmatrix} \tag{46}$$

we find

$$\begin{aligned} \alpha_{n-1} &= [D_nc_{n-1}\ldots D_2c_1] \\ \beta_{n-1} &= [D_nc_{n-1}\ldots D_2c_1D_1] \\ \gamma_{n-1} &= [c_{n-1}D_{n-1}\ldots c_1] \\ \delta_{n-1} &= [c_{n-1}D_{n-1}\ldots c_1D_1]. \end{aligned} \tag{47}$$

A proof of this result by induction is given in Hawkes (1967a). c_i denotes $-1/f_i$. Dymnikov (1968) has also discussed the use of Gaussian brackets in this context.

b. *Regenstreif matrices.* We now set $z_1 = z_{Fo}$, $z_2 = z_{Fi}$, $z_3 = z_{Go}$ and $z_4 = z_{Gi}$, giving

$$\mathbf{T}_x = \begin{pmatrix} 0 & f \\ -1/f & 0 \end{pmatrix} \quad \mathbf{T}_y = \begin{pmatrix} 0 & g \\ -1/g & 0 \end{pmatrix}. \tag{48}$$

Regenstreif (1966b, 1966c, 1967a, 1967c) has made a detailed study of a range of multiplets with the aid of these focal matrices, deriving rules for writing down the transfer matrix of a quadrupole sequence. He was particularly concerned to obtain the transfer matrix between the terminal planes of a series of quadrupoles the potential of which could be adequately represented by the rectangular model. This entails multiplying a set of quadrupole matrices and drift space matrices and this, as Regenstreif remarks, "leads to unmanageable expressions"; by transferring to the focal matrices, however, very much simpler expressions can be obtained. For the electron lens designer, Dušek matrices are more convenient than transfer matrices between somewhat arbitrary terminal planes, and Regenstreif's procedure may be adopted virtually unaltered to simplify the matrix products.

We write the Dušek matrix (in either the x–z or y–z plane, the suffix is to be understood) of the ith quadrupole of a sequence of n quadrupoles as

$$\mathbf{T}_i = \begin{pmatrix} a_i & b_i \\ c_i & d_i \end{pmatrix} \tag{49a}$$

and the separation between the ith and $(i+1)$th quadrupoles is characterized by

$$\mathbf{L}_{i,i+1} = \begin{pmatrix} 1 & L_{i,i+1} \\ 0 & 1 \end{pmatrix}. \tag{49b}$$

Following Regenstreif, we introduce the parameter

$$X_{i,i+1} = \frac{a_i}{c_i} + L_{i,i+1} + \frac{d_{i+1}}{c_{i+1}} \tag{50}$$

which is equal to the distance from the image focus of the i-th quadrupole to the object focus of the $(i+1)$-th quadrupole. At the ends of the sequence, it is convenient to write

$$\begin{aligned} \bar{X}_{1,2} &= X_{1,2} - \frac{1}{c_1 d_1} = \frac{b_1}{d_1} + L_{12} + \frac{d_2}{c_2} \\ \bar{X}_{n-1,n} &= X_{n-1,n} - \frac{1}{a_n c_n} = \frac{a_{n-1}}{c_{n-1}} + L_{n-1,n} + \frac{b_n}{a_n}. \end{aligned} \tag{51}$$

In order to reduce the number of suffices, we use the following notation for doublets, triplets, quadruplets and quintuplets; Regenstreif (1966c, 1967a) continues as far as sextuplets:

$$\begin{array}{llll} \bar{X}_{1,2} = \bar{X} & \bar{X}_{3,4} = \bar{Z} & X_{1,2} = X & X_{3,4} = Z \\ \bar{X}_{2,3} = \bar{Y} & \bar{X}_{4,5} = \bar{T} & X_{2,3} = Y & X_{4,5} = T. \end{array} \tag{52}$$

For a *doublet*, the Dušek transfer matrix between the mid-planes of the individual components is

$$\mathbf{T}^{(2)} = \begin{pmatrix} a_2 & b_2 \\ c_2 & d_2 \end{pmatrix} \begin{pmatrix} 1 & L_{12} \\ 0 & 1 \end{pmatrix} \begin{pmatrix} a_1 & b_1 \\ c_1 & d_1 \end{pmatrix} \tag{53}$$

and with

$$X = L_{12} + \frac{a_1}{c_1} + \frac{d_2}{c_2}, \quad \bar{X} = L_{12} + \frac{b_1}{d_1} + \frac{d_2}{c_2} \tag{54}$$

we have

$$\mathbf{T}^{(2)} = \begin{pmatrix} c_1 a_2 (X - 1/a_2 c_2) & d_1 a_2 (\bar{X} - 1/a_2 c_2) \\ c_1 c_2 X & d_1 c_2 \bar{X} \end{pmatrix}. \tag{55}$$

For a *triplet*, we write

$$\begin{aligned} \mathbf{T}^{(3)} &= \begin{pmatrix} a_3 & b_3 \\ c_3 & d_3 \end{pmatrix} \begin{pmatrix} 1 & L_{23} \\ 0 & 1 \end{pmatrix} \mathbf{T}^{(2)} \\ &= \begin{pmatrix} c_1 c_2 a_3 \left(X\bar{Y} - 1/c_2^2 \right) & d_1 c_2 a_3 \left(\bar{X}\bar{Y} - 1/c_2^2 \right) \\ c_1 c_2 c_3 \left(XY - 1/c_2^2 \right) & d_1 c_2 c_3 \left(\bar{X}Y - 1/c_2^2 \right) \end{pmatrix}. \end{aligned} \tag{56}$$

For a *quadruplet*,

$$\begin{aligned} \mathbf{T}^{(4)} &= \begin{pmatrix} a_4 & b_4 \\ c_4 & d_4 \end{pmatrix} \begin{pmatrix} 1 & L_{34} \\ 0 & 1 \end{pmatrix} \mathbf{T}^{(3)} \\ &= \begin{pmatrix} c_1 c_2 c_3 a_4 \left(XY\bar{Z} - \dfrac{\bar{Z}}{c_2^2} - \dfrac{X}{c_3^2} \right) & d_1 c_2 c_3 a_4 \left(\bar{X}Y\bar{Z} - \dfrac{\bar{Z}}{c_2^2} - \dfrac{\bar{X}}{c_3^2} \right) \\ c_1 c_2 c_3 c_4 \left(XYZ - \dfrac{Z}{c_2^2} - \dfrac{X}{c_3^2} \right) & d_1 c_2 c_3 c_4 \left(\bar{X}YZ - \dfrac{Z}{c_2^2} - \dfrac{\bar{X}}{c_3^2} \right) \end{pmatrix} \end{aligned} \tag{57}$$

and for a *quintuplet*

$$\begin{aligned}\mathbf{T}^{(5)} &= \begin{pmatrix} a_5 & b_5 \\ c_5 & d_5 \end{pmatrix}\begin{pmatrix} 1 & L_{45} \\ 0 & 1 \end{pmatrix}\mathbf{T}^{(4)} \\ &= \begin{pmatrix} A^{(5)} & B^{(5)} \\ C^{(5)} & D^{(5)} \end{pmatrix}\end{aligned} \tag{58a}$$

$$\begin{aligned}A^{(5)} &= c_1c_2c_3c_4a_5\left\{\left(XY-\frac{1}{c_2^2}\right)\left(Z\bar{T}-\frac{1}{c_4^2}\right)-\frac{X\bar{T}}{c_3^2}\right\} \\ B^{(5)} &= d_1c_2c_3c_4a_5\left\{\left(\bar{X}Y-\frac{1}{c_2^2}\right)\left(Z\bar{T}-\frac{1}{c_4^2}\right)-\frac{\bar{X}\bar{T}}{c_3^2}\right\} \\ C^{(5)} &= c_1c_2c_3c_4c_5\left\{\left(XY-\frac{1}{c_2^2}\right)\left(ZT-\frac{1}{c_4^2}\right)-\frac{XT}{c_3^2}\right\} \\ D^{(5)} &= d_1c_2c_3c_4c_5\left\{\left(\bar{X}Y-\frac{1}{c_2^2}\right)\left(ZT-\frac{1}{c_4^2}\right)-\frac{\bar{X}T}{c_3^2}\right\}\end{aligned} \tag{58b}$$

For a *multiplet* of n quadrupoles (Regenstreif, 1967d), we have

$$\mathbf{T}^{(n)} = \begin{pmatrix} A^{(n)} & B^{(n)} \\ C^{(n)} & D^{(n)} \end{pmatrix} \tag{59}$$

where

$$\begin{aligned}A^{(n)} &= a_n\left(\prod_{i=1}^{n-1} c_i\right)\alpha_n \\ B^{(n)} &= d_1a_n\left(\prod_{i=2}^{n-1} c_i\right)\beta_n \\ C^{(n)} &= \left(\prod_{i=1}^{n} c_i\right)\gamma_n \\ D^{(n)} &= d_1\left(\prod_{i=2}^{n} c_i\right)\delta_n\end{aligned} \tag{60}$$

and

$$\begin{aligned}\alpha_n &= \bar{X}_{n-1,n}\gamma_{n-1} - \frac{\gamma_{n-2}}{c_{n-1}^2}\\ \beta_n &= \bar{X}_{n-1,n}\delta_{n-1} - \frac{\delta_{n-2}}{c_{n-1}^2}\\ \gamma_n &= X_{n-1,n}\gamma_{n-1} - \frac{\gamma_{n-2}}{c_{n-1}^2}\\ \delta_n &= X_{n-1,n}\delta_{n-1} - \frac{\delta_{n-2}}{c_{n-1}^2}\end{aligned} \tag{61a}$$

with

$$\begin{aligned}\gamma_0 &= 0 & \delta_0 &= 0\\ \gamma_1 &= 1 & \delta_1 &= 1\\ \gamma_2 &= X_{12} & \delta_2 &= \bar{X}_{12}\end{aligned} \tag{61b}$$

3. Aberration coefficients: Round lenses

We shall be concerned with three types of aberrations: geometrical aberrations, chromatic aberrations and parasitic or mechanical aberrations. The geometrical aberrations arise when we retain terms of third and if necessary fifth order in the equations of motion; the chromatic aberrations reflect the fact that the cardinal elements of electron lenses are strongly dependent upon electron energy; and the parasitic aberrations are caused by imperfect alignment or construction of the various components.

In electron optics, it is usually sufficient to consider the dominant aberrations of each lens; the relative importance of the numerous aberrations will depend upon the role played by the lens. In a microscope, the angular aperture of the pencils of rays is small except at the objective lens and the spherical and chromatic aberration of the objective will therefore be dominant; at intermediate and projector lenses, the distortion and the chromatic aberration of magnification will be important and the spherical aberration negligible. Parasitic aberrations may be harmful wherever they originate; in round lenses, however, the most likely aberration to arise is astigmatism and this can be annulled with the aid of a multipole, optically equivalent in its simplest form to a weak quadrupole of variable orientation. The control of parasitic aberrations in round lenses is well understood, and we shall say no more about them here.

Just as we distinguish between the real cardinal elements of objective lenses and the asymptotic cardinal elements of projector or probe-forming lenses, so we must divide aberration coefficients into real and asymptotic classes. For objective lenses, the object is real and may be immersed in the lens field, so that only a part of the latter contributes to the imagery. The object imaged by other lenses is virtual, and hence the entire field is effective.

3.1 Real aberrations (objective lenses)

The spherical aberration coefficient is defined in terms of the angular aperture at the specimen plane; we write

$$\Delta x_i = x_i - Mx_o = MC_s^*\alpha^3. \tag{62}$$

C_s^* is referred to as the real spherical aberration coefficient. A variety of equivalent formulae are available, giving C_s^* in terms of $\Phi(z)$ and $B(z)$. The class of all such formulae is discussed by Seman (1955, 1958) and Hawkes (1967b); one celebrated form of the expression, obtained by Scherzer (1936), shows that the sign of C_s^* cannot change in normal circumstances. Rose (1967b) has recently proved that this remains true at relativistic voltages. For a correction, see Preikszas and Rose (1995).

The chromatic aberration is characterized by three coefficients:

$$\begin{aligned}\Delta x_i = x_i - Mx_o = M\left(C_c^*\alpha + C_M^* x_o - C_\theta^* y_o\right)\frac{\Delta\Phi_0}{\Phi_0}\\ \Delta y_i = y_i - My_o = M\left(C_c^*\gamma + C_M^* y_o + C_\theta^* x_o\right)\frac{\Delta\Phi_0}{\Phi_0}\end{aligned} \tag{63}$$

where as usual, the axes in the image plane are rotated with respect to those in the object plane through an angle Θ,

$$\Theta = \frac{\eta}{2}\int_{z_o}^{z_i}\frac{B(z)dz}{V^{1/2}} = \int_{z_o}^{z_i}\Omega(z)dz. \tag{64}$$

C_c^* characterizes a chromatic aberration that, like spherical aberration, depends only upon the angles α and γ at the object plane and is hence the most important chromatic effect in objective lenses; C_M^* corresponds to a chromatic change of magnification, and is analogous to distortion, while C_θ^* describes a chromatic change in the image rotation.

At high voltage, it is convenient to modify the definition of the chromatic aberration coefficients as follows:

$$\Delta x_i = M\left(\bar{C}_c^* \alpha + \bar{C}_M^* x_o - \bar{C}_\theta^* y_o\right)\frac{\Delta V_0}{V_0}$$
$$\Delta y_i = M\left(\bar{C}_c^* \gamma + \bar{C}_M^* y_o + \bar{C}_\theta^* x_o\right)\frac{\Delta V_0}{V_0}. \tag{65}$$

The reason for this can be seen from the expressions for the various coefficients in terms of $B(z)$ or $\Omega(z) = \eta B(z)/2V^{1/2}$ (we consider only magnetic lenses). We find that

$$\bar{C}_c^* = -\int_{z_o}^{z_i} \Omega^2 h^2 dz \quad \bar{C}_M^* = -\int_{z_o}^{z_i} \Omega^2 ghdz. \tag{66}$$

Since the paraxial properties of a lens are characterized by $\Omega(z)$, these coefficients will be independent of accelerating voltage if the cardinal elements are $B(z)$ are held constant by increasing the lens excitation, provided the dependence of $B(z)$ upon z does not alter (Dupouy, Perrier, & Trinquier, 1968).

The chromatic aberration may alternatively be defined in terms of the aperture radius instead of the slope at the specimen:

$$\Delta x_i = M\left(c_c^* x_a + c_M^* x_o - c_\theta^* y_o\right)\frac{\Delta\Phi}{\Phi} \tag{67}$$

($\Delta V/V$ may again be used). The rays $g(z)$ and $h(z)$ that were used in Eq. (66) satisfy the boundary conditions

$$g(z_o) = h'(z_o) = 1 \quad g'(z_o) = h(z_o) = 0 \tag{68a}$$

but the chromatic aberration coefficients appearing in (67) involve a different pair of rays, satisfying

$$s(z_o) = t(z_a) = 1 \quad s(z_a) = t(z_o) = 0 \tag{68b}$$

where $z = z_a$ is the aperture plane. It is clear that the dependence of the chromatic aberration of magnification upon the aperture position will not be simple. A detailed discussion is to be found in Hanszen and Lauer (1965–66); here we select only the salient points. The chromatic aberration c_c^* corresponds to a shift in the image plane conjugate to the object, so that a disk will be seen in the original image plane where a point would previously have been observed. The aberration c_M^* shifts the center of the disk laterally; it is, however, possible to select the beam that forms the image, by adopting a suitable position for the aperture, in such a way that there is no such lateral shift (Fig. 6A). If the angular aperture of the beam is limited at the source and not by a stop in the lens field, we use C_c^* and C_M^*; C_c^* again corresponds to an

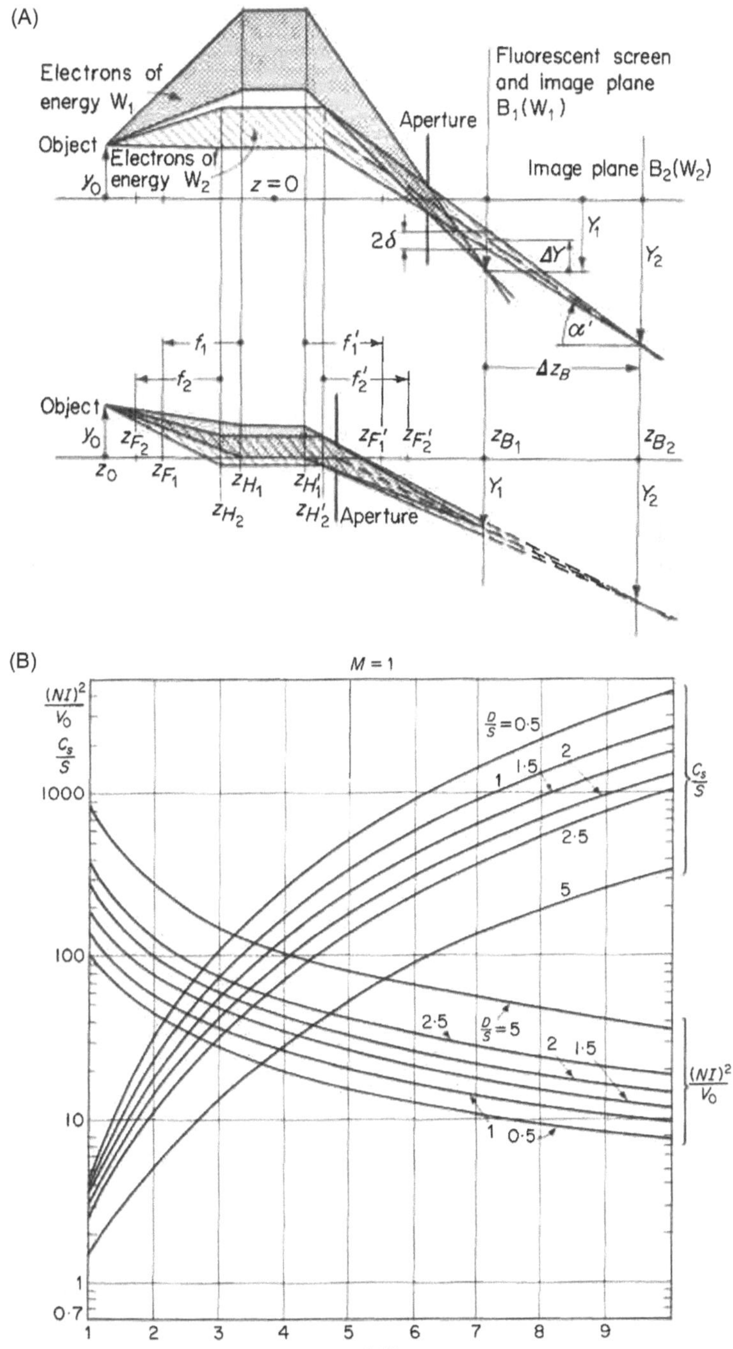

Fig. 6 (A) The dependence of the chromatic aberration of magnification upon aperture position (after Hanszen & Lauer, 1965–66). (B)–(F) The dependence of objective lens aberrations upon magnification or object position (after Der-Shvarts & Makarova, 1967). Curves (B)–(E) show C_s/S and the excitation parameter $(NI)^2/V_0$ against object distance/S and (F) shows the same quantities scaled with respect to D for a range of values of S/D. (B) $M=1$, (C) $M=5$, (D) $M=10$, (E) $M=100$ and (F) $M=100$.

(Continued)

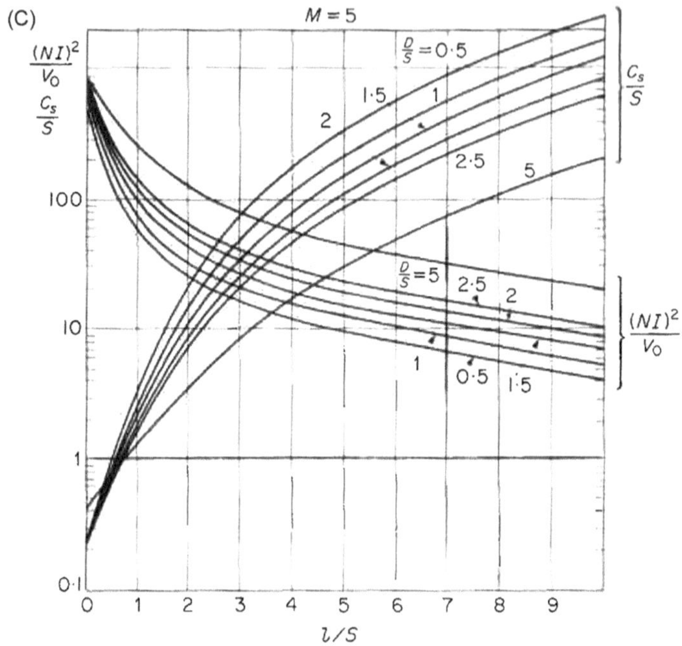
(C)
M = 5
(NI)²/V₀
Cs/S
D/S = 0·5
1·5
1
2
2·5
5
100
D/S = 5
2·5
2
10
1
0·5
1·5
1
0·1
0 1 2 3 4 5 6 7 8 9
l/S

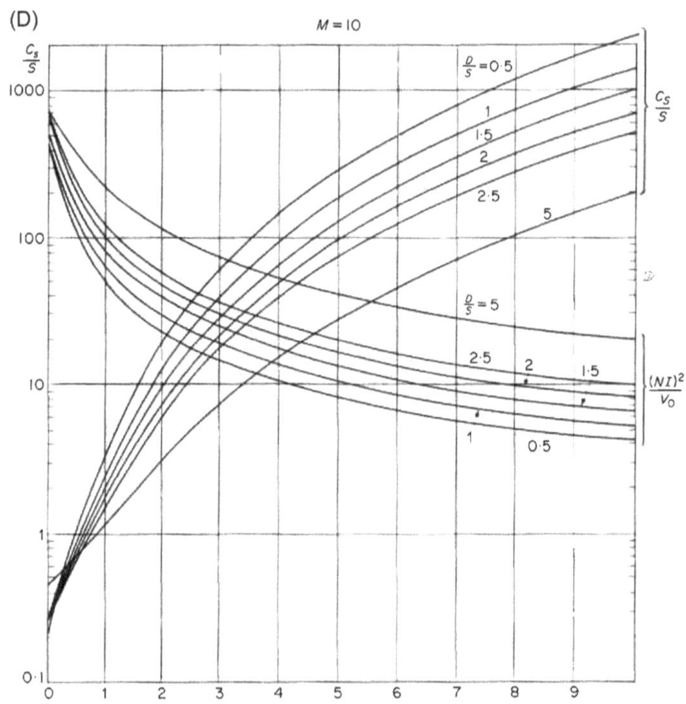
(D)
M = 10
Cs/S
1000
D/S = 0·5
1
1·5
2
2·5
5
100
D/S = 5
2·5
2
1·5
(NI)²/V₀
10
1
0·5
1
0·1
0 1 2 3 4 5 6 7 8 9
l/S

Fig. 6—cont'd

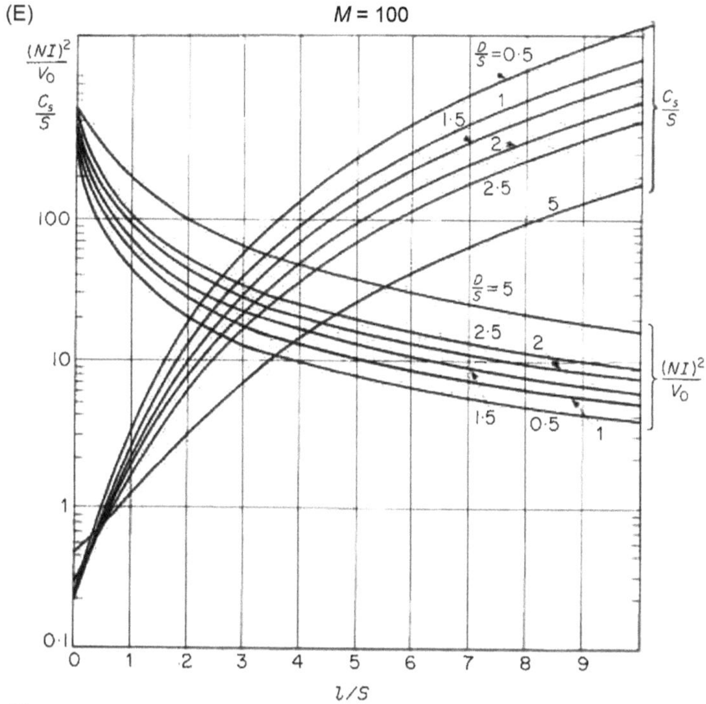
(E)
M = 100
(NI)²/V₀
Cₛ/S
D/S = 0·5
1
1·5
2
2·5
5
D/S = 5
2·5
2
1·5
0·5
1
100
10
1
0·1
0 1 2 3 4 5 6 7 8 9
l/S

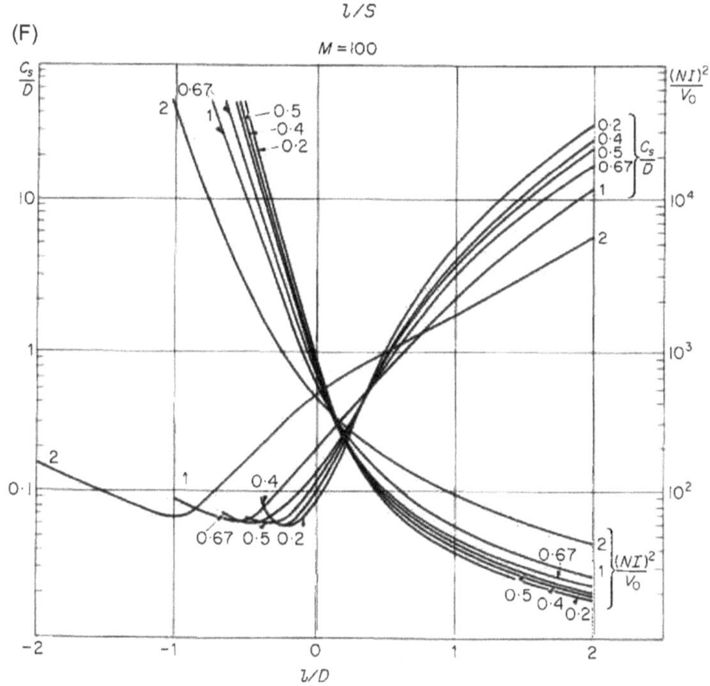
(F)
M = 100
Cₛ/D
(NI)²/V₀
0·67
2
1
0·5
0·4
0·2
0·2
0·4
0·5
0·67
1
2
10
10⁴
1
10³
0·1
10²
2
1
0·4
0·67 0·5 0·2
0·67
0·5 0·4 0·2
−2 −1 0 1 2
l/D

Fig. 6—cont'd

axial shift in Gaussian image plane and C_M^* to a lateral shift in the central point and if the object is not immersed in the field, we may use the analysis of Glaser (1956a). In the more common situation in which the real object is immersed, we may still use C_c^* and C_M^*; the effect of the field between the effective source and the object when the accelerating voltage varies must not be overlooked, however.

These coefficients of spherical and chromatic aberration vary with object position. Since objective lenses are normally operated at high magnification, this is of little practical importance, and the limiting values of these coefficients as the magnification tends to infinity are sometimes referred to as the aberration constants. In general, the dependence upon object position is not simple (Der-Shvarts & Makarova, 1967; see Fig. 6B–F); the simplicity of Newtonian fields extends to their real aberrations, however, and for these fields the spherical aberration coefficient can be written.

$$C_s^* = c_4^* z_o^4 + c_3^* z_o^3 + c_2^* z_o^2 + c_1^* z_o + c_0^* \tag{69}$$

or, writing $m = 1/M$,

$$C_s^* = C_4^* m^4 + C_3^* m^3 + C_2^* m^2 + C_1^* m + C_0. \tag{70}$$

If the lens is symmetric, $C_4^* = C_0^*$ and $C_3^* = C_1^*$ (Hawkes, 1968a). The thin-lens approximation to Eq. (70) has been given by Petrie (1962), who found that

$$C_s^* = kz_o^4/f = kf^3\left(m^4 - 4m^3 + 6m^2 - 4m + 1\right). \tag{71}$$

We postpone discussion of the significance of these variations with object position to the section on asymptotic aberrations, for it is here they become of real importance.

3.2 Asymptotic aberrations (lenses for which the object is virtual)

The entire lens field now contributes to the aberrations, irrespective of the object position. Of the geometrical aberrations, we consider only distortion and spherical aberration, the formulae for which are expressed in terms of the rays $G(z)$ and $H(z)$ defined in Eq. (19). The spherical aberration coefficient is defined by

$$\Delta x_i = MC_s\alpha^3 \tag{72}$$

and the distortion coefficient, D, by

$$\Delta x_i = MDx_o^3. \tag{73}$$

The parameters α and x_o are measured at the virtual object plane and Δx_i, M correspond to the asymptotic image plane. In probe-forming lenses, the magnification is very small, and these lenses are therefore characterized by the spherical aberration coefficient C_s', defined by

$$\Delta x_i = C_s' \alpha_i^3 \tag{74}$$

where α_i is the extreme angle at the image plane.

Since $H(z)$ appears raised to the fourth power in C_s and to the first power only in D, and since the virtual object position appears only in $H(z)$, it is clear that C_s may be written as a fourth degree polynomial in z_o while D will be linearly related to z_o. The linear relation between z_o and $m = 1/M$ ($z_o = z_{F_o} + fm$) implies that C_s may equally well be written as a fourth degree polynomial in m and that D and m are linearly related. These relationships are themselves interesting, but we may go further and establish the physical significance of the coefficients of the various powers of z_o or m. (The work of Verster, 1963, on electrostatic round lenses is extended to magnetic round lenses and quadrupoles in Hawkes, 1968b.)

The least laborious of the various methods of establishing relationships and symmetries among aberration coefficients is that of characteristic functions. The same is true of the coefficients that figure in the "aberration polynomials." The mathematical details may be found in Hawkes (1968b). Briefly, the chain of reasoning is as follows: all the properties of an electron optical system may be derived from one of various characteristic functions, which are integrals along the optic axis of the "projection" of the refractive index onto that axis. If n denotes the refractive index, and $\bar{\mu}$ is the projection of n on Oz, then Fermat's principle states that

$$\delta \int \bar{\mu} dz = 0 \tag{75a}$$

where

$$\bar{\mu} = \left\{\phi(1 + \varepsilon\phi)\left(1 + x'^2 + y'^2\right)\right\}^{1/2} - \eta\left(A_x x' + A_y y' + A_z\right) \tag{75b}$$

and $\mathbf{A}$ is the magnetic vector potential, $\mathbf{B} = \text{curl}\,\mathbf{A}$. The function $\bar{\mu}$ is then expanded as a power series in x, y and their derivatives and the terms of n-th degree are denoted $\bar{\mu}^{(n)}$. The paraxial properties are derived from $\bar{\mu}^{(2)}$ and the geometrical aberrations from the characteristic function $V(z_1, z_2)$, which we normally write V_{12}:

$$V(z_1, z_2) = V_{12} = \int_{z_1}^{z_2} \mu^{(4)} dz \tag{76}$$

where $\mu^{(4)} = \bar{\mu}^{(4)} / V_0^{1/2}$. It is not difficult to derive formulae for the asymptotic aberrations in an arbitrary "current" plane $z = z_c$. We denote the values of the asymptotes to $G(z)$ and $H(z)$ in image space at the plane $z = z_c$ by G_c^* and H_c^* respectively. The function V_{oc} represents $V(z_o, z_c)$ taken over three regions: $V_{oc} = V(z_o, -\infty) + V(-\infty, \infty) + V(\infty, z_c)$; in $V(z_o, -\infty)$ and $V(\infty, z_c)$, the asymptotic values of $G(z)$ and $H(z)$ are to be substituted into the integrands, the magnetic field vanishes and the electrostatic potential is constant and equal to the accelerating voltage (Fig. 7). We find

$$\begin{aligned} \Delta x(z_c) &= H_c^* \frac{\partial V_{oc}}{\partial x_o} - G_c^* \frac{\partial V_{oc}}{\partial \alpha} - \frac{1}{2} H_c^* \alpha \left(\alpha^2 + \gamma^2\right) \\ \Delta x'(z_c) &= H_c^{*\prime} \frac{\partial V_{oc}}{\partial x_o} - G_c^{*\prime} \frac{\partial V_{oc}}{\partial \alpha} - \frac{1}{2} H_c^{*\prime} \alpha \left(\alpha^2 + \gamma^2\right) + \frac{1}{2} \alpha_c \left(\alpha_c^2 + \gamma_c^2\right) \end{aligned} \tag{77}$$

with similar expressions for $\Delta y(z_c)$ and $\Delta y'(z_c) \cdot \alpha_c$ and γ_c denote $x'(z_c)$ and $y'(z_c)$. Let us now select the object focal plane as the object plane and the image focal plane as the current plane $z = z_c : z_o = z_{Fo}$ and $z_c = z_{Fi}$. This yields

$$\begin{aligned} \Delta x(z_{Fi}) &= f \left\{ \frac{\partial V_F}{\partial x_o} - \frac{1}{2} \alpha \left(\alpha^2 + \gamma^2\right) \right\} \\ \Delta x'(z_{Fi}) &= \frac{1}{f} \frac{\partial V_F}{\partial \alpha} - \frac{1}{2f} \frac{x_o x_o^2 + y_o^2}{f^2} \end{aligned} \tag{78}$$

in which V_F is the focal characteristic, $V_F = V(z_{Fo}, z_{Fi})$; we assume that there are no accelerating lenses.

The focal characteristic, V_F, must be invariant with respect to rotation in round lens systems, and we can ensure that this condition is satisfied by grouping the coordinates x_o, y_o, α and γ as system invariants:

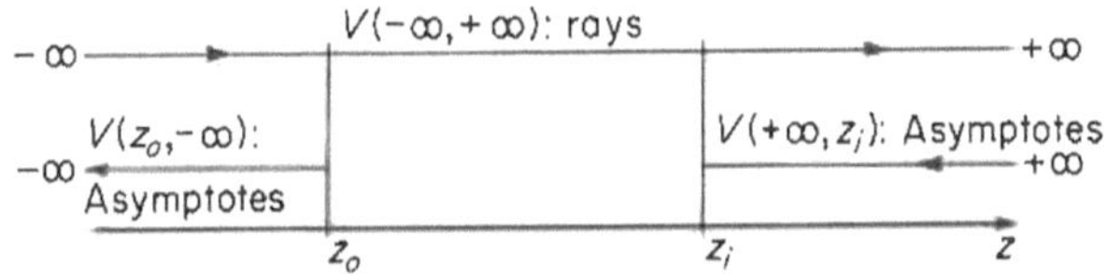

Fig. 7 The regions in which the contributions to V_{oi} are calculated.

$$\begin{aligned} \bar{r} &= x_o^2 + y_o^2 \qquad s = \alpha^2 + \gamma^2 \\ \bar{u} &= x_o\alpha + y_o\gamma \quad \bar{v} = x_o\gamma - y_o\alpha. \end{aligned} \tag{79}$$

The algebra is simplified if we use dimensionless groups, however, and we write

$$\xi_o = x_o/f \quad \eta_o = y_o/f \tag{80}$$

and

$$\begin{aligned} r &= \xi_o^2 + \eta_o^2 \qquad s = \alpha^2 + \gamma^2 \\ u &= \xi_o\alpha + \eta_o\gamma \quad v = \xi_o\gamma - \eta_o\alpha. \end{aligned} \tag{81}$$

The focal characteristic may be written

$$V_F = \begin{pmatrix} r \\ s \\ u \end{pmatrix}' \begin{pmatrix} F_{11} & F_{12} & F_{13} & F_{14} \\ 0 & F_{22} & F_{23} & F_{24} \\ 0 & 0 & F_{33} & F_{34} \end{pmatrix} \begin{pmatrix} r \\ s \\ u \\ v \end{pmatrix} \tag{82}$$

giving

$$\begin{aligned} x(z_{Fi}) &= f\alpha + 4F_{11}\xi_o r + 2F_{12}\xi_o s + 2F_{13}\xi_o u \\ &\quad + 2F_{14}\xi_o v + F_{13}\alpha r + \left(F_{23} - \frac{1}{2}f\right)\alpha s \\ &\quad + 2F_{33}\alpha u + F_{34}\alpha v + F_{14}\gamma r + F_{24}\gamma s + F_{34}\gamma u \\ x'(z_{Fi}) &= -\xi_o + \frac{1}{f}\left\{\left(F_{13} - \frac{1}{2}f\right)\xi_o r + F_{23}\xi_o s\right. \\ &\quad + 2F_{33}\xi_o u + F_{34}\xi_o v + 2F_{12}\alpha r + 4F_{22}\alpha s \\ &\quad \left. + 2F_{23}\alpha u + 2F_{24}\alpha v - F_{14}\eta_o r - F_{24}\eta_o s - F_{34}\eta_o u\right\} \end{aligned} \tag{83}$$

In purely electrostatic lenses, $F_{14} = F_{24} = F_{34} = 0$ and by considering $\eta_o = 0$, the inter-relations derived by Verster (1963, Eq. 1.99) may be confirmed immediately.

When the lens, electrostatic or magnetic, is symmetric, several of the F_{ij} are equal. Let us suppose that the electrons are incident from image space and travel towards object space; we substitute

$$\begin{aligned} \xi_o &= -\alpha_i \quad \alpha = x_i/f = \xi_i \\ \eta_o &= -\gamma_i \quad \gamma = y_i/f = \eta_i \end{aligned} \tag{84}$$

and write

$$\begin{aligned} r^* &= \xi_i^2 + \eta_i^2 & s^* &= \alpha_i^2 + \gamma_i^2 \\ u^* &= \xi_i\alpha_i + \eta_i\gamma_i & v^* &= \xi_i\gamma_i - \eta_i\alpha_i \end{aligned} \tag{85}$$

so that the focal characteristic becomes

$$V_F = \begin{pmatrix} r^* \\ s^* \\ u^* \end{pmatrix}' \begin{pmatrix} F_{22} & F_{12} & F_{23} & F_{24} \\ 0 & F_{11} & F_{13} & F_{14} \\ 0 & 0 & F_{33} & F_{34} \end{pmatrix} \begin{pmatrix} r^* \\ s^* \\ u^* \\ v^* \end{pmatrix} \tag{86}$$

and hence

$$\begin{aligned} F_{11} &= F_{22} \\ F_{13} &= F_{23} \\ F_{14} &= F_{24} \quad \text{(symmetric lens)} \end{aligned} \tag{87}$$

as already found by Verster. The aberration coefficients corresponding to any pair of conjugate planes between which the magnification is M may now be derived by writing

$$\begin{aligned} x(z_o) &= x(z_{Fo}) + \alpha f/M \\ x(z_i) &= x(z_{Fi}) - \alpha_i fM \end{aligned} . \tag{88}$$

With

$$\begin{aligned} &\xi = x(z_o)/f = \xi_o + \alpha/M \\ &\eta = y(z_o)/f = \eta_o + \gamma/M \\ &R = \xi^2 + \eta^2 \qquad S = s = \alpha^2 + \gamma^2 \\ &U = \xi\alpha + \eta\gamma \quad V = \xi\gamma - \eta\alpha \end{aligned} \tag{89}$$

we find

$$\begin{aligned} \frac{x(z_i)}{M} &= f\xi + \xi(KS + 2AU + DR + aV) \\ &\quad + \alpha(CS + 2KU + ER + 2kV) \\ &\quad - \eta(kS + aU + dR) \\ \frac{y(z_i)}{M} &= f\eta + \eta(KS + 2AU + DR + aV) \\ &\quad + \gamma(CS + 2KU + ER + 2kV) \\ &\quad + \xi(kS + aU + dR) \end{aligned} \tag{90}$$

(Expressions for $x'(z_i)$ and $y'(z_i)$ are given in Hawkes, 1968b.) With $m = 1/M$, the aberration coefficients A..... K are as follows:

$$\begin{aligned} C &= -4F_{11}m^4 + (4F_{13} - f/2)m^3 - 4(F_{12} + F_{33})m^2 \\ &\quad + (4F_{23} - f/2)m - 4F_{22} \qquad \text{(Spherical aberration)} \end{aligned} \tag{91a}$$

$$K = 4F_{11}m^3 - (3F_{13} - f/2)m^2 + 2(F_{12} + F_{33})m - F_{23} \qquad \text{(coma)} \tag{91b}$$

$$A = -4F_{11}m^2 + (2F_{13} - f/2)m - F_{33} \qquad \text{(astigmatism)} \tag{91c}$$

$$E = -4F_{11}m^2 + (2F_{13} - f/2)m - 2F_{12} \qquad \text{(field curvature)} \tag{91d}$$

$$D = 4F_{11}m + f/2 - F_{13} \qquad \text{(distortion)} \tag{91e}$$

$$k = -F_{14}m^2 + F_{34}m - F_{24} \qquad \text{(anisotropic coma)} \tag{91f}$$

$$a = 2F_{14}m - F_{34} \qquad \text{(anisotropic astigmatism)} \tag{91g}$$

$$d = -F_{14} \qquad \text{(anisotropic distortion)} \tag{91h}$$

The most interesting of these expressions are (91a) and (91e), for the spherical aberration and distortion.[a] These have the general form

$$\begin{aligned} C &= C_4m^4 + C_3m^3 + C_2m^2 + C_1m + C_0 \\ D &= D_1m + D_0 \end{aligned} \tag{92}$$

where

$$\begin{aligned} D_1 &= -C_4 \\ D_0 &= (3f - 2C_3)/8 \end{aligned} \tag{93}$$

If the spherical aberration functions C_4 and C_3 are known, the distortion may hence be derived immediately. If, furthermore, the lens is symmetric, then $C_4 = C_0$ and $C_3 = C_1$ and so

$$\begin{aligned} C &= C_4m^4 + C_3m^3 + C_2m^2 + C_3m + C_4 \quad \text{(symmetric lens)} \\ D &= -C_4m + (3f - 2C_3)/8 \end{aligned} \tag{94}$$

In practice, some of the coefficients C_i may not be required. In projector lenses, the magnification is normally high and hence m is small; we have

$$C_s = C \simeq C_0 \quad \text{(projector).} \tag{95}$$

In probe-forming lenses, the spherical aberration is characterized by $C_s' = M^4C_s$ [see Eqs. (72) and (74)]. We now have

$$C_s' = CM^4 = C_4 + C_3M + C_2M^2 + C_1M^3 + C_0M^4 \tag{96}$$

[a] It should not be forgotten that the distortion is here defined in terms of ξ.

and in probe-forming lenses, M is very small:

$$C_s' \simeq C_4. \quad \text{(probe-forming lens)} \tag{97}$$

If the lens is symmetric, $C_4 = C_0$ and hence $C_s' = C_s$.

If a lens is not symmetric, the coefficients for the reversed lens are obtained by making the interchanges

$$F_{11} \rightleftharpoons F_{22}, \quad F_{13} \rightleftharpoons F_{23}, \quad F_{14} \rightleftharpoons F_{24} \tag{98}$$

so that

$$C_s = C_0 m^4 + C_1 m^3 + C_2 m^2 + C_3 m + C_4 \quad \text{(lens reversed)} \tag{99}$$

If this *reversed* lens is used to form a probe, $C_s' = C_0$; it is thus equal to the spherical aberration of the original lens, before reversal, when used as a projector. Figures for a range of symmetric and asymmetric lenses are tabulated in Hawkes (1968d).

From Eqs. (92) and (99), it is clear that when the conjugate planes are interchanged by reversing the lens, so that $m \to M$, the spherical aberration is multiplied by M^4. This has been underlined by Archard (1958b), who derived the ratio of the radii of the spherical aberration disks from the elementary lens relation; briefly, he defines C_s by writing $\delta = MC_s\alpha^3$ (Fig. 8) and then considers the radius δ' of the spherical aberration disk produced at O_1 by a point object (O_2) placed at I_1 (Fig. 11B). The lens equation $1/v - 1/u = 1/f$ is differentiated to give $dv/du = v^2/u^2 = M^2$ and δ/δ' is set equal to $\alpha'\delta v/\alpha\delta u$, giving $\delta' = \delta/M = C_s\alpha^3$. This last step seems questionable, however, for the quantity $\delta v/\delta u$ merely gives the shift of the image plane, δv, when the object is shifted a distance δu such that the plane $v + \delta v$ is conjugate to $u + \delta u$. Why should the projections on the original object and image planes be equal to the radii of the spherical aberration disks? The relation between δ and δ' could have been derived without using the lens equation[b] by considering the height δ' of the object that must be placed at O_1 if the ray from its tip is to intersect the axis at I_1; from the definition of δ, $M(-\delta') + MC_s\alpha^3 = 0$ and hence $\delta' = C_s\alpha^3$.

In the preceding section, on the real aberrations of objective lenses, we mentioned that the real spherical aberration coefficients of lenses in which the fields may be regarded as Newtonian can also be written in the form

[b] Dr. J. C. E. Jennings drew my attention to this neat derivation.

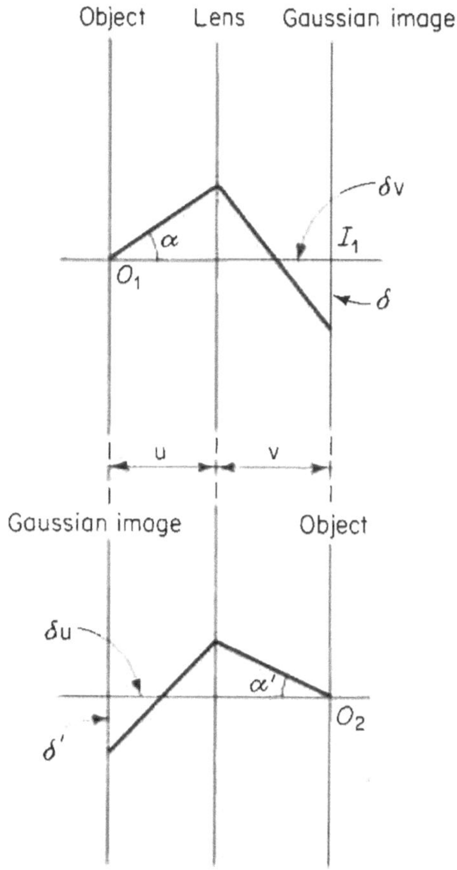

Fig. 8 The notation employed in Archard's derivation of the relation between "forward" and "backward" aberration coefficients.

$\sum_{i=0}^{4} c_i^* z_o^i$ or $\sum_{i=0}^{4} C_i^* m^i$. The reason why this should be the case for Newtonian fields but not for others is now apparent. The analysis involving the focal characteristic may be applied to any field; only with Newtonian fields, for which the foci are stationary and the focal lengths invariant and for which we again have a linear relation between $x(z_o)$, $x(z_{Fo})$ and $m = 1/M$ and between $x(z_i)$, $x(z_{Fi})$ and M, can we perform the final stage of the analysis and derive polynomial expressions for the aberration coefficients. (Strictly speaking, the term Newtonian is reserved for fields that satisfy Newton's lens equation but for which the magnification equations are not necessarily valid. For a careful discussion of this point, see Hutter, 1945, and for a full account of Newtonian fields, see Glaser, 1952, 1956a, 1956b.)

The chromatic aberrations of projector and probe-forming lenses must also be calculated using the "asymptotic" rays $G(z)$ and $H(z)$. We write

$$
\begin{aligned}
\Delta x_i &= M(\bar{C}_c\alpha + \bar{C}_M x_o - \bar{C}_\theta y_o)\frac{\Delta V_0}{V_0} \\
\Delta y_i &= M(\bar{C}_c\gamma + \bar{C}_M y_o + \bar{C}_\theta x_0)\frac{\Delta V_0}{V_0}.
\end{aligned}
\tag{100}
$$

The coefficients have the general form

$$
\begin{aligned}
\bar{C}_c = -\int_{-\infty}^{\infty} \Omega^2 H^2 dz; \quad \bar{C}_M = -\int_{-\infty}^{\infty} \Omega^2 GH dz \\
\bar{C}_\theta = \frac{1}{2}\int_{-\infty}^{\infty} \Omega dz.
\end{aligned}
\tag{101}
$$

3.3 Aberration coefficients: Quadrupoles

For a variety of reasons, the distinction between real and asymptotic aberrations is less important in quadrupoles than in round lenses. In practice, quadrupoles seem most likely to be used either to replace probe-forming or projector lenses or to correct the spherical aberration of round objective or probe-forming lenses, where the object is always virtual, so far as the quadrupoles are concerned. Even if an objective lens were composed wholly of quadrupoles and octopoles, it would probably be most convenient to place the object outside the lens field; if the object was immersed, the illumination would be affected nonuniformly by the quadrupolar pre-field, and although this could perhaps be matched to a noncircular aperture, it would inevitably render an already complex unit still more difficult to use. Furthermore, quadrupole multiplets are usually designed with some degree of geometrical and electrical symmetry, but if the object is immersed in the lens field, the optical benefit of such symmetries will be lost. The fact that quadrupoles are almost always used in combination provides another reason for concentrating attention on the asymptotic aberrations; provided the fields of the individual elements do not overlap too much, the aberrations of the whole multiplet may be calculated by adding the aberrations of the individual components, suitably weighted.

We shall therefore deal mainly with asymptotic aberrations, giving references to papers in which the real coefficients are discussed but only rarely reproducing their results here.

3.4 Geometrical aberrations

The geometrical aberrations of round lenses are conventionally divided into four broad classes; those that are independent of ray position in the object plane (spherical aberration), those that depend only on ray position in the object plane (distortion), those that depend linearly on the position coordinates in the object plane (coma) and those that depend quadratically on these coordinates (astigmatism and field curvature). Quadrupole aberrations are divided into the same classes but each class now contains more members so that the number of aberration coefficients required to give a complete specification is greater. Furthermore, several different definitions of the aberrations are current, based on assumptions that are applicable to particular situations, and this has led to some over-hasty comparisons (Tanguy & Joly, 1967). The main distinctions are these: (i) Are the aberrations expressed in terms of angle at the object plane (α, γ) or in terms of aperture coordinates (x_a, y_a) and is the aperture in field-free image space? (ii) Do the aberrations express a relationship between two pairs of conjugate planes, in the x–z and y–z planes respectively, or between a general pair of planes or between some specially selected pair? (iii) Is the system regular, by which we mean does it produce a stigmatic orthomorphic image of the object for all positions of the object plane? Any ambiguity about the answers to these questions can lead to confusion.

Let us first consider a system of quadrupoles and electrostatic round lenses forming line images of a real object point (x_o, y_o) situated in the real object plane, $z = z_o$ (for the next few paragraphs, we shall deal with real aberrations, since they illustrate the points more clearly; we then return to asymptotic aberrations). The beam is bounded by an aperture in the plane $z = z_a$, which in general lies within the lens field. We shall require the pairs of linearly independent solutions of the paraxial equations of motion that satisfy the boundary conditions

$$\begin{aligned} s_x(z_o) = s_y(z_o) = t_x(z_a) = t_y(z_a) = 1 \\ s_x(z_a) = s_y(z_a) = t_x(z_o) = t_y(z_o) = 0 \end{aligned}. \tag{102}$$

The aberrations are expressed in terms of the point characteristic $\bar{V}_{12} \equiv \bar{V}(z_1, z_2)$, defined by

$$\bar{V}_{12} = \int_{z_1}^{z_2} \bar{\mu}^{(4)} dz \tag{103}$$

in which $\bar{\mu}^{(4)}$ denotes the terms of fourth degree in x, y and their derivatives in the refractive index. ($\bar{V}_{12}$ differs slightly from the V_{12} of Eq. (76) because we include round electrostatic lenses and the object is now real.) We find that the ray positions in some arbitrary current plane, $z = z_c$, correct to third order, are given by

$$\begin{aligned} x(z_c) &= s_{xc}x_o + t_{xc}x_a + \left(t_{xc}\frac{\partial \bar{V}_{ac}}{\partial x_o} - s_{xc}\frac{\partial \bar{V}_{oc}}{\partial x_a} \right) / k_x \\ y(z_c) &= s_{yc}y_o + t_{yc}y_a + \left(t_{yc}\frac{\partial \bar{V}_{ac}}{\partial y_o} - s_{yc}\frac{\partial \bar{V}_{oc}}{\partial y_a} \right) / k_y \end{aligned} \tag{104}$$

in which k_x and k_y are constants, defined by

$$k_x = V^{1/2}\left(s_x t'_x - s'_x t_x\right); \quad k_y = V^{1/2}\left(s_y t'_y - s'_y t_y\right). \tag{105}$$

(We note that in the absence of round electrostatic lenses, we could have used $V_{12} = \bar{V}_{12}/V^{1/2}$ and the factors $V^{1/2}$ would then disappear from Eq. (105).) For the aberrations, therefore, we have

$$\begin{aligned} k_x \Delta x_c &= t_{xc}\frac{\partial \bar{V}_{ac}}{\partial x_o} - s_{xc}\frac{\partial \bar{V}_{oc}}{\partial x_a} \\ k_y \Delta y_c &= t_{yc}\frac{\partial \bar{V}_{ac}}{\partial y_o} - s_{yc}\frac{\partial \bar{V}_{oc}}{\partial y_a} \end{aligned}. \tag{106}$$

Writing the characteristic in terms of system invariants, we find

$$\bar{V}_{ac} = \begin{pmatrix} x_o^2 \\ y_o^2 \\ x_a^2 \\ y_a^2 \\ x_o x_a \end{pmatrix} \begin{pmatrix} V_{11} & V_{12} & V_{13} & V_{14} & V_{15} & V_{16} \\ 0 & V_{22} & V_{23} & V_{24} & V_{25} & V_{26} \\ 0 & 0 & V_{33} & V_{34} & V_{35} & V_{36} \\ 0 & 0 & 0 & V_{44} & V_{45} & V_{46} \\ 0 & 0 & 0 & 0 & 0 & V_{56} \end{pmatrix} \begin{pmatrix} x_o^2 \\ y_o^2 \\ x_a^2 \\ y_a^2 \\ x_o x_a \\ y_o y_a \end{pmatrix} \tag{107}$$

with a similar expression for $\bar{V}_{ac}$; we denote the matrix elements of $\bar{V}_{ac}$ by v_{ij}, reserving V_{ij} for those of $\bar{V}_{oc}$. If the aperture lies in field-free image space, most of the v_{ij} vanish. The aberrations can thus be written in the form

$$
\begin{aligned}
\Delta x(z_c) = {} & \{(3000)x_o^2 + (1200)y_o^2\}x_o && \text{(distortions)} \\
& + \{(0030)x_a^2 + (0012)y_a^2\}x_a && \text{(aperture aberrations)} \\
& + \{(1020)x_a^2 + (1002)y_a^2\}x_o && \text{(comas} \\
& + \{(2010)x_o^2 + (0210)y_o^2\}x_a && \text{and} \\
& + \{(1101)x_o + (0111)x_a\}y_o y_a && \text{astigmatisms)}
\end{aligned} \tag{108a}
$$

$$
\begin{aligned}
\Delta y(z_c) = {} & \{(2100)x_o^2 + (0300)y_o^2\}y_o && \text{(distortions)} \\
& + \{(0021)x_a^2 + (0003)y_a^2\}y_a && \text{(aperture aberrations)} \\
& + \{(0120)x_a^2 + (0102)y_a^2\}y_o && \text{(comas} \\
& + \{(2001)x_o^2 + (0201)y_o^2\}y_a && \text{and} \\
& + \{(1110)y_o + (1011)y_a\}x_o x_a && \text{astigmatisms)}
\end{aligned} \tag{108b}
$$

where

$$
\begin{aligned}
k_x(3000) &= 4t_{xc}v_{11} - s_{xc}V_{15} \\
k_x(1200) &= 2t_{xc}v_{12} - s_{xc}V_{25} \\
k_y(0300) &= 4t_{yc}v_{22} - s_{yc}V_{26} \\
k_y(2100) &= 2t_{yc}v_{12} - s_{yc}V_{16} \\
\\
k_x(2010) &= 3t_{xc}v_{15} - 2s_{xc}V_{13} \\
k_x(0210) &= t_{xc}v_{25} - 2s_{xc}V_{23} \\
k_x(1101) &= 2t_{xc}v_{16} - s_{xc}V_{56} \\
k_y(2001) &= t_{yc}v_{16} - 2s_{yc}V_{14} \\
k_y(0201) &= 3t_{yc}v_{26} - 2s_{yc}V_{24} \\
k_y(1110) &= 2t_{yc}v_{25} - s_{yc}V_{56} \\
\\
k_x(1020) &= 2t_{xc}v_{13} - 3s_{xc}V_{35} \\
k_x(1002) &= 2t_{xc}v_{14} - s_{xc}V_{45} \\
k_x(0111) &= t_{xc}v_{56} - 2s_{xc}V_{36} \\
k_y(0120) &= 2t_{yc}v_{23} - s_{yc}V_{36} \\
k_y(0102) &= 2t_{yc}v_{24} - 3s_{yc}V_{46} \\
k_y(1011) &= t_{yc}v_{56} - 2s_{yc}V_{45} \\
\\
k_x(0030) &= t_{xc}v_{35} - 4s_{xc}V_{33} \\
k_x(0012) &= t_{xc}v_{45} - 2s_{xc}V_{34} \\
k_y(0003) &= t_{yc}v_{46} - 4s_{yc}V_{44} \\
k_y(0021) &= t_{yc}v_{36} - 2s_{yc}V_{34}.
\end{aligned} \tag{109}
$$

We notice that only 16 of the 19 matrix elements V_{ij} appear, and similarly only 16 of the 19v_{ij}.

If the system produces a real, stigmatic image at $z = z_i$, then $t_{xi} = t_{yi} = 0$ and only the contributions to the aberrations from the V_{ij} remain. If the system creates real line images at $z = z_i$ and $z = z_j$, then $\Delta x(z_i)$ and $\Delta y(z_j)$ are again wholly specified by the V_{ij}. In both cases, 16 third-order parameters completely determine the aberrations.

We now turn to the asymptotic aberration coefficients, which we shall express in terms of ray position and slope at the virtual object planes; with quadrupoles, the virtual object plane in the x–z plane is frequently different from that in the y–z plane, since the asymptotic image of one lens forms the virtual object of the next and the image formed by a quadrupole is normally a pair of lines separated by the astigmatic distance: the object is astigmatic.

The asymptotic aberrations are expressed in terms of the rays $G_x(z)$, $G_y(z)$, $H_x(z)$ and $H_y(z)$ that satisfy the boundary conditions

$$\begin{aligned} \lim_{z\to-\infty} G_x(z) &= \lim_{z\to-\infty} G_y(z) = 1 \\ \lim_{z\to-\infty} H_x(z) &= z - z_{ox} \\ \lim_{z\to-\infty} H_y(z) &= z - z_{oy} \end{aligned} \tag{110}$$

where $z = z_{ox}$ and $z = z_{oy}$ contain the virtual objects in the x–z and y–z planes respectively (Fig. 9A). We require two values of the characteristic function: $V(z_{ox}, z_i)$ between the object plane $z = z_{ox}$ and the plane $z = z_i$ conjugate to it in the x–z plane and $V(z_{oy}, z_j)$ between $z = z_{oy}$ and the plane $z = z_j$ conjugate to it in the y–z plane. Each of these values of the characteristic consists of three sections:

$$\begin{aligned} V(z_{ox}, z_i) &= V(z_{ox}, -\infty) + V(-\infty, \infty) + V(\infty, z_i) \\ V\left(z_{oy}, z_j\right) &= V\left(z_{oy}, -\infty\right) + V(-\infty, \infty) + V\left(\infty, z_j\right). \end{aligned} \tag{111}$$

The term $V(-\infty,\infty)$, common to both, represents $\int_{-\infty}^{\infty} \mu^{(4)} dz$, into which the general solutions $x(z) = x_o G_x(z) + \alpha H_x(z)$, $y(z) = y_0 G_y(z) + \gamma H_y(z)$ are substituted. The terms $V(z_{ox}, -\infty)$ and $V(z_{oy}, -\infty)$ represent integrals of the *constant term* in $\mu^{(4)}$ (Φ= constant, $D = Q = 0$) into which the asymptotes in object space are substituted: $G_x^* = G_y^* = 1$; $H_x^* = z - z_{ox}$, $H_y^* = z - z_{ay}$. (We use the asterisk to indicate asymptotic forms of the G- and H-rays; the distinction between asymptotes in object and image space is not shown explicitly but will be clear from the context.) Partial integration reduces these terms to very simple functions of the object coordinates. The terms

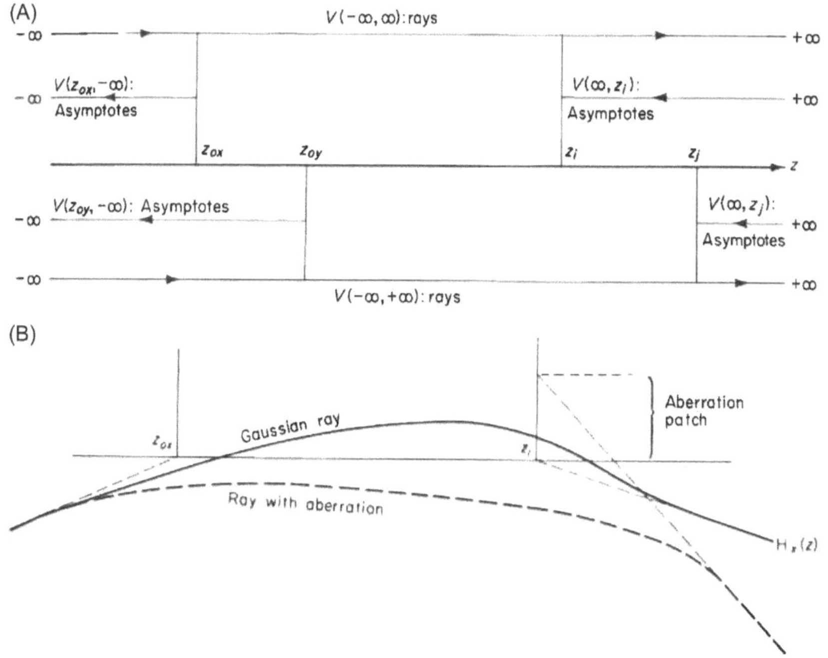

Fig. 9 (A) The regions in which the contributions to $V(z_{ox}, z_i)$ and to $V(z_{oy}, z_j)$ are calculated. (B) The meaning of asymptotic aberrations in one of the principal sections.

$V(\infty, z_i)$ and $V(\infty, z_j)$ likewise represent integrals of the constant term in $\mu^{(4)}$, into which the asymptotes in image space are substituted. Collecting together these pairs of terms, we find that

$$V(z_1, -\infty) + V(\infty, z_2) = -\frac{1}{8}\left[(xx' + yy')(x'^2 + y'^2)\right]_{z_1}^{z_2}. \tag{112}$$

The asymptotic aberrations (the physical significance of which is illustrated in Fig. 9B) are now given by

$$\begin{aligned}\Delta x(z_c) &= H^*_{xc}\frac{\partial V_x}{\partial x_o} - G^*_{xc}\frac{\partial V_x}{\partial \alpha} - \frac{1}{2}H^*_{xc}\alpha(\alpha^2 + \gamma^2)\\ \Delta y(z_d) &= H^*_{yd}\frac{\partial V_y}{\partial y_o} - G^*_{yd}\frac{\partial V_y}{\partial \gamma} - \frac{1}{2}H^*_{yd}\gamma(\alpha^2 + \gamma^2)\end{aligned} \tag{113}$$

in which $V_x = V(z_{ox}, z_c)$ and $V_y = V(z_{oy}, z_d)$; H^*_{xc}, H^*_{yd}, ... denote the heights of the asymptotes to the corresponding rays in image space, evaluated in the planes z_c or z_d.

The analysis that revealed the dependence of the asymptotic aberrations of round lenses upon object position or magnification can easily be extended to include quadrupoles. The consequences are particularly interesting because quadrupole multiplets are frequently designed with some degree of symmetry, and this is reflected in the coefficients that figure in the various polynomials.

In general, the characteristic function between two planes z_o and z_c is of the form

$$V_{oc} = \begin{pmatrix} \xi_o^2 \\ \eta_o^2 \\ \alpha_o^2 \\ \gamma_o^2 \\ \xi_o\alpha_o \end{pmatrix} \begin{pmatrix} 4000 & 2200 & 2020 & 2002 & 3010 & 2101 \\ 0 & 0400 & 0220 & 0202 & 1210 & 0301 \\ 0 & 0 & 0040 & 0022 & 1030 & 0121 \\ 0 & 0 & 0 & 0004 & 1012 & 0103 \\ 0 & 0 & 0 & 0 & 0 & 1111 \end{pmatrix} \begin{pmatrix} \xi_o^2 \\ \eta_o^2 \\ \alpha_o^2 \\ \gamma_o^2 \\ \xi_o\alpha_o \\ \eta_o\gamma_o \end{pmatrix} \tag{114}$$

in which

$$\xi_o = x_o/f \quad \eta_0 = y_o/g. \tag{115}$$

If the lens or multiplet is *symmetric*, several of the coefficients are equal; to show this, we substitute $\xi_o = -\alpha_c$, $\eta_o = -\gamma_c$, $\alpha_o = x_c/f = \xi_c$ and $\gamma_o = y_c/g = \eta_c$ giving

$$V_{oc} = \begin{pmatrix} \xi_c^2 \\ \eta_c^2 \\ \alpha_c^2 \\ \gamma_c^2 \\ \xi_c\alpha_c \end{pmatrix} \begin{pmatrix} 0040 & 0022 & 2020 & 0220 & 1030 & 0121 \\ 0 & 0004 & 2002 & 0202 & 1012 & 0103 \\ 0 & 0 & 4000 & 2200 & 3010 & 2101 \\ 0 & 0 & 0 & 0400 & 1210 & 0301 \\ 0 & 0 & 0 & 0 & 0 & 1111 \end{pmatrix} \begin{pmatrix} \xi_c^2 \\ \eta_c^2 \\ \alpha_c^2 \\ \gamma_c^2 \\ \xi_c\alpha_c \\ \eta_c\gamma_c \end{pmatrix} \tag{116}$$

so that

$$\begin{aligned} (4000) &= (0040) \quad & (0121) &= (2101) \\ (2200) &= (0022) \quad & (0400) &= (0004) \\ (2002) &= (0220) \quad & (1210) &= (1012) \\ (1030) &= (3010) \quad & (0103) &= (0301) \end{aligned} \tag{117}$$

or

$$(pqrs) = (rspq)(\text{symmetric quadrupole or system of quadrupoles}). \quad (118)$$

Another combination that has been thoroughly studied is the *antisymmetric* quadrupole multiplet. A multiplet is said to be antisymmetric if its central plane is a plane of geometrical symmetry and electrical antisymmetry (Fig. 10A). Here again, the coefficients $(pqrs)$ are inter-related; we rearrange Eq. (116) into the form

$$V_{oc} = \begin{pmatrix} \eta_c^2 \\ \xi_c^2 \\ \gamma_c^2 \\ \alpha_c^2 \\ \eta_c\gamma_c \end{pmatrix} \begin{pmatrix} 0004 & 0022 & 0202 & 2002 & 0103 & 1012 \\ 0 & 0040 & 0220 & 2020 & 0121 & 1030 \\ 0 & 0 & 0400 & 2200 & 0301 & 1210 \\ 0 & 0 & 0 & 4000 & 2101 & 3010 \\ 0 & 0 & 0 & 0 & 0 & 1111 \end{pmatrix} \begin{pmatrix} \eta_c^2 \\ \xi_c^2 \\ \gamma_c^2 \\ \alpha_c^2 \\ \eta_c\gamma_c \\ \xi_c\alpha_c \end{pmatrix} \quad (119)$$

and deduce that

$$\begin{aligned} (4000) &= (0004) \quad (2101) = (1012) \\ (0022) &= (2200) \quad (0400) = (0040) \\ (2020) &= (0202) \quad (1210) = (0121) \\ (3010) &= (0103) \quad (0301) = (1030) \end{aligned} \quad (120)$$

or

$$(pqrs) = (srqp). \quad (\text{antisymmetric multiplet}) \quad (121)$$

Let us now consider the asymptotic aberrations at the image foci in the two orthogonal planes when the astigmatic object is situated at the two object foci. We write

$$\begin{aligned} V_F &= V(z_{Fo}, z_{Fi}) \\ V_G &= V(z_{Go}, z_{Gi}) \\ V &= V(-\infty, \infty). \end{aligned} \quad (122)$$

The H-rays are defined by the initial conditions

$$\begin{aligned} \lim_{z\to-\infty} H_x(z) &= z - z_{Fo} \\ \lim_{z\to-\infty} H_y(z) &= z - z_{Go}. \end{aligned} \quad (123)$$

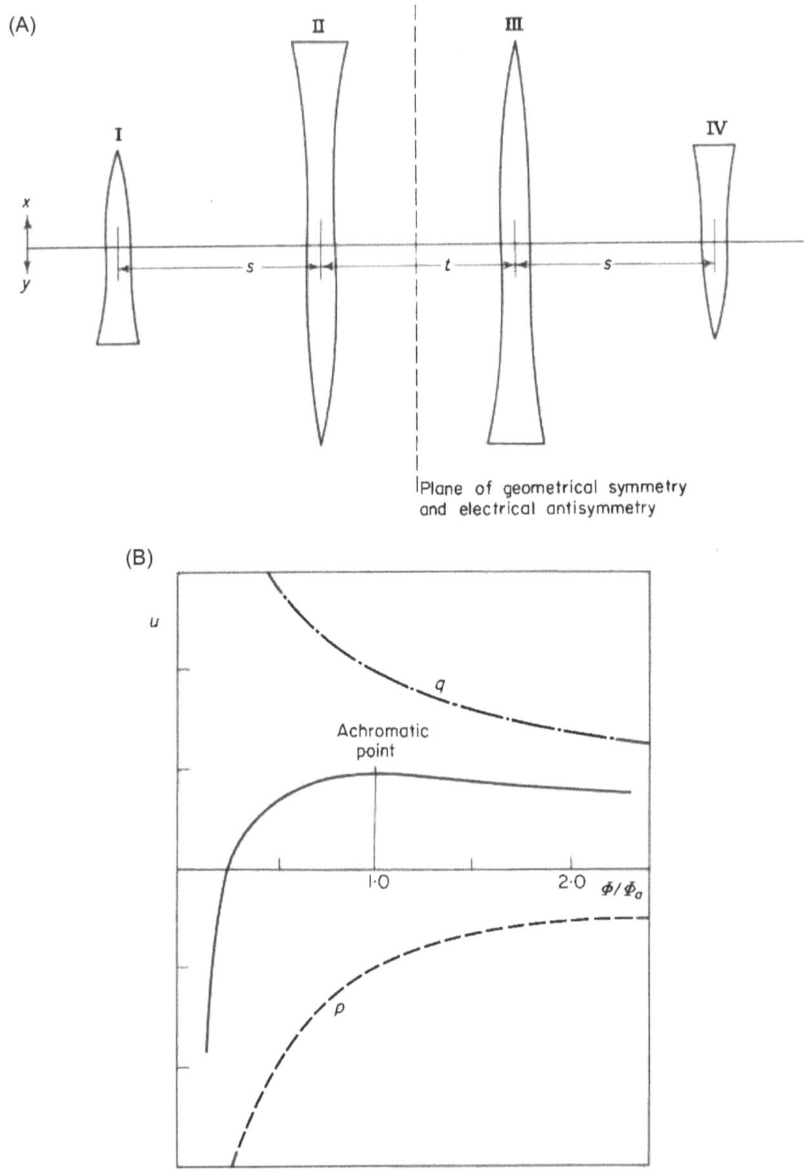

Fig. 10 (A) Schematic representation of an antisymmetric quadruplet; the center pair of lenses may be interchanged. (B) The variation of the excitation parameter $\beta^2(= u)$ with Φ/Φ_0; the dotted curves show the behavior of $\beta_M^2(= q)$ and $-\beta_E^2(= p)$. Φ_a denotes the accelerating voltage at which the lens is achromatic. *After Hardy, D.F. (1967).* Combined magnetic and electrostatic quadrupole electron lenses. *Dissertation, Cambridge University.*

Each of the functions V_F, V_G and V can of course be put in the general form of Eq. (114), but the matrix elements of V_F and V_G differ from those of V only by constants. Denoting the matrix elements that appear in V_F, V_G and V by $(pqrs)_F$, $(pqrs)_G$ and $(pqrs)$ respectively, we find that

$$
\begin{array}{ll}
(4000)_F = (4000) & (0400)_G = (0400) \\
(0040)_F = (0040) & (0004)_G = (0004) \\
(2200)_F = (2200) - d_i/8 & (2200)_G = (2200) + d_i/8 \\
(2020)_F = (2020) & (0202)_G = (0202) \\
(2002)_F = (2002)_G = (2002) & (0220)_F = (0202)_G = (0202) \\
(0022)_F = (0022) + d_o/8 & (0022)_G = (0022) + d_o/8 \\
(3010)_F = (3010) + f/8 & (0301)_G = (0301) + g/8 \\
(1030)_F = (1030) + f/8 & (0103)_G = (0103) + g/8 \\
(2101)_F = (2101)_G = (2101) + g/8 & \\
(1210)_F = (1210)_G = (1210) + f/8 & \\
(0121)_F = (0121)_G = (0121) + g/8 & \\
(1012)_F = (1012)_G = (1012) + f/8 & \\
(1111)_F = (1111)_G = (1111) &
\end{array}
\tag{124}
$$

in which

$$d_o = z_{Fo} - z_{Go};\ d_i = z_{Fi} - z_{Gi}. \tag{125}$$

We shall also need the relations

$$
\begin{aligned}
z_{ox} - z_{oy} &= d_o + \frac{f}{M_x} - \frac{g}{M_y} \\
z_i - z_j &= d_i - fM_x + gM_y
\end{aligned}
\tag{126}
$$

in which z_i is conjugate to z_{ox} with magnification M_x and z_j to z_{oy} with magnification M_y.

We now derive $\Delta x(z_{Fi})$ and $\Delta y(z_{Gi})$ with the aid of Eq. (113) and transfer the results to the arbitrary pairs of conjugate planes z_{ox} and z_i in the x–z plane and z_{oy} and z_j in the y–z plane (Hawkes, 1968b, 1968c). We write

$$\xi = x(z_{ox})/f \quad \eta = y(z_{oy})/g \tag{127}$$

and obtain

$$\begin{aligned}\Delta x(z_i) = M_x\{&(3000)\xi^3 + (1200)\xi\eta^2 + (0030)\alpha^3 + (0012)\alpha\gamma^2 + (2010)\xi^2\alpha \\ &+ (0210)\eta^2\alpha + (1101)\xi\eta\gamma + (1020)\xi\alpha^2 + (1002)\xi\gamma^2 + (0111)\eta\alpha\gamma\} \\ \Delta y(z_j) = M_y\{&(0300)\eta^3 + (2100)\xi^2\eta + (0003)\gamma^3 + (0021)\alpha^2\gamma + (2001)\xi^2\gamma \\ &+ (0201)\eta^2\gamma + (1110)\xi\eta\alpha + (0120)\eta\alpha^2 + (0102)\eta\gamma^2 + (1011)\xi\alpha\gamma\}.\end{aligned} \tag{128}$$

The coefficients are now polynomials in $1/M_x$ and $1/M_y$; writing

$$m_x = 1/M_x \quad m_y = 1/M_y \tag{129}$$

we find

$$\left.\begin{aligned}(3000) &= 4(4000)m_x - (3010) + 3f/8 \\ (1200) &= \{2(2200) - d_i/4\}m_x - (1210) + 3f/8 \\ (0300) &= 4(0400)m_y - (0301) + 3g/8 \\ (2100) &= \{2(2200) + d_i/4\}m_y - (2101) + 3g/8\end{aligned}\right\} \text{distortions}$$

$$\left.\begin{aligned}(2010) &= -12(4000)m_x^2 + \{6(3010) - 3f/4\}m_x - 2(2020) \\ (0210) &= -\{2(2200) - d_i/4\}m_x^2 + \{2(1210) - f/4\}m_x - 2(0220) \\ (1101) &= -\{4(2200) - d_i/2\}m_x m_y + \{2(2101) + g/4\}m_x \\ &\quad + \{2(1210) - 3f/4\}m_y - (1111) \\ (2001) &= -\{2(2200) + d_i/4\}m_y^2 + \{2(2101) - g/4\}m_y - 2(2002) \\ (0201) &= -12(0400)m_y^2 + \{6(0301) - 3g/4\}m_y - 2(0202) \\ (1110) &= -\{4(2200) + d_i/2\}m_x m_y + \{2(2101) - 3g/4\}m_x \\ &\quad + \{2(1210) + f/4\}m_y - (1111)\end{aligned}\right\} \text{astigmatisms}$$

$$
\left.\begin{aligned}
(1020) &= 12(4000)m_x^3 - \{9(3010) - 3f/8\}m_x^2 + 6(2020)m_x \\
&\quad -3(1030) - 3f/8 \\
(1002) &= \{2(2200) - d_i/4\}m_x m_y^2 - \{(1210) - 3f/8\}m_y^2 - \{2(2101) \\
&\quad + g/4\}m_x m_y + 2(2002)m_x + (1111)m_y - (1012) - f/8 \\
(0111) &= \{4(2200) - d_i/2\}m_x^2 m_y - \{2(2101) + g/4\}m_x^2 - \{4(1210) \\
&\quad - f/2\}m_x m_y + 2(1111)m_x + 4(0220)m_y - 2(0121) - g/4 \\
(0120) &= \{2(2200) + d_i/4\}m_x^2 m_y - \{(2101) - 3g/8\}m_x^2 - \{2(1210) \\
&\quad + f/4\}m_x m_y + (1111)m_x + 2(0220)m_y - (0121) - g/8 \\
(0102) &= 12(0400)m_y^3 - \{9(0301) - 3g/8\}m_y^2 + 6(0202)m_y \\
&\quad -3(0103) - 3g/8 \\
(1011) &= \{4(2200) + d_i/2\}m_x m_y^2 - \{2(1210) + f/4\}m_y^2 - \{4(2101) \\
&\quad - g/2\}m_x m_y + 4(2002)m_x + 2(1111)m_y - 2(1012) - f/4
\end{aligned}\right\} \text{coma}
$$

$$
\left.\begin{aligned}
(0030) &= -4(4000)m_x^4 + 4(3010)m_x^3 - 4(2020)m_x^2 + 4(1030)m_x \\
&\quad -4(0040) \\
(0012) &= -\{2(2200) - d_i/4\}m_x^2 m_y^2 + \{2(2101) + g/4\}m_x^2 m_y \\
&\quad + \{2(1210) - f/4\}m_x m_y^2 - 2(2002)m_x^2 - 2(0220)m_y^2 \\
&\quad -2(1111)m_x m_y + \{2(1012) - f/4\}m_x + \{2(0121) \\
&\quad + g/4\}m_y - 2(0022) - d_o/4 \\
(0003) &= -4(0400)m_y^4 + 4(0301)m_y^3 - 4(0202)m_y^2 + 4(0103)m_y \\
&\quad -4(0004) \\
(0021) &= -\{2(2200) + d_i/4\}m_x^2 m_y^2 + \{2(2101) - g/4\}m_x^2 m_y \\
&\quad + \{2(1210) + f/4\}m_x m_y^2 - 2(2002)m_x^2 - 2(0220)m_y^2 \\
&\quad -2(1111)m_x m_y + \{2(1012) + f/4\}m_x + \{2(0121) \\
&\quad - g/4\}m_y - 2(0022) + d_o/4
\end{aligned}\right\} \text{aperture aberrations}
$$

(130)

Of these coefficients, the most important in practice are the four aperture aberrations (00*rs*) and the four distortions (*pq*00). It is clear from Eq. (130) that these are related; if we write

$$\begin{aligned}
(3000) &= \sum_{i=0}^{1} (3000)_i m_x^i & (0300) &= \sum_{i=0}^{1} (0300)_i m_y^i \\
(1200) &= \sum_{i=0}^{1} (1200)_i m_x^i & (2100) &= \sum_{i=0}^{1} (2100)_i m_y^i
\end{aligned} \tag{131}$$

and

$$\begin{aligned}
(0030) &= \sum_{i=0}^{4} (0030)_i m_x^i & (0003) &= \sum_{i=0}^{4} (0003)_i m_y^i \\
(0012) &= \sum_{i,j=0}^{2} (0012)_{ij} m_x^i m_y^j & (0021) &= \sum_{i,j=0}^{2} (0021)_{ij} m_x^i m_y^j
\end{aligned} \tag{132}$$

then

$$\begin{aligned}
(3000)_1 &= -(0030)_4 & (0300)_1 &= -(0003)_4 \\
(3000)_0 &= \frac{3f - 2(0030)_3}{8} & (0300)_0 &= \frac{3g - 2(0003)_3}{8} \\
(1200)_1 &= -(0012)_{22} & (2100)_1 &= -(0021)_{22} \\
(1200)_0 &= \frac{f - 2(0012)_{12}}{4} & (2100)_0 &= \frac{g - 2(0021)_{21}}{4}.
\end{aligned} \tag{133}$$

So far, we have considered only aberrations of position, $\Delta x(z_i)$ and $\Delta y(z_j)$; Meads (1963) has also considered aberrations of slope, $\Delta x'$ and $\Delta y'$; the corresponding coefficients are defined by

$$\begin{aligned}
f\Delta x'(z_i) &= \sum_{p,q,r,s} [pqrs]\xi^p \eta^q \alpha^r \gamma^s & p+q+r+s &= 3 \\
& & p+r \quad & \text{odd} \\
g\Delta y'(z_j) &= \sum_{p,q,r,s} [pqrs]\xi^p \eta^q \alpha^r \gamma^s & p+q+r+s &= 3 \\
& & q+s \quad & \text{odd.}
\end{aligned} \tag{134}$$

The coefficients $[pqrs]$ are listed in the form of polynomials in the reciprocal magnifications m_x and m_y in Hawkes (1968b). Meads has established relations between the coefficients $(pqrs)$ and $[pqrs]$ which offer a very convenient means of checking calculations of the various coefficients; in the present notation, these relations take the following form (we have generalized the relation between (0021) and (0012) to the case of an astigmatic object):

$$
\begin{aligned}
[2010] + 3(3000)m_x + (2010) - 3fm_x/2 &= 0\\
2[1020] + 2(2010)m_x + 2(1020) + 3fm_x^2 &= 0\\
3[0030] + (1020)m_x + 3(0030) + 3fm_x\left(1 - m_x^2\right)/2 &= 0\\
[0210] + (1200)m_x + (0210) - fm_x/2 &= 0\\
[0111] + (1101)m_x + (0111) + fm_x m_y &= 0\\
[0012] + (1002)m_x + (0012) + fm_x\left(1 - m_y^2\right)/2 &= 0
\end{aligned}
$$

$$
\begin{aligned}
[1200]m_y - [2100]m_x + (1200)m_y - (2100)m_x - gm_x/2 + fm_y/2 + Lm_x m_y/2 &= 0\\
[1101]m_y - 2[2001]m_x + (1101)m_y - 2(2001)m_x + gm_x m_y - fm_y^2 - Lm_x m_y^2 &= 0\\
2[0210]m_y - [1110]m_x + 2(0210)m_y - (1110)m_x + gm_x^2 - fm_x m_y - Lm_x^2 m_y &= 0\\
[0111]m_y - [1011]m_x + (0111)m_y - (1011)m_x - gm_x^2 m_y + fm_x m_y^2 + Lm_x^2 m_y^2 &= 0
\end{aligned}
$$

$$
\begin{aligned}
[1101] + 2(2100)m_x + (1101) - fm_y - Lm_x m_y &= 0\\
2[1002] + 2(2001)m_x + 2(1002) + fm_y^2 + Lm_x m_y^2 &= 0\\
[0111] + (1110)m_x + (0111) + fm_x m_y + Lm_x^2 m_y &= 0\\
2[0012] + (1011)m_x + 2(0012) + fm_x\left(1 - m_y^2\right) - Lm_x^2 m_y^2 &= 0
\end{aligned}
$$

$$
\begin{aligned}
[1110] + 2(1200)m_y + (1110) - gm_x + Lm_x m_y &= 0\\
[1011] + (1101)m_y + (1011) + gm_x m_y - Lm_x m_y^2 &= 0\\
2[0120] + 2(0210)m_y + 2(0120) + gm_x^2 - Lm_x^2 m_y &= 0\\
2[0021] + (0111)m_y + 2(0021) + gm_y\left(1 - m_x^2\right) + Lm_x^2 m_y^2 &= 0
\end{aligned}
$$

$$
\begin{aligned}
(1110) - (1101) + Lm_x m_y &= 0\\
(1011) - 2(1002) - Lm_x m_y^2 &= 0\\
2(0120) - (0111) - Lm_x^2 m_y &= 0\\
(0021) - (0012) + Lm_x^2 m_y^2/2 - L_0/2 &= 0
\end{aligned}
$$

$$
\begin{aligned}
[2001] + (2100)m_y + (2001) - gm_y/2 &= 0\\
[1011] + (1110)m_y + (1011) + gm_x m_y &= 0\\
[0021] + (0120)m_y + (0021) + gm_y\left(1 - m_x^2\right)/2 &= 0\\
[0201] + 3(0300)m_y + (0201) - 3gm_y/2 &= 0\\
[0102] + (0201)m_y + (0102) + 3gm_y^2/2 &= 0\\
3[0003] + (0102)m_y + 3(0003) + 3gm_y\left(1 - m_y^2\right)/2 &= 0.
\end{aligned}
\tag{135}
$$

In these relations, $L = d_i + gM_y - fM_x = z_i - z_j$ and $L_0 = d_o + fm_x - gm_y = z_{ox} - z_{oy}$ (Eq. 126).

Let us now consider the effect of symmetry and antisymmetry on the coefficients $(0030)_i$, $(0003)_i$, $(0012)_{ij}$ and $(0021)_{ij}$. In the case of a *symmetric* lens or system of lenses, Eq. (117) show that the aperture aberrations take the form

$$\begin{aligned}
(0030) &= -4(4000)m_x^4 + 4(3010)m_x^3 - 4(2020)m_x^2 + 4(3010)m_x - 4(4000)\\
(0003) &= -4(0400)m_y^4 + 4(0301)m_y^3 - 4(0202)m_y^2 + 4(0301)m_y - 4(0400)\\
(0012) &= -\{2(2200) - d/4\}m_x^2m_y^2 + \{2(2101) + g/4\}m_x^2m_y\\
&\quad + \{2(1210) - f/4\}m_xm_y^2 - 2(2002)m_x^2 - 2(2002)m_y^2\\
&\quad -2(1111)m_xm_y + \{2(1210) - f/4\}m_x + \{2(2101) + g/4\}m_y\\
&\quad -\{2(2200) - d/4\}\\
(0021) &= -\{2(2200) + d/4\}m_x^2m_y^2 + \{2(2101) - g/4\}m_x^2m_y\\
&\quad + \{2(1210) + f/4\}m_xm_y^2 - 2(2002)m_x^2 - 2(2002)m_y^2\\
&\quad -2(1111)m_xm_y + \{2(1210) + f/4\}m_x + \{2(2101) - g/4\}m_y\\
&\quad -\{2(2200) + d/4\}
\end{aligned} \tag{136}$$

where

$$d = d_i = -d_o. \tag{137}$$

Hence

$$\begin{aligned}
&(0030)_4 = (0030)_0 \quad (0003)_4 = (0003)_0\\
&(0030)_3 = (0030)_1 \quad (0003)_3 = (0003)_1\\
&(0012)_{22} = (0012)_{00} \quad (0021)_{22} = (0021)_{00}\\
&(0012)_{21} = (0012)_{01} \quad (0021)_{21} = (0021)_{01}\\
&(0012)_{12} = (0012)_{10} \quad (0021)_{12} = (0021)_{10}\\
&(0012)_{20} = (0012)_{02} = (0021)_{20} = (0021)_{02} \quad \text{(symmetric}\\
&(0012)_{11} = (0021)_{11}. \qquad\qquad\qquad\qquad\qquad \text{system)}
\end{aligned} \tag{138}$$

When the system is *antisymmetric*, Eq. (120) give

$$
\begin{aligned}
(0030) &= -4(4000)m_x^4 + 4(3010)m_x^3 - 4(2020)m_x^2 + 4(1030)m_x - 4(0040)\\
(0003) &= -4(0040)m_y^4 + 4(1030)m_y^3 - 4(2020)m_y^2 + 4(3010)m_y - 4(4000)\\
(0012) &= -\{2(2200) - d/4\}m_x^2m_y^2 + \{2(2101) + f^*\}m_x^2m_y\\
&\quad + \{2(1210) - f^*\}m_xm_y^2 - 2(2002)m_x^2 - 2(0220)m_y^2\\
&\quad -2(1111)m_xm_y + \{2(2101) - f^*\}m_x + \{2(1210) + f^*\}m_y\\
&\quad -\{2(2200) + d/4\}\\
(0021) &= -\{2(2200) + d/4\}m_x^2m_y^2 + \{2(2101) - f^*\}m_x^2m_y\\
&\quad + \{2(1210) + f^*\}m_xm_y^2 - 2(2002)m_x^2 - 2(0220)m_y^2\\
&\quad -2(1111)m_xm_y + \{2(2101) + f^*\}m_x + \{2(1210) - f^*\}m_y\\
&\quad -\{2(2200) - d/4\}
\end{aligned}
\tag{139}
$$

where

$$
\begin{aligned}
f^* &= f/4 = g/4\\
d &= d_i = d_o
\end{aligned}
\tag{140}
$$

Hence

$$
\begin{aligned}
&(0030)_4 = (0003)_0 \quad (0003)_4 = (0030)_0\\
&(0030)_3 = (0003)_1 \quad (0003)_3 = (0030)_1\\
&\qquad (0030)_2 = (0003)_2\\
&(0012)_{22} = (0021)_{00} \quad (0012)_{00} = (0021)_{22}\\
&(0012)_{21} = (0021)_{10} \quad (0012)_{12} = (0021)_{01}\\
&(0012)_{20} = (0021)_{20} \quad (0012)_{02} = (0021)_{02}\\
&(0012)_{10} = (0021)_{21} \quad (0012)_{01} = (0021)_{12} \quad \text{(antisymmetric}\\
&\qquad (0012)_{11} = (0021)_{11}. \qquad \text{system)}
\end{aligned}
\tag{141}
$$

If the antisymmetric system is regular, then $d = 0$ and so $(0012)_{22} = (0012)_{00} = (0021)_{22} = (0021)_{00}$.

If the object of a regular system is stigmatic, the expressions for (0012) and (0021) collapse to

$$
\begin{aligned}
(0012) = (0021) = &-2(2200)m^4 + 2\{(2101) + (1210)\}m^3 \\
&-2\{(2002) + (0220) + (1111)\}m^2 + 2\{(1012) + (0121)\}m - 2(0022)
\end{aligned}
\tag{142}
$$

and the coefficients $(0012)_i$ and $(0021)_i$ in

$$
(0012) = \sum_{i=0}^{4} (0012)_i m^i, \quad (0021) = \sum_{i=0}^{4} (0021)_i m^i
$$

satisfy $(0012)_i = (0021)_i$; we have set $m_x = m_y = m$. If the system is antisymmetric or symmetric, $(0012)_4 = (0012)_0$ and $(0012)_3 = (0012)_1$.

We conclude this section with a list of formulae for the aperture aberration and distortion coefficients of systems of electrostatic round lenses, quadrupoles and octopoles. These expressions can be cast into a variety of forms (Hawkes, 1967b): derivatives of the potential functions may be eliminated, which is an advantage if these are obtained by relaxation methods; the number of terms may be reduced, which minimizes the labor of obtaining analytical expressions; and rearrangement may reveal that some coefficient cannot be made to vanish. The following list includes most of the expressions that have been found useful with some indication of pedigree.

a. *General formulae.* Electrostatic round lenses $\Phi(z)$, quadrupoles $D(z)$ and $Q(z)$ and octopoles $D_1(z)$ and $Q_1(z)$ or $\Xi(z)$. (Real aberrations, aperture coordinates, relativistically correct form.)

$$
\begin{aligned}
k_x \Delta x(z_c) = &\left(t_{xc}\int_a^c \sigma_{x4}\,dz - s_{xc}\int_o^c \sigma_{x3}\,dz\right)x_o^3 + \left(t_{xc}\int_a^c \sigma_{xy}\,dz - s_{xc}\int_o^c \sigma_{12}\,dz\right)x_o y_o^2 \\
&+ \left(t_{xc}\int_a^c \tau_{x3}\,dz - s_{xc}\int_o^c \tau_{x4}\,dz\right)x_a^3 + \left(t_{xc}^c\int_a^c \tau_{12}\,dz - s_{xc}\int_o^c \tau_{xy}\,dz\right)x_a y_a^2 \\
k_y \Delta y(z_c) = &\left(t_{yc}\int_a^c \sigma_{y4}\,dz - s_{yc}\int_o^c \sigma_{y3}\,dz\right)y_o^3 + \left(t_{yc}\int_a^c \sigma_{xy}\,dz - s_{yc}\int_o^c \sigma_{21}\,dz\right)x_o^2 y_o \\
&+ \left(t_{yc}\int_a^c \tau_{y3}\,dz - s_{yc}\int_o^c \tau_{y4}\,dz\right)y_a^3 + \left(t_{yc}\int_a^c \tau_{21}\,dz - s_{yc}\int_o^c \tau_{xy}\,dz\right)x_a^2 y_a
\end{aligned}
\tag{143}
$$

with

$$
\begin{aligned}
\sigma_{x4} &= X_1 s_x^4 + 2X_2 s_x^2 s_x'^2 + Z s_x'^4 \\
\sigma_{y4} &= Y_1 s_y^4 + 2Y_2 s_y^2 s_y'^2 + Z s_y'^4 \\
\tau_{x4} &= X_1 t_x^4 + 2X_2 t_x^2 t_x'^2 + Z t_x'^4 \\
\tau_{y4} &= Y_1 t_y^4 + 2Y_2 t_y^2 t_y'^2 + Z t_y'^4 \\
\sigma_{x3} &= X_1 s_x^3 t_x + X_2 s_x s_x' (s_x t_x)' + Z s_x'^3 t_x' \\
\sigma_{y3} &= Y_1 s_y^3 t_y + Y_2 s_y s_y' (s_y t_y)' + Z s_y'^3 t_y' \\
\tau_{x3} &= X_1 s_x t_x^3 + X_2 (s_x t_x)' t_x t_x' + Z s_x' t_x'^3 \\
\tau_{y3} &= Y_1 s_y t_y^3 + Y_2 (s_y t_y)' t_y t_y' + Z s_y' t_y'^3 \\
\sigma_{xy} &= V^\dagger s_x^2 s_y^2 + W\left\{ s_x^2 \left(s_y^2\right)' - \left(s_x^2\right)' s_y^2 \right\} + X_2 s_x^2 s_y'^2 + Y_2 s_x'^2 s_y^2 + Z s_x'^2 s_y'^2 \\
\tau_{xy} &= V^\dagger t_x^2 t_y^2 + W\left\{ t_x^2 \left(t_y^2\right)' - \left(t_x^2\right)' t_y^2 \right\} + X_2 t_x^2 t_y'^2 + Y_2 t_x'^2 t_y^2 + Z t_x'^2 t_y'^2 \\
\sigma_{21} &= V^\dagger s_x^2 s_y t_y + W\left\{ s_x^2 (s_y t_y)' - \left(s_x^2\right)' s_y t_y \right\} + X_2 s_x^2 s_y' t_y' + Y_2 s_x'^2 s_y t_y + Z s_x'^2 s_y' t_y' \\
\tau_{21} &= V^\dagger t_x^2 s_y t_y + W\left\{ \left(t_x^2 (s_y t_y)' - \left(t_x^2\right)' s_y t_y \right\} + X_2 t_x^2 s_y' t_y' + Y_2 t_x'^2 s_y t_y + Z t_x'^2 s_y' t_y' \\
\sigma_{12} &= V^\dagger s_x t_x s_y^2 + W\left\{ s_x t_x \left(s_y^2\right)' - (s_x t_x)' s_y^2 \right\} + X_2 s_x t_x s_y'^2 + Y_2 s_x' t_x' s_y^2 + Z s_x' t_x' s_y'^2 \\
\tau_{12} &= V^\dagger s_x t_x t_y^2 + W\left\{ s_x t_x \left(t_y^2\right)' - (s_x t_x)' t_y^2 \right\} + X_2 s_x t_x t_y'^2 + Y_2 s_x' t_x' t_y^2 + Z s_x' t_x' t_y'^2
\end{aligned}
\tag{144}
$$

in which

$$
\begin{aligned}
X_1 &= \frac{\sigma}{32} \frac{\Phi^{(\text{iv})}}{V^{1/2}} - \frac{(\Phi'' - D)^2}{32 V^{3/2}} - \frac{\sigma D'' - 2\eta Q'' V^{1/2}}{24 V^{1/2}} + \frac{2\sigma D_1}{V^{1/2}} - 4\eta Q_1; \\
Y_1 &= \frac{\sigma}{32} \frac{\Phi^{(\text{iv})}}{V^{1/2}} - \frac{(\Phi'' + D)^2}{32 V^{3/2}} + \frac{\sigma D'' - 2\eta Q'' V^{1/2}}{24 V^{1/2}} + \frac{2\sigma D_1}{V^{1/2}} - 4\eta Q_1; \\
X_2 &= -\frac{\sigma}{8} \frac{\Phi'' - D}{V^{1/2}}; \quad Y_2 = -\frac{\sigma}{8} \frac{\Phi'' + D}{V^{1/2}}; \\
V^\dagger &= \frac{\sigma}{32} \frac{\Phi^{(\text{iv})}}{V^{1/2}} + \frac{D^2 - \Phi''^2}{32 V^{3/2}} - 3\left(\frac{2\sigma D_1}{V^{1/2}} - 4\eta Q_1 \right); \\
W &= \frac{1}{4} \eta Q'; \quad Z = -\frac{1}{2} V^{1/2}
\end{aligned}
\tag{145}
$$

and we recall that the s- and t-rays satisfy the boundary conditions $s_x(z_o) = s_y(z_o) = t_x(z_a) = t_y(z_a) = 1$ and $s_x(z_a) = s_y(z_a) = t_x(z_o) = t_y(z_o) = 0$.

These formulae were given in Hawkes (1963, 1965b) together with relativistically correct formulae for all the geometrical aberration coefficients of orthogonal systems with straight axes and in particular, of systems of round electrostatic lenses, quadrupoles and octopoles. Simplified (nonrelativistic) formulae, in which the number of terms in the integrands is very considerably reduced and many of the derivatives of $S_x(z)$, $s_y(z)$, $t_x(z)$ and $t_y(z)$ are eliminated, are to be found in Hawkes (1965c); the relativistically correct formulae for the aperture aberrations of quadrupoles (no electrostatic round lens) are cast into a general form, from which any expression can be deduced straightforwardly, in Hawkes (1967b).

The aberration coefficients of quadrupoles were first derived by Melkich (1944); his formulae are shown to be equivalent to those of Hawkes (1965b) in an appendix to the latter.

b. *Quadrupoles and octopoles only, asymptotic aberrations.* (Aberrations expressed in terms of coordinates at the astigmatic object planes: x_o, α at $z = z_{ox}$ and y_o, γ at $z = z_{oy}$, and evaluated at the asymptotic image planes z_i and z_j conjugate to z_{ox} and z_{oy} respectively.)

$$
\begin{aligned}
x_o^3 : (3000) &= \int_{-\infty}^{\infty} F G_x^3 H_x dz + 1/2f^2 \\
y_o^3 : (0300) &= \int_{-\infty}^{\infty} \overline{F} G_y^3 H_y dz + 1/2g^2 \\
x_o y_o^2 : (1200) &= \int_{-\infty}^{\infty} \left[K G_x H_x G_y^2 + L\left\{ G_x H_x \left(G_y^2\right)' - (G_x H_x)' G_y^2 \right\}\right] dz + \frac{3}{8g^2} - \frac{d_i}{4fg^2} m_x \\
x_o^2 y_o : (2100) &= \int_{-\infty}^{\infty} \left[K G_x^2 G_y H_y - L\left\{ (G_x^2)' G_y H_y - G_x^2 \left(G_y H_y\right)' \right\}\right] dz + \frac{3}{8f^2} + \frac{d_i}{4f^2 g} m_y \\
\alpha^3 : (0030) &= \int_{-\infty}^{\infty} F H_x^4 dz \\
\gamma^3 : (0003) &= \int_{-\infty}^{\infty} \overline{F} H_y^4 dz \\
\alpha\gamma^2 : (0012) &= \int_{-\infty}^{\infty} \left[K H_x^2 H_y^2 + L\left\{ H_x^2 \left(H_y^2\right)' - (H_x^2)' H_y^2 \right\}\right] dz + \frac{1}{4}\left(d_i m_x^2 m_y^2 - d_o\right) \\
\alpha^2\gamma : (0021) &= \int_{-\infty}^{\infty} \left[K H_x^2 H_y^2 + L\left\{ H_x^2 \left(H_y^2\right)' - (H_x^2)' H_y^2 \right\}\right] dz - \frac{1}{4}\left(d_i m_x^2 m_y^2 - d_o\right)
\end{aligned}
\tag{146}
$$

where

$$
\begin{aligned}
F &= \frac{D^2 - 4\eta\sigma DQV^{1/2} + 6\eta^2 Q^2 V}{12V^2} + \frac{5}{24}\varepsilon\frac{D^2}{V} - \frac{\sigma D'' - 4\eta Q'' V^{1/2}}{96V} - \Xi \\
\bar{F} &= \frac{D^2 - 4\eta\sigma DQV^{1/2} + 6\eta^2 Q^2 V}{12V^2} + \frac{5}{24}\varepsilon\frac{D^2}{V} + \frac{\sigma D'' - 4\eta Q'' V^{1/2}}{96V} - \Xi \\
K &= -\frac{1}{8}\left(\frac{\sigma D - 2\eta Q V^{1/2}}{V}\right)^2 + \frac{\varepsilon D^2}{8V} + 3\Xi \\
L &= \frac{3}{32}\frac{\sigma D' - 4\eta Q' V^{1/2}}{V}
\end{aligned}
\tag{147}
$$

and we recall that

$$
\begin{aligned}
\Xi(z) &= \frac{2\sigma D_1 - 4\eta Q_1 V^{1/2}}{V} \\
d_i &= z_i - z_j \quad d_o = z_{ox} - z_{oy}.
\end{aligned}
\tag{148}
$$

c. *Quadrupoles and octopoles only.* The formulae of §b expressed in terms of the quantities β^2, $k(z)$ and n defined in Eqs. (11a–11f). ($n = -1$ corresponds to purely electrostatic quadrupoles, $n = 0$ to magnetic quadrupoles and, as we shall see later, $n = \sigma^2$ or nonrelativistically, $n = 1$ corresponds to achromatic quadrupoles.)

$$
\begin{aligned}
F &= \frac{1}{6}\beta^4 k^2\left(\frac{3n^2}{\sigma^2} - 2n + 3\right) + \frac{1}{24}\beta^2 k'' - \Xi \\
\bar{F} &= \frac{1}{6}\beta^4 k^2\left(\frac{3n^2}{\sigma^2} - 2n + 3\right) - \frac{1}{24}\beta^2 k'' - \Xi \\
K &= -\frac{1}{2}\beta^4 k^2\left(\frac{n^2}{\sigma^2} - 2n + 1\right) + 3\Xi \\
L &= -\frac{3}{8}\beta^2 k'.
\end{aligned}
\tag{149}
$$

d. *Quadrupoles and octopoles: asymptotic aperture aberrations.* (Formulae in which the potential functions appear undifferentiated.)

$$\alpha^3 : (0030) = \int_{-\infty}^{\infty} \left\{ \frac{1}{96} \left(\frac{7 + 16\varepsilon V}{V^2} D^2 - 24\eta\sigma \frac{DQ}{V^{3/2}} + 32\eta^2 \frac{Q^2}{V} \right) H_x^4 \right.$$
$$\left. - \frac{1}{8} \left(\frac{\sigma D}{V} - 4\eta \frac{Q}{V^{1/2}} \right) H_x^2 H_x'^2 \right\} dz - \int_{-\infty}^{\infty} \Xi H_x^4 dz$$
$$\gamma^3 : (0003) = \int_{-\infty}^{\infty} \left\{ \frac{1}{96} \left(\frac{7 + 16\varepsilon V}{V^2} D^2 - 24\eta\sigma \frac{DQ}{V^{3/2}} + 32\eta^2 \frac{Q^2}{V} \right) H_y^4 \right.$$
$$\left. + \frac{1}{8} \left\{ \frac{\sigma D}{V} - 4\eta \frac{Q}{V^{1/2}} \right) H_y^2 H_y'^2 \right\} dz - \int_{-\infty}^{\infty} \Xi H_y^4 dz$$
$$\alpha\gamma^2 : (0012) = -\frac{1}{32} \int_{-\infty}^{\infty} \left\{ \left(\frac{D^2}{V^2} + 8\eta\sigma \frac{DQ}{V^{3/2}} - 32\eta^2 \frac{Q^2}{V} \right) H_x^2 H_y^2 + 6 \left(\frac{\sigma D}{V} - 4\eta \frac{Q}{V^{1/2}} \right) \right.$$
$$\left. \times \left(H_x^2 H_y'^2 - H_x'^2 H_y^2 \right) \right\} dz + \frac{1}{4} \left(d_i m_x^2 m_y^2 - d_o \right) + 3 \int_{-\infty}^{\infty} \Xi H_x^2 H_y^2 dz.$$
$$\alpha^2\gamma : (0021) = \frac{1}{32} \int_{-\infty}^{2} \left\{ \left(\frac{D^2}{V^2} + 8\eta\sigma \frac{DQ}{V^{3/2}} - 32\eta^2 \frac{Q^2}{V} \right) H_x^2 H_y^2 + 6 \left(\frac{\sigma D}{V} - 4\eta \frac{Q}{V^{1/2}} \right) \right.$$
$$\left. \times \left(H_x^2 H_y'^2 - H_x'^2 H_y^2 \right) \right\} dz - \frac{1}{4} \left(d_i m_x^2 m_y^2 - d_o \right) + 3 \int_{-\infty}^{\infty} \Xi H_x^2 H_y^2 dz.$$
(150a)

e. *Quadrupoles and octopoles: asymptotic aperture aberrations.* [The formulae of §d in terms of the variables of Eqs. (11a–11f); cf. §c.]

$$\alpha^3 : (0030) = \int_{-\infty}^{\infty} \left\{ \frac{1}{6} \left(\frac{3n^2}{\sigma^2} - 2n + 2 \right) \beta^4 k^2 H_x^4 + \frac{1}{2} \beta^2 k H_x^2 H_x'^2 \right\} dz - \int_{-\infty}^{\infty} \Xi H_x^4 dz$$
$$\gamma^3 : (0003) = \int_{-\infty}^{\infty} \left\{ \frac{1}{6} \left(\frac{3n^2}{\sigma^2} - 2n + 2 \right) \beta^4 k^2 H_y^4 - \frac{1}{2} \beta^2 k H_y^2 H_y'^2 \right\} dz - \int_{-\infty}^{\infty} \Xi H_y^4 dz$$
$$\alpha\gamma^2 : (0012) = -\int_{-\infty}^{\infty} \left\{ \left(\frac{n^2}{2\sigma^2} - n - 1 \right) \beta^4 k^2 H_x^2 H_y^2 + \frac{3}{4} \beta^2 k \left(H_x^2 H_y'^2 - H_x'^2 H_y^2 \right) \right\} dz$$
$$+ 3 \int_{-\infty}^{\infty} \Xi H_x^2 H_y^2 dz + \frac{1}{4} \left(d_i m_x^2 m_y^2 - d_o \right)$$
$$\alpha^2\gamma : (0021) = -\int_{-\infty}^{\infty} \left\{ \left(\frac{n^2}{2\sigma^2} - n - 1 \right) \beta^4 k^2 H_x^2 H_y^2 + \frac{3}{4} \beta^2 k \left(H_x^2 H_y'^2 - H_x'^2 H_y^2 \right) \right\} dz$$
$$+ 3 \int_{-\infty}^{\infty} \Xi H_x^2 H_y^2 dz - \frac{1}{4} \left(d_i m_x^2 m_y^2 - d_o \right).$$
(150b)

f. *Quadrupoles: real aperture aberrations.* (Formulae showing which coefficients can change sign.)

At $z = z_i$, the real line image in the x–z plane, we have

$$\begin{aligned}
\alpha^3 : (0030) &= \frac{1}{6}\int_o^i h_x'^4 dz + \frac{1}{6}\int_o^i \left\{ 4\left(\frac{3}{2}\frac{\sigma D}{V} - 4\eta\frac{Q}{V^{1/2}}\right)^2 + \frac{D^2}{V^2}\left(6-\sigma^2\right)\right\} h_x^4 dz \\
&= \frac{1}{6}\int_o^i h_x'^4 dz + \frac{1}{6}\left(3n^2/\sigma^2 - 2n + 2\right)\beta^4 \int_o^i k^2 h_x^4 dz \\
\alpha\gamma^2 : (0012) &= -\frac{1}{2} h_{xi}'^2 h_{yi} h_{yi}' + \frac{3}{2}\int_o^i h_x'^2 h_y'^2 dz - \frac{1}{32}\int_o^i \left(\frac{D^2}{V^2} + 8\eta\frac{\sigma D Q}{V^{3/2}} - 32\eta^2\frac{Q^2}{V}\right) h_x^2 h_y^2 dz \\
&= -\frac{1}{2} h_{xi}'^2 h_{yi} h_{yi}' + \frac{3}{2}\int_o^i h_x'^2 h_y'^2 dz - \frac{1}{2}\left(n^2/\sigma^2 - 2n - 2\right)\beta^4 \int_o^i k^2 h_x^2 h_y^2 dz.
\end{aligned} \tag{151a}$$

At $z = z_j$, the real line image in the y–z plane, we have

$$\begin{aligned}
\gamma^3 : (0003) &= \frac{1}{6}\int_o^j h_y'^4 dz + \frac{1}{6}\int_o^j \left\{ 4\left(\frac{3}{2}\frac{\sigma D}{V} - 4\eta\frac{Q}{V^{1/2}}\right)^2 + \frac{D^2}{V^2}\left(6-\sigma^2\right)\right\} h_y^4 dz \\
&= \frac{1}{6}\int_o^j h_y'^4 dz + \frac{1}{6}\left(3n^2/\sigma^2 - 2n + 2\right)\beta^4 \int_o^j k^2 h_y^4 dz \\
\alpha^2\gamma : (0021) &= -\frac{1}{2} h_{xj} h_{xj}' h_{yj}'^2 + \frac{3}{2}\int_o^j h_x'^2 h_y'^2 dz - \frac{1}{32}\int_o^j \left(\frac{D^2}{V^2} + 8\eta\frac{\sigma D Q}{V^{3/2}} - 32\eta^2\frac{Q^2}{V}\right) h_x^2 h_y^2 dz \\
&= -\frac{1}{2} h_{xj} h_{xj}' h_{yj}'^2 + \frac{3}{2}\int_o^j h_x'^2 h_y'^2 dz - \frac{1}{2}\left(n^2/\sigma^2 - 2n - 2\right)\beta^4 \int_o^j k^2 h_x^2 h_y^2 dz.
\end{aligned} \tag{151b}$$

When a stigmatic image is formed at $z = z_i$ (and hence $z_i = z_j$), we have

$$\begin{aligned}
(0012) = (0021) &= \frac{3}{2}\int_o^i h_x'^2 h_y'^2 dz - \frac{1}{32}\int_o^i \left(\frac{D^2}{V^2} + 8\eta\frac{\sigma D Q}{V^{3/2}} - 32\eta^2\frac{Q^2}{V}\right) h_x^2 h_y^2 dz \\
&= \frac{3}{2}\int_o^i h_x'^2 h_y'^2 dz - \frac{1}{2}\left(n^2/\sigma^2 - 2n - 2\right)\beta^4 \int_o^i k^2 h_x^2 h_y^2 dz.
\end{aligned} \tag{151c}$$

(These formulae are of course not valid if either of the line foci is virtual.) It is clear that (0030) and (0003) cannot change sign in the nonrelativistic case, and that the mixed aberration coefficient of a stigmatic magnetic quadrupole system cannot change sign (Hawkes, 1966–67; Moses, 1966; Ovsyannikova & Yavor, 1965); in the nonrelativistic situation, (0012) = (0021) is always positive in a stigmatic system provided $1 - \sqrt{3} < n < 1 + \sqrt{3}$—this range includes magnetic and achromatic lens systems. When the relativistic modification becomes important, (0030) and (0003) can vanish however (Rose, 1967b); a negative term appears in the integrand when $6 - \sigma^2 < 0$, that is, when the accelerating voltage exceeds about 0.7 MV.

g. *Quadrupoles only; thin-lens approximation.* Dymnikov, Fishkova, and Yavor (1965b, 1965e) have calculated approximate formulae for the aperture aberrations on the assumption that the lens is thin (stigmatic object). These formulae are derived from expressions in which the function $k(z)$ appears undifferentiated, and Dymnikov et al. find

$$\begin{aligned}(0030) &= \frac{\kappa}{6}\left(3n^2 - 2n + 2\right)\frac{z_o^4}{f^2 L}\\ (0003) &= \frac{\kappa}{6}\left(3n^2 - 2n + 2\right)\frac{z_o^4}{g^2 L}\\ (0012) = (0021) &= \frac{\kappa}{6}\left(n^2 - 2n - 2\right)\frac{z_o^4}{fgL}\end{aligned} \qquad (152a)$$

and of course $f = -g$. Here, L denotes the effective length, $L = \int_{-\infty}^{\infty} k(z)dz$ and κ is a form factor,

$$\kappa = \int_{-\infty}^{\infty} k^2 dz \Big/ \int_{-\infty}^{\infty} kdz = \frac{1}{L}\int_{-\infty}^{\infty} k^2 dz. \qquad (152b)$$

For the class of bell-shaped fields for which

$$k(z) = 1/\left\{1 + (z/d)^2\right\}^n \qquad (152c)$$

we have (Dymnikov et al., 1965b)

$$\kappa = \frac{(2n-1)(2n+1)\ldots(4n-3)}{2n(2n+2)\ldots(4n-2)} \qquad (152d)$$

so that for $n = 2, \kappa = \frac{5}{8}$.

If, however, we consider the expressions for (0030) and (0003) from which derivatives of the h-rays have been eliminated (Eq. 149), we find

$$\begin{aligned}(0030) &= \frac{\kappa}{6}\left(3n^2 - 2n + 3\right)\frac{z_o^4}{f^2 L}\\ (0003) &= \frac{\kappa}{6}\left(3n^2 - 2n + 3\right)\frac{z_o^4}{g^2 L}\end{aligned} \qquad (152e)$$

since $\int_{-\infty}^{\infty} k'' dz$ vanishes. Different expressions for the mixed aberrations are also obtained. This may well explain why the thin-lens curves in Dymnikov, Fishkova, and Yavor (1965c) and Dymnikov et al. (1965e) are separated from the curves obtained using the full formula even in the region where one might expect the thin-lens approximation to be adequate.

3.5 Chromatic aberrations

Quadrupole lens systems differ from round lenses in that their chromatic aberration coefficients can be reduced to zero or made negative, provided that both electrostatic and magnetic quadrupoles are present. It is thus possible, in principle at least, to construct achromatic quadrupole multiplets and to design quadrupole correctors that can cancel the chromatic aberration of round lenses.

The chromatic aberrations of quadrupoles are easily obtained by considering the chromatic characteristic $\bar{K}(z_o, z_c)$, which we also write $\bar{K}_{oc}$, defined by

$$\bar{K}(z_o, z_c) = \int_{z_o}^{z_c} \bar{\mu}^{(c)} dz \tag{153a}$$

in which $\bar{\mu}^{(c)} = \partial\bar{\mu}^{(2)}/\partial\Phi$. The asymptotic formulae will be obtained if we regard $\bar{K}_{oc}$ as being composed of three sections,

$$\bar{K}_{oc} = \bar{K}(z_o, -\infty) + \bar{K}(-\infty, \infty) + \bar{K}(\infty, z_c). \tag{153b}$$

In a system of round electrostatic lenses, quadrupoles and octopoles, $\bar{\mu}^{(2)}$ has the form

$$\begin{aligned}\bar{\mu}^{(2)} = & -\frac{1}{8}\frac{\sigma}{V^{1/2}}\Phi''\left(x^2+y^2\right) + \left(\frac{1}{8}\frac{\sigma D}{V^{1/2}} - \frac{1}{2}\eta Q\right)\left(x^2-y^2\right) \\ & + \frac{1}{2}V^{1/2}\left(x'^2+y'^2\right)\end{aligned} \tag{154}$$

so that

$$\begin{aligned}\bar{\mu}^{(c)} = & \frac{1}{16}\frac{\Phi''}{V^{3/2}}\left(x^2+y^2\right) - \frac{1}{16}\frac{D}{V^{3/2}}\left(x^2-y^2\right) \\ & + \frac{\sigma}{4V^{1/2}}\left(x'^2+y'^2\right).\end{aligned} \tag{155}$$

Integrating the final term, we obtain

$$\bar{K} = \frac{1}{2}\int \bar{K}_\phi(z)\left(x^2+y^2\right)dz + \frac{1}{2}\int \bar{K}_q(z)\left(x^2-y^2\right)dz + \left[\bar{K}(z)\right] \tag{156}$$

in which

$$\begin{aligned}\bar{K}_\phi(z) &= \frac{\sigma(5+\sigma^2)}{16V^{5/2}}\Phi'^2\\ \bar{K}_q(z) &= -\frac{(1+\sigma^2)D - 4\eta\sigma Q V^{1/2}}{8V^{3/2}}\\ \bar{K}(z) &= \frac{\sigma}{4V^{1/2}}(xx'+yy') + \frac{(1+\sigma^2)\Phi'}{16V^{3/2}}\left(x^2+y^2\right).\end{aligned} \tag{157}$$

When there is no round lens component, we write

$$K_q(z) = \bar{K}_q(z)/V^{1/2} = -\frac{(1+\sigma^2)D - 4\eta\sigma Q V^{1/2}}{8V^2}$$

and

$$k_q(z) = VK_q(z) = -\frac{(1+\sigma^2)D - 4\eta\sigma Q V^{1/2}}{8V}.$$

a. *Real chromatic aberrations.* When the beam is confined within an aperture at $z = z_a$, the chromatic aberrations are of the form

$$\begin{aligned}k_x\Delta x(z_c) &= \left(\kappa_{cx}x_a + \kappa_{Mx}x_o\right)\Delta\Phi\\ k_y\Delta y(z_c) &= \left(\kappa_{cy}y_a + \kappa_{My}y_o\right)\Delta\Phi\end{aligned} \tag{158}$$

in which

$$\begin{aligned}\kappa_{cx} &= t_{xc}\int_a^c\left(\bar{K}_\phi + \bar{K}_q\right)s_x t_x dz - s_{xc}\int_o^c\left(\bar{K}_\phi + \bar{K}_q\right)t_x^2 dz\\ \kappa_{cy} &= t_{yc}\int_a^c\left(\bar{K}_\phi - \bar{K}_q\right)s_y t_y dz - s_{yc}\int_o^c\left(\bar{K}_\phi - \bar{K}_q\right)t_y^2 dz\\ \kappa_{Mx} &= t_{xc}\int_a^c\left(\bar{K}_\phi + \bar{K}_q\right)s_x^2 dz - s_{xc}\int_o^c\left(\bar{K}_\phi + \bar{K}_q\right)s_x t_x dz\\ \kappa_{My} &= t_{yc}\int_a^c\left(\bar{K}_\phi - \bar{K}_q\right)s_y^2 dz - s_{yc}\int_o^c\left(\bar{K}_\phi - \bar{K}_q\right)s_y t_y dz\end{aligned} \tag{159}$$

where as usual, $k_x = V^{1/2}(g_x h'_x - g'_x h_x)$ and $k_y = V^{1/2}(g_y h'_y - g'_y h_y)$. It is clear that, in the absence of any round lens component $\left(\bar{K}_\phi = 0\right)$, all the chromatic aberration coefficients vanish if $\bar{K}_q(z) = 0$, and hence if

$$\left(1+\sigma^2\right)D = 4\eta\sigma Q V^{1/2} \tag{160}$$

or introducing the notation of Eqs. (11a–11f),

$$\left(1+\sigma^2\right)\beta_E^2=\sigma^2\beta_M^2. \tag{161}$$

Nonrelativistically, these conditions reduce to

$$D=2\eta QV^{1/2},\quad 2\beta_E^2=\beta_M^2. \tag{162}$$

This result shows that if an electrostatic and a magnetic quadrupole are superimposed and $D(z)\propto Q(z)$, so that the function $k(z)$ describes both potential distributions, then the combination will be achromatic if the lenses act in opposition and the magnetic lens strength is double that of the electrostatic lens. Fig. 10B shows how this condition arises; it can be seen that whereas β_E^2 and β_M^2 vary monotonically with accelerating voltage, the difference between them, β^2, passes through a maximum where $d(\beta^2)/d\Phi$ vanishes and hence the lens is achromatic.

The possibility of achieving achromatism in this way was first pointed out by Kel'man and Yavor (1961), and the condition was later independently rediscovered by Septier (1963); a fuller account of the various papers is to be found in QO[c] and in Hawkes (1965c). Experimental work on achromatic quadrupoles and quadrupole multiplets has been carried out in Leningrad (Kel'man, Yavor, Dymnikov, & Ovsyannikova, 1963; Yavor, Fishkova & Shpak, 1964) and Cambridge (Hardy, 1967). We postpone further discussion of this to the sections on Quadrupole systems and Quadrupole data.

b. *Asymptotic aberrations.* We now consider the chromatic aberration between a pair of conjugate planes, $z=z_{ox}$ and $z=z_i$, in the x–z plane and another pair, $z=z_{oy}$ and $z=z_j$, in the y–z plane. Writing

$$\begin{aligned} V_0^{1/2}\Delta x_i &= M_x\left(\bar{C}_{cx}\alpha+\bar{C}_{Mx}x_o\right)\Delta\Phi \\ V_0^{1/2}\Delta y_j &= M_y\left(\bar{C}_{cy}\gamma+\bar{C}_{M_y}y_o\right)\Delta\Phi \end{aligned} \tag{163}$$

we find that the coefficients $\bar{C}_{cx},\bar{C}_{cy},\bar{C}_{Mx}$ and $\bar{C}_{My}$ are given by

[c] In QO, however, it is wrongly stated that the derivation of the achromatic condition in Kel'man and Yavor (1961) is based upon the "rectangular model"; in fact the proof is general (cf. Hawkes, 1965c).

$$
\begin{aligned}
\bar{C}_{cx} &= -\int_{-\infty}^{\infty} \left(\bar{K}_\phi + \bar{K}_q\right) H_x^2 dz \\
\bar{C}_{cy} &= -\int_{-\infty}^{\infty} \left(\bar{K}_\phi - \bar{K}_q\right) H_y^2 dz \\
\bar{C}_{Mx} &= -\int_{-\infty}^{\infty} \left(\bar{K}_\phi + \bar{K}_q\right) G_x H_x dz \\
\bar{C}_{My} &= -\int_{-\infty}^{\infty} \left(\bar{K}_\phi - \bar{K}_q\right) G_y H_y dz.
\end{aligned}
\tag{164}
$$

In pure quadrupole systems, we write

$$
\begin{aligned}
\Delta x_i &= M_x\left(C_{cx}\alpha + C_{Mx}x_o\right)\frac{\Delta\Phi}{V} \\
\Delta y_j &= M_y\left(C_{cy}\gamma + C_{My}y_o\right)\frac{\Delta\Phi}{V}
\end{aligned}
\tag{165}
$$

where

$$
\begin{aligned}
C_{cx} &= -\int_{-\infty}^{\infty} k_q H_x^2 dz \\
C_{cy} &= \int_{-\infty}^{\infty} k_q H_y^2 dz \\
C_{Mx} &= -\int_{-\infty}^{\infty} k_q G_x H_x dz \\
C_{My} &= \int_{-\infty}^{\infty} k_q G_y H_y dz
\end{aligned}
\tag{166}
$$

and nonrelativistically,

$$
\begin{aligned}
C_{cx} &= \frac{n-1}{2}\beta^2\int_{-\infty}^{\infty} k(z) H_x^2 dz; & C_{cy} &= -\frac{n-1}{2}\beta^2\int_{-\infty}^{\infty} k(z) H_y^2 dz \\
C_{Mx} &= \frac{n-1}{2}\beta^2\int_{-\infty}^{\infty} k(z) G_x H_x dz; & C_{My} &= -\frac{n-1}{2}\beta^2\int_{-\infty}^{\infty} k(z) G_y H_y dz.
\end{aligned}
\tag{167}
$$

c. *Higher order effects in achromatic lenses.* This topic is discussed by Hardy (1967), whom we follow closely except that Hardy's treatment is nonrelativistic. Let us consider the variation of some first order parameter, Λ—a focal length or the position of a focus, for example—with the accelerating voltage. If $\Lambda = \Lambda(\Phi)$, then

$$\Lambda(\Phi + \delta\Phi) = \Lambda(\Phi) + \frac{d\Lambda}{d\Phi}\delta\Phi + \frac{1}{2}\frac{d^2\Lambda}{d\Phi^2}(\delta\Phi)^2 + \frac{1}{6}\frac{d^3\Lambda}{d\Phi^3}(\delta\Phi)^3 + \ldots \quad (168)$$

Rewriting derivatives with respect to Φ in terms of derivatives with respect to β^2 and $d(\beta^2)/d\Phi$, we obtain

$$\begin{aligned}\delta\Lambda &= \Lambda(\Phi + \delta\Phi) - \Lambda(\Phi)\\ &= \Lambda_1\frac{\delta V}{V} + \Lambda_2\left(\frac{\delta V}{V}\right)^2 + \Lambda_3\left(\frac{\delta V}{V}\right)^3 + \ldots\end{aligned} \quad (169)$$

where $\delta V = \sigma\delta\Phi$ and

$$\begin{aligned}
\Lambda_1 &= \frac{1}{2}\left(\frac{\sigma^2+1}{\sigma^2}\beta_E^2 - \beta_M^2\right)\Lambda'\\
&= \frac{n-\sigma^2}{2\sigma^2}\beta^2\Lambda'\\
2\Lambda_2 &= \frac{1}{4}\left(\frac{\sigma^2+1}{\sigma^2}\beta_E^2 - \beta_M^2\right)^2\Lambda'' + \left(\frac{1+2\sigma^2}{4\sigma^2}\beta_M^2 - \frac{3+\sigma^2}{2\sigma^2}\beta_E^2\right)\Lambda'\\
&= \left(\frac{n-\sigma^2}{2\sigma^2}\beta^2\right)^2\Lambda'' + \frac{1-5n+2\sigma^2}{4\sigma^2}\beta^2\Lambda'\\
6\Lambda_3 &= \frac{1}{8}\left(\frac{\sigma^2+1}{\sigma^2}\beta_E^2 - \beta_M^2\right)^3\Lambda'' + \frac{3}{2\sigma^2}\left(\frac{1+\sigma^2}{\sigma^2}\beta_E^2 - \beta_M^2\right)\left(\frac{1+2\sigma^2}{4}\beta_M^2 - \frac{3+\sigma^2}{2}\beta_E^2\right)\Lambda''\\
&\quad - \frac{3}{8\sigma^2}\left\{(3+2\sigma^2)\beta_M^2 - 2\frac{\sigma^4+6\sigma^2+1}{\sigma^2}\beta_E^2\right\}\Lambda'\\
&= \left(\frac{n-\sigma^2}{2\sigma^2}\beta^2\right)^3\Lambda''' + \frac{3}{8\sigma^4}(n-\sigma^2)(2\sigma^2-5n+1)\beta^4\Lambda''\\
&\quad - \frac{3}{8\sigma^4}\{2(\sigma^4-n) + 3\sigma^2(1-3n)\}\beta^2\Lambda'
\end{aligned} \quad (170)$$

where primes denote differentiation with respect to β^2. If the lens is achromatic, then $n = \sigma^2$ and

$$\begin{aligned}
\Lambda_1 &= 0\\
\Lambda_2 &= \frac{1-3\sigma^2}{8\sigma^2}\beta^2\Lambda' \simeq -\frac{1}{4}\beta^2\Lambda'\\
\Lambda_3 &= \frac{7\sigma^2-1}{16\sigma^2}\beta^2\Lambda' \simeq \frac{3}{8}\beta^2\Lambda'
\end{aligned} \quad (171)$$

where the approximate values are the nonrelativistic expressions.

3.6 Parasitic aberrations

Under this heading, we include all the defects that arise from misalignment, inaccurate machining, inhomogeneity of magnetic materials, asymmetries in the magnetic circuitry and similar practical imperfections. In quadrupoles, these aberrations are of paramount importance, for there is no device as simple as the stigmator to correct them and they can stem from such varied sources. No account of experimental work with quadrupoles is complete without a section on this problem, which is most fully discussed by Meyer (1961), Deltrap (1964a), Hardy (1967), for whom the problem was aggravated by the fact that his electrostatic and magnetic quadrupoles had to be combined into single units, Meads (1967) and Kawakatsu, Vosburgh, and Siegel (1968b); specific systems are discussed by Bauer (1965–66), Crewe, Eggenberger, Welter, and Wall (1967) and Keck, Liebich, and Pose (1967).

It is not easy to generalize about these aberrations, for their relative importance varies widely with lens design. By considering several individual treatments, we shall try to give a reasonably full account of the methods that have been used to study them. In order to avoid repetition, we do not describe the work of Meyer (1961) and Deltrap (1964a) here, as it is covered in some detail in *QO*, concentrating instead on more recent papers.

If a quadrupole does not possess the planes of symmetry and antisymmetry of a symmetrically excited lens in the standard orientation, the potential distribution will contain parasitic terms, included in Eqs. (7a) and (7b). We can relate each additional term to a particular imperfection and derive formulae for the resulting aberrations: Meyer (1961) does this very thoroughly. This is often too elaborate a procedure, however, for by their very nature parasitic effects are accidental and every lens will be characterized by its own set of parasitic potentials; moreover, it is exceedingly difficult to relate a physical defect to its potential quantitatively. If a two-dimensional approximation is adequate, the problem can be solved by the methods of conformal mapping; Shukeilo (1959) investigated the effect of replacing hyperbolic pole-pieces by circular ones, using this technique, and a very thorough analysis is to be found in the work of Doinikov (1966a, 1966b); Doinikov and Titov (1966); Doinikov and Samsonov (1967).

Let us now return to Eq. (7a); it is convenient to rewrite this in the form

$$\begin{aligned}\phi(r,\theta,z) = {} & \left(\Phi_0 - \frac{1}{4}\Phi_0''r^2 + \frac{1}{64}\Phi_0^{(iv)}r^4 - \ldots\right) \\ & + r\cos\theta\left(\Phi_{1c} - \frac{1}{9}\Phi_{11}''r^2 + \ldots\right) \\ & + r\sin\theta\left(\Phi_{1s} - \frac{1}{9}\Phi_{1s}''r^2 + \ldots\right) \\ & + r^2\cos 2\theta\left(\Phi_{2c} - \frac{1}{12}\Phi_{2c}''r^2 + \ldots\right) \\ & + r^2\sin 2\theta\left(\Phi_{2s} - \frac{1}{12}\Phi_{2s}''r^2 + \ldots\right) \\ & + r^3\cos 3\theta(\Phi_{3c} - \ldots) + r^3\sin 3\theta(\Phi_{3s} - \ldots) \\ & + r^4\cos 4\theta(\Phi_{4c} - \ldots) + r^4\sin 4\theta(\Phi_{4s} - \ldots)\end{aligned} \tag{172}$$

with a similar expression for $\psi(r,\theta,z)$. Fig. 11 shows the electrode or pole-piece shift that would create each of these potentials. The functions Φ_0 and Ψ_0 correspond to round lens effects; Φ_0 will merely impose a weak electrostatic lens effect on the system, which will normally be of no importance. The function Ψ_0 represents a round magnetic lens effect, and if this were strong, the resulting optical behavior would be of the very complex kind analyzed by Gullstrand (1915) and Carathéodory (1937), see Hawkes (1965a); a case that is amenable to analytical treatment has been analyzed by Schiske (1966–67). In practice, however, the field will be weak and will cause a slight rotation of the image. All the functions with subscript 1 represent deflecting fields, which will displace the image and can cause second-order aberrations. The functions Φ_{2s} and Ψ_{2c} represent quadrupoles inclined to the standard orientation at 45°.These potentials often occur in systems of quadrupoles in which the components are slightly twisted with respect to one another (Bauer, 1965–66; Crewe, 1965; Crewe et al., 1967; Kartashev & Kotov, 1966) and can be canceled by inserting a weak quadrupole between the components at 45° to the standard orientation. The functions with subscript 3 introduce second-order aberrations and Φ_{4s}, Ψ_{4c} correspond to octopoles inclined to the standard orientation at $22\frac{1}{2}^\circ$.

So far, we have spoken only of mechanical imperfections; in magnetic quadrupoles, however, considerable asymmetries in excitation may occur, since the "magnetostatic potential of the pole-piece face depends upon the ratio between the gap reluctance and the reluctance of the iron of the pole-piece shank and the junctions between it and the 'earth rail' and the pole-pieces themselves. Consequently any asymmetry in these reluctances

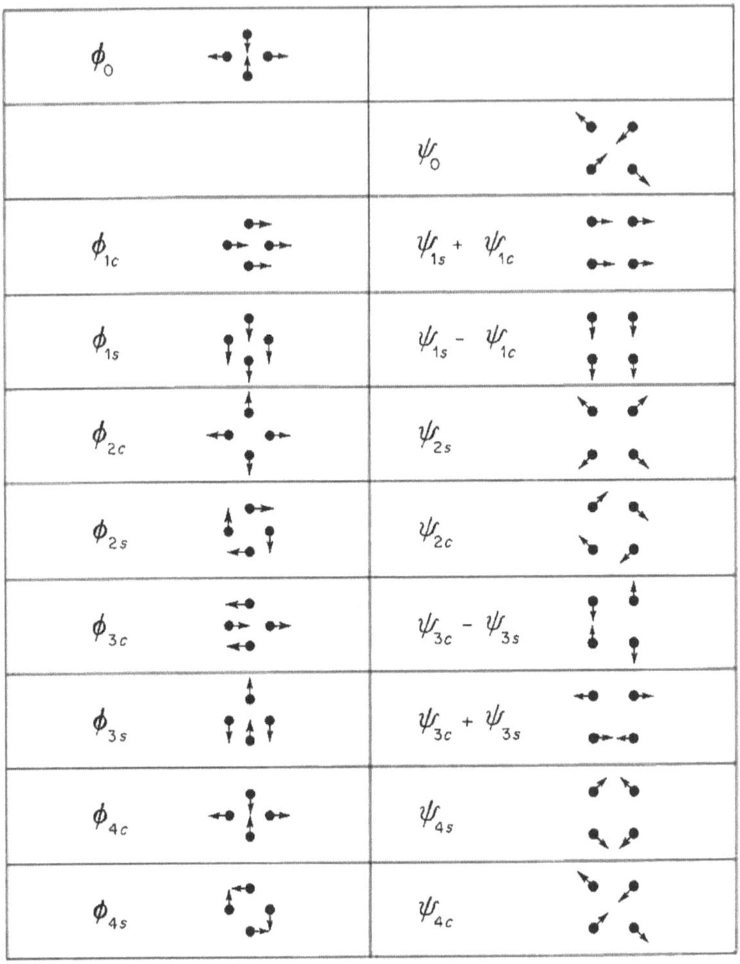

Fig. 11 Distortions of the geometry of electrostatic (left) and magnetic (right) quadrupoles that correspond to the potential functions shown. The optical effects of adjacent perturbations are the same. *After Hardy, D. F. (1967).* Combined magnetic and electrostatic quadrupole electron lenses. *Dissertation, Cambridge University.*

must result in an asymmetry in the magnetostatic potentials of the four pole-pieces" (Hardy, 1967). Small potential inequalities can be counteracted by moving the pole-piece towards or away from the optic axis, however; Hardy found that "it was sometimes. .. necessary to displace the pole-piece up to 0.08 cm from what appeared to be its ideal position from the point of view of mechanical alignment" with a bore-radius of 1.1 cm. Where strong-focusing is concerned, asymmetries may be deliberately introduced into the excitation in order to counteract small misalignments (Keck et al., 1967; Koltay & Szabó, 1965).

We now turn to the work of Kawakatsu et al. (1968b), to which we shall return in the section on quadruplet projectors. These authors consider how the potential of a magnetic quadrupole is affected by displacement from the true optic axis, rotation and tilt. When the lens is perfectly aligned, the magnetostatic potential (Eq. 6a) is given by $\psi(x, y, z) = Q(z)xy$ to a first order approximation, and we write this as $\psi(x, y, z) = 2(Q_0/a^2)k(z)xy$ in which $k(z)$ is normalized to rise to a maximum value of unity and a is the bore radius. If the quadrupole is shifted bodily through a distance δx, δy, rotated about its axis through an angle $\delta\Theta$ and tilted so that its axis is inclined to the true optic axis, ψ becomes

$$\begin{aligned}\psi(x, y, z) = 2\frac{Q_0}{a^2}k(z)[xy - x\{\delta y + \delta\lambda(z - z_1)\sin\theta_1 + x\delta\Theta\} \\ -y\{\delta x + \delta\lambda(z - z_1)\cos\theta_1 - y\delta\Theta\}]\end{aligned} \tag{173}$$

in which $z = z_1$ is the point about which the tilt, characterized by ($\cos\theta_1\delta\lambda$, $\sin\theta_1\delta\lambda$), occurs. Kawakatsu et al. (1968b) then make the assumption that for each type of misalignment, the potential can be obtained by modulating the solution of the two-dimensional problem with the function $k(z)$. They consider radial shift and rotation of a single pole (Fig. 12A), and find that the equations of motion become, to a first order approximation,

$$\begin{aligned}x'' + \beta^2 k(z)x = \beta^2 k(G - R + Hx - Sy) \\ y'' - \beta^2 k(z)y = \beta^2 k(G - R - Hy - Sx)\end{aligned} \tag{174}$$

where

$$\left.\begin{aligned}G = 2\delta a/3\pi \\ H = \delta a/2a\end{aligned}\right\}\ \text{radial error} \qquad \left.\begin{aligned}R = 8a\delta\theta/9\sqrt{2}\pi^2 \\ S = \delta\theta/4\end{aligned}\right\}\ \text{rotational error.} \tag{175}$$

Similar equations are obtained when the magnetic material is inhomogeneous:

$$\begin{aligned}x'' + \beta^2 kx = \beta^2 k\left|\frac{a}{3\pi}\left(\frac{\Delta B}{B}\right)_P + \frac{1}{4}\left(\frac{\Delta B}{B}\right)_P x + \left(\frac{\Delta B}{B}\right)_Q x\right| \\ y'' - \beta^2 ky = \beta^2 k\left|\frac{a}{3\pi}\left(\frac{\Delta B}{B}\right)_P - \frac{1}{4}\left(\frac{\Delta B}{B}\right)_P y - \left(\frac{\Delta B}{B}\right)_Q y\right|\end{aligned} \tag{176}$$

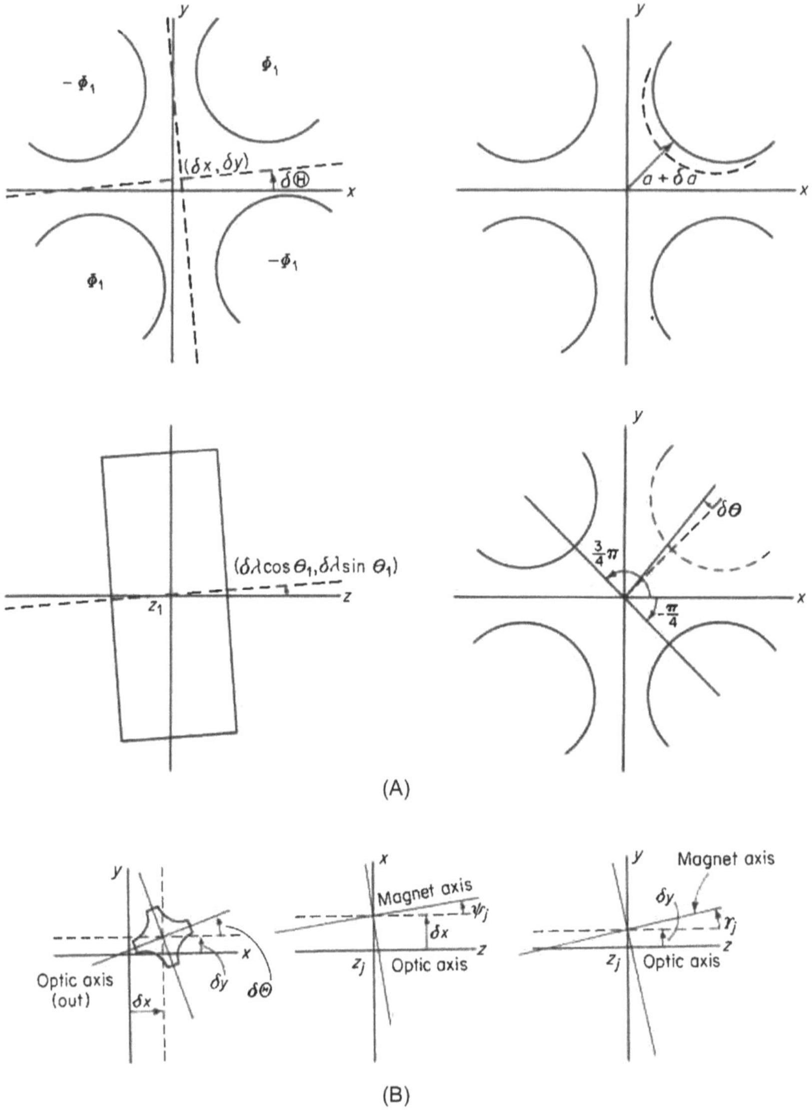

Fig. 12 (A) The distortions of the geometry of a magnetic quadrupole considered by Kawakatsu et al. (1968b). (B) The distortions of the geometry of a magnetic quadrupole considered by Meads (1967).

in which subscript P denotes error arising from a single pole and Q that arising from the entire lens. The aberrations, referred back to the object plane, are then given by

$$
\begin{aligned}
\Delta x_o = {} & (0000)_A\left|\delta x + G - R + \frac{a}{3\pi}\left(\frac{\Delta B}{B}\right)_P\right| + (0000)_D\delta\lambda\cos\theta_1 \\
& + \left|(1000)_H x_o + (0010)_H x_a\right|\left|H + \frac{1}{4}\left(\frac{\Delta B}{B}\right)_P + \left(\frac{\Delta B}{B}\right)_Q\right| \\
& + \left|(0100)_C y_o + (0001)_C y_a\right|\left|\delta\Theta + \frac{1}{2}S\right| \\
\Delta y_o = {} & (0000)_B\left|\delta y - G + R - \frac{a}{3\pi}\left(\frac{\Delta B}{B}\right)_P\right| + (0000)_E\delta\lambda\sin\theta_1 \\
& + \left|(0100)_H y_o + (0001)_H y_a\right|\left|H + \frac{1}{4}\left(\frac{\Delta B}{B}\right)_P + \left(\frac{\Delta B}{B}\right)_Q\right| \\
& + \left|(1000)_C x_0 + (0010)_C x_a\right|\left|\delta\Theta + \frac{1}{2}S\right|.
\end{aligned}
\tag{177}
$$

The coefficients $(\alpha\beta\gamma\delta)$ for a stigmatic system are listed in Kawakatsu et al. (1968b) and we shall not reproduce them here. The authors have calculated the tolerances that have to be met in a quadruplet projector with the following dimensions: bore radius $a = 1.5$ cm; pole-piece length, 3 cm; center-to-center spacing between quadrupoles, 4.7 cm; projector focal length varying from 0.75 to 3.6 mm. If the radial and rotational error in the magnification is to be less than 3%, the following requirements must be fulfilled: $|\delta\Theta| < 2.7\times10^{-3}$ rad, for a doublet (quadruplet with center pair of quadrupoles switched off) or $|\delta\Theta| < 1.7\times10^{-3}$ rad for the quadruplet; $|\Delta a/a| < 3.3\times10^{-2}$ (doublet) or 6.3×10^{-2} (quadruplet); $|\delta\theta| < 2.2\times10^{-2}$ rad (doublet) or 1.3×10^{-2} rad (quadruplet); $|(\Delta B/B)_P| < 6.6\times10^{-2}$ or 1.2×10^{-1}; $|(\Delta B/B)_Q| < 1.6\times10^{-2}$ or 3.1×10^{-2}; in order to keep $|\Delta M/M| < 10^{-4}$, where M is the magnification, fluctuations in lens current, I, must satisfy $|\Delta I/I| < 8\times10^{-5}$ or 1.6×10^{-4}. The deflection caused by tilt and displacement is comparatively harmless, and Kawakatsu et al. (1968b) ensured that $|\delta x| \le 0.5$ mm, $|\delta y| \le 0.5$ mm and $|\delta\lambda| \le 10^{-2}$ rad.

Another detailed study of these parasitic aberrations has been made by Meads (1963, 1967). Meads is concerned with the effects of misalignment of perfectly constructed magnetic quadrupoles and he assumes that the potential distribution does not depart far from the rectangular model. He therefore considers displacement, rotation and tilt of the whole quadrupole: in the notation of the foregoing paragraphs, he considers δx, δy, $\delta\Theta$, $\cos\theta_1\delta\lambda$ and $\sin\theta_1\delta\lambda$. The aberrations are deduced by constructing the appropriate Green's functions; Meads shows that the aberrations in an arbitrary plane $z = z_c$ can be expressed in terms of those at the centers of the individual quadrupoles, so that if $T(z_j, z_c)$ is the transfer matrix from the center of the j-th quadrupole, $z = z_j$, to the current plane, then

$$\begin{pmatrix} \Delta x \\ \Delta x' \end{pmatrix}_c = \sum_j T(z_j, z_c) \begin{pmatrix} \Delta x \\ \Delta x' \end{pmatrix}_j. \tag{178}$$

in which Δx_j and $\Delta x'_j$ are "the apparent displacement and change in slope at z_j, as seen by an observer at $[z_c]$, that result from the misalignment of only that magnet centered at z_j" (Meads, 1963). These apparent aberrations are given by the following formulae, in which we have retained Meads' notation for the tilt (Fig. 12B); the rectangular model, discussed in detail in the chapter on Quadrupole data, is used and $\Lambda_j = \beta_j L_j/2$.

$$\begin{aligned}
\beta_j \Delta x_j &= 2\left(\sin\Lambda_j \cosh\Lambda_j - \cos\Lambda_j \sinh h\Lambda_j\right)\delta\Theta \cdot y'_j + 2\left(\Lambda_j \cos\Lambda_j - \sin\Lambda_j\right)\psi_j \\
\Delta x'_j &= -2\beta_j\left(\sin\Lambda_j \cosh\Lambda_j + \cos\Lambda_j \sinh\Lambda_j\right)\delta\Theta \cdot y_j + 2\beta_j \sin\Lambda_j \delta x_j \\
\beta_j \Delta y_j &= -2\left(\cos\Lambda_j \sinh\Lambda_j - \sin\Lambda_j \cosh\Lambda_j\right)\delta\Theta \cdot x'_j + 2\left(\Lambda_j \cosh\Lambda_j - \sinh\Lambda_j\right)\Upsilon_j \\
\Delta y'_j &= -2\beta_j\left(\cos\Lambda_j \sinh\Lambda_j + \sin\Lambda_j \cosh\Lambda_j\right)\delta\Theta \cdot x_j - 2\beta_j \sinh\Lambda_j \delta y_j.
\end{aligned} \tag{179}$$

The magnetic doublet has been studied in detail by Kartashev and Kotov (1966), who consider the effects of altering the lens separation from the correct value, displacing the lens axis parallel to the optic axis, rotation of one lens with respect to the other and rotation of a lens about transverse axes; they derive transfer matrices for each of these situations.

The papers in which difficulties encountered experimentally are described usually stress the aberrations that arise when adjacent quadrupoles are twisted with respect to one another (Bauer, 1965–66; Crewe, 1965; Crewe et al., 1967); Crewe, we recall, overcomes this defect by inserting a weak quadrupole between the others, inclined at 45° to the standard orientation. For Thomson (1967), however, the most serious problem was that mechanical and magnetic alignment did not coincide: "From measurements of the deflection in single weak quadrupoles, it was calculated that the misalignment was several tenths of a millimetre. This is much larger than could be caused by constructional errors, or by deformation of the centre jig; it cannot be accounted for by twist as this does not give rise to a deflection of the paraxial rays.

The presence of external magnetic fields was also ruled out. Mumetal screening was placed around the complete lens unit, to no effect, and in any case when the lens unit was rotated the direction of the deflection moved with it.

We have noted earlier that the precise position of the coils does not influence the imaging, but to make sure that they are not responsible, those of a single quadrupole were each moved round one pole. No change was observed.

We are thus forced to the conclusion that the magnetic equipotential does not coincide with the physical pole-piece surface, that is to say, there must be irregularities in the magnetic circuit. The various parts are all push fits together, so an error as large as a few tenths of a millimetre can only arise from irregularities in the magnetic material itself. It is possible that this could be corrected by more thorough heat treatment than is usual, but it is likely that the metal itself is not homogeneous. Although the former can be tried, the latter can only be investigated by devising sufficiently sensitive experiments, or by reconstructing the whole lens out of more reliable material, possibly an alloy other than mumetal" (Thomson, 1967).

Enough has been said to show that the problems posed by parasitic aberrations are frustrating ones, which each experimentalist must solve for his particular system; they probably form the most formidable obstacle to the widespread use of quadrupoles at the present time.

4. Fifth-order aberrations

Rather little is known about these aberrations, but they cannot be entirely neglected because of their importance if third-order spherical aberration can be overcome; furthermore, the shape of the electrodes or poles of quadrupoles strongly affects their fifth-order effects, and it is therefore wise to design quadrupoles with this in mind.

We shall not deal at all with the fifth-order aberrations of round lenses; they are considered in Scherzer (1949), My Chzhen (1957), Archard, Mulvey, and Petrie (1960), Meyer (1961) and Hawkes (1965b). The fifth-order aperture aberrations of quadrupoles contain terms in α^5, $\alpha^3\beta^2$ and $\alpha\beta^4$ in the x–z plane, and in β^5, $\alpha^2\beta^3$ and $\alpha^4\beta$ in the y–z plane. Some estimate of the coefficients is obtained from the shadow-projection method of measuring these quantities (see Appendix 2) and this is pursued by Deltrap (1964a) and Hardy (1967). Dhuicq (1968a) discusses the effect of these aberrations upon probe-size in a probe-forming system.

The nature of the fifth-order aberrations is discussed qualitatively in Hawkes (1965a); if we consider the asymptotic aperture aberrations between z_{ox} and z_i in the x–z plane and between z_{oy} and z_j in the y–z plane, we have

$$x(z_i) = M_x\{x_o + (0030)\alpha^3 + (0012)\alpha\beta^2 + (0050)\alpha^5 + (0032)\alpha^3\beta^2 + (0014)\alpha\beta^4\}$$
$$y(z_j) = M_y\{y_o + (0003)\beta^3 + (0021)\alpha^2\beta + (0005)\beta^5 + (0023)\alpha^2\beta^3 + (0041)\alpha^4\beta\}. \tag{180}$$

Each fifth-order coefficient contains a contribution from ϕ_6 and another from ϕ_2: it is shown in Hawkes (1965a) that the ϕ_6 contributions are inter-related, but the work of Foster (1968) indicates that the entire coefficients obey the same equalities. The result is that if $z_{ox} = z_{oy}$ and $z_i = z_j$, then

$$(0032) = 2(0041) \text{ and } (0023) = 2(0014). \tag{181}$$

We may anticipate that these coefficients will be expressible as sixth degree polynomials in m_x and m_y.

Foster (1968) is concerned with the design of a triplet of air-cored solenoid quadrupoles producing unit magnification, for use as a lens spectrometer working on the focal separation principle. Two ray configurations are considered (Figs. 13A and B) and the effects of the octopole correctors are imposed in two different ways: either the octopoles are placed at the center lens, at an outer lens and midway between the two or the quadrupoles are constructed from eight solenoids, which together produce quadrupole and octopole effects. Numerous points of general interest emerge from the analysis, and we summarize some of these here. The aberrations are obtained by a variety of methods; by considering a perturbation of the paraxial rays, $x = x_1(z)$, $y = y_1(z)$, such that $x(z)$ and $y(z)$ become

$$x(z) = \lambda x_1 + \lambda^3 x_3 + \lambda \nu^2 x_{12} + \lambda^5 x_5 + \lambda^3 \nu^2 x_{32} + \lambda \nu^4 x_{14} + \ldots$$
$$y(z) = \nu y_1 + \nu^3 y_3 + \lambda^2 \nu y_{21} + \nu^5 y_5 + \lambda^2 \nu^3 y_{23} + \lambda^4 \nu y_{41} + \ldots, \tag{182}$$

the refractive index may be written

$$\mu = \mu_0 + \lambda^2 \mu_{2x} + \nu^2 \mu_{2y} + \lambda^4 \mu_{4x} + \nu^4 \mu_{4y} + \lambda^2 \nu^2 \mu_{4xy} + \lambda^6 \mu_{6x} + \nu^6 \mu_{6y}$$
$$+ \lambda^2 \nu^4 \mu_{6yx} + \lambda^4 \nu^2 \mu_{6xy} \tag{183}$$

in which μ_{2x} and μ_{2y} contain terms in x_1^2 and y_1^2 respectively, μ_{4x} contains terms in x_1^4 and x_3x_1, μ_{4y} terms in y_1^4 and y_3y_1 and μ_{4xy} terms in $x_1^2y_1^2$, x_1x_{12} and y_1y_{12}; m_{6x} contains terms in x_1^6, $x_1^3x_3$, x_5x_1 and x_3^2, and m_{6xy} terms in $x_1^4y_1^2$, $x_1y_1^2x_3$, y_{21}^2, $x_1^2y_1y_{21}$, $x_1^3x_{12}$, x_1x_{32}, y_1y_{41} and x_3x_{12} with similar patterns for m_{6y} and m_{6yx}. Foster writes down the Euler equations for each variational equation of the form $\delta \int \mu_n dz = 0$; μ_{2x} and μ_{2y} yield the familiar

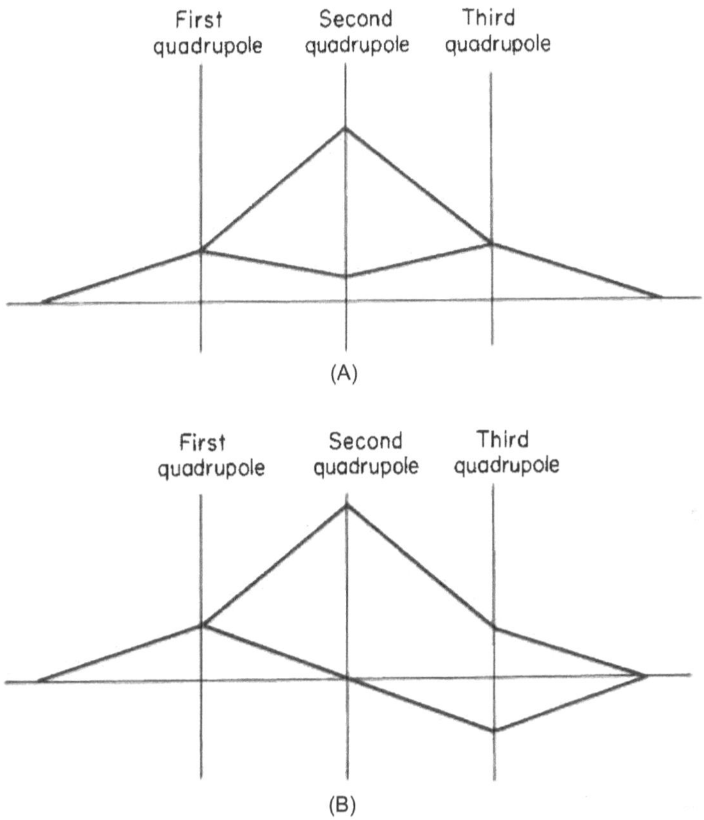

Fig. 13 Schematic diagram of the ray configurations in the triplets studied by Foster (1968).

paraxial equations of motion, μ_{4x}, μ_{4y} and μ_{4xy} yield equations from which the primary geometrical aberrations may be derived and the terms in μ_6 yield equations for the secondary (fifth-order) aberrations.

Foster has also devised an original and very powerful method that depends upon exact ray tracing and is "suitable for finding the aperture aberrations of any system of lenses whose field functions are known, whether or not the system is orthogonal. Furthermore, exact expressions for the field functions may be used ..." (Foster, 1968). The exact ray equation is first solved numerically by a refined predictor–corrector method (Burfoot, 1952; Fox & Goodwin, 1949; Jennings & Pratt, 1955). The aberrations are deduced by differentiating the departure from paraxial imagery with respect to slope in the object plane; in the cases considered by Foster, for example, where the system is stigmatic and orthomorphic with unit magnification, we know that (to the third-order approximation) $\Delta x_i = Ax_i'^3 + Bx_i'y_i'^2$ and $\Delta y_i = Cy_i'^3 + Dx_i'^2 y_i$ so that for $y_o' = 0$,

$$\frac{d^3(\Delta x_i)}{dx_o'^3} = -3!A \tag{184}$$

for $x_0' = 0$

$$\frac{d^3(\Delta y_i)}{dy_o'^3} = -3!C \tag{185}$$

and for $x_o' = y_o'$

$$\frac{d^3(\Delta x_i)}{dx_o^3} = -3!(A + B) \text{ and } \frac{d^3(\Delta y_i)}{dy_o^3} = -3!(C + D). \tag{186}$$

Higher order aberrations can be derived by further differentiation, and Foster uses this method to obtain the fifth-order aberration coefficients of triplets with and without octopole correctors and to show that the equalities between the coefficients [Eq. (181)] are satisfied even when the third-order aberrations do not vanish.

The fifth-order aberration coefficients vary little among the several systems considered, with one exception: when the primary aperture aberrations are corrected with the aid of eight-pole elements excited to provide both octopole and quadrupole fields, all the fifth-order coefficients change in sign. If, therefore, each eight-pole element is replaced by a dodecapole—a quadrupole superimposed upon an eight-pole element—each providing half the required potential, we may expect the fifth-order aberrations to cancel, and Foster shows that this is indeed the case. Much of the analysis is performed on the assumption that the air-cored solenoids (of length l and radius b) can be replaced by point poles of strength $p = \pi b^2 nI$, where n is the number of turns/cm and I (e.m.u.) is the current; calculations using more exact formulae for the field of the solenoid proved that the point pole approximation was very adequate with the dimensions used ($l = 6$ cm, $b = 1.7$ cm, $n = 100$ turns/cm, bore radius = 6.5 cm, source-to-screen distance = 128 cm, quadrupole separation = 30 cm, electron accelerating voltage $\simeq$ 8kV). With the aid of this point-pole approximation, Foster explains how this change in the sign of the fifth-order aberrations comes about.

A preliminary attempt has been made in Leningrad (Fishkova, Ovsyannikova, & Yavor, 1966; Ovsyannikova & Yavor, 1967b) to obtain analytical expressions for the secondary aperture aberration coefficients.

These authors consider two-dimensional fields, for which $k(z) = 1$ for all z; in the equations of motion, we set $k' = k'' = 0$ and with the origin at the object plane, the first order trajectories take the form

$$\begin{aligned} x &= x_0 \cos \beta z + \alpha \frac{\sin \beta z}{\beta} \\ y &= y_o \cosh \beta z + \gamma \frac{\sinh \beta z}{\beta} \end{aligned} \tag{187}$$

so that in the converging (x–z) plane, the first line image occurs at $\beta z = \pi$. (This two-dimensional procedure is also employed in connection with the third-order aberration by Strashkevich (1963, 1964, 1966) and Dymnikov et al. (1964a), for example.) The two-dimensional potential distribution is

$$\begin{aligned} \phi(x, y) &= -\frac{\Phi_E k_1}{r_E^2}\left[\left(x^2 - y^2\right) + \frac{k_2}{k_1}\frac{1}{r_E^4}\left\{x^6 - y^6 - 15x^2y^2\left(x^2 - y^2\right)\right\}\right] \\ \psi(x, y) &= -\frac{2\Psi_M N_1}{r_M^2}\left[xy + \frac{N_2}{N_1}\frac{1}{r_M^4}\left\{xy\left(x^4 + y^4\right) - \frac{10}{3}x^3y^3\right\}\right] \end{aligned} \tag{188}$$

in which k_1, k_2, N_1, and N_2 are constants determined by the shape of the electrodes or pole-pieces and r_E^2, r_M^2 are the bore-radii of the electrodes and poles, respectively; Φ_E and Ψ_M are the potentials at their surfaces. The trajectory equation in the x–z plane has the form

$$x'' + \beta^2 x = F_3 + F_5 \tag{189a}$$

in which

$$\begin{aligned} F_3 =\; & x'\left(x'x'' + y'y''\right) + \beta^2\left\{\frac{1}{2}(n-1)x\left(x'^2 + y'^2\right) - nx'\left(xx' - yy'\right)\right\} \\ & -\frac{1}{2}\beta^4 n(n-1)x\left(x^2 - y^2\right) \\ F_5 =\; & -x'\left(x'^2 + y'^2\right)\left(x'x'' + y'y''\right) + \frac{1}{8}\beta^2(n+1)x\left(x'^2 + y'^2\right)^2 \\ & + \beta^4 n\left\{\frac{1}{4}(n+1)x\left(x'^2 + y'^2\right) - ny'\left(xy' + x'y\right)\right\}\left(x^2 - y^2\right) \\ & + \frac{1}{8}\beta^6 n^2(5n-3)x\left(x^2 - y^2\right)^2 + \beta^2\left(\frac{3nk_2}{r_E^4 k_1} - \frac{n+1}{r_M^4}\frac{N_2}{N_1}\right)x\left(x^4 - 10x^2y^2 + 5y^4\right). \end{aligned} \tag{189b}$$

In the line image plane, Ovsyannikova and Yavor (1967b) find

$$\Delta x_i = -\left[(30)\alpha^3 + (12)\alpha\beta^2 + \left\{(50)_2 + (50)_6\right\}\alpha^5 + \left\{(32)_2 + (32)_6\right\}\alpha^3\beta^2 + \left\{(14)_2 + (14)_6\right\}\alpha\beta^4\right] \tag{190}$$

in which we have written (pq) for $(00pq)$; $(pq)_2$ arises from terms in ϕ_2 and $(pq)_6$ from those in ϕ_6. The coefficients are given by

$$\begin{aligned}
(30)/L &= \left(3n^2 - 2n + 3\right)/16 \\
(12)/L &= \left\{4(n-1)^2 - (n+1)(n-3)\sinh 2\pi/\pi\right\}/32 \\
(50)_2/L &= -\left(105n^4 + 60n^3 + 126n^2 - 84n + 129\right)/1024 \\
(32)_2/L &= -\left\{65n^4 + 4n^3 + 78n^2 - 108n + 137 + (n+1)\left(-51n^3 + 47n^2 - 113n\right.\right. \\
&\quad \left.\left. + 109\right)\sinh 2\pi/5\pi + (n+1)\left(n^3 - 21n^2 + 27n - 15\right)\cosh 2\pi\right\}/512 \\
(14)_2/L &= -\left\{29n^4 + 8n^3 + 18n^2 - 96n + 57\right. \\
&\quad + (n+1)\left(-23n^3 + 23n^2 + 3n + 85\right)\sinh 2\pi/2\pi \\
&\quad + (n+1)\left(n^3 + 43n^2 - 77n - 119\right)\sinh 4\pi/20\pi \\
&\quad \left. + 4(n+1)\left(n^3 - 5n^2 + 5n - 5\right)\cosh 2\pi\right\}/512 \\
(50)_6/L &= 5\left\{(n+1)(L/r_M)^4 N_2/N_1 - 3n(L/r_E)^4 k_2/k_1\right\}/16\pi^4 \\
(32)_6 &= 6(1 - \sinh 2\pi/5\pi)(50)_6 \\
(14)_6 &= (3 - \sinh 2\pi/\pi + \sinh 4\pi/20\pi)(50)_6
\end{aligned} \tag{191}$$

in which

$$L = \pi/\beta. \tag{192}$$

Ovsyannikova and Yavor give figures for the various coefficients for the case $N_2/N_1 = k_2/k_1 = 0.033$, L/r_E or L/r_M equal to 4 and 10 and $n = -1$ (electrostatic lens) and $n = 0$ (magnetic lens).

CHAPTER THREE

Quadrupole potential functions

Contents

1. Calculation by analog and numerical methods

All the requisite properties of electron lenses can be calculated once the potential or field functions are known. In the case of round lenses, $\Phi(z)$ and $B(z)$ may be obtained either by measurement (Germain, 1967) or by relaxation, since the problem is essentially two-dimensional. For the same reason, such analog devices as the resistance network and the electrolytic tank are very suitable (see Francken, 1967; Weber, 1967a, 1967b; for a more general approach, see Fifer, 1961; Karplus, 1958; Karplus & Soroka, 1959). The situation is more complex for the quadrupole functions $D(z)$ and $Q(z)$, however; let us consider an electrostatic quadrupole in the standard orientation, and attempt to find a separated solution of $\nabla^2\phi = 0$ in the form

$$\phi(r, \theta, z) = R(r, z)\Theta(\theta). \tag{1}$$

We find

$$\frac{\partial^2\Theta}{\partial\theta^2} = -n^2\Theta \tag{2a}$$

$$\frac{\partial^2 R}{\partial r^2} + \frac{1}{r}\frac{\partial R}{\partial r} + \frac{\partial^2 R}{\partial z^2} - \frac{n^2 R}{r} = 0 \tag{2b}$$

We retain only the solutions.

Advances in Imaging and Electron Physics, Volume 224
ISSN 1076-5670
https://doi.org/10.1016/B978-0-323-98865-0.00003-0

$$\Theta_n(\theta) = \cos n\theta \quad n = 2j, j \quad \text{a positive integer or zero} \tag{3}$$

and so the general solution is

$$\phi(r,\theta,z) = \sum_{j=0}^{\infty} R_{2j}(r,z)\cos 2j\theta. \tag{4}$$

For round lenses, $j=0$ and Eq. (2b) collapses to the familiar two-dimensional equation for $R(r,z)$:

$$\frac{\partial^2 R_0}{\partial r^2} + \frac{1}{r}\frac{\partial R_0}{\partial r} + \frac{\partial^2 R_0}{\partial z^2} = 0. \tag{5}$$

For quadrupole lenses with planes of symmetry and antisymmetry, however, $j=1,3,5, \ldots 2m+1, \ldots$ and a family of functions R_j are involved. Similarly for octopoles, $j=2,6,10, \ldots 4m+2, \ldots$ At this stage, it is usual to discard all the quadrupole contributions other than $\cos 2\theta$ and all the octopole terms except $\cos 4\theta$. This is equivalent to selecting the electrode cross-section to be approximately hyperbolic in form. With this assumption then, that one azimuthal term is adequate to describe the lens, we have.

$$\phi(r,\theta,z) = R_2(r,z)\cos 2\theta \quad \text{(quadrupole)} \tag{6}$$

$$\phi(r,\theta,z) = R_4(r,z)\cos 4\theta \quad \text{(octopole)} \tag{7}$$

and

$$\frac{\partial^2 R_2}{\partial r^2} + \frac{1}{r}\frac{\partial R_2}{\partial r} + \frac{\partial^2 R_2}{\partial z^2} - \frac{4R_2}{r^2} = 0 \tag{8}$$

$$\frac{\partial^2 R_4}{\partial r^2} + \frac{1}{r}\frac{\partial R_4}{\partial r} + \frac{\partial^2 R_4}{\partial z^2} - \frac{16R_4}{r^2} = 0. \tag{9}$$

These are the partial differential equations for which Thomson (1967) obtains iteration formulae, as we explain below. The functions R_2 and R_4 both vanish on the axis ($r=0$), however, since

$$R_2(r,z) \propto D(z)r^2 - \frac{1}{12}D''r^4 + \ldots \tag{10}$$

and

$$R_4(r,z) = D_1(z)r^4 - \ldots \tag{11}$$

and the functions of interest, $D(z)$ and $D_1(z)$, are not easy to extract from tables of values of $R_2(r,z)$ and $R_4(r,z)$. Following a suggestion of H.E.R. Preston, we may write

$$R_2(r, z) = r^2 \rho_2(r, z) \tag{12}$$
$$R_4(r, z) = r^4 \rho_4(r, z) \tag{13}$$

The functions $\rho_2(r, z)$, $\rho_4(r, z)$ satisfy

$$\begin{aligned} \frac{\partial^2 \rho_2}{\partial r^2} + \frac{5}{r}\frac{\partial \rho_2}{\partial r} + \frac{\partial^2 \rho_2}{\partial z^2} = 0 \\ \frac{\partial^2 \rho_4}{\partial r^2} + \frac{9}{r}\frac{\partial \rho_4}{\partial r} + \frac{\partial^2 \rho_4}{\partial z^2} = 0 \end{aligned} \tag{14}$$

and in general

$$\frac{\partial^2 \rho_n}{\partial r^2} + \frac{2n+1}{r}\frac{\partial \rho_n}{\partial r} + \frac{\partial^2 \rho_n}{\partial z^2} = 0 \tag{15}$$

$D(z)$ and $D_1(z)$ can now be obtained immediately once the values of ρ_2 and ρ_4 are known on the axis.

1.1 Relaxation

Let us now divide the region within which the potential is to be calculated into a rectangular network of points, mesh-length h. We have

$$\begin{aligned} R(r, z) = {} & R(r_0, z_0) + \left(\frac{\partial R}{\partial z}\right)_{r_0, z_0}(z - z_0) + \left(\frac{\partial R}{\partial r}\right)_{r_0, z_0}(r - r_0) \\ & + \frac{1}{2!}\left\{\left(\frac{\partial^2 R}{\partial z^2}\right)_{r_0, z_0}(z - z_0)^2 + 2\left(\frac{\partial^2 R}{\partial r \partial z}\right)_{r_0, z_0}(r - r_0)(z - z_0) \right. \\ & \left. + \left(\frac{\partial^2 R}{\partial r^2}\right)_{r_0, z_0}(r - r_0)^2\right\} + \ldots \end{aligned} \tag{16}$$

so that with

$$\begin{aligned} R_1 = R(r_0, z_0 + h) \quad R_2 = R(r_0 + h, z_0) \\ R_3 = R(r_0, z_0 - h) \quad R_4 = R(r_0 - h, z_0) \end{aligned} \tag{17}$$

Eq. (2b) becomes

$$\frac{R_2 + R_4 - 2R_0}{h^2} + \frac{1}{r}\frac{R_2 - R_4}{2h} + \frac{R_1 + R_3 - 2R_0}{h^2} - \frac{4j^2 R_0}{r^2} = 0. \tag{18}$$

Writing $b = r/h$, we find

$$\mathcal{R} = -4R_0\left(1 + j^2/b^2\right) + R_1 + R_3 + R_2(1 + 1/2b) + R_4(1 - 1/2b) = 0. \quad (19)$$

The quantity $\mathcal{R}$ is called the "residual" at the point (r_0, z_0). The process of relaxation consists in guessing the potentials at the mesh-points and then altering them in such a way that the residuals become zero. The points on the boundaries and on the axis must be considered separately. The residuals are evaluated at each point of the mesh in turn, usually by scanning along the rows of mesh points in succession if a computer is used, and the potential at each mesh-point is altered to make the residual there zero or, as we explain below, to alter its sign. These alterations may be made immediately, so that the residual at a point $(r, z+h)$ is calculated using the new value of the potential at (r, z); this is called successive displacement. Alternatively, new values for the potentials may be calculated for all the points with a fixed set of values, after which the potentials of all the points are altered simultaneously and the process repeated; this is known as simultaneous displacement. The latter occupies considerably more storage space in a computer than successive displacement, and the rate of convergence is slower.

With successive displacement, the convergence rate is appreciably increased if the residual at each point is altered so that instead of becoming zero, its sign is changed: if R at some point must be changed to $R+\varepsilon$ to make R zero, we change it to $R+\omega\varepsilon$, where $1<\omega<2$. This is known as over-relaxation. It is extremely difficult to calculate the over-relaxation factor ω for all but the very simplest cases; the most relevant discussion is to be found in the work of Weber (1967a, 1967b), Carré (1961) and Carré and Wreathall (1964) and more general but correspondingly more complicated analysis is given by Forsythe and Wasow (1960), Varga (1962) and Durand (1966). Thomson (1966, 1967) found that the solution appropriate to a $p \times q$ mesh used for solving $\partial^2 V/\partial x^2 + \partial^2 V/\partial y^2 = 0$, namely

$$\omega = 2/\left\{1 + \left(1 - \lambda^2\right)^{1/2}\right\} \quad (20a)$$

with

$$\lambda = 1 - \frac{\pi^2}{4}\left(\frac{1}{p^2} + \frac{1}{q^2}\right) \quad (20b)$$

gave satisfactory results. Eq. (3.32) of Weber (1967a) shows that for the two-dimensional Laplace equation in (r, z),

$$\lambda = 1 - \pi^2/4p^2 - \mu^2/4q^2 \quad (20c)$$

would be better, where μ is the first zero of the zero-order Bessel function $J_0(x)$. Hardisty (1964) varies λ during the relaxation in such a way that the rate of convergence is always as high as possible, but Thomson (1967) comments that the gain may in practice be very slight.

1.2 Analog methods

Both the electrolytic tank and the resistance network have been used to obtain the quadrupole potential functions. The essence of these analog devices is explained in detail by Francken (1967) and Weber (1967a, 1967b) and here we consider only their special application to quadrupoles. Gershtein's induced current method could presumably be used also, but we know of no practical attempt (Gershtein, 1958, 1959a, 1959b, 1960, 1964, 1967). The electrolytic tank has been employed by Gregory and Sander (1962), and we refer the reader to their paper for details.

The resistance network has only recently been used to obtain quadrupole functions, by Belonozhko, Gaidukova, and Der-Shvarts (1967). A network for round lenses (Fig. 1A) was already available (Der-Shvarts & Netrebenko, 1959) with row resistances $h_i = v_0/2i$ (with $h_0 = 4v_0$) and column resistances $v_i = v_0/(2i+1)$ where v_0 was $80\,\Omega$ (cf. Francken, 1959, 1967 or Liebmann, 1950). "Leakage" resistors are connected to the nodes of this network (Fig. 1B), the values of which are selected in such a way that Eq. (5) is converted into Eqs. (8) or (9). For quadrupoles, the leakage resistors are given the values

$$c_i = iv_0/8 \tag{21}$$

and for octopoles,

$$c_i = iv_0/32 \tag{22}$$

For the reasons we have set out in connection with relaxation, it might seem more convenient to design a network to solve the equations for $\rho_2(r, z)$ or $\rho_4(r, z)$ (Eq. 14). In general, we have

$$\frac{\partial^2 \rho_n}{\partial r^2} + \frac{(2n+1)}{r}\frac{\partial \rho_n}{\partial r} + \frac{\partial^2 \rho_n}{\partial z^2} = 0. \tag{23}$$

For $n = 0$, we obtain the familiar rotationally symmetric equation, requiring resistances $v_i = v_0/(2i+1)$, $h_i = v_0/2i$; more generally, however, we have

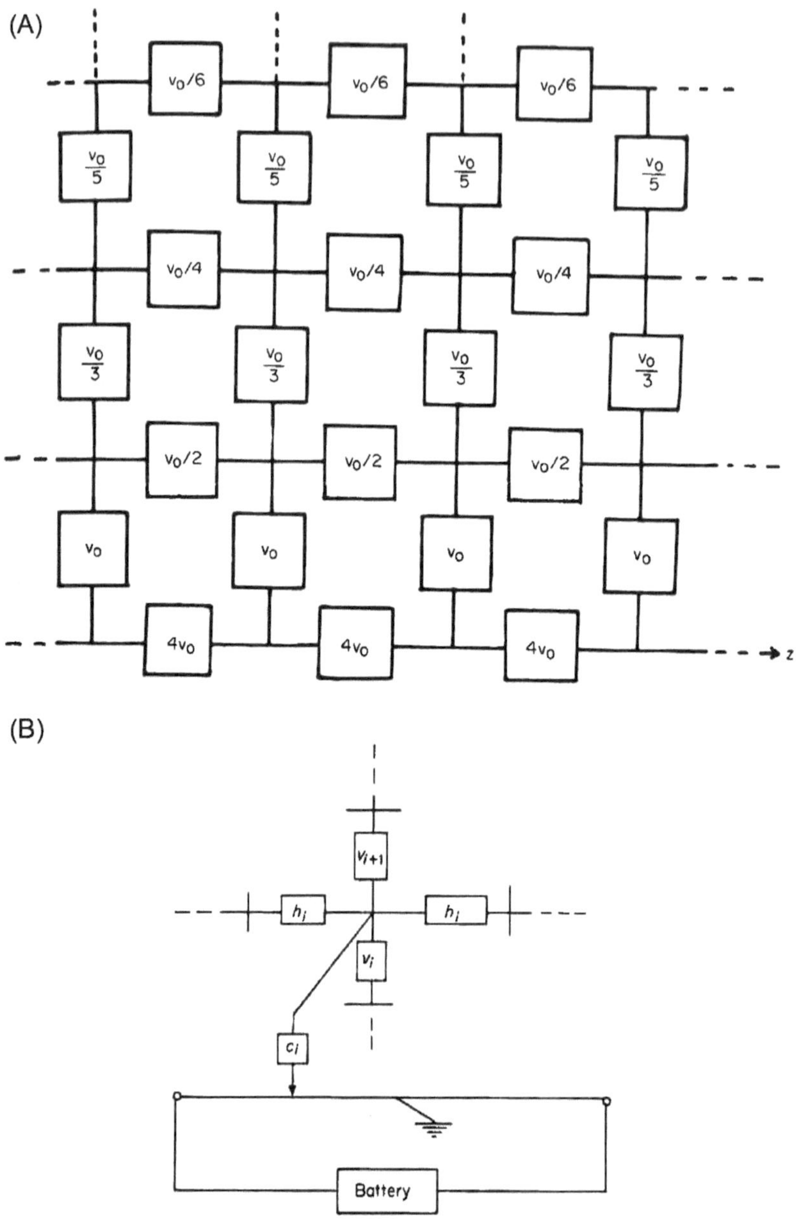

Fig. 1 (A) The standard resistance network, used for calculating the potential in a round lens. (B) The modified network described by Belonozhko et al. (1967), used for calculating the potential in quadrupoles and octopoles.

$$\begin{aligned} v_i &= \frac{2i - (2n+1)}{2i + (2n+1)} v_{i-1} \\ h_i &= 1 + \frac{2n+1}{2i} v_i \end{aligned}, \tag{24}$$

so that some resistances would have to be negative. Weber (1967a, 1967b) has pointed out, in connection with the rotationally symmetric case, that the resistances in the neighborhood of the axis can be given values different from those derived above; this has the advantage that the resistances close to the axis need not be too large compared with those near the periphery. It is conceivable that the problem of the negative resistance could be avoided by some similar device, but we shall not pursue this further here; nor have we made any attempt to consider stability and convergence (cf. Tschopp, 1961).

2. Curve-fitting and models

2.1 Best-fit curves

Once the potential functions have been established by measurement, relaxation or an analog method, it is usual to fit them to suitable analytical functions. These may be divided into two groups: functions that permit the equations of motion to be solved and the aberration integrals to be evaluated in terms of tabulated functions; and functions that have been selected because they fit the results particularly well. The first group of functions must be used in any theoretical study of the general behavior of systems, and two such functions, the rectangular model and the bell-shaped model, are particularly satisfactory for very long quadrupoles and short quadrupoles respectively. For a computer study of a particular lens, the best-fitting function from the second group is of course better.

A number of functions have been used to obtain good curve-fitting. In Cornell, Vosburgh (1967), following a suggestion of H. Kawakatsu, used

$$k(z) = \operatorname{sech}\{a(z^2 + bz)\} \tag{25}$$

for magnetic quadrupoles having bore radius 1.5 cm, and poles in the form of segments of truncated circular cylinders, radius 1.725 cm and length 3.0 cm; for these dimensions,

$$a = 0.238 \quad b = 0.5. \tag{26}$$

Fig. 2A shows how well this function matches the measured values. Vosburgh (1967) also measured the field in a doublet system, and established

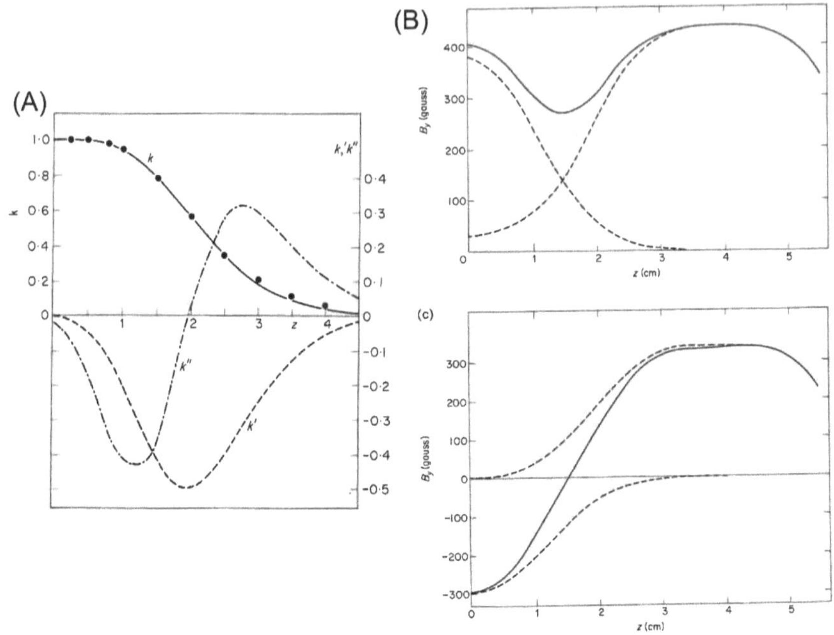

Fig. 2 (A) The function $k(z) = \text{sech}\ a(z^2 + bz)$ and the derivatives $k'(z)$ and $k''(z)$. Dots indicate the measured distribution. (B) and (C) The function $k(z)$ in a quadrupole doublet (continuous line) and in the two components (broken line), showing that the total $k(z)$ may be obtained by superposition of the functions for each component. (B) Lenses excited in the same sense; (C) in opposite senses. *(C) After Kawakatsu, H., Vosburgh, K. G., & Siegel, B. M. (1968a).* Journal of Applied Physics, 39, *245. With the permission of AIP Publishing, doi:10.1063/1.1655740.*

that this could be obtained accurately by superposition of the fields of the individual lenses (Fig. 2B and C).

At the Institut d'Electronique Fondamentale at Orsay, P.A. Möller has measured the potential function within long magnetic lenses (bore radius = 4 cm, length = 15 cm) and between pairs of these lenses; the doublet measurements confirm that the potential is obtained accurately by superposition. For these long lenses, Möller suggested a composite model, consisting of a flat central plateau terminated by functions

$$k(z) = \frac{1}{1 + [(z \pm 5)/3.8]^{3.7}}. \tag{27}$$

This model was adopted by Dhuicq (1968a) for his numerical work on multiplets.

Thomson (1967) was concerned only with short lenses, the length of which never exceeded the bore radius; as trial functions, he selected

$$k(z) = A \exp\left(-z^2/b^2\right) \tag{28}$$

and

$$k(z) = A/\left(1 + z^2/b^2\right)^n \quad n = 2,4,6\ldots12 \tag{29}$$

The constant A was selected by matching the functions at $z=0$, and b was set by minimizing the root mean square difference between the function and the tabulated values. Examples are given in Fig. 3.

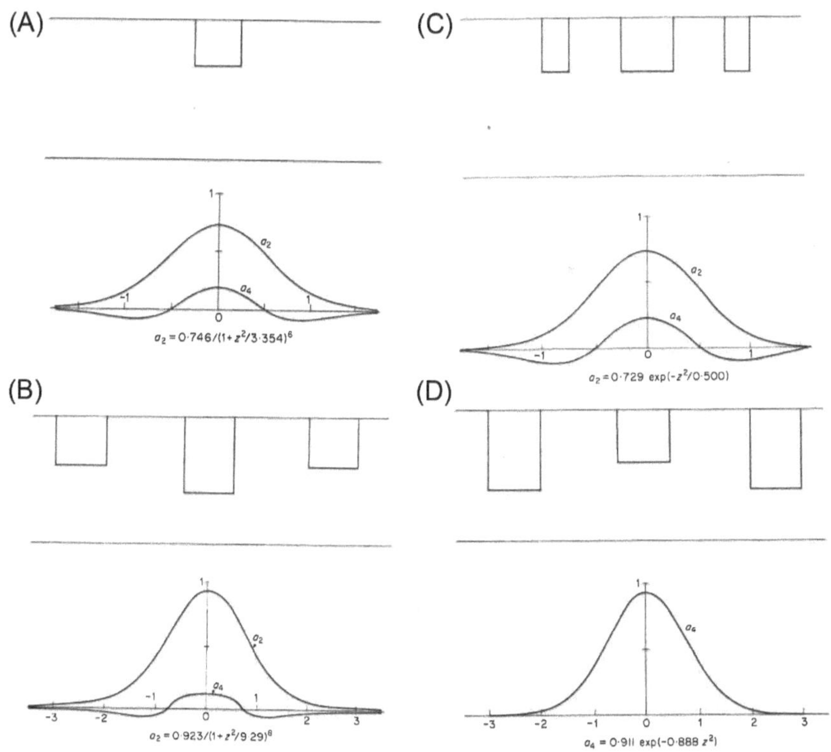

Fig. 3 The functions $k(z)(=a_2)$ and the coefficient of the term in r^4 in the series expansion of the potential (a_4) for a range of geometries, fitted to analytic test functions ; (A)–(C) correspond to quadrupoles and (D) to an octopole. *After Thomson, M. G. R. (1967).* The aberrations of quadrupole electron lenses. *Dissertation, Cambridge University.*

2.2 Electrode shape and fifth-order effects

Hitherto, we have accepted the restriction to lens potentials described by a $\cos 2\theta$ (quadrupole) or $\cos 4\theta$ (octopole) variation in azimuth. The effect of this is illustrated by a set of curves (Fig. 4) taken from Thomson (1967), which represent the pole-piece profiles that would correspond to his relaxation procedure. If the pole-piece cross-section is appreciably different from these profiles—if, that is, it does not bear a marked resemblance to a hyperbola—neglect of the $\cos 6\theta$ and higher terms in the quadrupole potential expansion will be unjustifiable. The additional terms that must be included will affect only fifth-order aberrations, which are caused both by the terms in $r^6 \cos 2\theta$ and those in $r^6 \cos 6\theta$. The magnitude of these aberrations will therefore vary with the electrode shape. This point is discussed by Fishkova, Ovsyannikova, and Yavor (1966), Ovsyannikova and Yavor (1967b) and by Hardy (1967) and the effect of fifth-order aberrations on probe size is considered by Dhuicq (1968a).

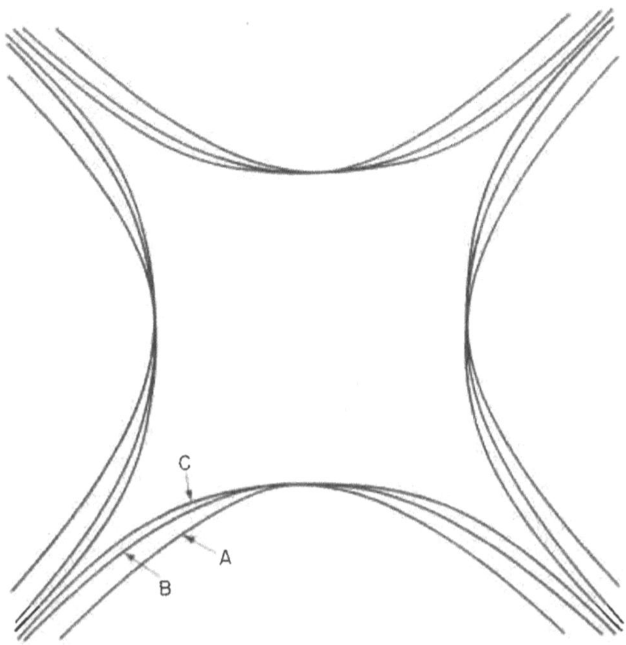

Fig. 4 Equipotentials at the center and end of the central element of Fig. 3A. (A) rectangular hyperbola; (B) section at the center; (C) section at the end. *After Thomson, M. G. R. (1967).* The aberrations of quadrupole electron lenses. *Dissertation, Cambridge University.*

It is extremely difficult to discuss fifth-order aberrations analytically, except in the thin lens approximations, and it is difficult to estimate the magnitude of ϕ_6 except in the two-dimensional approximation (e.g., Bernard, 1953; Fishkova, Siladi, & Yavor, 1967; Siladi, Shpak, & Yavor, 1968). Hardy (1967), using the two-dimensional formula

$$\phi(r, \theta) = \phi_2 r^2 \cos 2\theta + \phi_6 r^6 \cos 6\theta, \tag{30}$$

has generated the electrode profiles that correspond to various values of ϕ_6 (Fig. 5) for $\phi_2 = 16$ (arbitrary units); the curves represent $\phi = \pm 1$. Hardy found experimentally that the fifth-order aberrations of quadrupoles with hemispherical electrodes depend upon the ratio of electrode radius (R) to bore radius (a), the optimum value being $R/a = 1.7$. In fact, however, Hardy did not use hemispherical electrodes in his achromatic lenses, but

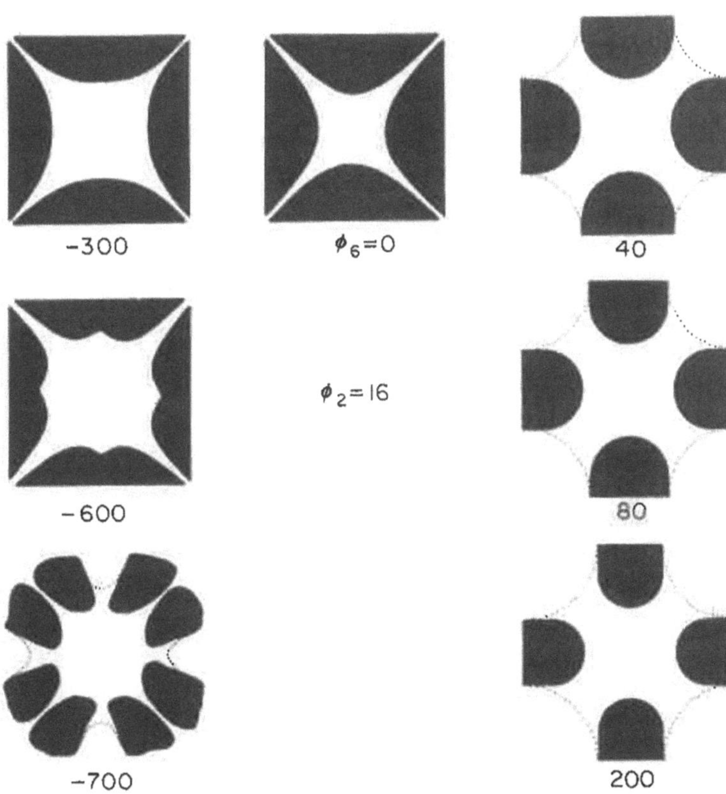

Fig. 5 Computer-generated electrode profiles, showing fifth-order effects of varying magnitude. *After Hardy, D. F. (1967).* Combined magnetic and electrostatic quadrupole electron lenses. *Dissertation, Cambridge University.*

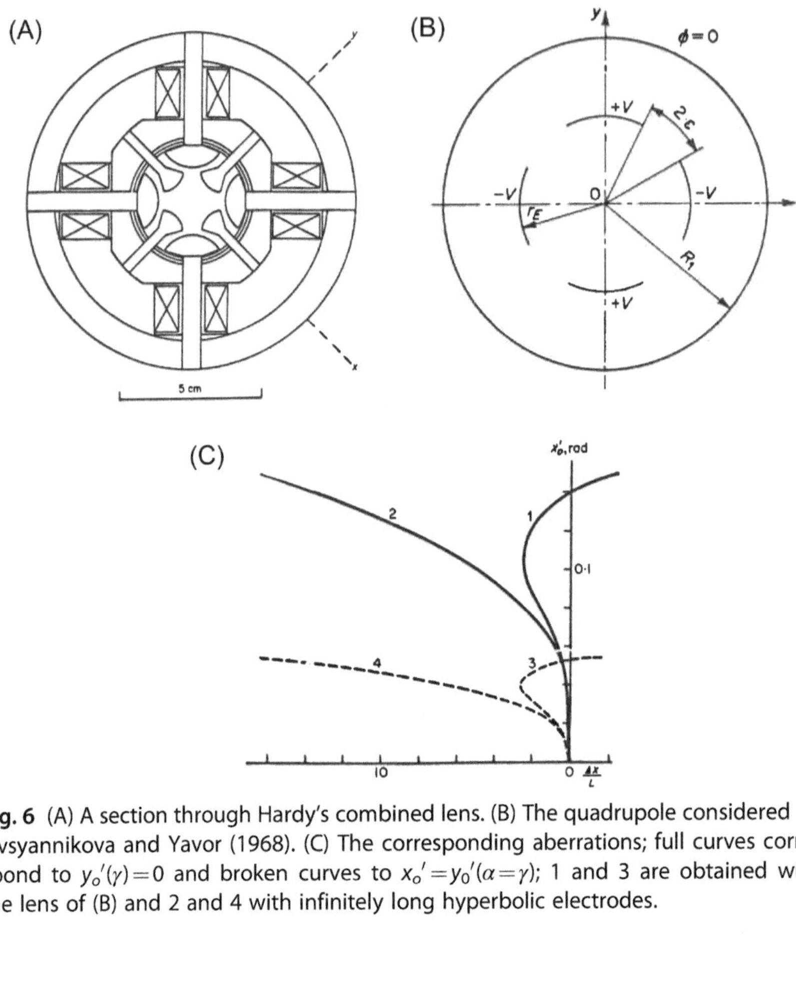

Fig. 6 (A) A section through Hardy's combined lens. (B) The quadrupole considered by Ovsyannikova and Yavor (1968). (C) The corresponding aberrations; full curves correspond to $y_o'(\gamma)=0$ and broken curves to $x_o'=y_o'(\alpha=\gamma)$; 1 and 3 are obtained with the lens of (B) and 2 and 4 with infinitely long hyperbolic electrodes.

truncated cones, their bases closed by spherical caps centered on the apices of the cones (Fig. 6A); for these, the best value of R/a proved to be 1.35.

Ovsyannikova and Yavor (1968) have examined the possibility of balancing third and fifth-order aberrations (Fig. 6B and C).

2.3 Analytical models for which the equations of motion are soluble

We now turn to the potential models that are amenable to theoretical treatment. The two that have been most extensively studied are the rectangular model, for which

$$k(z) = 1 \quad \text{for} \quad |z| < \frac{1}{2}L$$
$$k(z) = 0 \quad \text{for} \quad |z| > \frac{1}{2}L \tag{31}$$

and the bell-shaped model, for which

$$k(z) = \frac{1}{\{1 + (z/d)^2\}^2}. \tag{32}$$

The rectangular model may be made more realistic by attaching half-bells to its extremities, so that

$$\begin{aligned} k(z) &= 1 && \text{for } |z| \leqslant \frac{1}{2}L \\ k(z) &= \frac{1}{\{1 + (z \pm L/2)^2/d^2\}^2} && \text{for } z \left\{ \begin{array}{l} \leqslant -\tfrac{1}{2}L \\ \geqslant \tfrac{1}{2}L. \end{array} \right. \end{aligned} \tag{33}$$

The bell-shaped model may be modified to describe asymmetric quadrupoles:

$$\begin{aligned} k(z) &= \frac{1}{\{1 + (z/p)^2\}^2} && z \leq 0 \\ k(z) &= \frac{1}{\{1 + (z/q)^2\}^2} && z \geq 0. \end{aligned} \tag{34}$$

These are not the only functions $k(z)$ for which the equations of motion can be solved in terms of tabulated functions; with the triangular potential (Hawkes, 1965–66)

$$\begin{aligned} k(z) &= mz + 1 && z > 0 \\ k(z) &= -mz + 1 && z < 0 \end{aligned} \tag{35}$$

the equations of motion can be solved in terms of Airy integral functions although the aberrations cannot be calculated explicitly. For

$$k(z) = \operatorname{sech}(z/d) \tag{36}$$

we have

$$\begin{aligned} x'' + \beta^2 \operatorname{sech}^2(z/d)x &= 0 \\ y'' - \beta^2 \operatorname{sech}^2(z/d)y &= 0 \end{aligned} \tag{37}$$

and writing

$$u = -\tanh z/d \tag{38}$$

we obtain

$$\begin{aligned}(1-u^2)\ddot{x} - 2u\dot{x} + \beta^2 d^2 x &= 0\\(1-u^2)\ddot{y} - 2u\dot{y} - \beta^2 d^2 y &= 0\end{aligned} \tag{39}$$

where dots signify differentiation with respect to u. These equations have solutions in terms of Legendre functions of order n, where $n(n+1) = \pm\beta^2 d^2$. Other suitable forms of the function $k(z)$ can be obtained by extending the work of Grümm (1956) but with the advent of high-speed computers, there is little incentive to study any but the simplest models in detail. The rectangular and bell-shaped models give a very satisfactory indication of the behavior of real lenses, and since the other models inevitably involve a considerable amount of computation, one might just as well use the best-fit curve for the lens in question.

3. Note added in proof

We have recently used a computer Library Routine that permits division by a banded matrix to solve Eq. (15) for any value of n without relaxation (cf. McKeown & Ferrari, 1967). The number of points is still acceptable, and the functions $\Phi(z)$, $D(z)$ and $D_1(z)$ are obtained immediately; it is of considerable interest that the same program can be used to obtain the function $B(z)$ that characterizes a round magnetic lens, since $\nabla^2 B = 0$ [we recall that in the present notation, $B(z)$ denotes $B_z(0,0,z)$]. We have employed both the five-and nine-point formulae, which are to be found in Durand (1966, vol. II, pp. 27–34).

The advantage of using cubic splines (Ahlberg, Nilson, & Walsh, 1967) in solving paraxial ray equations has been discussed by Worster (1969).

CHAPTER FOUR

Quadrupole systems: Their suitability for specific tasks

Contents

The number of reasons for introducing quadrupoles and octopoles into practical systems is limited and it is difficult to foresee any but the following roles for these elements:

(i) As corrector units, to reduce the spherical and chromatic aberrations of good round lenses (Archard–Deltrap correctors).

(ii) As objective lenses with lower aberrations than the best conventional round lens available. It is very difficult to gain any improvement at ordinary accelerating voltages (up to a few hundred kilovolts); at higher voltages, however, the optical quality of conventional lenses deteriorates, and they become extremely cumbersome (weight estimates are given by Cohen, 1966; Hart, 1966; Tadano, 1966; a recent announcement in *Le Monde*, June 13, 1968 that a 3.5 MV microscope is to be built at Toulouse mentions a gross weight of 20 metric tons) and it is in this domain that quadrupoles may well prove to be superior.

(iii) As projector (and intermediate) lenses with less distortion and chromatic aberration of magnification than a comparable round lens, and less variation of the various parameters with magnification.

(iv) As probe-forming lenses combining long working-distance with modest spherical aberration.

Advances in Imaging and Electron Physics, Volume 224
ISSN 1076-5670
https://doi.org/10.1016/B978-0-323-98865-0.00004-2

(v) As a simple means of obtaining a good line image or a stigmatic image with unequal or unit magnifications in the principal directions.

The remainder of this chapter will be devoted to these various applications, with some account of the efforts that have been made to design and build suitable multiplets. Numerical data are for the most part collected together in "Quadrupole Data"; if information is sought about a specific system, the present chapter and "Quadrupole Data" should both be consulted, however, as the dividing-line between them is often imprecise.

1. Quadrupole multiplets as corrector units

In his classic paper on the "Sphärische und Chromatische Korrektur von Elektronenlinsen," Scherzer (1947) suggested various ways of using non-rotationally symmetric lenses to combat the spherical and chromatic aberrations of round lenses, and one of these proposals was pursued experimentally by Seeliger (1948–49, 1949, 1951, 1953) and Möllenstedt (1956a, 1956b). Archard (1954) pointed out that the cylindrical lenses of Scherzer's designs could be combined into quadrupoles, thus reducing considerably the number of separate components, and in the following year, he suggested "Two new simplified systems for the correction of spherical aberration in electron lenses" (Archard, 1955). The first of these, "system A," is essentially a quadrupole triplet system; it consists of four elements, to be placed immediately behind a high magnification objective. The first is a quadrupole, the second an octopole and the third and fourth create mixed quadrupole–octopole fields. The second of Archard's proposals, "system B," is essentially the Archard–Deltrap telescopic corrector. It consists of an antisymmetric quadruplet, modified so that three octopole potentials affect the beam. The first component after the objective is a quadrupole, excitation +1 (arbitrary units); next comes a combined quadrupole–octopole, with quadrupole excitation −2. This is followed by an octopole and after this comes another combined quadrupole–octopole, excitation +2; the beam is finally rendered stigmatic by a fourth quadrupole, excitation −1. Here, we have all the essentials of the corrector that at the present time seems the most promising: four quadrupoles, geometrical symmetry and electrical antisymmetry, afocal imagery.

In 1959, Archard (1960a) pointed out that microanalysis could benefit from quadrupole—octopole triplets, designed to improve the probe, and a triplet system, resembling "system A," is described by Archard, Mulvey, and Petrie (1960); here, the first lens is a quadrupole–octopole, the second element is an octopole, the third is again a combined unit and the fourth is a quadrupole.

The next stage in the design of corrector combinations represented a major advance; this was an extended experimental and theoretical study by Deltrap (1964a, 1964b) of the design of these units, culminating in the construction and testing of a telescopic quadruplet corrector. Deltrap did not impose magnetic antisymmetry on his designs, and they are in consequence slightly more complex than perhaps they need be. Apart from his successful attempt to demonstrate the practical viability of such a corrector, a number of important general points emerge from this work. Deltrap examines with some care a large family of triplets and quadruplets, in order to see what type of imagery each would be capable of. A thin-lens study of triplets suggests that high magnification can be achieved only if the object is virtual and large demagnification only if the image is virtual. It seems probable that this remains true in the general case, because "the last element will always be converging in one plane and diverging in the other, while for a real focus the rays must be converging in both planes at the exit of the system" (Deltrap, 1964a). This difficulty is still present in quadruplets, of course, but is more easily overcome. After preliminary approximate calculations giving the octopole excitations necessary to correct various systems, Deltrap comes to the conclusion that the most convenient is a system with two separate internal "line foci" (Fig. 1) at which two of the octopoles act. A third octopole may be inserted anywhere in the system provided that its effect is not equivalent to that of either of the other octopoles: provided, that is, that the beam asymmetry is different in all three octopoles.

Let us now turn to the details of Deltrap's final system which is specifically designed to correct a probe-forming lens. The four elements of the

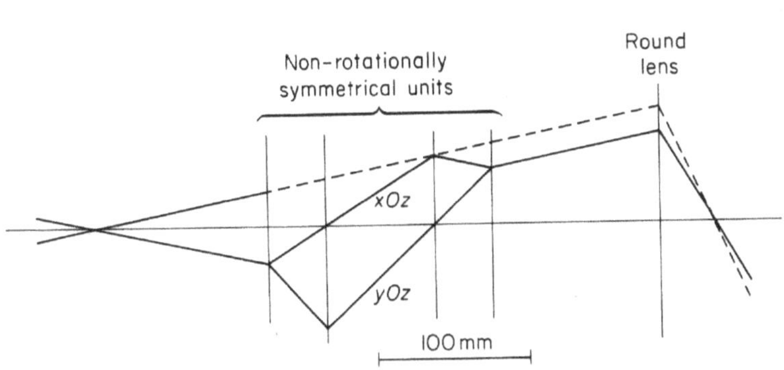

Fig. 1 Ray diagram of the system employed by Deltrap (1964a, 1964b).

corrector are all capable of producing octopole and quadrupole potentials, and the pole-tips are shaped with a view to minimizing the fifth-order aberration (Deltrap, 1964a). The poles are made out of mumetal, which is more difficult to machine than the Swedish iron used in Deltrap's prototypes and has lower saturation magnetization, but has far less remanent magnetization. The geometry is shown in Fig. 1; the principal plane of the round lens is 11.3 cm from the fourth element and the object point of the whole assembly is 37 cm in front of the round lens. When the quadrupoles are adjusted to give the desired ray path, and the octopoles are adjusted for correction, the excitations are as follows:

Q_1 : 50 amp-turns	O_1 : 0 amp-turns
Q_2 : 43	O_2 : 5·5
Q_3 : 30	O_3 : 8·5
Q_4 : 23	O_4 : 16

Further details are to be found in Hawkes (1966a) and in Septier (1966). Deltrap's work is being applied in a scanning microscope by Crewe, Cohen, and Meads (1968). A microscope corrector, operating on this principle, is planned by Welton and Allison (1966, 1967) and Worsham (1969).

We now turn to the work of Möller, Dhuicq, and Septier (1966a, 1966b) and Regenstreif (1966a, 1966b, 1966c, 1966d, 1967b, 1967c, 1967e) on the quadrupole combinations that allow us to achieve specific forms of transfer matrix. These authors are particularly concerned with systems for channeling beams of high energy particles in accelerators, but it so happens that such systems are normally telescopic, like Archard–Deltrap correctors. Of the various questions considered by Möller et al. and Regenstreif, we are concerned with the following: what combinations of quadrupoles can give transfer matrices of the form

$$\mathbf{T} = \begin{pmatrix} 1 & b \\ 0 & 1 \end{pmatrix} = \mathbf{T}_{+b} \tag{1a}$$

and

$$\mathbf{T} = \begin{pmatrix} -1 & b \\ 0 & -1 \end{pmatrix} = \mathbf{T}_{-b} \tag{1b}$$

between the terminal planes of the first and last quadrupoles? Since Regenstreif considers the question of obtaining

$$\mathbf{T} = \begin{pmatrix} \pm 1 & 0 \\ c & \pm 1 \end{pmatrix} = \mathbf{T}_{\pm c} \tag{1c}$$

at the same time, we shall also include this here. The rectangular model is employed throughout, and Regenstreif restricts the variable $\theta = \beta L$ to the range $0 \leqslant \theta \leqslant \pi/2$.

1.1 Doublets

"No AG [alternating gradient] doublet can achieve any of the four simple transfer matrices considered" (Regenstreif, 1967b). Möller et al. (1966a) had likewise shown that a doublet cannot give $\mathbf{T}_{\pm b}$.

1.2 Triplets

Neither $\mathbf{T}_{+c}$ nor $\mathbf{T}_{-c}$ can be attained with a triplet. $\mathbf{T}_{+b}$ can be obtained but electrons incident with a finite slope will emerge with a finite slope of the same sign. T_{-b} can never be obtained in the DCD (diverging–converging–diverging) principal section of a triplet (Regenstreif, 1967b).

1.3 Quadruplets

Regenstreif (1967c) is devoted wholly to the problem of selecting quadruplets such that $\mathbf{T}_x = \mathbf{T}_y = \mathbf{T}_{-b}$, after which he investigates whether they fulfill various subsidiary requirements. We postpone discussion of this work to Section 4 of Quadrupole Data. Möller et al. (1966a) show that $\mathbf{T}_{+b}$ cannot be attained with a symmetric quadruplet; with an antisymmetric quadruplet, however (Möller et al., 1966b), both $\mathbf{T}_{\pm b}$ can be achieved.

2. Quadrupole multiplets as objective lenses or as probe-forming lenses

The lenses that stand to gain most from reduction of the spherical and chromatic aberration coefficients are the objective lenses of microscopes and the probe-forming lenses of microanalyzers. The corrector system that we have discussed in the preceding section is designed to follow an objective lens or to precede a probe-forming lens, and in either case the fact that the contribution of the corrector to the aberrations is effectively reduced by a factor M^4 where M is the objective magnification or probe-lens demagnification has advantages. Nevertheless, it is interesting to see whether the round lens can be dispensed with, the quadrupole multiplet providing both focusing and aberration correction.

This problem has been studied by Dymnikov, Baranova, Fishkova and Yavor in Leningrad, by Dhuicq and Septier in Paris and by Bauer in Dresden. From this work, two promising systems have emerged: the antisymmetric quadruplet (Dymnikov et al., 1965a, 1965b, 1965c, 1965d, 1965e, 1965f; Dhuicq and Septier, 1966a, 1966b; Dhuicq, 1968a, 1968c) and the quintuplet consisting of an antisymmetric doublet and a symmetric triplet (Bauer, 1964, 1965–66; Dhuicq, 1968a). Each of these systems exploits the properties of antisymmetry in multiplets; in the quintuplet, the magnification is provided by the doublet and the triplet converts the real object into a virtual object close to the asymptotic object focus of the doublet (Figs. 2 and 3A). The quadruplet is sufficiently flexible to need no such auxiliary lenses.

Further details of the design parameters of these systems will be given in "Quadrupole Data". Here we mention only a few of the experimental results obtained by Bauer and Dhuicq. With his quintuplet objective, Bauer achieved a resolution of about 200 Å and this figure could doubtless be reduced by increasing the stability and incorporating a better stigmator; the theoretical value is about 70 Å at 50 kV. We should stress that there are no octopoles in the system: these figures represent the performance of quadrupoles alone. Bauer points out that the triplet is the main source of aberration, and that it could be most simply reduced by exciting the lenses asymmetrically (Section 6 of Quadrupole Data).

Dhuicq (1968a) uses the quintuplet as a probe-forming lens, and the role of the triplet is now to convey the "virtual probe," formed near the image focus of the doublet, into free space (Fig. 3B). Dhuicq discusses in considerable detail the effect of the third- and fifth-order aberrations on the quality of the probe; we cannot reproduce this here, but we recapitulate his general conclusions. With two independent octopoles, Dhuicq obtains a probe comparable with one used by Castaing (1951), but no better:

> *Nous sommes ici dans la même situation que les expérimentateurs qui ont tenté de corriger l'aberration sphérique d'un objectif de microscope: malgré tout le soin et les efforts apportés dans la réalisation du système optique et dans la tentative de correction, le résultat final n'est. pas meilleur que celui obtenu avec le système non corrigé. Cela tient au fait que l'ensemble optique nécessaire à la correction possède un nombre élevé d'électrodes n'ayant pas la symétrie de révolution, de réalisation mécanique beaucoup plus délicate que les lentilles rondes et possédant lui-même des aberrations (tout au moins en y) beaucoup plus fortes que celles des lentilles de révolution.*

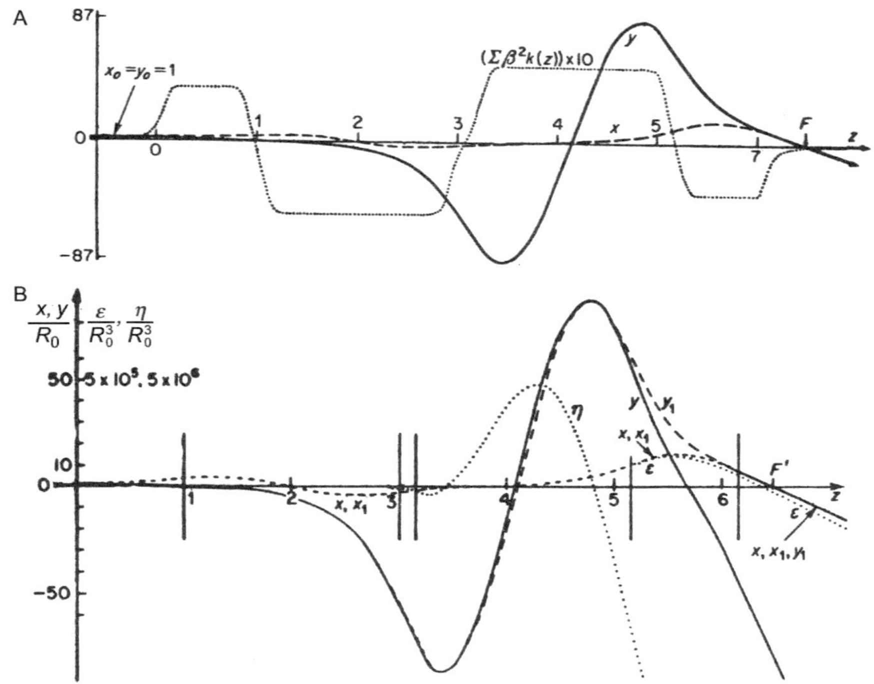

Fig. 2 The quadrupole quadruplet as a probe-forming lens. (A) The function $\beta^2 k(z)$ (*dotted curve*) and $g_x(z)$, $g_y(z)$. (B) The trajectories $R_o g_x(z)$, $R_o g_y(z)$ $(=X_1, Y_1)$, their perturbation caused by third-order aberrations (X, Y) and the aberrations $\varepsilon(=X - X_1)$ and $\eta(=Y - Y_1)$. *After Dhuicq, D., & Septier, A. (1966a).* Comptes Rendus, B263, *280; Dhuicq, D., & Septier, A. (1966b).* Comptes Rendus, B263, *364.*

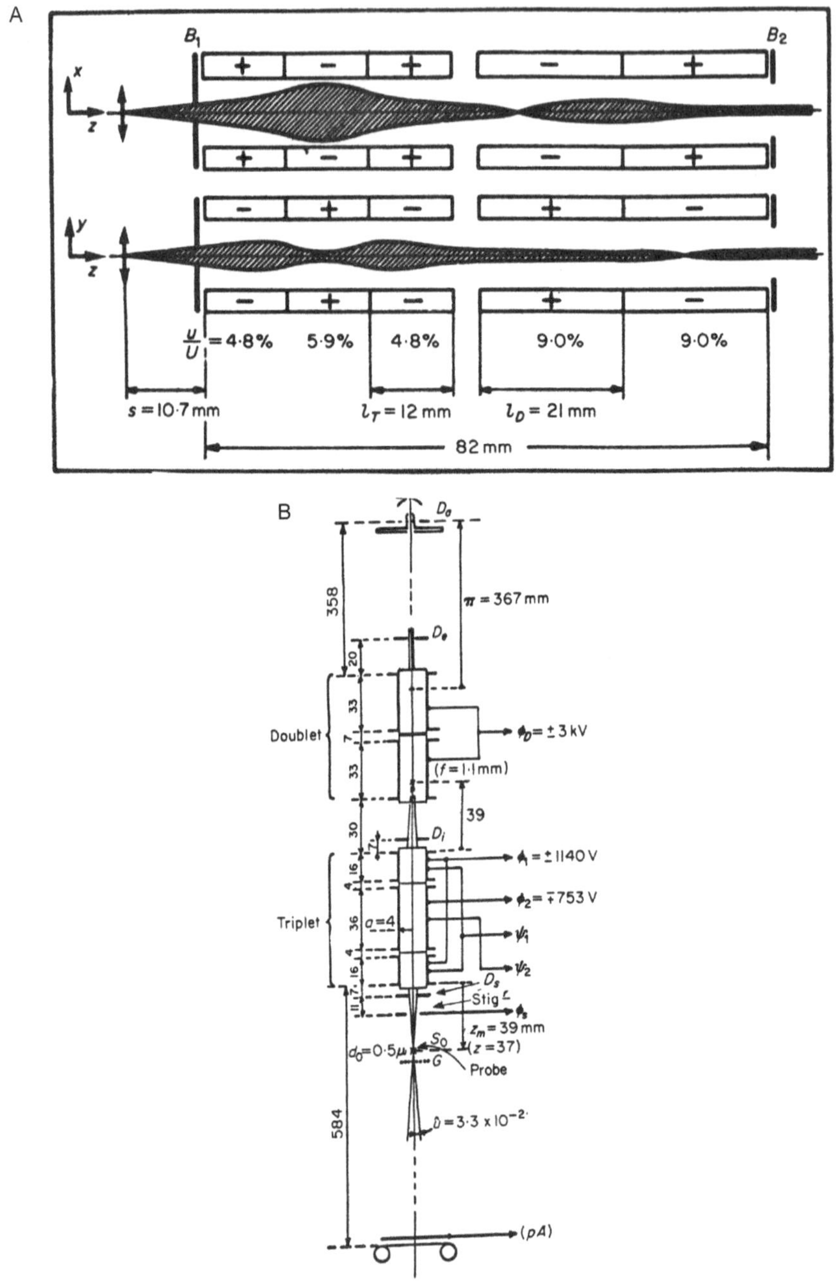

Fig. 3 (A) The quintuplet objective. (B) The quintuplet probe-forming lens. *(A) After Bauer, H.-D. (1965–66).* Optik, 23, *596. (B) After Dhuicq, D. (1968a).* Contribution à la correction des aberrations d'ouverture du 3ème ordre dans les systèmes de lentilles quadrupolaires stigmatiques au ler ordre. *Thèse, Paris-Orsay.*

Since the system could not correct all the aperture aberrations, the fact that it was as good as that of Castaing is in itself encouraging and Dhuicq is now improving his optical system by including a third octopole and a dodecapole designed to compensate the worst of the fifth-order aberrations. He points out that the system is suitable for producing very small probes and that the long working distance and low lens excitations are indisputably attractive features.

The characteristics of the antisymmetric quadruplet are collected together in Section 4 of Quadrupole Data. We shall not consider it further here.

3. Quadrupoles as projector lenses

The doublet and the quadruplet have been considered for this role by several authors, and the triplet has also been studied, though to a lesser extent. The earliest discussion is to be found in a series of remarkably complete, unpublished reports by Glaser (1955), the essence of which was subsequently included in a patent specification (Glaser, 1956b); Glaser carefully analyses the symmetric doublet and describes a two-lens "electron microscope" consisting of a conventional round objective and a regular doublet, which acts as a projector and as a corrector for the spherical aberration of the objective. Two octopoles are incorporated in the doublet.

Experimental as well as theoretical work on a doublet projector was performed by Reisman (1957) and Reisman and Siegel, 1957); Septier (1958a, 1958b) published the theory of the doublet using the rectangular model and Dhuicq (1961) pursued this experimentally. Reisman investigated both magnetic and electrostatic quadrupoles: in the experimental work, the magnetic lenses were 7.3 or 2.0 cm long with bore radius 1.06 cm; the poles were in the form of right cylinders of truncated hyperbolic section with 420 turns/pole. The electrostatic lenses were 2.03 cm long with bore radius 0.24 cm; the electrodes were in the form of flat fins, eight in number so that either quadrupole or octopole fields could be generated. Reisman showed that a magnetic doublet can provide a wide range of projector magnifications, and his measurements of the distortion, although beset with uncertainties of experimental origin, are by no means discouraging.

Dhuicq (1961) employed electrostatic quadrupoles (for reasons of practical convenience) with right cylindrical electrodes of radius 4.5 mm, length 3.3 cm and bore radius 4 mm; the gap between the end-faces of the lenses was 7 mm. With these long narrow quadrupoles, he found good agreement with the predictions of the rectangular model.

Quadruplet projectors have been studied by Dhuicq (1961) and by Vosburgh (1967) and Kawakatsu, Vosburgh, and Siegel (1968a). Dhuicq studies mainly the symmetric quadruplet, and we shall postpone discussion of his results to Section 4 of Quadrupole Data. Kawakatsu et al. have made a detailed theoretical and experimental examination of the antisymmetric quadruplet as a projector lens, two converging lenses following two diverging lenses and conversely, in the two principal directions; they consider the distortion coefficients and the chromatic aberration and find that these compare very favorably with those of a good round lens. The focal length can be short with a large bore radius and can be varied over a wide range; "throughout this change in focal length, the distortion and chromatic aberrations change very little and are always lower than even the best set of rotationally symmetric lenses designed to cover this range of focal lengths" (Kawakatsu et al., 1968a). The pole-pieces consisted of segments of circular cylinders of radius $1.15a$, where $a = 1.5\,\text{cm}$ is the bore radius, and 3.0 cm long. The gaps between the lenses were 1.7 cm wide (so that if we consider the "effective length," the lenses are in contact). The maximum magnetic flux required for 100 kV electrons is about 1 kg, and thus "a quadruplet projector with the same pole piece dimensions could be used in electron microscopes with accelerating voltages of several million volts, while the focal lengths and aberrations of the lens system would still have the same values." Further details of this system are given in Section 4 of Quadrupole Data.

Dhuicq (1961) has also investigated a range of triplets; none was suitable for a probe-forming lens or objective but some could have been employed as projectors. Triplets are further discussed in Section 3 of Quadrupole Data.

4. Formation of a line image and pseudo-stigmatic imagery

It is easy to produce a good line image with a quadrupole doublet; Le Poole (1964), Bok, Kramer, and Le Poole (1964) and Bok (1968) have described a scanning microscope and a mirror microscope each employing a line image produced by a doublet. The mirror microscope system is interesting in that the incident electrons pass through the doublet, are reflected close to the specimen and return through the doublet; since the latter is magnetic, however, the sign of the excitations is reversed, so that the converging–diverging plane for the incident electrons is still the converging–diverging plane for the emergent beam: the optical system possesses the characteristics of an antisymmetric quadruplet.

Crewe, Eggenberger, Welton, and Wall (1967) have also used a quadrupole doublet in a scanning device, to produce not a line but a spot; the magnification is different in the two principal sections and the system is pseudo-stigmatic. Crewe et al. conclude that "a scanning electron microscope can use a quadrupole doublet to focus the beam into a spot at least as small as 150 by 450Å"; the size of their probe was in fact limited not by the aperture aberrations but by astigmatism, which can be corrected.

5. Octopoles alone

Hitherto, we have assumed that the spherical aberration of a round lens can be corrected only if quadrupoles and octopoles are present simultaneously. The quadrupoles create an asymmetry in the electron beam, without which the octopoles cannot effect the desired correction uniformly. If only round lenses and octopoles are present, the circular aberration disc produced by the spherical aberrations of a round lens can at best be reduced to a rosette. In an extremely interesting paper, Scherzer and Typke (1967) have considered whether such partial correction might nevertheless be worthwhile, and they estimate that the resolving power of the combination could be as much as three times greater than that of the round lens alone. The importance of this conclusion scarcely needs emphasizing, and we therefore recapitulate their discussion in some detail. The types of lens considered are shown in Fig. 4; in the electrostatic lens, four metal electrodes are incorporated in the central electrode, and in the magnetic lens, four soft iron poles are inserted through the image-side pole-piece. In both cases, the modification will alter the round lens strength and introduce an octopole field component; astigmatism is canceled with the aid of an eight-pole stigmator.

In the Gaussian image plane, we have[a]

$$\begin{aligned} x_i &= -M(C_s \cos\theta_a + C_o \cos 3\theta_a)\alpha_o^3 \\ y_i &= -M(C_s \sin\theta_a - C_o \sin 3\theta_a)\alpha_o^3 \end{aligned} \qquad (2)$$

for a ray that intersects the optic axis at the object plane, making an angle α_o with the axis. The azimuthal angle θ_a is measured in the aperture plane, which is assumed to be just behind the objective; C_o is the aperture aberration coefficient of the octopole field, and is proportional to the coefficient (0003) of Hawkes (1965a, 1965b). When $C_o = -C_s$,

[a] We have retained the minus sign that Scherzer and Typke include in their definition in order to make the aberration coefficients positive.

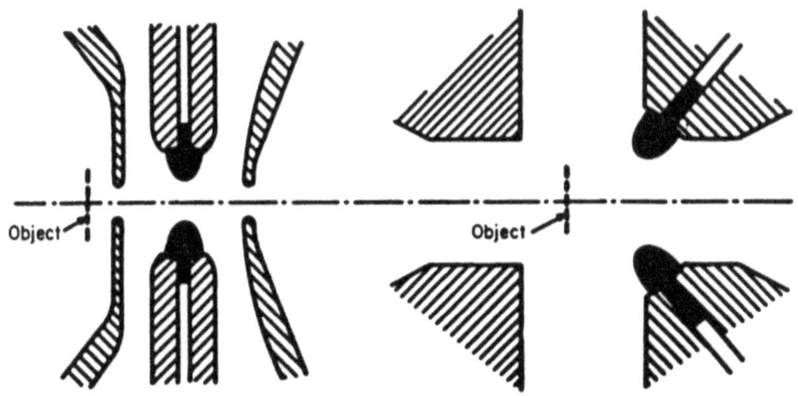

Fig. 4 Electrostatic (left) and magnetic (right) round lenses incorporating octopole correctors. *After Scherzer, O., & Typke, D. (1967).* Optik, 26, *564.*

$$\begin{aligned} x_i &= -4M\alpha_o^3 C_s \cos\theta_a \sin^2\theta_a \\ y_i &= -4M\alpha_o^3 C_s \sin\theta_a \cos^2\theta_a \end{aligned} \tag{3}$$

Fig. 5 illustrates the current density distribution in the image plane when the aperture is uniformly filled; in the lower half of the figure, corresponding isophotes are shown for the round lens alone ($C_o = 0$) in the Gaussian image plane (left) and in the optimum plane (right). Close to the axis in the image plane, the current density is higher in the corrected system and the intensity thus falls off more steeply; it is this concentration of the pencil that suggests that the resolution may be improved.

The important parameter, so far as the point resolution is concerned, is the change in the phase, γ, of the electron wave as it passes through the lens; if we consider the Gaussian image plane, we have

$$\gamma = \frac{2\pi}{\lambda}\left(\frac{1}{4}C_s\alpha^4 + \frac{1}{4}C_o\cos 4\theta_a\alpha^4 + \frac{1}{6}C_5\alpha^6\right) \tag{4}$$

in which C_5 is the fifth-order aberration coefficient and λ denotes wavelength in the object plane. We need consider only the contribution from within the first Fresnel zone, and Scherzer and Typke restrict their integration to the region for which $\gamma < \pi/3$. The wave function in the image plane is obtained from that in the aperture plane, and the region of integration proves to be a star-shaped area bounded by hyperbolae if the effect of C_5 is neglected (Fig. 6); in reality, C_5 limits the region so that

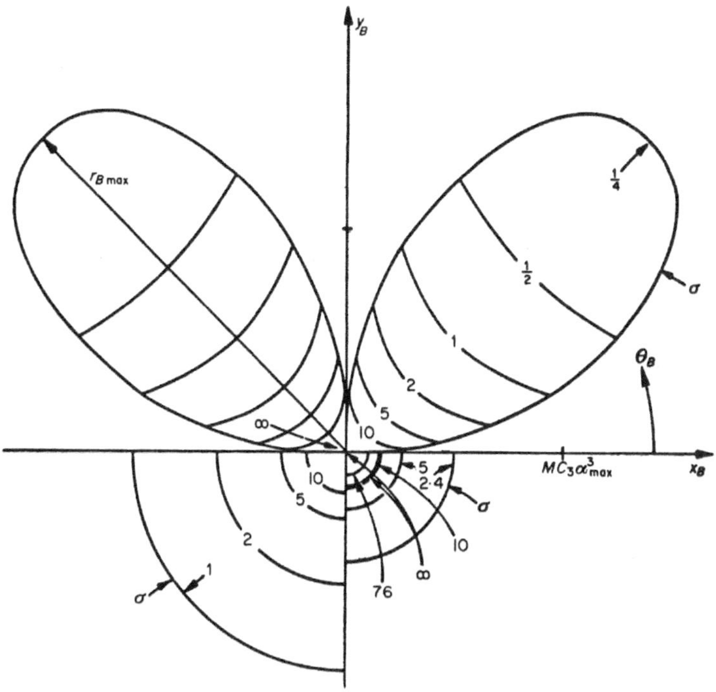

Fig. 5 The intensity distribution at the image produced by an objective lens with an octopole corrector. *Above*: in the Gaussian image plane; *Below*: the intensity distribution in the Gaussian image plane (left) and in the optimum image plane (right) without the octopole corrector. *After Scherzer, O., & Typke, D. (1967).* Optik, 26, *564.*

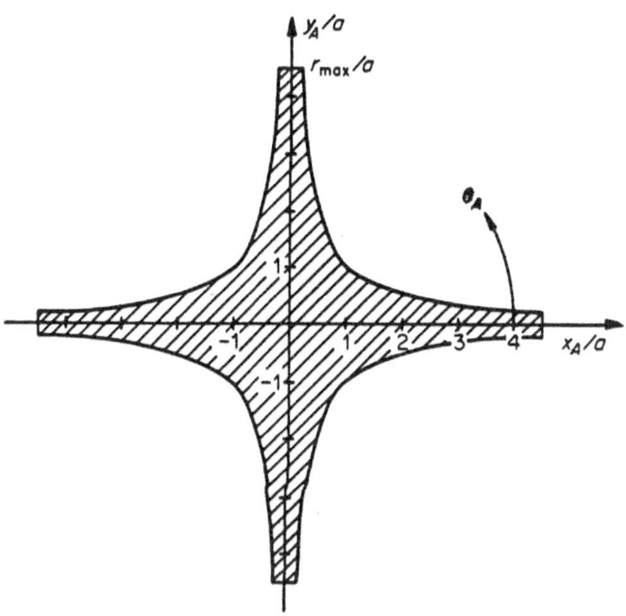

Fig. 6 The useful aperture. *After Scherzer, O., & Typke, D. (1967).* Optik, 26, *564.*

$$r_{\max} = f(\lambda/C_5)^{1/6} = a\left(\frac{2^6 \cdot 3^3 \cdot C_s^3}{\lambda C_5^2}\right)^{1/12} \tag{5}$$

where

$$a = f(\lambda/12C_s)^{1/4}. \tag{6}$$

The regions outside the hyperbolae serve no useful purpose in the imagery and the physical aperture should therefore be star-shaped; the parameter a, which defines the dimensions of the aperture, will in practice be of the order of 10 μm. Scherzer and Typke then evaluate the intensity distribution in the directions in which the octopole acts against the round lens aberration (*Korrekturschnitte*) and in the directions, at 45° to these, in which it reinforces it; the results are reproduced in Fig. 7A. These reveal that two of the features of such a correcting device that might have proved troublesome are in fact tolerable: firstly, the resolution between two object points is hardly dependent upon the relative orientation of the line joining the object points and the octopole (Fig. 7B); and secondly, optical artifacts should not render the image difficult to interpret (Fig. 8). The final conclusion of Scherzer and Typke is that in comparison with the round lens alone, the resolution of an objective with an octopole corrector should be improved by two to four times.

6. Chromatic correction

Hitherto, we have tended to bracket the chromatic and spherical aberrations of round lenses together, as the pair of defects that quadrupoles can conveniently correct. The chromatic effects of quadrupoles are often considered separately from the aperture aberrations, however, and it is convenient to accord chromatic correction a separate section here; this reflects the intermediate state of development of the subject and when the design of quadrupole lenses and correctors has become more sophisticated, such a division will doubtless seem unnatural.

The chromatic aberration of quadrupoles and the correction of that of a round lens with the aid of quadrupoles and octopoles is considered by several authors (Hardy, 1967; Kawakatsu et al., 1968a; Kel'man & Yavor, 1961, 1963a; Scherzer, 1947; Schiske, 1967; Vosburgh, 1967; Yavor et al., 1964). We shall devote most of this section to Hardy's work, since this is concerned with the many practical problems that can arise, and to that of Schiske.

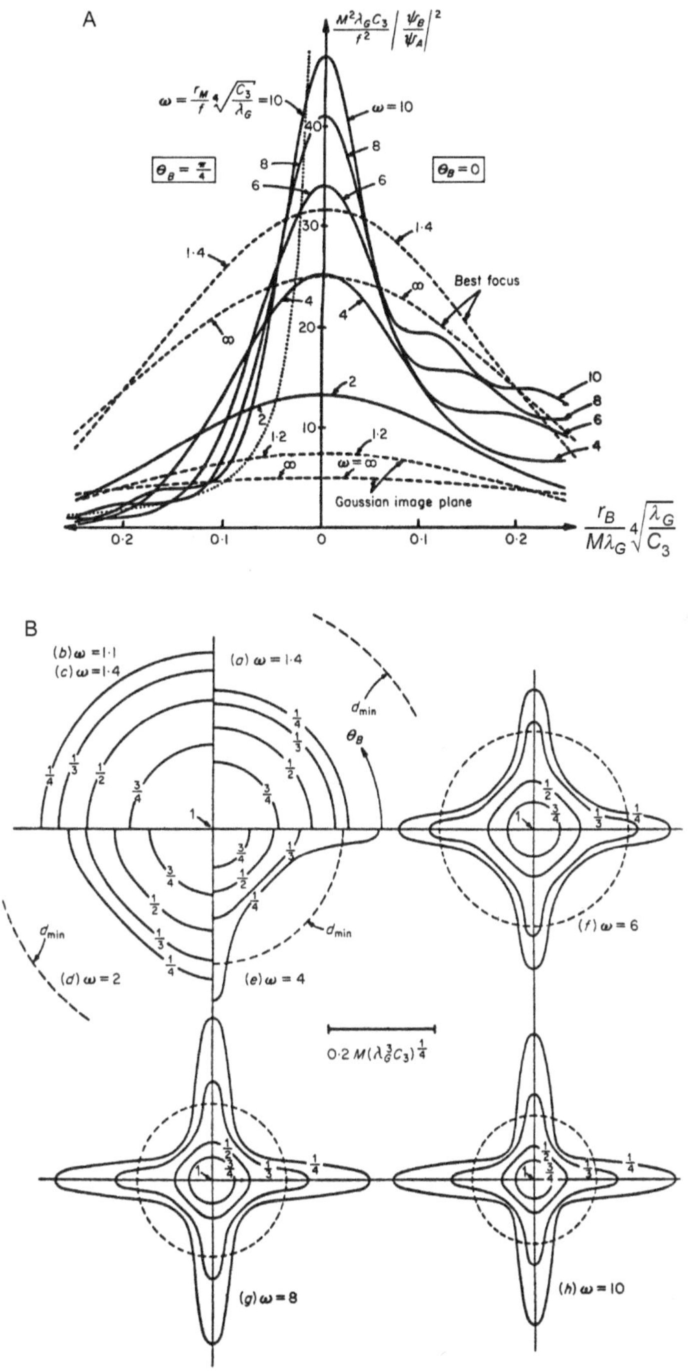

Fig. 7 (A) The radial intensity distribution in the Gaussian image plane for various apertures, *left* in the correction plane and *right* in the plane at 45° to this; the value given by geometrical optics is dotted and the uncorrected value is shown by a broken line. (B) Lines of equal current density for different apertures, $\alpha=(\lambda/C_3)^{1/4}\omega$. (A) Uncorrected round lens, optimum focusing and optimum aperture; (B) as (a) in the Gaussian image plane; (c)–(h) corrected round lens, Gaussian image plane. The resolving power, d_{min} is shown by a dashed line as a function of the angle θ_a. Note that $C_3 \equiv C_s$ and $\lambda_G \equiv \lambda$. *(A) After Scherzer, O., & Typke, D. (1967).* Optik, 26, *564; (C)–(H) After Scherzer, O., & Typke, D. (1967).* Optik, 26, 564.

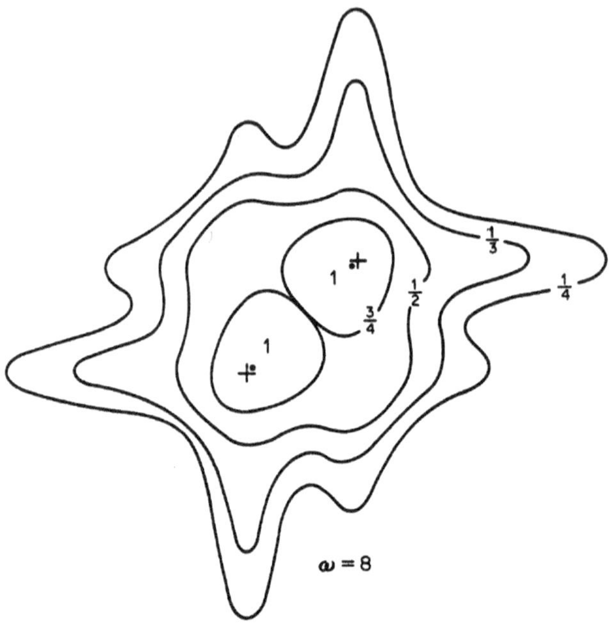

Fig. 8 Lines of equal current density produced by superposition of two diffraction patterns; + signs indicate the geometrical optical image points. *After Scherzer, O., & Typke, D. (1967).* Optik, *26, 564.*

We have not yet mentioned the extensive work of Regenstreif (1964, 1965), a thesis by Gouiran (1967) and a paper by Lenormand (1968), Regenstreif (1965), Gouiran (1967) and Lenormand (1968) are concerned with periodic alternating gradient quadrupoles (the geometrical aberrations of such sequences are discussed by Tanguy, 1967). Regenstreif (1964), however, gives a systematic thin-lens treatment of the chromatic aberrations of quadrupole multiplets, and we summarize his results here. Let us consider a system of n quadrupoles separated by $n-1$ drift spaces; each lens has transfer matrix $\begin{pmatrix} 1 & 0 \\ \delta & 1 \end{pmatrix}$, adopting Regenstreif's notation, so that $\delta = -1/f$ or $-1/g$. The transfer matrix $\mathbf{M}$ between the terminal lenses is then $\begin{pmatrix} a & b \\ c & d \end{pmatrix}$; if the beam momentum ($p$) is increased by an amount Δp, $\mathbf{M}$ is transformed into $\mathbf{M} + \overline{\mathbf{M}}(\Delta p/p)$, which is written $\begin{pmatrix} \overline{a} & \overline{b} \\ \overline{c} & \overline{d} \end{pmatrix}$. When the object plane lies at a distance u from the first lens ($z_0 = -u$ if the origin is taken at the first lens), the longitudinal chromatic aberration Δv is given by

$$\Delta v = \frac{\alpha(x_o + ux'_o)^2 + 2\beta(x_o + ux'_o)x'_o + \gamma x'^2_o}{\{cx_o + (cu + d)x'_o\}^2} \frac{\Delta p}{p} \tag{7}$$

in which $\alpha = a\bar{c} - \bar{a}c, 2\beta = (a\bar{d} - \bar{a}d) + (b\bar{c} - \bar{b}c)$ and $\gamma = b\bar{d} - \bar{b}d$; Δv represents the shift in the point at which the ray with initial parameters (x_0, x'_0) intersects the axis in image space. The transverse aberration in the image plane is given by

$$\Delta x = -\frac{(\alpha u + \beta)(x_o + ux'_o) + (\beta u + \gamma)x'_o}{cu + d} \frac{\Delta p}{p}.$$

Regenstreif then considers three special cases: systems that focus a parallel beam, systems that focus a point into a point and systems that render parallel a beam emerging from a point. The remainder of this report contains a systematic discussion of the possibility of minimizing the chromatic aberrations in single quadrupoles, doublets and triplets. The reader is referred to the original work for details of this.

The aspect of Hardy's work that we consider here concerns quadrupole multiplets: is it possible, by replacing the electrostatic or magnetic quadrupoles of a multiplet by combined "electromagnetic" quadrupoles, to make the multiplet achromatic or, if its function is to correct spherical aberration, to make it cancel chromatic aberration as well? By investigating some typical situations, Hardy gives a good indication of the extent to which we may answer these questions affirmatively. The cases considered are (i) the quadrupole triplet of achromatic quadrupoles; (ii) the round lens with an achromatic quadruplet corrector.

(i) The triplet consists of three equally spaced electromagnetic quadrupoles; these are not perfectly achromatic since the functions $k(z)$ for the electrostatic and magnetic contributions are not quite the same, but this should be unimportant at reasonably low excitations. Each quadrupole has a (magnetic) effective length of 1.84 cm, the lenses are 4.40 cm apart (center-to-center) and the accelerating voltage is 15 kV; Hardy obtains an achromatic stigmatic image a distance 3.25 cm from the center of the final quadrupole when the object is 19.5 cm in front of this plane with the following excitations:

Excitation	First	Second	Third lens
Magnetic (amp-turns/pole)	63(36)	−114(−52)	156(60)
Electrostatic (kV/electrode)	−0.55	1.05	−1.80.

(The figures in brackets are the excitations required to produce a stigmatic image at the same position when the electrostatic component is switched off.) Fig. 9 represents the movement of the line images as the accelerating voltage is varied; clearly, both image distances pass through minima at the achromatic working point, and with a more sensitive method of achieving achromatism, they would have coincided exactly. Furthermore, the magnification remains nearly constant in the vicinity of the achromatic point, so that the triplet is completely achromatic.

So far as the aperture aberrations are concerned, however, the achromatic triplet is rather worse than the magnetic triplet; for the magnetic triplet, $(30)_M = 10\,\text{cm}$, $(03)_M = 247\,\text{cm}$ and $(12)_M = (21)_M = 77\,\text{cm}$, while for the achromatic triplet $(30)_A = 16\,\text{cm} = 1{\cdot}6(30)_M$, $(03)_A = 726\,\text{cm} \simeq 3(03)_M$ but $(12)_A = (21)_A = 32\,\text{cm} \simeq 0{\cdot}4(12)_M$.

(ii) The round lens is an asymmetric probe-forming lens ($D_1 = 1\,\text{cm}$, $D_2 = 1.25\,\text{cm}$, $S = 2\,\text{cm}$ and the corrector consists of four equally spaced

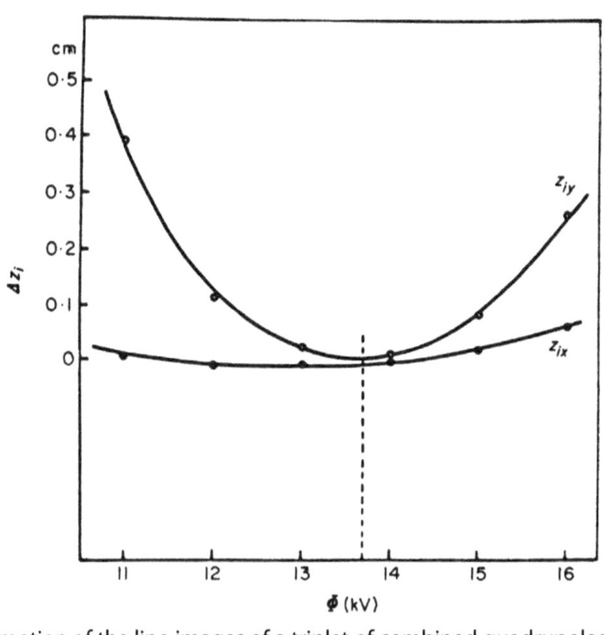

Fig. 9 The motion of the line images of a triplet of combined quadrupoles as the accelerating voltage is varied about the achromatic point, measured by Hardy (1967).

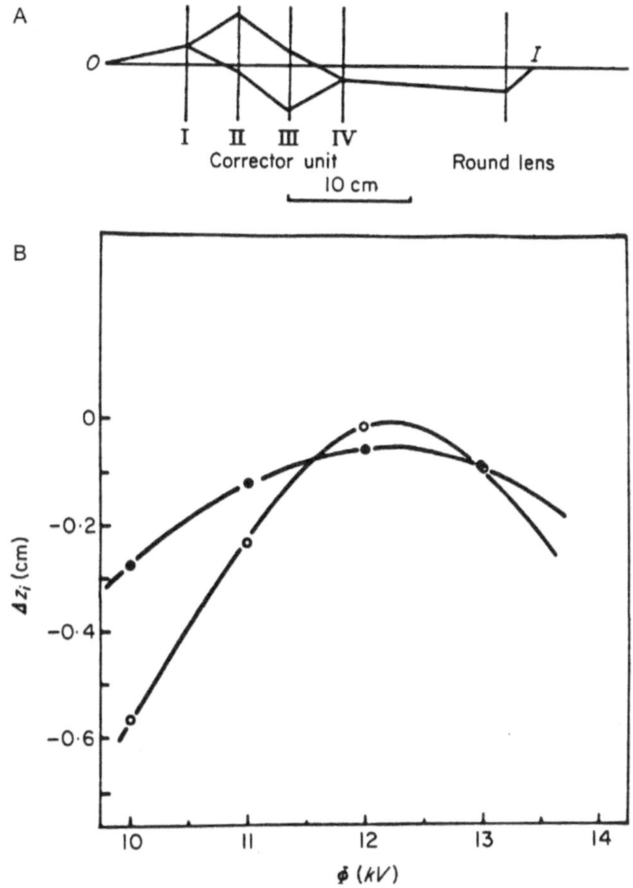

Fig. 10 (A) The optical arrangement used for the correction of the paraxial chromatic aberration of a round lens. (B) Experimental measurements of the chromatic change in position of the line images formed by a round lens together with the "chromatic corrector." *After Hardy, D. F. (1967).* Combined magnetic and electrostatic quadrupole electron lenses. *Dissertation, Cambridge University.*

quadrupoles (Fig. 10A); when the focal length and chromatic aberration of the round lens are 3.5 and 4.2 cm, respectively, the combination of corrector and round lens is achromatic when the excitations are as follows:

Excitation	First	Second	Third	Fourth quadrupole
Magnetic (amp-turns/pole)	72	−200	189	−42
Electrostatic (kV/electrode)	0	2.50	−3.00	0.

Fig. 10B shows the behavior of the line images in the vicinity of the achromatic working point. Here, the positions of these images reach a maximum, so that the secondary chromatic aberration coefficients must be negative; furthermore, the magnification varies rapidly. The aperture aberration now increases considerably, however:

	(30)	(03)	(12) = (21)	(cm)
Round lens	24	24	24	
Round lens and multiplet	39	35	27	
Round lens and multiplet in achromatic condition	190	55	27	

Hardy then considers whether the aperture aberrations can be corrected by exciting the electrostatic components asymmetrically; he excites all but the first component asymmetrically by applying the same voltage to all four electrodes. The aperture aberration is reasonably well corrected (Fig. 11) under the following conditions:

Excitation	First	Second	Third	Fourth quadrupole
Magnetic (amp-turns/pole)	78	−63	57	−36
Electrostatic (kV/electrode)	0	1.25	1·10	−3.00

Round lens: focal length $\simeq$4 cm, round lens distant 12 cm from the nearest quadrupole (center-to-center). Accelerating voltage: 15 kV.

> *In view of the high excitations found necessary to correct the aperture aberrations alone, it was impossible to correct the chromatic aberrations at the same time using this system. Although it is possible to envisage a more sophisticated corrector that could achieve correction of both aberrations of a round lens, the adjustment of the excitations would be a formidable problem*
>
> ***(Hardy, 1967).***

We now turn briefly to the work of Schiske (1967), [b] who considers how combined electromagnetic quadrupole multiplets may be used to correct the chromatic aberration of a round lens. The correctors considered are the Archard–Deltrap telescopic quadruplet and the doublet, the analysis for the latter being based on the work of Junior and Antony-Spies (1967; see "Quadrupole Data"). For each of these, Schiske discusses how the two

[b] The text of this conference lecture was kindly sent to me by Dr. Schiske.

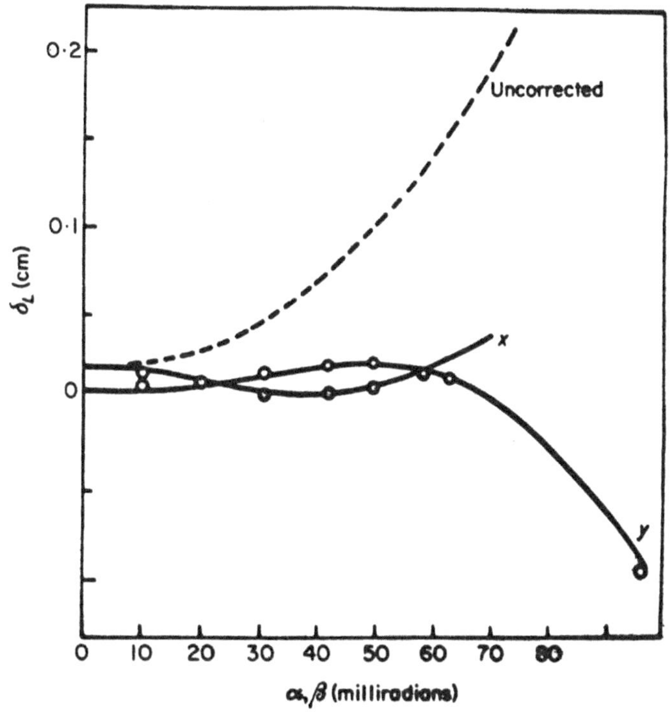

Fig. 11 The longitudinal aperture aberration, δ_L, as a function of angle (α,β) in the principal directions of a spherically corrected system. *After Hardy, D. F. (1967).* Combined magnetic and electrostatic quadrupole electron lenses. *Dissertation, Cambridge University.*

necessary conditions may be satisfied: (i) the axial chromatic aberration must be equal in the two principal sections and equal and opposite to that of the round lens to be corrected; (ii) the chromatic aberration of magnification must be the same in the two principal sections but it can be compensated by that of a round lens. Schiske points out that the quadruplet can be simplified to a doublet by making the first and fourth quadrupoles electrostatic and the two central quadrupoles electromagnetic; if the electrostatic excitation of the first two lenses is the same, the electrodes can be united and the net excitation of the second lens can be negative with respect to that of the first by making the magnetic component strong enough. The remainder of the paper is concerned with doublets; Schiske discusses how the two requirements may be reconciled with the conditions for stigmatic and orthomorphic imagery. If the object of the doublet is virtual and formed within it and the image falls at the end of the doublet, "können die beiden

Vergrösserungsfarbfehler völlig ausgeglichen werden. Der axiale Farbfehler ist aber dann ebenfalls fast völlig kompensiert, so dass eine Anwendung als Korrekturstück kaum möglich erscheint" (Schiske, 1967). Parallel incident rays are then considered, but neither arrangement is immediately inviting.

CHAPTER FIVE

Quadrupole data

Contents

1. The single quadrupole

1.1 Long quadrupoles: The rectangular model

The function $k(z)$ is now of the form

$$\begin{aligned} k(z) &= 1 \quad |z| < L/2 \\ k(z) &= 0 \quad |z| > L/2. \end{aligned} \tag{1}$$

The Gaussian equations of motion, in the form $x'' + \beta^2 k(z)x = 0$, $y'' - \beta^2 k(z)y = 0$ (Eq. 10 in Definitions, Notation, and Methods of analysis) for $|z| < L/2$ have solutions

Advances in Imaging and Electron Physics, Volume 224
ISSN 1076-5670
https://doi.org/10.1016/B978-0-323-98865-0.00005-4

$$x = A_x \cos\beta z + B_x \sin\beta z$$
$$y = A_y \cosh\beta z + B_y \sinh\beta z. \tag{2}$$

The G-rays are given by

$$G_x(z) = \cos\beta(z + L/2)$$
$$G_y(z) = \cosh\beta(z + L/2) \tag{3a}$$

and in image space ($z > L/2$)

$$G_x(z) = -\theta\sin\theta\left(\zeta - \frac{1}{2}\right) + \cos\theta$$
$$G_y(z) = \theta\sinh\theta\left(\zeta - \frac{1}{2}\right) + \cosh\theta \tag{3b}$$

where

$$\theta = \beta L \quad \zeta = z/L \tag{4}$$

so that the asymptotic cardinal elements are given by

$$\begin{aligned}
\frac{z_{Fi}}{L} &= -\frac{z_{Fo}}{L} = \frac{\cot\theta}{\theta} + \frac{1}{2}; & \frac{z_{Gi}}{L} &= -\frac{z_{Go}}{L} = -\frac{\coth\theta}{\theta} + \frac{1}{2}\\
\frac{z_{Pi}}{L} &= -\frac{z_{Po}}{L} = -\frac{\tan\theta/2}{\theta} + \frac{1}{2}; & \frac{z_{Qi}}{L} &= -\frac{z_{Qo}}{L} = -\frac{\tanh\theta/2}{\theta} + \frac{1}{2}\\
\frac{f}{L} &= \frac{1}{\theta\sin\theta}; & \frac{g}{L} &= -\frac{1}{\theta\sinh\theta}.
\end{aligned} \tag{5}$$

If θ is small (thin or weak lens approximation),

$$\begin{aligned}
\frac{z_{Fi}}{L} &\simeq \frac{1}{6} + \frac{1}{\theta^2}; & \frac{z_{Gi}}{L} &\simeq \frac{1}{6} - \frac{1}{\theta^2}\\
z_{Pi} &\simeq 0; & z_{Qi} &\simeq 0\\
\frac{f}{L} &\simeq \frac{1}{\theta^2} + \frac{1}{6}; & \frac{g}{L} &\simeq \frac{1}{6} - \frac{1}{\theta^2}.
\end{aligned} \tag{6}$$

The constant $\frac{1}{6}$ can normally be neglected.

In practice, the length L is the "effective length" found experimentally to be given by

$$L = l + 1{\cdot}1a \tag{7}$$

where l is the electrode length and a is the bore-radius. The coefficient 1.1, measured by Septier (Grivet & Septier, 1958, 1960), has been justified theoretically by Reisman (1957) who points out that if the rectangle is terminated by half-bells and reduced to $l - 2a$ in length, then $k\ (z)$ will have fallen to $\frac{1}{2}$ at the edge of the rectangle; the areas under the modified and unmodified curves are the same if

$$L = l + (\pi - 2)a. \tag{8}$$

[The terminating half-bells here are of the form $1/\{1 + (z \pm \bar{z})^2/a^2\}$.] The parameter θ is given by

$$\theta^2 = \left(\frac{L}{a}\right)^2 \frac{U}{V_0} \quad \text{(electrostatic quadrupole)} \tag{9a}$$

or

$$\theta^2 = \mu_0 \eta \left(\frac{L}{a}\right)^2 \frac{NI}{V_0^{1/2}} \quad \text{(magnetic quadrupole)} \tag{9b}$$

($\mu_0 = 4\pi \times 10^{-7}$) where $+U$ denotes the potential on the electrodes and NI is the number of ampere-turns per pole. We recall that $V_0 = \Phi_0(1 + \varepsilon\Phi_0)$ where Φ_0 (volts) is the accelerating voltage.

The focal lengths (f, g) and the focal distances (z_{Fi}, z_{Gi}) are plotted as a function of θ in Fig. 1 and tabulated in Table 1.

Junior and Antony-Spies (1967) have pointed out that stigmatic but not distortion-free imagery can be achieved with a single quadrupole, although the image is never real. Figs. 2A and B illustrate this condition.

The aperture aberrations are defined by

$$\begin{aligned} \Delta x(z_i) &= M_x(0030)\alpha^3 + M_x(0012)\alpha\gamma^2 \\ \Delta y(z_j) &= M_y(0003)\gamma^3 + M_y(0021)\alpha^2\gamma \end{aligned} \tag{10}$$

and the distortions by

$$\begin{aligned} \Delta x(z_i) &= M_x(3000)x_o^3 + M_x(1200)x_o y_o^2 \\ \Delta y(z_j) &= M_y(0300)y_o^3 + M_y(2100)x_o^2 y_o \end{aligned} \tag{11}$$

where x_o is measured in the plane $z = z_{ox}$ and y_o in $z = z_{oy}$; z_i is conjugate to z_{ox} and z_j to z_{oy}. The coefficients may be written

$$\begin{aligned} \frac{(0030)}{L} &= \sum_{i=0}^{4} \langle 0030 \rangle_i \zeta_x^i; \qquad \frac{(0003)}{L} = \sum_{i=0}^{4} \langle 0003 \rangle_i \zeta_y^i \\ \frac{(0012)}{L} &= \sum_{i=0}^{2}\sum_{j=0}^{2} \langle 0012 \rangle_{ij} \zeta_x^i \zeta_y^j; \quad \frac{(0021)}{L} = \sum_{i=0}^{2}\sum_{j=0}^{2} \langle 0021 \rangle_{ij} \zeta_x^i \zeta_y^j \end{aligned} \tag{12}$$

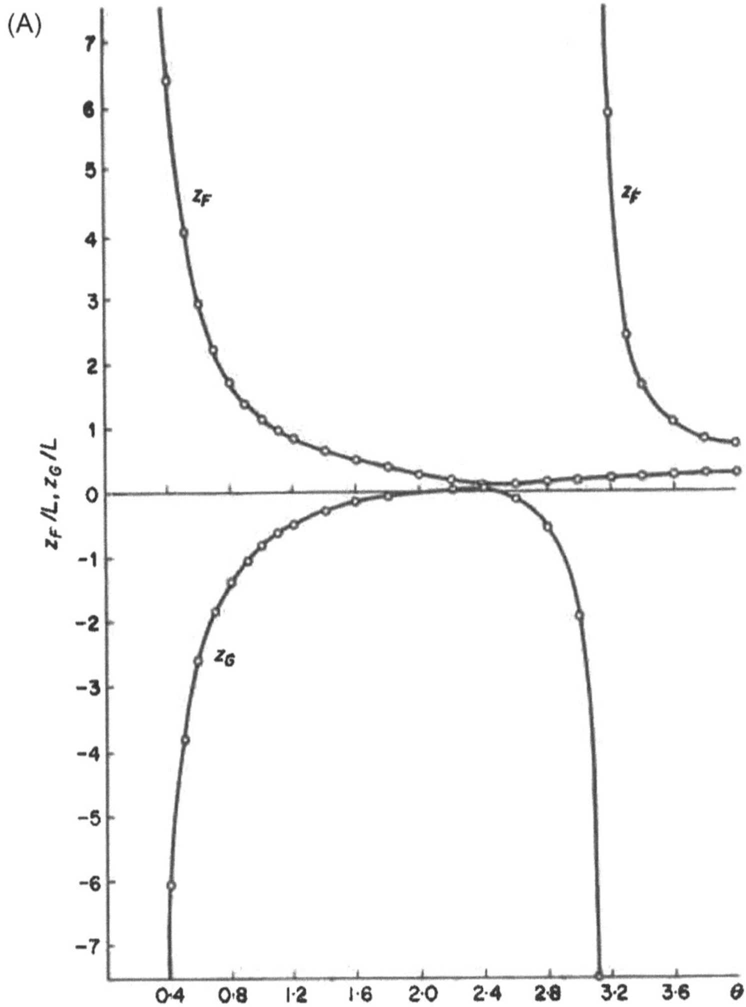

Fig. 1 The cardinal elements of a quadrupole lens: rectangular model.

and

$$
\begin{aligned}
(3000)L^2 &= \sum_{i=0}^{1} \langle 3000 \rangle_i \zeta_x^i; \quad (0300)L^2 = \sum_{i=0}^{1} \langle 0300 \rangle_i \zeta_y^i \\
(1200)L^2 &= \sum_{i=0}^{1} \langle 1200 \rangle_i \zeta_x^i; \quad (2100)L^2 = \sum_{i=0}^{1} \langle 2100 \rangle_i \zeta_y^i
\end{aligned}
\tag{13}
$$

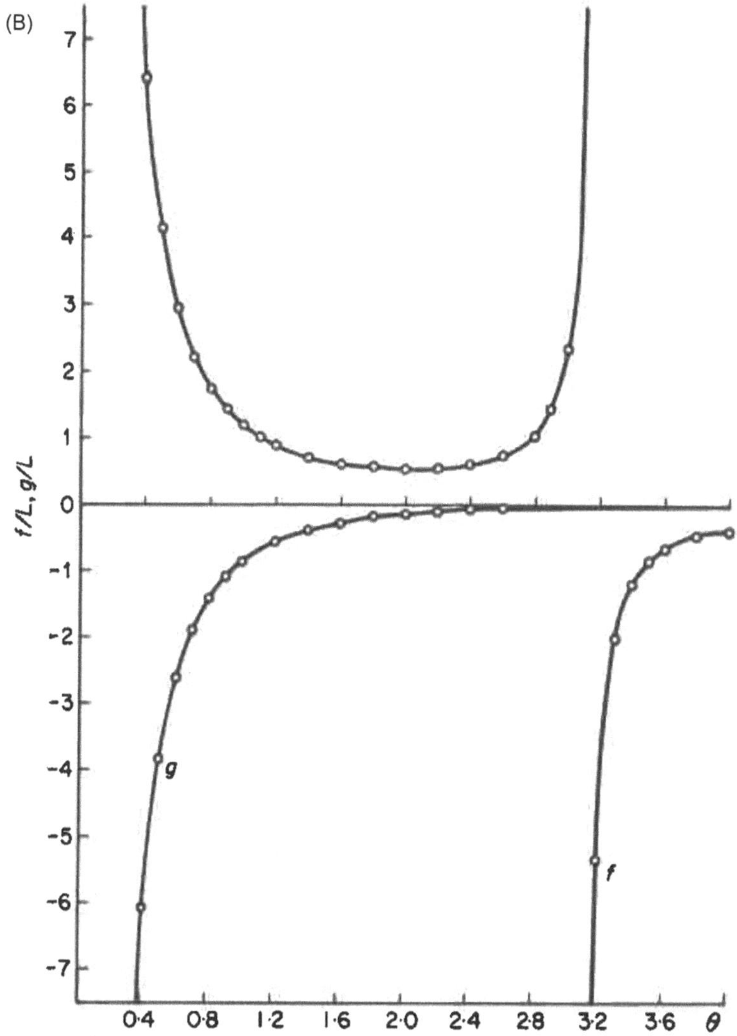

Fig. 1—Cont'd

or

$$\frac{(0030)}{L} = \sum_{i=0}^{4} (0030)_i m_x^i; \qquad \frac{(0003)}{L} = \sum_{i=0}^{4} (0003)_i m_y^i$$

$$\frac{(0012)}{L} = \sum_{i=0}^{2} \sum_{j=0}^{2} (0012)_{ij} m_x^i m_y^j; \quad \frac{(0021)}{L} = \sum_{i=0}^{2} \sum_{j=0}^{2} (0021)_{ij} m_x^i m_y^j \tag{14}$$

Table 1 The rectangular model. Cardinal elements as a function of $\theta = \beta L$. *NI* is defined to be $\theta^2/0.7453$.

θ	θ^2	*NI*	f/L	g/L	z_{Fi}/L	z_{Gi}/L
1.000⋆-1	1.000⋆-2	1.342⋆-2	1.002⋆2	−9.983⋆1	1.002⋆2	−9.983⋆1
2.000⋆-1	4.000⋆-2	5.367⋆-2	2.517⋆1	−2.483⋆1	2.517⋆1	−2.483⋆1
3.000⋆-1	9.000⋆-2	1.208⋆-1	1.128⋆1	−1.095⋆1	1.128⋆1	−1.094⋆1
4.000⋆-1	1.600⋆-1	2.147⋆-1	6.420	−6.086	6.413	−6.080
5.000⋆-1	2.500⋆-1	3.354⋆-1	4.172	−3.838	4.161	−3.828
6.000⋆-1	3.600⋆-1	4.830⋆-1	2.952	−2.618	2.936	−2.603
7.000⋆-1	4.900⋆-1	6.575⋆-1	2.216	−1.883	2.196	−1.864
8.000⋆-1	6.400⋆-1	8.587⋆-1	1.743	−1.407	1.714	−1.382
9.000⋆-1	8.100⋆-1	1.087	1.418	−1.082	1.382	−1.051
1.000	1.000	1.342	1.188	−8.509⋆-1	1.142	−8.130⋆-1
1.100	1.210	1.624	1.020	−6.806⋆-1	9.627⋆-1	−6.357⋆-1
1.200	1.440	1.932	8.941⋆-1	−5.521⋆-1	8.240⋆-1	−4.996⋆-1
1.300	1.690	2.268	7.983⋆-1	−4.529⋆-1	7.136⋆-1	−3.927⋆-1
1.400	1.960	2.630	7.248⋆-1	−3.751⋆-1	6.232⋆-1	−3.068⋆-1
1.500	2.250	3.019	6.683⋆-1	−3.131⋆-1	5.473⋆-1	−2.365⋆-1
1.600	2.560	3.435	6.253⋆-1	−2.631⋆-1	4.817⋆-1	−1.781⋆-1
1.700	2.890	3.878	5.932⋆-1	−2.223⋆-1	4.236⋆-1	−1.289⋆-1
1.800	3.240	4.347	5.705⋆-1	−1.888⋆-1	3.704⋆-1	−8.677⋆-2
1.900	3.610	4.844	5.562⋆-1	−1.610⋆-1	3.202⋆-1	−5.040⋆-2
2.000	4.000	5.367	5.499⋆-1	−1.379⋆-1	2.712⋆-1	−1.866⋆-2
2.100	4.410	5.917	5.517⋆-1	−1.184⋆-1	2.215⋆-1	9.311⋆-3
2.200	4.840	6.494	5.622⋆-1	−1.020⋆-1	1.691⋆-1	3.415⋆-2
2.300	5.290	7.098	5.830⋆-1	−8.807⋆-2	1.115⋆-1	5.639⋆-2
2.400	5.760	7.728	6.169⋆-1	−7.623⋆-2	4.513⋆-2	7.642⋆-2
2.500	6.250	8.386	6.684⋆-1	−6.611⋆-2	−3.546⋆-2	9.457⋆-2
2.600	6.760	9.070	7.461⋆-1	−5.745⋆-2	−1.393⋆-1	1.111⋆-1
2.700	7.290	9.781	8.666⋆-1	−5.001⋆-2	−2.835⋆-1	1.263⋆-1
2.800	7.840	1.052⋆1	1.066	−4.360⋆-2	−5.045⋆-1	1.402⋆-1

Table 1 The rectangular model. Cardinal elements as a function of $\theta = \beta L$. *NI* is defined to be $\theta^2/0.7453$.—cont'd

θ	θ^2	*NI*	f/L	g/L	z_{Fi}/L	z_{Gi}/L
2.900	8.410	1.128⋆1	1.441	−3.806⋆-2	−8.994⋆-1	1.531⋆-1
3.000	9.000	1.208⋆1	2.362	−3.327⋆-2	−1.838	1.650⋆-1
3.100	9.610	1.289⋆1	7.758	−2.912⋆-2	−7.251	1.761⋆-1
3.200	1.024⋆1	1.374⋆1	−5.353	−2.552⋆-2	5.844	1.865⋆-1
3.300	1.089⋆1	1.461⋆1	−1.921	−2.238⋆-2	2.397	1.961⋆-1
3.400	1.156⋆1	1.551⋆1	−1.151	−1.965⋆-2	1.613	2.052⋆-1
3.500	1.225⋆1	1.644⋆1	−8.145⋆-1	−1.727⋆-2	1.263	2.138⋆-1
3.600	1.296⋆1	1.739⋆1	−6.277⋆-1	−1.519⋆-2	1.063	2.218⋆-1
3.700	1.369⋆1	1.837⋆1	−5.101⋆-1	−1.337⋆-2	9.326⋆-1	2.294⋆-1
3.800	1.444⋆1	1.937⋆1	−4.301⋆-1	−1.178⋆-2	8.402⋆-1	2.366⋆-1
3.900	1.521⋆1	2.041⋆1	−3.728⋆-1	−1.038⋆-2	7.706⋆-1	2.434⋆-1
4.000	1.600⋆1	2.147⋆1	−3.303⋆-1	−9.161⋆-3	7.159⋆-1	2.498⋆-1

In all the computer-printed Tables, *n signifies $\times 10^n$.

and

$$
\begin{aligned}
(3000)L^2 &= \sum_{i=0}^{1} (3000)_i m_x^i; \quad (0300)L^2 = \sum_{i=0}^{1} (0300)_i m_y^i \\
(1200)L^2 &= \sum_{i=0}^{1} (1200)_i m_x^i; \quad (2100)L^2 = \sum_{i=0}^{1} (2100)_i m_y^i
\end{aligned}
\tag{15}
$$

where $m_x = 1/M_x$ and $m_y = 1/M_y$; it is convenient to set $\zeta_x = z_{ox}/L + \frac{1}{2}$ and $\zeta_y = z_{oy}/L + \frac{1}{2}$ so that the scaled object distances, ζ_x and ζ_y, are measured from the beginning of the rectangle.

Formulae for the aperture aberration coefficients have been derived by Dymnikov, Fishkova, and Yavor (1965d, 1965f) for the case of a stigmatic object ($z_{ox} = z_{oy}$) and by Ovsyannikova and Yavor (1967a) and Fishkova et al. (1968a) for the astigmatic situation. We have rearranged their formulae into the polynomial expressions to give the coefficients $\langle 00pq \rangle_i$ and $\langle 00pq \rangle_{ij}$ and then substituted $z_{ox} = z_{Fo} + fm_x$, $z_{oy} = z_{G_0} + gm_y$ to give the coefficients $(00pq)_i$ and $(00pq)_{ij}$, with the following results:

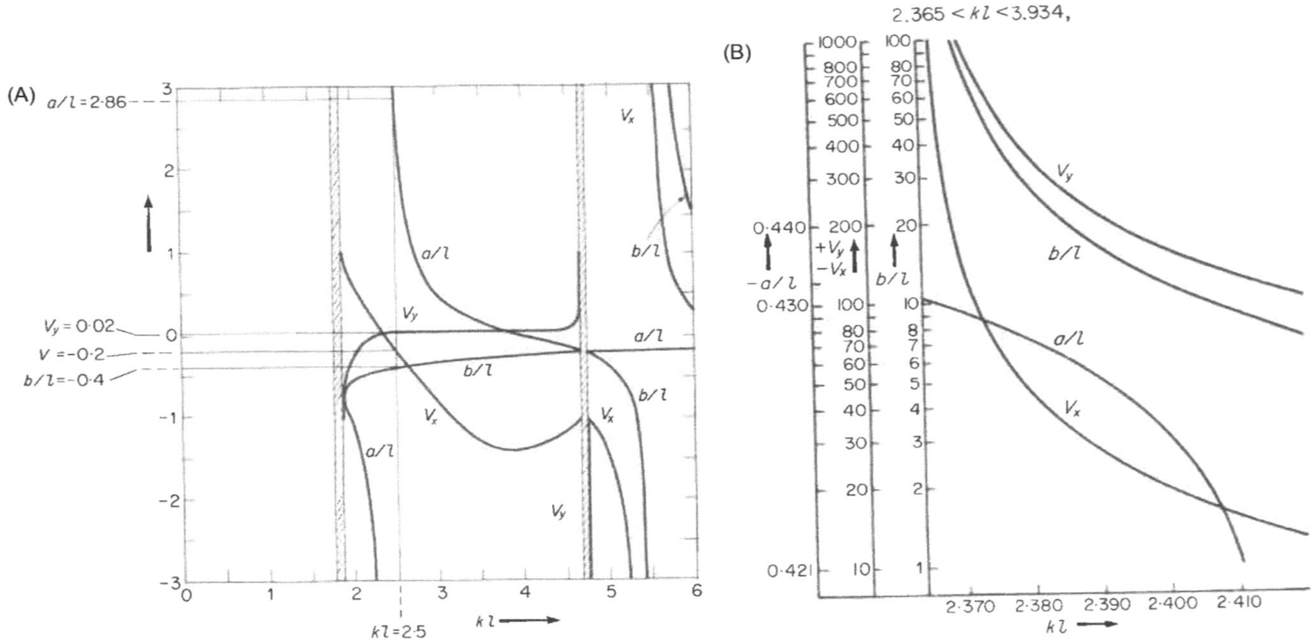

Fig. 2 (A) Stigmatic imagery with a single quadrupole (rectangular model). $l(=L)$ is the lens-length a and b are the object and image distances and $kl(=\theta)$ is the excitation; V_x and V_y ($=M_x$ and M_y) are magnifications. Pseudo-stigmatic imagery occurs with $kl=\theta=2.5$, $a/l=2.86$, $b/l=-0.4$, $V_x=-0.2$ and $V_y=0.02$, for example. (After Junior & Antony-Spies, 1967.) (B) The single quadrupole producing a real image of a virtual object. Notation as Fig. 2(A). (After Antony-Spies & Junior, 1967.) (C) The dependence of the aberration coefficients of a single quadrupole on the dimensionless excitation parameter $\theta=\beta L$ (rectangular model). Full curves correspond to the x–z plane and broken curves to the y–z plane. 1 and 4: electrostatic; 2 and 5: magnetic; 3 and 6: achromatic. The object is at the focus and the aperture is in the center of the lens. The aberrations are expressed in terms of position in the object and aperture planes. (i) Aperture aberrations: (0030) L^2 and (0003) L^2. (ii) Aperture aberrations: (0012) L^2 and (0021) L^2. (iii) Astigmatism: (2010) L^2 and (0201) L^2. (iv) Astigmatism: (0210) L^2 and (2001) L^2. (v) Astigmatism: (1101) L^2 and (1110) L^2 (vi) Coma: (1020) L^2 and (0102) L^2. (vii) Coma: (1002) L^2 and (0120) L^2. (viii) Coma: (0111) $L^2$and (1011) L^2. (ix) Distortions: (3000) L^2 and (0300) L^2 (x) Distortions: (1200) L^2 and (2100) L^2. (After Ovsyannikova, 1968.)

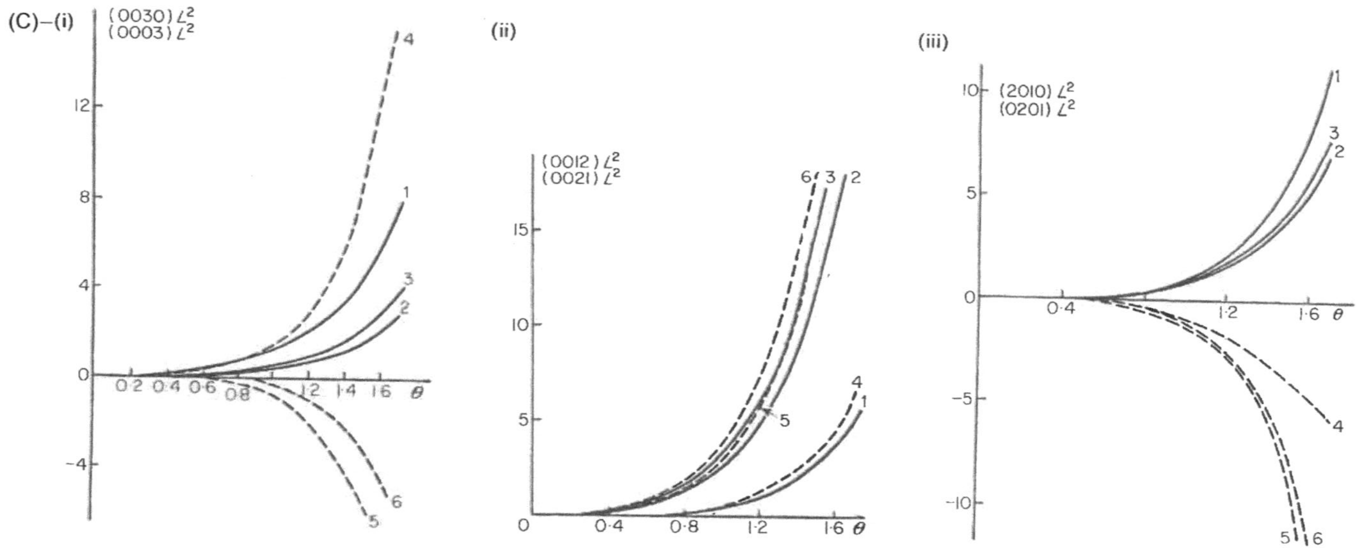
(C)–(i)
(0030) L^2
(0003) L^2
12
8
4
0
−4
0.2 0.4 0.6 0.8 1.2 1.4 1.6 θ
1
2
3
4
5
6
(ii)
(0012) L^2
(0021) L^2
15
10
5
0
0.4 0.8 1.2 1.6 θ
1
2
3
4
5
6
(iii)
(2010) L^2
(0201) L^2
10
5
0
−5
−10
0.4 1.2 1.6 θ
1
2
3
4
5
6

Fig. 2—Cont'd

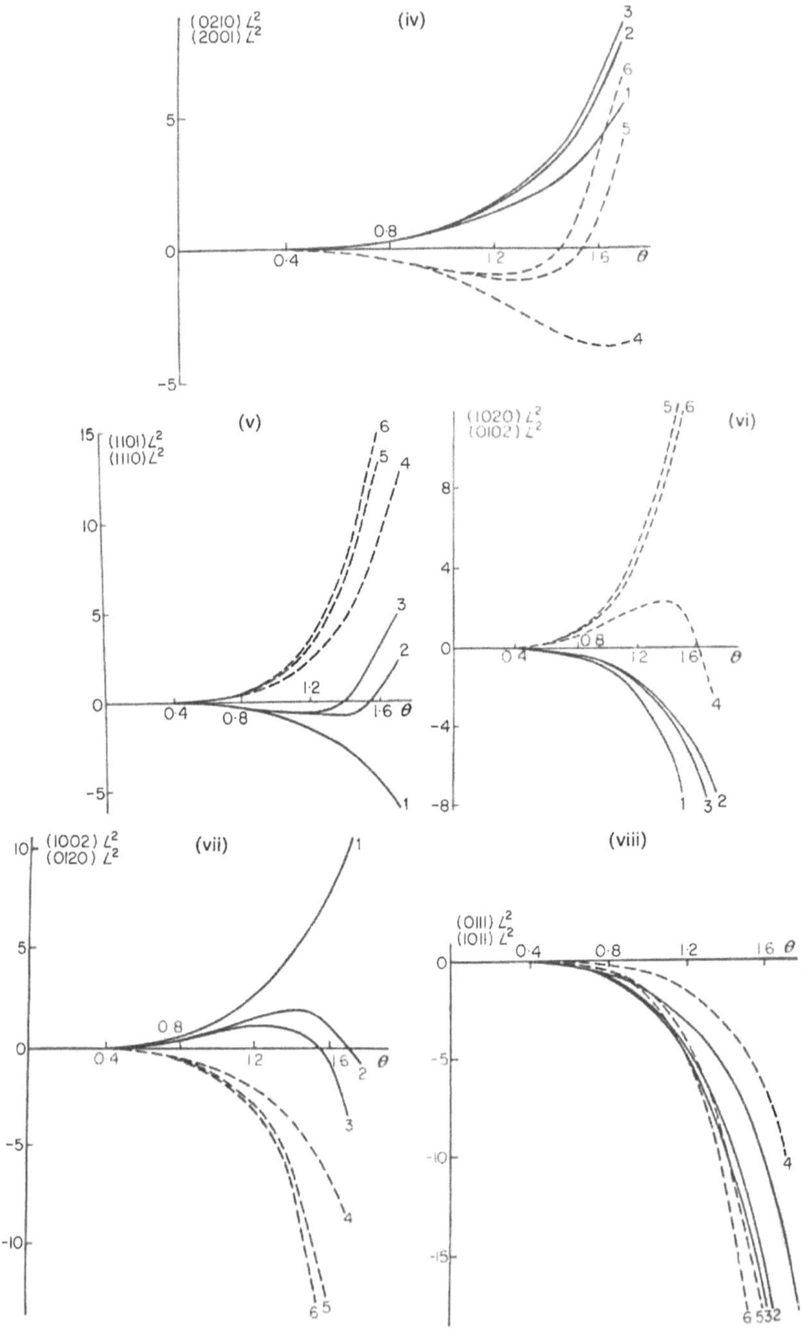
(0210) L2
(2001) L2
(iv)
(v)
(1101) L2
(1110) L2
(vi)
(1020) L2
(0102) L2
(vii)
(1002) L2
(0120) L2
(viii)
(0111) L2
(1011) L2

Fig. 2—Cont'd

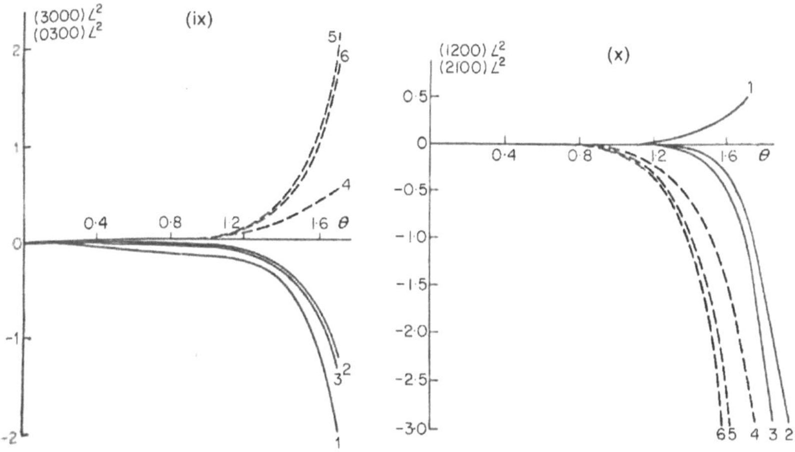

Fig. 2—Cont'd

$$
\begin{aligned}
\langle 0030 \rangle_4 &= \frac{\theta^4}{192}\left\{12(1+u) + 8u\frac{\sin 2\theta}{\theta} - (3-u)\frac{\sin 4\theta}{\theta}\right\} \\
\langle 0030 \rangle_3 &= \frac{\theta^2}{48}\{3 - 5u + 4u\cos 2\theta - (3-u)\cos 4\theta\} \\
\langle 0030 \rangle_2 &= \frac{\theta^2}{32}\left\{4(1+u) + (3-u)\frac{\sin 4\theta}{\theta}\right\} \\
\langle 0030 \rangle_1 &= \frac{1}{48}\{-3(1+u) + 4u\cos 2\theta + (3-u)\cos 4\theta\} \\
\langle 0030 \rangle_0 &= \frac{1}{192}\left\{12(1+u) - 8u\frac{\sin 2\theta}{\theta} - (3-u)\frac{\sin 4\theta}{\theta}\right\}
\end{aligned}
\tag{16}
$$

and

$$
\begin{aligned}
\langle 0012 \rangle_{22} = \frac{\theta^4}{16}\Bigg\{ &- 2(3+\nu) - (1+\nu)\frac{\sin 2\theta}{\theta} + \frac{5-\nu}{2}\frac{\sin 2\theta}{\theta}\cosh 2\theta \\
&+ (1-\nu)\frac{\sinh 2\theta}{\theta} + \frac{1-\nu}{2}\frac{\sinh 2\theta}{\theta}\cos 2\theta\Bigg\}
\end{aligned}
$$

$$
\begin{aligned}
\langle 0012 \rangle_{21} = -\frac{\theta^2}{16}\{&(5-\nu)\sin 2\theta \sinh 2\theta - 3(1-\nu) + 2(1-\nu)\cosh 2\theta \\
&+ (1-\nu)\cos 2\theta \cosh 2\theta\}
\end{aligned}
$$

$$
\begin{aligned}
\langle 0012\rangle_{12} &= -\frac{\theta^2}{16}\{-(5-\nu)\cos 2\theta\cosh 2\theta + 3(1-\nu) + 2(1+\nu)\cos 2\theta \\
&\quad + (1-\nu)\sin 2\theta\sinh 2\theta\} \\
\langle 0012\rangle_{20} &= \frac{\theta^2}{16}\left\{2(3+\nu) + (1+\nu)\frac{\sin 2\theta}{\theta} + (1-\nu)\frac{\sinh 2\theta}{\theta} + \frac{5-\nu}{2}\frac{\sin 2\theta}{2}\right. \\
&\quad \left.\cosh 2\theta + \frac{1-\nu}{2}\frac{\sinh 2\theta}{\theta}\cos 2\theta\right\} \\
\langle 0012\rangle_{11} &= \frac{\theta^2}{8}\left\{(1-\nu)\frac{\sin 2\theta}{\theta}\cosh 2\theta - (5-\nu)\frac{\sinh 2\theta}{\theta}\cos 2\theta\right\} \\
\langle 0012\rangle_{02} &= \frac{\theta^2}{16}\left\{-2(3+\nu) + (1+\nu)\frac{\sin 2\theta}{\theta} + (1-\nu)\frac{\sinh 2\theta}{\theta}\right. \\
&\quad \left.-\frac{5-\nu}{2}\frac{\sin 2\theta}{2}\cosh 2\theta - \frac{1-\nu}{2}\frac{\sinh 2\theta}{\theta}\cos 2\theta\right\} \\
\langle 0012\rangle_{10} &= -\frac{1}{16}\{7+\nu-2(1+\nu)\cos 2\theta - (5-\nu)\cos 2\theta\cosh 2\theta \\
&\quad + (1-\nu)\sin 2\theta\sinh 2\theta\} \\
\langle 0012\rangle_{01} &= -\frac{1}{16}\{-(1-\nu) + 2(1-\nu)\cosh 2\theta - (1-\nu)\cos 2\theta\cosh 2\theta \\
&\quad -(5-\nu)\sin 2\theta\sinh 2\theta\} \\
\langle 0012\rangle_{00} &= \frac{1}{16}\left\{2(3+\nu) - (1+\nu)\frac{\sin 2\theta}{\theta} + (1-\nu)\frac{\sinh 2\theta}{\theta}\right. \\
&\quad \left.-\frac{5-\nu}{2}\frac{\sin 2\theta}{2}\cosh 2\theta - \frac{1-\nu}{2}\frac{\sinh 2\theta}{\theta}\cos 2\theta\right\} \qquad (17)
\end{aligned}
$$

and

$$(0030)_4 = \frac{1}{192\sin^4\theta}\left\{12(1+u) + 8u\frac{\sin 2\theta}{\theta} + (u-3)\frac{\sin 4\theta}{\theta}\right\} = (0030)_0$$

$$(0030)_3 = \frac{1}{48\sin^4\theta}\left\{-12(1+u)\cos\theta + (3-11u)\frac{\sin\theta}{\theta} + 3(1-u)\frac{\sin 3\theta}{\theta}\right\} = (0030)_1$$

$$(0030)_2 = \frac{1}{16\sin^4\theta}\left\{2(1+u)(\cos 2\theta + 2) + (5u-3)\frac{\sin 2\theta}{\theta}\right\}$$

$$(0012)_{22} = \frac{1}{16\sin^2\theta\sinh^2\theta}\left\{-2(3+v) - (1+v)\frac{\sin 2\theta}{\theta} + (1-v)\frac{\sinh 2\theta}{\theta}\right.$$
$$\left. + \frac{1}{2}(5-v)\frac{\sin 2\theta}{\theta}\cosh 2\theta + \frac{1}{2}(1-v)\frac{\sinh 2\theta}{\theta}\cos 2\theta\right\} = (0012)_{00}$$

$$(0012)_{21} = -\frac{1}{16\sin^2\theta\sinh^2\theta}\left\{-4(3+v)\cosh\theta + 5(1-v)\frac{\sinh\theta}{\theta} + 3(1-v)\frac{\sin 2\theta}{\theta}\cosh\theta\right.$$
$$\left. + (1-v)\cos 2\theta\frac{\sinh\theta}{\theta}\right\} = (0012)_{01}$$

$$(0012)_{12} = \frac{1}{16\sin^2\theta\sinh^2\theta}\left\{4(3+v)\cos\theta + (5v-1)\frac{\sin\theta}{\theta} - (5-v)\frac{\sin\theta}{\theta}\cosh 2\theta\right.$$
$$\left. -3(1-v)\frac{\sinh 2\theta}{\theta}\cos\theta\right\} = (0012)_{10}$$

$$(0012)_{20} = \frac{1}{32\sin^2\theta\sinh^2\theta}\left\{-4(3+v) + 3(1-v)\left(\frac{\sin 2\theta}{\theta} + \frac{\sinh 2\theta}{\theta}\right)\right\} = (0012)_{02}$$

$$(0012)_{11} = -\frac{1}{4\sin^2\theta\sinh^2\theta}\left\{2(3+v)\cos\theta\cosh\theta - 3(1-v)\left(\frac{\sin\theta}{\theta}\cosh\theta + \frac{\sinh\theta}{\theta}\cos\theta\right)\right\} \tag{18}$$

in which

$$u = 3n^2 - 2n + 2, \quad v = n^2 - 2n - 2 \tag{19}$$

The coefficients $\langle 0003\rangle_i$, $(0003)_i$, $\langle 0021\rangle_{ij}$ and $(0021)_{ij}$ are obtained from these formulae by writing $i\theta$ in place of θ, so that $\theta^2 \to -\theta^2, \sin k\theta/\theta \rightleftharpoons \sinh k\,\theta/\theta$ and $\cos k\theta \rightleftharpoons \cosh k\theta$. For the distortions, we find

$$\langle 3000\rangle_1 = \frac{\theta^4}{192}\left\{-12(1+u) - 8u\frac{\sin 2\theta}{\theta} + (3-u)\frac{\sin 4\theta}{\theta}\right\}$$

$$\langle 3000\rangle_0 = \frac{\theta^2}{192}\{33 + 5u - 4(9+u)\cos 2\theta + (3-u)\cos 4\theta\}$$

$$\langle 1200\rangle_1 = \frac{\theta^4}{32}\left\{4(3+v) + 2(1+v)\frac{\sin 2\theta}{\theta} - 2(1-v)\frac{\sinh 2\theta}{\theta}\right.$$
$$\left. -(5-v)\cosh 2\theta\frac{\sin 2\theta}{\theta} - (1-v)\cos 2\theta\frac{\sinh 2\theta}{\theta}\right\}$$

$$\langle 1200\rangle_0 = \frac{\theta^2}{32}\{-1 - 3v + 4\cosh 2\theta + 2(1+v)\cos 2\theta - (5-v)\cos 2\theta\cosh 2\theta$$
$$+ (1-v)\sinh 2\theta\sin 2\theta\} \tag{20}$$

and

$$(3000)_1 = \frac{\theta^2}{192}\cdot\frac{\theta}{\sin\theta}\left\{-12(1+u) - 8u\frac{\sin 2\theta}{\theta} + (3-u)\frac{\sin 4\theta}{\theta}\right\}$$

$$(3000)_0 = \frac{\theta^2}{192}\cdot\frac{\theta}{\sin\theta}\left\{12(1+u)\cos\theta + (51+11u)\frac{\sin\theta}{\theta} + 3(u-7)\frac{\sin 3\theta}{\theta}\right\}$$

$$(1200)_1 = \frac{\theta^4}{16}\cdot\frac{\theta}{\sin\theta}\left\{2(3+v) + (1+v)\frac{\sin 2\theta}{\theta} - (1-v)\frac{\sinh 2\theta}{\theta}\right.$$
$$\left. -\frac{1}{2}(5-v)\frac{\sin 2\theta}{\theta}\cosh 2\theta - \frac{1}{2}(1-v)\frac{\sinh 2\theta}{\theta}\cos 2\theta\right\}$$

$$(1200)_0 = \frac{\theta^2}{16}\cdot\frac{\theta}{\sin\theta}\left\{-2(3+v)\cos\theta - \frac{1}{2}(3+v)\frac{\sin\theta}{\theta} + \frac{3}{2}(1-v)\frac{\sinh 2\theta}{\theta}\cos\theta\right.$$
$$\left. + \frac{1}{2}(9-v)\frac{\sin\theta}{\theta}\cosh 2\theta\right\} \tag{21}$$

We have checked that $(1200)_i$ and $(3000)_i$ are related to $(0012)_{22}$, $(0012)_{12}$, $(0030)_4$ and $(0030)_3$ as predicted by Eq. (133) in Definitions. Writing $i\theta$ for θ again gives the coefficients in the other principal section.

The chromatic aberration coefficients, defined by

$$\begin{aligned}\Delta x(z_i) &= M_x(C_{cx}\alpha + C_{Mx}x_o)\frac{\Delta\Phi}{\Phi}\\ \Delta y(z_j) &= M_y(C_{cy}\gamma + C_{My}y_o)\frac{\Delta\Phi}{\Phi}\end{aligned} \tag{22}$$

are given by

$$
\begin{aligned}
\frac{C_{cx}}{L} &= \frac{1}{4}(n-1)\theta^2\left\{\left(1+\frac{\sin 2\theta}{2\theta}\right)\zeta_x^2 + \frac{\cos 2\theta - 1}{\theta^2}\zeta_x + \frac{1-\sin 2\theta/2\theta}{\theta^2}\right\} \\
\frac{C_{cy}}{L} &= -\frac{1}{4}(n-1)\theta^2\left\{\left(1+\frac{\sinh 2\theta}{2\theta}\right)\zeta_y^2 - \frac{\cosh 2\theta - 1}{\theta^2}\zeta_y - \frac{1-\sinh 2\theta/2\theta}{\theta^2}\right\} \\
C_{Mx} &= -\frac{1}{4}(n-1)\theta^2\left\{\left(1+\frac{\sin 2\theta}{2\theta}\right)\zeta_x + \frac{\cos 2\theta - 1}{2\theta^2}\right\} \\
C_{My} &= \frac{1}{4}(n-1)\theta^2\left\{\left(1+\frac{\sinh 2\theta}{2\theta}\right)\zeta_y - \frac{\cosh 2\theta - 1}{2\theta^2}\right\}
\end{aligned} \tag{23}
$$

(where as before $\zeta_x = z_{ox}/L + 1/2$, $\zeta_y = z_{oy}/L + 1/2$). Replacing object distance by magnification, we find

$$
\begin{aligned}
\frac{C_{cx}}{L} &= \frac{n-1}{4\sin^2\theta}\left\{\left(1+\frac{\sin 2\theta}{2\theta}\right)\left(m_x^2+1\right) - 2\left(\cos\theta + \frac{\sin\theta}{\theta}\right)m_x\right\} \\
\frac{C_{cy}}{L} &= -\frac{n-1}{4\sinh^2\theta}\left\{\left(1+\frac{\sinh 2\theta}{2\theta}\right)\left(m_y^2+1\right) - 2\left(\cosh\theta + \frac{\sinh\theta}{\theta}\right)m_y\right\} \\
C_{Mx} &= -\frac{n-1}{4}\frac{\theta}{\sin\theta}\left\{\left(1+\frac{\sin 2\theta}{2\theta}\right)m_x - \left(\cos\theta + \frac{\sin\theta}{\theta}\right)\right\} \\
C_{My} &= -\frac{n-1}{4}\frac{\theta}{\sinh\theta}\left\{\left(1+\frac{\sinh 2\theta}{2\theta}\right)m_y - \left(\cosh\theta + \frac{\sinh\theta}{\theta}\right)\right\}
\end{aligned} \tag{24}
$$

The aperture aberration coefficients $\langle 00rs\rangle_i$, $\langle 00rs\rangle_{ij}$, $(00rs)_i$ and $(00rs)_{ij}$ are listed in Tables 2–9. The dependence of the aperture aberration coefficients (0030), (0003), (0012) and (0021) on object position is illustrated graphically in Dymnikov, Fishkova, and Yavor (1966); their curves are reproduced in QO as Fig. 5. Ovsyannikova (1968) has plotted curves showing the variation of all the geometrical aberration coefficients with excitation (Fig. 2C). The coefficients of m_x^2, m_x, m_y^2 and m_y in the chromatic aberration are listed in Table 10; the quantities tabulated are

$$
\begin{aligned}
A_x &= \frac{1}{4}\operatorname{cosec}^2\theta(1+\sin 2\theta/2\theta); & A_y &= -\frac{1}{4}\operatorname{cosech}^2\theta(1+\sinh 2\theta/2\theta) \\
B_x &= -\frac{1}{2}\operatorname{cosec}^2\theta(\cos\theta+\sin\theta/\theta); & B_y &= \frac{1}{2}\operatorname{cosech}^2\theta(\cosh\theta+\sinh\theta/\theta) \\
C_x &= -\frac{1}{4}\theta\operatorname{cosec}\theta(1+\sin 2\theta/2\theta); & C_y &= -\frac{1}{4}\theta\operatorname{cosech}\theta(1+\sinh 2\theta/2\theta) \\
D_x &= \frac{1}{4}\theta\operatorname{cosec}\theta(\cos\theta+\sin\theta/\theta); & D_y &= \frac{1}{4}\theta\operatorname{cosech}\theta(\cosh\theta+\sinh\theta/\theta).
\end{aligned} \tag{25}
$$

Table 2 The rectangular model. The aperture aberration coefficients $<0030>_i$ as a function of $\theta = \beta L$

Electrostatic

θ	$<30>_4$	$<30>_3$	$<30>_2$	$<30>_1$	$<30>_0$
2.000★-1	1.826★-3	2.853★-3	2.207★-2	2.077★-2	6.823★-3
4.000★-1	2.752★-2	4.203★-2	1.100★-1	9.115★-2	2.899★-2
6.000★-1	1.266★-1	1.860★-1	3.093★-1	2.272★-1	7.038★-2
8.000★-1	3.534★-1	4.908★-1	6.458★-1	4.338★-1	1.341★-1
1.000	7.494★-1	9.639★-1	1.095	6.883★-1	2.190★-1
1.200	1.341	1.569	1.589	9.374★-1	3.185★-1
1.400	2.153	2.257	2.070	1.114	4.208★-1
1.600	3.217	2.986	2.537	1.165	5.122★-1
1.800	4.593	3.691	3.061	1.074	5.809★-1
2.000	6.399	4.240	3.753	8.692★-1	6.207★-1
2.200	8.887	4.422	4.679	6.117★-1	6.317★-1
2.400	1.252★1	4.019	5.812	3.669★-1	6.195★-1
2.600	1.862★1	2.975	7.029	1.799★-1	5.925★-1
2.800	2.624★1	1.547	8.183	6.450★-2	5.585★-1
3.000	3.800★1	3.262★-1	9.201	1.022★-2	5.234★-1
3.200	5.370★1	6.389★-2	1.015★1	1.712★-3	4.909★-1
3.400	7.316★1	1.351	1.119★1	3.549★-2	4.629★-1
3.600	9.572★1	4.321	1.253★1	1.235★-1	4.413★-1
3.800	1.203★2	8.562	1.421★1	2.806★-1	4.284★-1
4.000	1.461★2	1.330★1	1.614★1	5.051★-1	4.264★-1

Magnetic

θ	$<30>_4$	$<30>_3$	$<30>_2$	$<30>_1$	$<30>_0$
2.000★-1	5.297★-4	2.735★-4	1.948★-2	1.948★-2	6.561★-3
4.900★-1	8.293★-3	4.657★-3	7.249★-2	7.199★-2	2.504★-2
6.000★-1	4.032★-2	2.523★-2	1.477★-1	1.425★-1	5.219★-2
8.0000★-1	1.196★-1	8.314★-2	2.385★-1	2.132★-1	8.376★-2
1.000	2.672★-1	2.016★-1	3.513★-1	2.705★-1	1.157★-1

Table 2 The rectangular model. The aperture aberration coefficients $\langle 0030\rangle_i$ as a function of $\theta = \beta L$—cont'd

Magnetic

θ	$\langle 30\rangle_4$	$\langle 30\rangle_3$	$\langle 30\rangle_2$	$\langle 30\rangle_1$	$\langle 30\rangle_0$
1.200	4.950⋆-1	3.896⋆-1	5.026⋆-1	3.086⋆-1	1.449⋆-1
1.400	8.059⋆-1	6.253⋆-1	7.074⋆-1	3.284⋆-1	1.699⋆-1
1.600	1.206	8.522⋆-1	9.658⋆-1	3.332⋆-1	1.902⋆-1
1.800	1.729	9.978⋆-1	1.260	3.243⋆-1	2.057⋆-1
2.000	2.454	1.007	1.562	2.995⋆-1	2.165⋆-1
2.200	3.515	8.720⋆-1	1.855	2.556⋆-1	2.222⋆-1
2.400	5.086	6.378⋆-1	2.147	1.934⋆-1	2.225⋆-1
2.600	7.350	3.790⋆-1	2.468	1.211⋆-1	2.175⋆-1
2.800	1.048⋆1	1.631⋆-1	2.854	5.401⋆-2	2.081⋆-1
3.000	1.463⋆1	3.047⋆-2	3.325	9.891⋆-3	1.962⋆-1
3.200	1.994⋆1	5.835⋆-3	3.863	1.702⋆-3	1.841⋆-1
3.400	2.650⋆1	1.340⋆-1	4.426	3.194⋆-2	1.741⋆-1
3.600	3.434⋆1	5.058⋆-1	4.969	9.152⋆-2	1.677⋆-1
3.800	4.338⋆1	1.238	5.473	1.638⋆-1	1.656⋆-1
4.000	5.337⋆1	2.402	5.964	2.317⋆-1	1.673⋆-1

Achromatic

θ	$\langle 30\rangle_4$	$\langle 30\rangle_3$	$\langle 30\rangle_2$	$\langle 30\rangle_1$	$\langle 30\rangle_0$
2.000⋆-1	7.894⋆-4	7.894⋆-4	2.000⋆-2	1.973⋆-2	6.614⋆-3
4.000⋆-1	1.214⋆-2	1.213⋆-2	8.000⋆-2	7.582⋆-2	2.583⋆-2
6.000⋆-1	5.757⋆-2	5.739⋆-2	1.800⋆-1	1.594⋆-1	5.583⋆-2
8.000⋆-1	1.664⋆-1	1.647⋆-1	3.200⋆-1	2.573⋆-1	9.382⋆-2
1.000	3.637⋆-1	3.540⋆-1	5.000⋆-1	3.540⋆-1	1.363⋆-1
1.200	6.643⋆-1	6.255⋆-1	7.200⋆-1	4.343⋆-1	1.796⋆-1
1.400	1.075	9.517⋆-1	9.800⋆-1	4.856⋆-1	2.201⋆-1
1.600	1.609	1.279	1.280	4.996⋆-1	2.546⋆-1
1.800	2.302	1.536	1.620	4.742⋆-1	2.807⋆-1
2.000	3.243	1.654	2.000	4.134⋆-1	2.973⋆-1

Continued

Table 2 The rectangular model. The aperture aberration coefficients <0030>$_i$ as a function of $\theta = \beta L$—cont'd

Achromatic

θ	<30>$_4$	<30>$_3$	<30>$_2$	<30>$_1$	<30>$_0$
2.200	4.590	1.582	2.420	3.268⋆-1	3.041⋆-1
2.400	6.573	1.314	2.880	2.281⋆-1	3.019⋆-1
2.600	9.483	8.982⋆-1	3.380	1.329⋆-1	2.925⋆-1
2.800	1.363⋆1	4.399⋆-1	3.920	5.611⋆-2	2.782⋆-1
3.000	1.931⋆1	8.962⋆-2	4.500	9.957⋆-3	2.616⋆-1
3.200	2.669⋆1	1.745⋆-2	5.120	1.704⋆-3	2.454⋆-1
3.400	3.584⋆1	3.774⋆-1	5.780	3.265⋆-2	2.318⋆-1
3.600	4.662⋆1	1.269	6.480	9.791⋆-2	2.224⋆-1
3.800	5.877⋆1	2.703	7.220	1.872⋆-1	2.182⋆-1
4.000	7.191⋆1	4.582	8.000	2.864⋆-1	2.191⋆-1

Table 3 The rectangular model. The aperture aberration coefficients <0003>$_i$ as a function of $\theta = \beta L$

Electrostatic

θ	<03>$_4$	<03>$_3$	<03>$_2$	<03>$_1$	<03>$_0$
2.000⋆-1	1.906⋆-3	3.016⋆-3	−1.780⋆-2	−1.917⋆-2	−6.503⋆-3
4.000⋆-1	3.255⋆-2	5.253⋆-2	−4.122⋆-2	−6.538⋆-2	−2.385⋆-2
6.000⋆-1	1.845⋆-1	3.069⋆-1	4.997⋆-2	−9.314⋆-2	−4.397⋆-2
8.000⋆-1	6.902⋆-1	1.191	5.846⋆-1	2.037⋆-2	−4.719⋆-2
1.000	2.126	3.804	2.411	5.811⋆-1	1.071⋆-2
1.200	5.979	1.100⋆1	7.673	2.322	2.261⋆-1
1.400	1.621⋆1	3.021⋆1	2.170⋆1	6.954	8.054⋆-1
1.600	4.359⋆1	8.084⋆1	5.762⋆1	1.841⋆1	2.186
1.800	1.177⋆2	2.133⋆2	1.474⋆2	4.563⋆1	5.288
2.000	3.201⋆2	5.579⋆2	3.686⋆2	1.088⋆2	1.205⋆1

Table 3 The rectangular model. The aperture aberration coefficients $\langle 0003\rangle_i$ as a function of $\theta = \beta L$—cont'd

Electrostatic

θ	$\langle 03\rangle_4$	$\langle 03\rangle_3$	$\langle 03\rangle_2$	$\langle 03\rangle_1$	$\langle 03\rangle_0$
2.200	8.740⋆2	1.450⋆3	9.074⋆2	2.532⋆2	2.651⋆1
2.400	2.388⋆3	3.744⋆3	2.209⋆3	5.803⋆2	5.720⋆1
2.600	6.504⋆3	9.608⋆3	5.333⋆3	1.317⋆3	1.220⋆2
2.800	1.762⋆4	2.450⋆4	1.279⋆4	2.969⋆3	2.585⋆2
3.000	4.740⋆4	6.209⋆4	3.051⋆4	6.664⋆3	5.460⋆2
3.200	1.266⋆5	1.563⋆5	7.243⋆4	1.492⋆4	1.152⋆3
3.400	3.353⋆5	3.913⋆5	1.713⋆5	3.333⋆4	2.432⋆3
3.600	8.811⋆5	9.739⋆5	4.037⋆5	7.436⋆4	5.137⋆3
3.800	2.298⋆6	2.411⋆6	9.483⋆5	1.658⋆5	1.087⋆4
4.000	5.952⋆6	5.938⋆6	2.222⋆6	3.694⋆5	2.303⋆4

Magnetic

θ	$\langle 03\rangle_4$	$\langle 03\rangle_3$	$\langle 03\rangle_2$	$\langle 03\rangle_1$	$\langle 03\rangle_0$
2.000⋆-1	5.368⋆-4	2.593⋆-4	−2.055⋆-2	−2.054⋆-2	−6.775⋆-3
4.000⋆-1	8.745⋆-3	3.740⋆-3	−8.969⋆-2	−8.910⋆-2	−2.845⋆-2
6.000⋆-1	4.532⋆-2	1.446⋆-2	−2.375⋆-1	−2.300⋆-1	−6.960⋆-2
8.000⋆-1	1.455⋆-1	1.777⋆-2	−5.461⋆-1	−4.980⋆-1	−1.397⋆-1
1.000	3.476⋆-1	−8.772⋆-2	−1.228	−1.008	−2.569⋆-1
1.200	6.292⋆-1	−6.991⋆-1	−2.818	−2.004	−4.558⋆-1
1.400	6.611⋆-1	−3.111	−6.650	−4.005	−8.031⋆-1
1.600	−1.011	−1.118⋆1	−1.601⋆1	−8.130	−1.430
1.800	−9.488	−3.579⋆1	−3.889⋆1	−1.682⋆1	−2.597
2.000	−4. 091⋆1	−1.066⋆2	−9.465⋆1	−3.542⋆1	−4.831
2.200	−1.434⋆2	−3.023⋆2	−2.299⋆2	−7.571⋆1	−9.208
2.400	−4.553⋆2	−8.284⋆2	−5.558⋆2	−1.637⋆2	−1.794⋆1
2.600	−1.363⋆3	−2.213⋆3	−1.337⋆3	−3.572⋆2	−3.563⋆1
2.800	−3.922⋆3	−5.797⋆3	−3.202⋆3	−7.841⋆2	−7.185⋆1
3.000	−1.097⋆4	−1.496⋆4	−7.633⋆3	−1.729⋆3	−1.467⋆2

Continued

Table 3 The rectangular model. The aperture aberration coefficients $<0003>_i$ as a function of $\theta = \beta L$—cont'd

Magnetic

θ	$<03>_4$	$<03>_3$	$<03>_2$	$<03>_1$	$<03>_0$
3.200	−3.007⋆4	−3.812⋆4	−1.811⋆4	−3.823⋆3	−3.024⋆2
3.400	−8.102⋆4	−9.621⋆4	−4.283⋆4	−8.472⋆3	−6.283⋆2
3.600	−2.153⋆5	−2.408⋆5	−1.009⋆5	−1.880⋆4	−1.313⋆3
3.800	−5.659⋆5	−5.982⋆5	−2.371⋆5	−4.176⋆4	−2.758⋆3
4.000	−1.473⋆6	−1.477⋆6	−5.554⋆5	−9.281⋆4	−5.816⋆3

Achromatic

θ	$<03>_4$	$<03>_3$	$<03>_2$	$<03>_1$	$<03>_0$
2.000⋆-1	8.106⋆-4	8.107⋆-4	−2.000⋆-2	−2.027⋆-2	−6.720⋆-3
4.000⋆-1	1.350⋆-2	1.350⋆-2	−8.000⋆-2	−8.436⋆-2	−2.753⋆-2
6.000⋆-1	7.316⋆-2	7.296⋆-2	−1.800⋆-1	−2.027⋆-1	−6.447⋆-2
8.000⋆-1	2.544⋆-1	2.524⋆-1	−3.200⋆-1	−3.944⋆-1	−1.212⋆-1
1.000	7.034⋆-1	6.905⋆-1	−5.000⋆-1	−6.905⋆-1	−2.034⋆-1
1.200	1.699	1.641	−7.200⋆-1	−1.139	−3.194⋆-1
1.400	3.770	3.554	−9.800⋆-1	−1.813	−4.814⋆-1
1.600	7.908	7.223	−1.280	−2.822	−7.067⋆-1
1.800	1.595⋆1	1.402⋆1	−1.620	−4.328	−1.020
2.000	3.129⋆1	2.631⋆1	−2.000	−6.577	−1.456
2.200	6.005⋆1	4.808⋆1	−2.420	−9.933	−2.064
2.400	1.133⋆2	8.605⋆1	−2.880	−1.494⋆1	−2.914
2.600	2.105⋆2	1.515⋆2	−3.380	−2.241⋆1	−4.107
2.800	3.864⋆2	2.631⋆2	−3.920	−3.355⋆1	−5.786
3.000	7.010⋆2	4.516⋆2	−4.500	−5.018⋆1	−8.155
3.200	1.259⋆3	7.678⋆2	−5.120	−7.498⋆1	−1.150⋆1
3.400	2.239⋆3	1.295⋆3	−5.780	−1.120⋆2	−1.625⋆1
3.600	3.948⋆3	2.167⋆3	−6.480	−1.672⋆2	−2.300⋆1
3.800	6.905⋆3	3.603⋆3	−7.220	−2.495⋆2	−3.262⋆1
4.000	1.199⋆4	5.958⋆3	−8.000	−3.724⋆2	−4.633⋆1

Table 4 The rectangular model. The aperture aberration coefficients $<0012>_{ij}$ as a function of $\theta = \beta L$

Electrostatic

θ	$<12>_{22}$	$<12>_{21}$	$<12>_{12}$	$<12>_{20}$	$<12>_{11}$	$<12>_{02}$	$<12>_{10}$	$<12>_{01}$	$<12>_{00}$
2.000⋆-1	−7.684⋆-4	1.600⋆-3	−7.467⋆-4	4.026⋆-2	−3.783⋆-2	−2.079⋆-2	2.080⋆-2	−3.999⋆-2	−6.505⋆-3
4.000⋆-1	−1.086⋆-2	2.548⋆-2	−9.404⋆-3	1.638⋆-1	−1.237⋆-1	−9.210⋆-2	9.287⋆-2	−1.593⋆-1	−2.399⋆-2
6.000⋆-1	−4.440⋆-2	1.266⋆-1	−2.644⋆-2	3.765⋆-1	−1.641⋆-1	−2.367⋆-1	2.454⋆-1	−3.517⋆-1	−4.576⋆-2
8.000⋆-1	−1.939⋆-1	3.799⋆-1	7.370⋆-3	6.776⋆-1	2.775⋆-2	−4.777⋆-1	5.261⋆-1	−5.936⋆-1	−5.874⋆-2
1.000	−1.860⋆-1	8.245⋆-1	2.874⋆-1	1.041	7.547⋆-1	−8.140⋆-1	9.954⋆-1	−8.245⋆-1	−4.128⋆-2
1.200	−3.719⋆-1	1.329	1.210	1.384	2.418	−1.182	1.709	−9.231⋆-1	3.865⋆-2
1.400	−1.087	1.345	3.349	1.522	5.403	−1.405	2.680	−6.860⋆-1	2.233⋆-1
1.600	−3.614	−4.575⋆-1	7.211	1.125	9.780	−1.148	3.816	1.787⋆-1	5.606⋆-1
1.800	−1.083⋆1	−6.554	1.258⋆1	−3.029⋆-1	1.476⋆1	1.038⋆-1	4.830	2.023	1.093
2.000	−2.791⋆1	−2.065⋆1	1.720⋆1	−3.356	1.784⋆1	2.978	5.126	5.163	1.839
2.200	−6.204⋆1	−4.689⋆1	1.478⋆1	−8.501	1.377⋆1	7.977	3.706	9.687	2.756
2.400	−1.195⋆2	−8.715⋆1	−7.530	−1.558⋆1	−6.379	1.498⋆1	−8.510⋆-1	1.513⋆1	3.704
2.600	−1.968⋆2	−1.353⋆2	−7.098⋆1	−2.293⋆1	−5.520⋆1	2.236⋆1	−1.023⋆1	2.002⋆1	4.392
2.800	−2.632⋆2	−1.673⋆2	−2.040⋆2	−2.618⋆1	−1.468⋆2	2.573⋆1	−2.591⋆1	2.134⋆1	4.339

Continued

Table 4 The rectangular model. The aperture aberration coefficients $<0012>_{ij}$ as a function of $\theta = \beta L$—cont'd

Electrostatic

θ	$<12>_{22}$	$<12>_{21}$	$<12>_{12}$	$<12>_{20}$	$<12>_{11}$	$<12>_{02}$	$<12>_{10}$	$<12>_{01}$	$<12>_{00}$
3.000	−2.298⋆2	−1.268⋆2	−4.336⋆2	−1.674⋆1	−2.905⋆2	1.653⋆1	−4.816⋆1	1.409⋆1	2.860
3.200	9.075⋆1	8.978⋆1	−7.626⋆2	1.920⋆1	−4.782⋆2	−1.910⋆1	−7.447⋆1	−8.768	−8.746⋆-1
3.400	1.021⋆3	6.411⋆2	−1.125⋆3	1.003⋆2	−6.635⋆2	−9.984⋆1	−9.729⋆1	−5.545⋆1	−7.673
3.600	3.011⋆3	1.722⋆3	−1.318⋆3	2.460⋆2	−7.334⋆2	−2.453⋆2	−1.015⋆2	−1.329⋆2	−1.798⋆1
3.800	6.522⋆3	3.491⋆3	−9.053⋆2	4.670⋆2	−4.770⋆2	−4.661⋆2	−6.232⋆1	−2.418⋆2	−3.134⋆1
4.000	1.166⋆4	5.898⋆3	8.669⋆2	7.458⋆2	4.337⋆2	−7.448⋆2	5.475⋆1	−3.687⋆2	−4.561⋆1

Magnetic

θ	$<12>_{22}$	$<12>_{21}$	$<12>_{12}$	$<12>_{20}$	$<12>_{11}$	$<12>_{02}$	$<12>_{10}$	$<12>_{01}$	$<12>_{00}$
2.000⋆-1	1.631⋆-3	3.983⋆-3	1.669⋆-3	4.105⋆-2	−3.463⋆-2	−1.998⋆-2	2.199⋆-2	−3.878⋆-2	−6.025⋆-3
4.000⋆-1	2.747⋆-2	6.280⋆-2	2.995⋆-2	1.758⋆-1	−7.265⋆-2	−7.848⋆-2	1.114⋆-1	−1.394⋆-1	−1.631⋆-2
6.000⋆-1	1.483⋆-1	3.078⋆-1	1.778⋆-1	4.321⋆-1	9.255⋆-2	−1.624⋆-1	3.345⋆-1	−2.470⋆-1	−6.941⋆-3
8.000⋆-1	4.941⋆-1	9.144⋆-1	6.684⋆-1	8.320⋆-1	8.215⋆-1	−2.193⋆-1	7.870⋆-1	−2.457⋆-1	6.352⋆-2
1.000	1.219	1.998	1.918	1.355	2.604	−1.086⋆-1	1.564	7.405⋆-2	2.550⋆-1
1.200	2.351	3.410	4.521	1.891	5.921	4.674⋆-1	2.705	1.049	6.449⋆-1
1.400	3.466	4.459	9.010	2.200	1.091⋆1	2.042	4.111	3.165	1.322
1.600	3.234	3.667	1.515⋆1	1.915	1.668⋆1	5.444	5.420	7.033	2.372
1.800	−7.822⋆-1	−1.019	2.055⋆1	6.682⋆-1	2.036⋆1	1.177⋆1	5.868	1.330⋆1	3.853
2.000	−1.099⋆1	−1.082⋆1	1.832⋆1	−1.558	1.572⋆1	2.221⋆1	4.168	2.244⋆1	5.756

2.200	−2.504⋆1	−2.221⋆1	−5.749	−3.702	−7.887	3.756⋆1	−1.515	3.439⋆1	7.956
2.400	−2.368⋆1	−1.876⋆1	−7.558⋆1	−2.373	−6.564⋆1	5.735⋆1	−1.335⋆1	4.808⋆1	1.015⋆1
2.600	5.410⋆1	4.298⋆1	−2.237⋆2	9.980	−1.747⋆2	7.867⋆1	−3.322⋆1	6.087⋆1	1.185⋆1
2.800	3.539⋆2	2.545⋆2	−4.815⋆2	4.732⋆1	−3.465⋆2	9.487⋆1	−6.147⋆1	6.820⋆1	1.232⋆1
3.000	1.168⋆3	7.806⋆2	−8.537⋆2	1.321⋆2	−5.718⋆2	9.491⋆1	−9.487⋆1	6.380⋆1	1.078⋆1
3.200	3.006⋆3	1.881⋆3	−1.267⋆3	2.960⋆2	−7.948⋆2	6.504⋆1	−1.237⋆2	4.128⋆1	6.606
3.400	6.638⋆3	3.907⋆3	−1.488⋆3	5.769⋆2	−8.783⋆2	−4.746	−1.287⋆2	−2.067	−1.423⋆-1
3.600	1.305⋆4	7.251⋆3	−1.012⋆3	1.010⋆3	−5.658⋆2	−1.058⋆2	−7.820⋆1	−5.798⋆1	−7.888
3.800	2.316⋆4	1.219⋆4	1.040⋆3	1.607⋆3	5.434⋆2	−1.835⋆2	7.184⋆1	−9.568⋆1	−1.242⋆1
4.000	3.720⋆4	1.861⋆4	5.951⋆3	2.329⋆3	2.971⋆3	−9.335⋆1	3.717⋆2	−4.574⋆1	−5.554
					Achromatic				
θ	**$<12>_{22}$**	**$<12>_{21}$**	**$<12>_{12}$**	**$<12>_{20}$**	**$<12>_{11}$**	**$<12>_{02}$**	**$<12>_{10}$**	**$<12>_{01}$**	**$<12>_{00}$**
2.000⋆-1	2.431⋆-3	4.778⋆-3	2.474⋆-3	4.131⋆-2	−3.357⋆-2	−1.971⋆-2	2.239⋆-2	−3.837⋆-2	−5.865⋆-3
4.000⋆-1	4.025⋆-2	7.523⋆-2	4.307⋆-2	1.798⋆-1	−5.562⋆-2	−7.394⋆-2	1.175⋆-1	−1.328⋆-1	−1.375⋆-2
6.000⋆-1	2.126⋆-1	3.682⋆-1	2.459⋆-1	4.507⋆-1	1.781⋆-1	−1.376⋆-1	3.643⋆-1	−2.121⋆-1	5.999⋆-3
8.000⋆-1	6.934⋆-1	1.093	8.887⋆-1	8.835⋆-1	1.086	−1.332⋆-1	8.740⋆-1	−1.297⋆-1	1.043⋆-1
1.000	1.687	2.389	2.461	1.460	3.220	1.265⋆-1	1.753	3.736⋆-1	3.538⋆-1
1.200	3.258	4.104	5.625	2.060	7.089	1.017	3.038	1.707	8.470⋆-1
1.400	4.984	5.497	1.090⋆1	2.425	1.274⋆1	3.192	4.588	4.448	1.688

Continued

Table 4 The rectangular model. The aperture aberration coefficients $<0012>_{ij}$ as a function of $\theta = \beta L$—cont'd

Achromatic

θ	$<12>_{22}$	$<12>_{21}$	$<12>_{12}$	$<12>_{20}$	$<12>_{11}$	$<12>_{02}$	$<12>_{10}$	$<12>_{01}$	$<12>_{00}$
1.600	5.516	5.042	1.780⋆1	2.178	1.899⋆1	7.642	5.955	9.317	2.976
1.800	2.568	8.262⋆-1	2.321⋆1	9.919⋆-1	2.222⋆1	1.566⋆1	6.214	1.706⋆1	4.773
2.000	−5.349	−7.540	1.870⋆1	−9.588⋆-1	1.501⋆1	2.863⋆1	3.848	2.819⋆1	7.062
2.200	−1.271⋆1	−1.398⋆1	−1.259⋆1	−2.102	−1.510⋆1	4.742⋆1	−3.255	4.262⋆1	9.689
2.400	8.241	4.030	−9.826⋆1	2.028	−8.539⋆1	7.147⋆1	−1.752⋆1	5.906⋆1	1.230⋆1
2.600	1.377⋆2	1.024⋆2	−2.746⋆2	2.095⋆1	−2.145⋆2	9.745⋆1	−4.088⋆1	7.449⋆1	1.433⋆1
2.800	5.596⋆2	3.951⋆2	−5.740⋆2	7.182⋆1	−4.131⋆2	1.179⋆2	−7.333⋆1	8.382⋆1	1.498⋆1
3.000	1.634⋆3	1.083⋆3	−9.938⋆2	1.817⋆2	−6.656⋆2	1.210⋆2	−1.104⋆2	8.037⋆1	1.342⋆1
3.200	3.977⋆3	2.478⋆3	−1.435⋆3	3.883⋆2	−9.003⋆2	9.308⋆1	−1.402⋆2	5.796⋆1	9.099
3.400	8.511⋆3	4.996⋆3	−1.609⋆3	7.358⋆2	−9.499⋆2	2.695⋆1	−1.392⋆2	1.573⋆1	2.368
3.600	1.639⋆4	9.094⋆3	−9.102.2	1.264⋆3	−5.100⋆2	−5.933⋆1	−7.043⋆1	−3.301⋆1	−4.523
3.800	2.870⋆4	1.509⋆4	1.689⋆3	1.987⋆3	8.835⋆2	−8.925⋆1	1.166⋆2	−4.698⋆1	−6.117
4.000	4.571⋆4	2.284⋆4	7.646⋆3	2.856⋆3	3.817⋆3	1.238⋆2	4.773⋆2	6.190⋆1	7.799

Table 5 The rectangular model. The aperture aberration coefficients $<0012>_{ij}$ as a function of $\theta = \beta L$

Electrostatic

θ	$<21>_{22}$	$<21>_{21}$	$<21>_{12}$	$<21>_{20}$	$<21>_{11}$	$<21>_{02}$	$<21>_{10}$	$<21>_{01}$	$<21>_{00}$
2.000⋆-1	−8.324⋆-4	−8.534⋆-4	1.600⋆-3	1.919⋆-2	4.210⋆-2	−3.973⋆-2	3.999⋆-2	−1.920⋆-2	6.825⋆-3
4.000⋆-1	−1.495⋆-2	−1.623⋆-2	2.548⋆-2	6.653⋆-2	1.919⋆-1	−1.553⋆-1	1.593⋆-1	−6.731⋆-2	2.911⋆-2
6.000⋆-1	−9.079⋆-2	−1.039⋆-1	1.266⋆-1	1.078⋆-1	5.063⋆-1	−3.342⋆-1	3.517⋆-1	−1.167⋆-1	7.158⋆-2
8.000⋆-1	−3.613⋆-1	−4.244⋆-1	3.799⋆-1	7.551⋆-2	1.031	−5.506⋆-1	5.936⋆-1	−1.256⋆-1	1.397⋆-1
1.000	−1.142	−1.332	8.245⋆-1	−1.420⋆-1	1.710	−7.647⋆-1	8.245⋆-1	−4.914⋆-2	2.353⋆-1
1.200	−3.088	−3.476	1.329	−7.045⋆-1	2.252	−9.353⋆-1	9.231⋆-1	1.352⋆-1	3.505⋆-1
1.400	−7.378	−7.854	1.345	−1.804	1.935	−1.063	6.860⋆-1	3.808⋆-1	4.577⋆-1
1.600	−1.581⋆1	−1.571⋆1	−4.575⋆-1	−3.614	−5.738⋆-1	−1.284	−1.787⋆-1	4.948⋆-1	4.984⋆-1
1.800	−3.053⋆1	−2.814⋆1	−6.554	−6.184	−7.293	−2.045	−2.023	2.734⋆-2	3.689⋆-1
2.000	−5.313⋆1	−4.516⋆1	−2.065⋆1	−9.282	−2.067⋆1	−4.363	−5.163	−1.865	−9.075⋆-2
2.200	−8.257⋆1	−6.443⋆1	−4.689⋆1	−1.222⋆1	−4.264⋆1	−1.018⋆1	−9.687	−6.553	−1.103
2.400	−1.124⋆2	−7.984⋆1	−8.715⋆1	−1.375⋆1	−7.263⋆1	−2.270⋆1	−1.513⋆1	−1.602⋆1	−2.941
2.600	−1.287⋆2	−8.141⋆1	−1.353⋆2	−1.228⋆1	−1.041⋆2	−4.664⋆1	−2.002⋆1	−3.278⋆1	−5.899
2.800	−1.140⋆2	−5.948⋆1	−1.673⋆2	−6.701	−1.195⋆2	−8.795⋆1	−2.134⋆1	−5.952⋆1	−1.022⋆1
3.000	−6.762⋆1	−1.808⋆1	−1.268⋆2	1.487	−8.454⋆1	−1.528⋆2	−1.409⋆1	−9.835⋆1	−1.597⋆1

Continued

Table 5 The rectangular model. The aperture aberration coefficients <0012>$_{ij}$ as a function of $\theta = \beta L$—cont'd

Electrostatic

θ	<21>$_{22}$	<21>$_{21}$	<21>$_{12}$	<21>$_{20}$	<21>$_{11}$	<21>$_{02}$	<21>$_{10}$	<21>$_{01}$	<21>$_{00}$
3.200	−6.083⋆1	−5.250	8.978⋆1	4.300	5.612⋆1	−2.450⋆2	8.768	−1.494⋆2	−2.293⋆1
3.400	−3.549⋆2	−1.694⋆2	6.411⋆2	−1.914⋆1	3.771⋆2	−3.624⋆2	5.545⋆1	−2.093⋆2	−3.035⋆1
3.600	−1.614⋆3	−8.498⋆2	1.722⋆3	−1.116⋆2	9.568⋆2	−4.912⋆2	1.329⋆2	−2.688⋆2	−3.690⋆1
3.800	−5.235⋆3	−2.701⋆3	3.491⋆3	−3.401⋆2	1.837⋆3	−6.010⋆2	2.418⋆2	−3.120⋆2	−4.062⋆1
4.000	−1.379⋆4	−6.829⋆3	5.898⋆3	−8.457⋆2	2.949⋆3	−6.448⋆2	3.687⋆2	−3.179⋆2	−3.930⋆1

Magnetic

θ	<21>$_{22}$	<21>$_{21}$	<21>$_{12}$	<21>$_{20}$	<21>$_{11}$	<21>$_{02}$	<21>$_{10}$	<21>$_{01}$	<21>$_{00}$
2.000⋆-1	1.567⋆-3	1.530⋆-3	4.015⋆-3	1.998⋆-2	4.530⋆-2	−3.891⋆-2	4.118⋆-2	−1.799⋆-2	7.305⋆-3
4.000⋆-1	2.338⋆-2	2.109⋆-2	6.484⋆-2	7.852⋆-2	2.430⋆-1	−1.417⋆-1	1.778⋆-1	−4.744⋆-2	3.679⋆-2
6.000⋆-1	1.019⋆-1	7.728⋆-2	3.309⋆-1	1.635⋆-1	7.629⋆-1	−2.600⋆-1	4.409⋆-1	−1.200⋆-2	1.104⋆-1
8.000⋆-1	2.367⋆-1	1.101⋆-1	1.041	2.299⋆-1	1.824	−2.922⋆-1	8.545⋆-1	2.224⋆-1	2.619⋆-1
1.000	2.627⋆-1	−1.588⋆-1	2.455	1.717⋆-1	3.559	−5.937⋆-2	1.393	8.494⋆-1	5.316⋆-1
1.200	−3.656⋆-1	−1.394	4.641	−1.978⋆-1	5.755	7.138⋆-1	1.920	2.108	9.568⋆-1
1.400	−2.825	−4.740	7.006	−1.127	7.439	2.385	2.118	4.232	1.556
1.600	−8.958	−1.159⋆1	7.485	−2.824	6.331	5.308	1.425	7.349	2.310
1.800	−2.048⋆1	−2.260⋆1	1.419	−5.213	−1.695	9.626	−9.847⋆-1	1.130⋆1	3.129
2.000	−3.621⋆1	−3.532⋆1	−1.953⋆1	−7.484	−2.279⋆1	1.487⋆1	−6.121	1.541⋆1	3.827

2.200	−4.557⋆1	−3.975⋆1	−6.741⋆1	−7.420	−6.429⋆1	1.940⋆1	−1.491⋆1	1.815⋆1	4.097
2.400	−1.661⋆1	−1.146⋆1	−1.552⋆2	−5.468⋆-1	−1.319⋆2	1.967⋆1	−2.763⋆1	1.693⋆1	3.509
2.600	1.223⋆2	9.689⋆1	−2.880⋆2	2.064⋆1	−2.236⋆2	9.681	−4.300⋆1	8.077	1.555
2.800	5.031⋆2	3.623⋆2	−4.448⋆2	6.680⋆1	−3.192⋆2	−1.881⋆1	−5.690⋆1	−1.266⋆1	−2.234
3.000	1.330⋆3	8.894⋆2	−5.469⋆2	1.504⋆2	−3.658⋆2	−7.440⋆1	−6.080⋆1	−4.864⋆1	−8.051
3.200	2.854⋆3	1.786⋆3	−4.148⋆2	2.811⋆2	−2.604⋆2	−1.609⋆2	−4.051⋆1	−9.940⋆1	−1.545⋆1
3.400	5.263⋆3	3.097⋆3	2.786⋆2	4.575⋆2	1.623⋆2	−2.673⋆2	2.400⋆1	−1.559⋆2	−2.282⋆1
3.600	8.422⋆3	4.679⋆3	2.028⋆3	6.520⋆2	1.124⋆3	−3.517⋆2	1.562⋆2	−1.939⋆2	−2.681⋆1
3.800	1.140⋆4	5.999⋆3	5.436⋆3	7.916⋆2	2.858⋆3	−3.184⋆2	3.759⋆2	−1.659⋆2	−2.170⋆1
4.000	1.175⋆4	5.878⋆3	1.098⋆4	7.371⋆2	5.486⋆3	6.653	6.856⋆2	5.010	7.585⋆-1

Achromatic

θ	$<21>_{22}$	$<21>_{21}$	$<21>_{12}$	$<21>_{20}$	$<21>_{11}$	$<21>_{02}$	$<21>_{10}$	$<21>_{01}$	$<21>_{00}$
2.000⋆-1	2.367⋆-3	2.325⋆-3	4.821⋆-3	2.024⋆-2	4.637⋆-2	−3.864⋆-2	4.157⋆-2	−1.759⋆-2	7.465⋆-3
4.000⋆-1	3.616⋆-2	3.352⋆-2	7.796⋆-2	8.252⋆-2	2.600⋆-1	−1.372⋆-1	1.839⋆-1	−4.081⋆-2	3.934⋆-2
6.000⋆-1	1.662⋆-1	1.377⋆-1	3.990⋆-1	1.820⋆-1	8.485⋆-1	−2.352⋆-1	4.706⋆-1	2.289⋆-2	1.233⋆-1
8.000⋆-1	4.360⋆-1	2.882⋆-1	1.261	2.814⋆-1	2.089	−2.061⋆-1	9.415⋆-1	3.384⋆-1	3.027⋆-1
1.000	7.310⋆-1	2.322⋆-1	2.998	2.763⋆-1	4.176	1.757⋆-1	1.582	1.149	6.304⋆-1
1.200	5.420⋆-1	−7.006⋆-1	5.744	−2.891⋆-2	6.923	1.263	2.252	2.765	1.159

Continued

Table 5 The rectangular model. The aperture aberration coefficients $<0012>_{ij}$ as a function of $\theta = \beta L$—cont'd

Achromatic

θ	$<21>_{22}$	$<21>_{21}$	$<21>_{12}$	$<21>_{20}$	$<21>_{11}$	$<21>_{02}$	$<21>_{10}$	$<21>_{01}$	$<21>_{00}$
1.400	−1.307	−3.702	8.893	−9.014⋆-1	9.273	3.534	2.595	5.515	1.923
1.600	−6.676	−1.021⋆1	1.013⋆1	−2.561	8.632	7.506	1.960	9.633	2.914
1.800	−1.713⋆1	−2.075⋆1	4.076	−4.889	1.711⋆-1	1.352⋆1	−6.387⋆-1	1.506⋆1	4.049
2.000	−3.057⋆1	−3.204⋆1	−1.915⋆1	−6.885	−2.350⋆1	2.129⋆1	−6.441	2.117⋆1	5.132
2.200	−3.324⋆1	−3.152⋆1	−7.425⋆1	−5.821	−7.151⋆1	2.926⋆1	−1.665⋆1	2.638⋆1	5.830
2.400	1.532⋆1	1.134⋆1	−1.779⋆2	3.855	−1.516⋆2	3.379⋆1	−3.179⋆1	2.791⋆1	5.659
2.600	2.059⋆2	1.563⋆2	−3.389⋆2	3.161⋆1	−2.634⋆2	2.845⋆1	−5.067⋆1	2.169⋆1	4.039
2.800	7.088⋆2	5.029⋆2	−5.373⋆2	9.130⋆1	−3.858⋆2	4.235	−6.876⋆1	2.958	4.275⋆-1
3.000	1.796⋆3	1.192⋆3	−6.870⋆2	2.000⋆2	−4.596⋆2	−4.827⋆1	−7.637⋆1	−3.207⋆1	−5.410
3.200	3.826⋆3	2.383⋆3	−5.829⋆2	3.734.2	−3.660⋆2	−1.329⋆2	−5.694⋆1	−8.272⋆1	−1.296⋆1
3.400	7.135⋆3	4.186⋆3	1.578⋆2	6.164⋆2	9.069⋆1	−2.356⋆2	1.352⋆1	−1.381⋆2	−2.031⋆1
3.600	1.177⋆4	6.522⋆3	2.130⋆3	9.065⋆2	1.180⋆3	−3.052⋆2	1.640⋆2	−1.689⋆2	−2.344⋆1
3.800	1.694⋆4	8.899⋆3	6.085⋆3	1.171⋆3	3.198⋆3	−2.242⋆2	4.206⋆2	−1.173⋆2	−1.540⋆1
4.000	2.027⋆4	1.011⋆4	1.268⋆4	1.265⋆3	6.332⋆3	2.238⋆2	7.912⋆2	1.126⋆2	1.411⋆1

Table 6 The rectangular model. The aperture aberration coefficients $(0030)_i$ as a function of $\theta = \beta L$.

Electrostatic

θ	$(30)_4 = (30)_0$	$(30)_3 = (30)_1$	$(30)_2$
2.000⋆-1	7.335⋆2	−2.921⋆3	4.375⋆3
4.000⋆-1	4.675⋆1	−1.834⋆2	2.732⋆2
6.000⋆-1	9.607	−3.650⋆1	5.381⋆1
8.000⋆-1	3.259	−1.168⋆1	1.688⋆1
1.000	1.495	−4.848	6.786
1.200	8.572⋆-1	−2.364	3.165
1.400	5.942⋆-1	−1.264	1.629
1.600	4.917⋆-1	−6.724⋆-1	9.303⋆-1
1.800	4.864⋆-1	−2.431⋆-1	6.799⋆-1
2.000	5.850⋆-1	2.688⋆-1	8.624⋆-1
2.200	8.879⋆-1	1.304	1.937
2.400	1.813	4.405	6.040
2.600	5.583	1.790⋆1	2.533⋆1
2.800	3.391⋆1	1.259⋆2	1.846⋆2
3.000	1.183⋆3	4.680⋆3	6.994⋆3
3.200	4.411⋆4	1.761⋆5	2.641⋆5
3.400	1.284⋆2	4.987⋆2	7.411⋆2
3.600	1.486⋆1	5.438⋆1	7.952⋆1
3.800	4.117	1.371⋆1	1.970⋆1
4.000	1.748	5.028	7.161

Magnetic

θ	$(30)_4 = (30)_0$	$(30)_3 = (30)_1$	$(30)_2$
2.000⋆-1	2.125⋆2	−8.375⋆2	1.250⋆3
4.000⋆-1	1.409⋆1	−5.313⋆1	7.809⋆1
6.000⋆-1	3.060	−1.075⋆1	1.540⋆1
8.000⋆-1	1.103	−3.513	4.855

Continued

Table 6 The rectangular model. The aperture aberration coefficients $(0030)_i$ as a function of $\theta = \beta L$.—cont'd

Magnetic

θ	$(30)_4 = (30)_0$	$(30)_3 = (30)_1$	$(30)_2$
1.000	5.330⋆-1	−1.490	1.978
1.200	3.164⋆-1	−7.370⋆-1	9.538⋆-1
1.400	2.225⋆-1	−3.894⋆-1	5.316⋆-1
1.600	1.844⋆-1	−1.868⋆-1	3.603⋆-1
1.800	1.831⋆-1	−1.881⋆-2	3.404⋆-1
2.000	2.244⋆-1	2.061⋆-1	4.964⋆-1
2.200	3.512⋆-1	6.718⋆-1	1.043
2.400	7.364⋆-1	2.022	2.888
2.600	2.278	7.649	1.100⋆1
2.800	1.354⋆1	5.084⋆1	7.482⋆1
3.000	4.555⋆2	1.804⋆3	2.696⋆3
3.200	1.638⋆4	6.540⋆4	9.804⋆4
3.400	4.650⋆1	1.800⋆2	2.672⋆2
3.600	5.332	1.925⋆1	2.802⋆1
3.800	1.485	4.795	6.819
4.000	6.355⋆-1	1.748	2.450

Achromatic

θ	$(30)_4 = (30)_0$	$(30)_3 = (30)_1$	$(30)_2$
2.000⋆-1	3.167⋆2	−1.254⋆3	1.875⋆3
4.000⋆-1	2.062⋆1	−7.918⋆1	1.171⋆2
6.000⋆-1	4.370	−1.590⋆1	2.308⋆1
8.000⋆-1	1.534	−5.146	7.260
1.000	7.253⋆-1	−2.162	2.940
1.200	4.245⋆-1	−1.062	1.396
1.400	2.968⋆-1	−5.642⋆-1	7.511⋆-1
1.600	2.459⋆-1	−2.839⋆-1	4.743⋆-1

Table 6 The rectangular model. The aperture aberration coefficients $(0030)_i$ as a function of $\theta = \beta L$.—cont'd

Achromatic

θ	$(30)_4 = (30)_0$	$(30)_3 = (30)_1$	$(30)_2$
1.800	2.438⋆-1	−6.368⋆-2	4.083⋆-1
2.000	2.965⋆-1	2.186⋆-1	5.696⋆-1
2.200	4.586⋆-1	7.983⋆-1	1.222
2.400	9.517⋆-1	2.499	3.519
2.600	2.939	9.699	1.387⋆1
2.800	1.761⋆-1	6.586⋆1	9.678⋆1
3.000	6.010⋆2	2.379⋆3	3.556⋆3
3.200	2.192⋆4	8.754⋆4	1.312⋆5
3.400	6.289⋆1	2.438⋆2	3.620⋆2
3.600	7.238	2.628⋆1	3.832⋆1
3.800	2.011	6.577	9.394
4.000	8.563⋆-1	2.404	3.392

Table 7 The rectangular model. The aperture aberration coefficients $(0003)_i$ as a function of $\theta = \beta L$.

Electrostatic

θ	$(03)_4 = (03)_0$	$(03)_2 = (03)_1$	$(03)_2$
2.000⋆-1	7.251⋆2	−2.913⋆3	4.375⋆3
4.000⋆-1	4.466⋆1	−1.813⋆2	2.732⋆2
6.000⋆-1	8.665	−3.558⋆1	5.382⋆1
8.000⋆-1	2.709	−1.117⋆1	1.690⋆1
1.000	1.115	−4.537	6.824
1.200	5.554⋆-1	−2.172	3.211
1.400	3.208⋆-1	−1.166	1.670
1.600	2.088⋆-1	−6.888⋆-1	9.293⋆-1
1.800	1.497⋆-1	−4.243⋆-1	5.422⋆-1
2.000	1.156⋆-1	−2.782⋆-1	3.271⋆-1

Continued

Table 7 The rectangular model. The aperture aberration coefficients $(0003)_i$ as a function of $\theta = \beta L$.—cont'd

Electrostatic

θ	$(03)_4 = (03)_0$	$(03)_2 = (03)_1$	$(03)_2$
2.200	9.454⋆-2	−1.898⋆-1	2.021⋆-1
2.400	8.061⋆-2	−1.336⋆-1	1.270⋆-1
2.600	7.085⋆-2	−9.633⋆-2	8.082⋆-2
2.800	6.365⋆-2	−7.079⋆-2	5.191⋆-2
3.000	5.811⋆-2	−5.279⋆-2	3.358⋆-2
3.200	5.367⋆-2	−3.982⋆-2	2.183⋆-2
3.400	5.002⋆-2	−3.031⋆-2	1.425⋆-2
3.600	4.692⋆-2	−2.324⋆-2	9.337⋆-3
3.800	4.426⋆-2	−1.791⋆-2	6.131⋆-3
4.000	4.192⋆-2	−1.387⋆-2	4.035⋆-3

Magnetic

θ	$(03)_4 = (03)_0$	$(03)_2 = (03)_1$	$(03)_2$
2.000⋆-1	2.042⋆2	−8.292⋆2	1.250⋆3
4.000⋆-1	1.200⋆1	−5.105⋆1	7.809⋆1
6.000⋆-1	2.129	−9.834	1.540⋆1
8.000⋆-1	5.710⋆-1	−3.005	4.848
1.000	1.822⋆-1	−1.179	1.965
1.200	5.845⋆-2	−5.409⋆-1	9.298⋆-1
1.400	1.309⋆-2	−2.768⋆-1	4.870⋆-1
1.600	−4.844⋆-3	−1.537⋆-1	2.734⋆-1
1.800	−1.206⋆-2	−9.103⋆-2	1.610⋆-1
2.000	−1.478⋆-2	−5.689⋆-2	9.802⋆-2
2.200	−1.552⋆-2	−3.717⋆-2	6.112⋆-2
2.400	−1.537⋆-2	−2.521⋆-2	3.876⋆-2
2.600	−1.484⋆-2	−1.764⋆-2	2.489⋆-2
2.800	−1.417⋆-2	−1.265⋆-2	1.612⋆-2
3.000	−1.345⋆-2	−9.262⋆-3	1.051⋆-2

Table 7 The rectangular model. The aperture aberration coefficients $(0003)_i$ as a function of $\theta = \beta L$.—cont'd

Magnetic

θ	$(03)_4 = (03)_0$	$(03)_2 = (03)_1$	$(03)_2$
3.200	−1.275★-2	−6.889★-3	6.881★-3
3.400	−1.209★-2	−5.189★-3	4.523★-3
3.600	−1.147★-2	−3.947★-3	2.981★-3
3.800	−1.090★-2	−3.026★-3	1.969★-3
4.000	−1.037★-2	−2.334★-3	1.303★-3

Achromatic

θ	$(03)_4 = (03)_0$	$(03)_2 = (03)_1$	$(03)_2$
2.000★-1	3.084★2	−1.246★3	1.875★3
4.000★-1	1.853★1	−7.710★1	1.171★2
6.000★-1	3.436	−1.498★1	2.308★1
8.000★-1	9.985★-1	−4.638	7.259
1.000	3.687★-1	−1.851	2.937
1.200	1.578★-1	−8.671★-1	1.386
1.400	7.463★-2	−4.545★-1	7.235★-1
1.600	3.789★-2	−2.591★-1	4.045★-1
1.800	2.028★-2	−1.577★-1	2.372★-1
2.000	1.130★-2	−1.012★-1	1.438★-1
2.200	6.496★-3	−6.770★-2	8.931★-2
2.400	3.824★-3	−4.689★-2	6.641★-2
2.600	2.294★-3	−3.338★-2	3.607★-2
2.800	1.396★-3	−2.428★-2	2.328★-2
3.000	8.593★-4	−1.797★-2	1.512★-2
3.200	5.338★-4	−1.348★-2	9.871★-3
3.400	3.340★-4	−1.021★-2	6.469★-3
3.600	2.102★-4	−7.805★-3	4.252★-3
3.800	1.330★-4	−6.003★-3	2.802★3
4.000	8.443★-5	−4.642★-3	1.849★-3

Table 8 The rectangular model. The aperture aberration coefficients $(0012)_{ij}$ as a function of $\theta = \beta L$

Electrostatic

θ	$(12)_{22} = (12)_{00}$	$(12)_{21} = (12)_{01}$	$(12)_{12} = (12)_{10}$	$(12)_{20} = (12)_{02}$	$(12)_{11}$
2.000⋆-1	−3.002⋆2	6.376⋆2	6.000⋆2	−3.125⋆2	−1.250.3
4.000⋆-1	−1.659⋆1	4.225⋆1	3.279⋆1	−1.954⋆1	−7.784⋆1
6.000⋆-1	−2.651	9.174	4.911	−3.869	−1.514⋆1
8.000⋆-1	−6.249⋆-1	3.295	8.453⋆-1	−1.232	−4.591
1.000	−1.902⋆-1	1.578	−4.170⋆-2	−5.113⋆-1	−1.705
1.200	−9.062⋆-2	9.148⋆-1	−2.640⋆-1	−2.526⋆-1	−6.630⋆-1
1.400	−8.038⋆-2	6.108⋆-1	−3.142⋆-1	−1.420⋆-1	−2.076⋆-1
1.600	−9.780⋆-2	4.571⋆-1	−3.178⋆-1	−8.868⋆-2	2.670⋆-2
1.800	−1.257⋆-1	3.785⋆-1	−3.129⋆-1	−6.090⋆-2	1.720⋆-1
2.000	−1.604⋆-1	3.459⋆-1	−3.132⋆-1	−4.597⋆-2	2.879⋆-1
2.200	−2.039⋆-1	3.518⋆-1	−3.264⋆-1	−3.850⋆-2	4.140⋆-1
2.400	−2.641⋆-1	4.076⋆-1	−3.625⋆-1	−3.668⋆-2	6.012⋆-1
2.600	−3.616⋆-1	5.683⋆-1	−4.450⋆-1	−4.198⋆-2	9.740⋆-1
2.800	−5.687⋆-1	1.096	−6.582⋆-1	−6.640⋆-2	2.065

3.000	−1.419	5.037	−1.676	−2.502⋆-1	9.974
3.200	1.694	2.404⋆1	7.231⋆-1	−9.785⋆-1	4.801⋆1
3.400	5.222⋆-1	1.026	5.094⋆-1	−3.419⋆-2	1.983
3.600	2.738⋆-1	2.797⋆-1	3.002⋆-1	−7.636⋆-3	5.016⋆-1
3.800	1.674⋆-1	1.197⋆-1	2.108⋆-1	−2.676⋆-3	1.893⋆-1
4.000	1.068⋆-1	6.402⋆-2	1.636⋆-1	−1.172⋆-3	8.369⋆-2

Magnetic

θ	$(12)_{22} = (12)_{00}$	$(12)_{21} = (12)_{01}$	$(12)_{12} = (12)_{10}$	$(12)_{20} = (12)_{02}$	$(12)_{11}$
2.000⋆-1	6.373⋆2	−1.237⋆3	−1.275⋆3	6.252⋆2	2.500⋆3
4.000⋆-1	4.194⋆1	−7.493⋆1	−8.438⋆1	3.923⋆1	1.563⋆2
6.000⋆-1	8.857	−1.398⋆1	−1.822⋆1	7.889	3.089⋆1
8.000⋆-1	2.972	−4.041	−6.448	2.615	9.788
1.000	1.246	−1.445	−2.997	1.177	4.022
1.200	5.727⋆-1	−5.689⋆-1	−1.647	6.641⋆-1	1.951
1.400	2.562⋆-1	−2.234⋆-1	−1.006	4.509⋆-1	1.063
1.600	8.751⋆-2	−7.391⋆-2	−6.508⋆-1	3.578⋆-1	6.319⋆-1
1.800	−9.076⋆-3	−6.207⋆-3	−4.221⋆-1	3.244⋆-1	4.020⋆-1
2.000	−6.315⋆-2	2.423⋆-2	−2.441⋆-1	3.316⋆-1	2.697⋆-1

Continued

Table 8 The rectangular model. The aperture aberration coefficients $(0012)_{ij}$ as a function of $\theta = \beta L$—cont'd

Magnetic

θ	$(12)_{22} = (12)_{00}$	$(12)_{21} = (12)_{01}$	$(12)_{12} = (12)_{10}$	$(12)_{20} = (12)_{02}$	$(12)_{11}$
2.200	−8.232⋆-2	3.626⋆-2	−6.327⋆-2	3.819⋆-1	1.878⋆-1
2.400	−5.236⋆-2	3.771⋆-2	1.937⋆-1	5.045⋆-1	1.312⋆-1
2.600	9.940⋆-2	2.904⋆-2	7.212⋆-1	8.046⋆-1	8.022⋆-2
2.800	7.646⋆-1	−7.770⋆-3	2.417	1.778	−1.235⋆-2
3.000	7.214	−3.439⋆-1	1.652⋆1	9.386	−6.951⋆-1
3.200	5.609⋆1	−2.897	1.076⋆2	5.153⋆1	−5.807
3.400	3.397	−1.587⋆-1	5.906	2.533	−3.324⋆-1
3.600	1.186	−4.634⋆-2	1.981	7.981⋆-1	−1.085⋆-1
3.800	5.944⋆-1	−1.821⋆-2	1.002	3.955⋆-1	−5.203⋆-2
4.000	3.407⋆-1	−7.460⋆-3	6.103⋆-1	2.456⋆-1	−2.985⋆-2

Achromatic

θ	$(12)_{22} = (12)_{00}$	$(12)_{21} = (12)_{01}$	$(12)_{12} = (12)_{10}$	$(12)_{20} = (12)_{02}$	$(12)_{11}$
2.000⋆-1	9.497⋆2	−1.862⋆3	−1.900⋆3	9.377⋆2	3.759⋆3
4.000⋆-1	6.145⋆1	−1.140⋆2	−1.234⋆2	5.883⋆1	2.343⋆2
6.000⋆-1	1.269⋆1	−2.170⋆1	−2.593⋆1	1.181⋆1	4.623⋆1
8.000⋆-1	4.171	−6.487	−8.879	3.898	1.458⋆1

1.000	1.725	−2.453	−3.982	1.739	5.931
1.200	7.938⋆-1	−1.063	−2.108	9.697⋆-1	2.822
1.400	3.684⋆-1	−5.014⋆-1	−1.236	6.486⋆-1	1.487
1.600	1.493⋆-1	−2.509⋆-1	−7.617⋆-1	5.066⋆-1	8.337⋆-1
1.800	2.988⋆-2	−1.344⋆-1	−4.585⋆-1	4.528⋆-1	4.787⋆-1
2.000	−3.974⋆-2	−8.300⋆-2	−2.210⋆-1	4.574⋆-1	2.636⋆-1
2.200	−4.178⋆-2	−6.892⋆-2	2.444⋆-2	5.220⋆-1	1.124⋆-1
2.400	1.822⋆-2	−8.560⋆-2	3.791⋆-1	6.849⋆-1	−2.542⋆-2
2.600	2.531⋆-1	−1.507⋆-1	1.110	1.087	−2.177⋆-1
2.800	1.209	−3.757⋆-1	3.442	2.393	−7.048⋆-1
3.000	1.009⋆1	−2.138	2.258⋆1	1.260⋆1	−4.251
3.200	7.422⋆1	−1.188⋆1	1.432⋆2	6.904⋆1	−2.374⋆1
3.400	4.355	−5.534⋆-1	7.705	3.389	−1.104
3.600	1.491	−1.550⋆-1	2.541	1.067	−3.118⋆-1
3.800	7.367⋆-1	−6.417⋆-2	1.266	5.282⋆-1	−1.325⋆-1
4.000	4.186⋆-1	−3.129⋆-2	7.593⋆-1	3.278⋆-1	−6.770⋆-2

Table 9 The rectangular model. The aperture aberration coefficients $(0021)_{ij}$ as a function of $\theta = \beta L$.

Electrostatic

θ	$(21)_{22} = (21)_{00}$	$(21)_{21} = (21)_{01}$	$(21)_{12} = (21)_{10}$	$(21)_{20} = (21)_{02}$	$(21)_{11}$
2.000★-1	−3.252★2	6.500★2	6.126★2	−3.125★2	−1.250★3
4.000★-1	−2.283★1	4.530★1	3.600★1	−1.954★1	−7.784★1
6.000★-1	−5.421	1.048★1	6.387	−3.869	−1.514★1
8.000★-1	−2.173	3.999	1.717	−1.232	−4.591
1.000	−1.168	2.003	5.525★-1	−5.113★-1	−1.705
1.200	−7.524★-1	1.191	1.831★-1	−2.526★-1	−6.630★-1
1.400	−5.454★-1	7.983★-1	4.826★-2	−1.420★-1	−2.076★-1
1.600	−4.277★-1	5.887★-1	−5.179★-3	−8.868★-2	2.670★-2
1.800	−3.543★-1	4.729★-1	−2.7684★-2	−6.090★-2	1.720★-1
2.000	−3.053★-1	4.148★-1	−3.826★-2	−4.597★-2	2.879★-1
2.200	−2.714★-1	4.028★-1	−4.532★-2	−3.850★-2	4.140★-1
2.400	−2.485★-1	4.457★-1	−5.409★-2	−3.668★-2	6.012★-1
2.600	−2.364★-1	5.971★-1	−7.194★-2	−4.198★-2	9.740★-1
2.800	−2.463★-1	1.118	−1.251★-1	−6.640★-2	2.065
3.000	−4.1774★-1	5.054	−4.953★-1	−2.502★-1	9.974
3.200	−1.135	2.406★1	−1.954	−9.785★-1	4.801★1
3.400	−1.815★-1	1.035	−6.611★-2	−3.419★-2	1.983
3.600	−1.467★-1	2.873★-1	−1.370★-2	−7.636★-3	5.016★-1
3.800	−1.344★-1	1.256★-1	−4.234★-3	−2.676★-3	1.893★-1
4.000	−1.263★-1	6.860★-2	−1.532★-3	−1.172★-3	8.369★-2

Magnetic

θ	$(21)_{22} = (21)_{00}$	$(21)_{21} = (21)_{01}$	$(21)_{12} = (21)_{10}$	$(21)_{20} = (21)_{02}$	$(21)_{11}$
2.000★-1	6.123★2	−1.225★3	−1.262★3	6.252★2	2.500★3
4.000★-1	3.569★1	−7.189★1	−8.117★1	3.923★1	1.563★2
6.000★-1	6.087	−1.267★1	−1.674★1	7.889	3.089★1
8.000★-1	1.424	−3.338	−5.577	2.615	9.788
1.000	2.687★-1	−1.020	−2.403	1.177	4.022

Table 9 The rectangular model. The aperture aberration coefficients $(0021)_{ij}$ as a function of $\theta = \beta L$.—cont'd

			Magnetic		
θ	$(21)_{22} = (21)_{00}$	$(21)_{21} = (21)_{01}$	$(21)_{12} = (21)_{10}$	$(21)_{20} = (21)_{02}$	$(21)_{11}$
1.200	−8.907⋆-2	−2.928⋆-1	−1.200	6.641⋆-1	1.951
1.400	−2.088⋆-1	−3.584⋆-2	−6.435⋆-1	4.509⋆-1	1.063
1.600	−2.424⋆-1	5.763⋆-2	−3.381⋆-1	3.578⋆-1	6.319⋆-1
1.800	−2.377⋆-1	8.821⋆.2	−1.368⋆-1	3.244⋆-1	4.020⋆-1
2.000	−2.081⋆-1	9.316⋆-2	3.088⋆-2	3.316⋆-1	2.697⋆-1
2.200	−1.498⋆-1	8.725⋆-2	2.178⋆-1	3.819⋆-1	1.878⋆-1
2.400	−3.672⋆-2	7.582⋆-2	5.021⋆-1	5.045⋆-1	1.312⋆-1
2.600	2.246⋆-1	5.776⋆-2	1.094	8.046⋆-1	8.022⋆-2
2.800	1.087	1.403⋆-2	2.950	1.778	−1.235⋆-2
3.000	8.216	−3.273⋆-1	1.770⋆1	9.386	−6.951⋆-1
3.200	5.326⋆1	−2.884	1.049⋆2	5.153⋆1	−5.807
3.400	2.693	−1.488⋆-1	5.331	2.533	−3.324⋆-1
3.600	7.658⋆-1	−3.874⋆-2	1.667	7.981⋆-1	−1.085⋆-1
3.800	2.926⋆-1	−1.232⋆-2	7.873⋆-1	3.955⋆-1	−5.203⋆-2
4.000	1.076⋆-1	−2.880⋆-3	4.452⋆-1	2.456⋆-1	−2.985⋆-2
			Achromatic		
θ	$(21)_{22} = (21)_{00}$	$(21)_{21} = (21)_{01}$	$(21)_{12} = (21)_{10}$	$(21)_{20} = (21)_{02}$	$(21)_{11}$
2.000⋆-1	9.247⋆2	−1.850⋆3	−1.887⋆3	9.377⋆2	3.750⋆3
4.000⋆-1	5.520⋆1	−1.109⋆2	−1.202⋆2	5.883⋆1	2.343⋆2
6.000⋆-1	9.923	−2.039⋆1	−2.445⋆1	1.181⋆1	4.623⋆1
8.000⋆-1	2.623	−5.783	−8.008	3.898	1.458⋆1
1.000	7.475⋆-1	−2.028	−3.388	1.739	5.931
1.200	1.321⋆-1	−7.874⋆-1	−1.661	9.697⋆-1	2.822
1.400	−9.662⋆-2	−3.139⋆-1	−8.740⋆-1	6.486⋆-1	1.487
1.600	−1.807⋆-1	−1.194⋆-1	−4.491⋆-1	5.066⋆-1	8.337⋆-1
1.800	−1.988⋆-1	−4.004⋆-2	−1.732⋆-1	4.528⋆-1	4.787⋆-1
2.000	−1.757⋆-1	−1.407⋆-2	5.393⋆-2	4.574⋆-1	2.636⋆-1

Continued

Table 9 The rectangular model. The aperture aberration coefficients $(0021)_{ij}$ as a function of $\theta = \beta L$.—cont'd

Achromatic

θ	$(21)_{22} = (21)_{00}$	$(21)_{21} = (21)_{01}$	$(21)_{12} = (21)_{10}$	$(21)_{20} = (21)_{02}$	$(21)_{11}$
2.200	−1.093⋆-1	−1.792⋆-2	3.056⋆-1	5.220⋆-1	1.124⋆-1
2.400	3.386⋆-2	−4.749⋆-2	6.875⋆-1	6.849⋆-1	−2.542⋆-2
2.600	3.783⋆-1	−1.220⋆-1	1.483	1.087	−2.177⋆-1
2.800	1.531	−3.539⋆-1	3.975	2.393	−7.048⋆-1
3.000	1.109⋆1	−2.121	2.376⋆1	1.260⋆1	−4.251
3.200	7.140⋆1	−1.186⋆1	1.405⋆2	6.904⋆1	−2.374⋆1
3.400	3.651	−5.436⋆-1	7.129	3.389	−1.104
3.600	1.070	−1.474⋆-1	2.228	1.067	−3.118⋆-1
3.800	4.349⋆-1	−5.828⋆-2	1.051	5.282⋆-1	−1.325⋆-1
4.000	1.856⋆-1	−2.671⋆-2	5.941⋆-1	3.278⋆-1	−6.770⋆-2

1.2 The short quadrupole: The bell-shaped model

The best analytical model is now Glaser's bell-shaped model

$$k(z) = \frac{1}{\left\{1 + (z/d)^2\right\}^2} \tag{26}$$

or in the asymmetric case, two half-bells:

$$\begin{aligned} k(z) &= \frac{1}{\left\{1 + (z/p)^2\right\}^2} \qquad z \leqslant 0 \\ k(z) &= \frac{1}{\left\{1 + (z/q)^2\right\}^2} \qquad z \geqslant 0 \end{aligned} \tag{27}$$

The asymmetric case is considered separately in §1.3. The excitation parameter β^2 is now given by

$$\beta^2 = \frac{V}{a^2\Phi_0(1 + \varepsilon\Phi_0)}\left(1 - \frac{a^2}{3d^2}\right) \tag{28a}$$

or

$$\beta^2 = \mu_0\eta\frac{NI}{a^2}\frac{1}{\left\{\Phi_0(1 + \varepsilon\Phi_0)\right\}^{1/2}}\left(1 - \frac{a^2}{3d^2}\right) \tag{28b}$$

Table 10 The rectangular model. The chromatic aberration coefficients A_x, A_y, B_x, B_y, C_x, C_y, D_x and D_y.

θ	A_x	B_x	C_x	D_x	A_y	B_y	C_y	D_y
0.2	1.250⋆1	−2.500⋆1	−4.967⋆-1	4.967⋆-1	−1.250⋆1	2.500⋆1	−5.034⋆-1	5.033⋆-1
0.4	3.127	−6.247	−4.871⋆-1	4.865⋆-1	−3.127	6.247	−5.137⋆-1	5.132⋆-1
0.6	1.393	−2.770	−4.720⋆-1	4.693⋆-1	−1.393	2.771	−5.320⋆-1	5.293⋆-1
0.9	7.893⋆-1	−1.548	−4.530⋆-1	4.442⋆-1	−7.876⋆-1	1.552	−5.596⋆-1	5.512⋆-1
1.0	5.136⋆-1	−9.757⋆-1	−4.322⋆-1	4.105⋆-1	−5.093⋆-1	9.841⋆-1	−5.985⋆-1	5.783⋆-1
1.2	3.688⋆-1	−6.556⋆-1	−4.125⋆-1	3.666⋆-1	−3.596⋆-1	6.734⋆-1	−6.514⋆-1	6.099⋆-1
1.4	2.882⋆-1	−4.499⋆-1	−3.977⋆-1	3.104⋆-1	−2.706⋆-1	4.841⋆-1	−7.215⋆-1	6.453⋆-1
1.6	2.456⋆-1	−2.980⋆-1	−3.929⋆-1	2.383⋆-1	−2.138⋆-1	3.599⋆-1	−8.127⋆-1	6.840⋆-1
1.8	2.312⋆-1	−1.655⋆-1	−4.053⋆-1	1.450⋆-1	−1.756⋆-1	2.739⋆-1	−9.298⋆-1	7.253⋆-1
2.0	2.452⋆-1	−2.320⋆-2	−4.458⋆-1	2.117⋆-2	−1.487⋆-1	2.119⋆-1	−1.078	7.687⋆-1
2.2	2.997⋆-1	1.690⋆-1	−5.331⋆-1	−1.503⋆-1	−1.290⋆-1	1.660⋆-1	−1.265	8.137⋆-1
2.4	4.342⋆-1	4.997⋆-1	−7.039⋆-1	−4.050⋆-1	−1.143⋆-1	1.311⋆-1	−1.499	8.600⋆-1
2.6	7.809⋆-1	1.239	−1.047	−8.305⋆-1	−1.028⋆-1	1.042⋆-1	−1.789	9.072⋆-1
2.8	1.977	3.665	−1.854	−1.719	−9.387⋆-2	8.329⋆-2	−2.149	9.552⋆-1
3.0	1.197⋆1	2.367⋆1	−5.067	−5.011	−8.624⋆-2	6.680⋆-2	−2.592	1.004
3.2	7.470⋆1	1.492⋆2	1.395⋆1	1.393⋆1	−8.005⋆-2	5.373⋆-2	−3.137	1.053
3.4	4.107	7.978	3.568	3.466	−7.481⋆-2	4.331⋆-2	−3.806	1.102
3.6	1.417	2.604	2.258	2.074	−7.030⋆-2	3.498⋆-2	−4.627	1.151
3.8	7.528⋆-1	1.271	1.750	1.478	−6.636⋆-2	2.829⋆-2	−5.633	1.201
4.0	4.905⋆-1	7.358⋆-1	1.485	1.114	−6.288⋆-2	2.291⋆-2	−6.864	1.251

where as usual, a denotes the bore-radius (Hardy, 1967) and for a short lens, in which $a \simeq d$,

$$1 - a^2/3d^2 = 0.67 \tag{29}$$

which agrees well with Deltrap's experimental result that this quantity lies between 0.6 and 0.7. ($\mu_0\eta = 0.3726.$).

The characteristic length is now the "quarter-width," d, and we write

$$\omega_x^2 = 1 + \beta^2 d^2 \quad \omega_y^2 = 1 - \beta^2 d^2. \tag{30}$$

The asymptotic cardinal elements are given by

$$\begin{aligned} \frac{f}{d} &= -\frac{\omega_x}{\sin \pi\omega_x} \quad \frac{z_{Fi}}{d} = -\frac{z_{Fo}}{d} = \omega_x \cot \pi\omega_x \\ \frac{g}{d} &= -\frac{\omega_y}{\sin \pi\omega_y} \quad \frac{z_{Gi}}{d} = -\frac{z_{Go}}{d} = \omega_y \cot \pi\omega_y. \end{aligned} \tag{31}$$

When $\beta^2 d^2 > 1$, the cardinal elements in the $y-z$ plane become

$$\frac{g}{d} = -\frac{\sigma_y}{\sinh \pi\sigma_y} \quad \frac{z_{Gi}}{d} = -\frac{z_{Go}}{d} = \sigma_y \coth \pi\sigma_y \tag{32}$$

in which

$$\sigma_y^2 = \beta^2 d^2 - 1. \tag{33}$$

The real cardinal elements are given by

$$\begin{aligned} \frac{f^*}{d} &= \operatorname{cosec}(\pi/\omega_x) \quad \frac{z^*_{Fi}}{d} = -\frac{z^*_{Fo}}{d} = -\cot(\pi/\omega_x) \\ \frac{g^*}{d} &= \operatorname{cosec}(\pi/\omega_y) \quad \frac{z^*_{Gi}}{d} = -\frac{z^*_{Go}}{d} = -\cot(\pi/\omega_y). \end{aligned} \tag{34}$$

When $\beta^2 d^2$ is small and hence ω_x^2 and ω_y^2 are close to unity, we have

$$\frac{f}{d} \simeq \frac{2}{\pi\beta^2 d^2} = \frac{d}{L} \cdot \frac{1}{\beta^2 d^2}; \quad \frac{g}{d} \simeq -\frac{2}{\pi\beta^2 d^2} = -\frac{d}{L} \cdot \frac{1}{\beta^2 d^2} \tag{35}$$

so that

$$\frac{f}{L} = \frac{1}{\beta^2 L^2}; \quad \frac{g}{L} = -\frac{1}{\beta^2 L^2} \tag{36}$$

in which L is the effective length, $L = \pi d/2$; in the thin or weak lens approximation, therefore, we obtain the same formula as we do with the rectangular model, as we should expect. The real cardinal elements tend to the same limits.

The focal lengths and the positions of the foci are tabulated in Table 11 and plotted in Fig. 3; the table gives the values of the various quantities scaled with respect to d and with respect to the effective length L. [For any parameter X, $X/L = (2/\pi)X/d$.].

The asymptotic aberration coefficients may as usual be written in the form of polynomials in the object distances or the magnifications, so that

$$(0030)/d = \sum_{i=0}^{4} \langle 0030\rangle_i \zeta_x^i = \sum_{i=0}^{4} (0030)_i m_x^i$$

and

$$(0012)/d = \sum_{i,j=0}^{2} \langle 0012\rangle_{ij} \zeta_x^i \zeta_y^j = \sum_{i,j=0}^{2} (0012)_{ij} m_x^i m_y^j,$$

with corresponding expressions for the other coefficients; ζ_x and ζ_y now denote z_{ox}/d and z_{oy}/d respectively. The coefficients $\langle 00rs\rangle_i$ and $\langle 00rs\rangle_{ij}$ are listed in Hawkes (1967c, 1967d), Hawkes and Hardy (1967) and Hawkes and Cham (1967); the expressions are not reproduced here. We have not derived the formulae for the coefficients $(00rs)_i$ and $(00rs)_{ij}$ analytically; instead, we have modified the computer program giving numerical values of the coefficients $\langle 00rs\rangle_i$ and $\langle 00rs\rangle_{ij}$ so that it performs the matrix products that transform these into the coefficients $(00rs)_i$ and $(00rs)_{ij}$. Writing (rs) for $(00rs)$, these matrices are as follows:

$$\begin{pmatrix} (30)_4 \\ (30)_3 \\ (30)_2 \\ (30)_1 \\ (30)_0 \end{pmatrix} = \begin{pmatrix} f^4 & 0 & 0 & 0 & 0 \\ 4f^3\zeta_{Fo} & f^3 & 0 & 0 & 0 \\ 6f^2\zeta_{Fo}^2 & 3f^2\zeta_{Fo} & f^2 & 0 & 0 \\ 4f\zeta_{Fo}^3 & 3f\zeta_{Fo}^2 & 2f\zeta_{Fo} & f & 0 \\ \zeta_{Fo}^4 & \zeta_{Fo}^3 & \zeta_{Fo}^2 & \zeta_{Fo} & 1 \end{pmatrix} \begin{pmatrix} \langle 30\rangle_4 \\ \langle 30\rangle_3 \\ \langle 30\rangle_2 \\ \langle 30\rangle_1 \\ \langle 30\rangle_0 \end{pmatrix} \tag{37}$$

and

Table 11 The symmetric bell-shaped model. Cardinal elements as a function of βd. $\theta = \beta L$ where $L = \pi d/2$; the focal lengths and focal distances are scaled with respect to both d and L.

βd	θ	f/d	g/d	z_{Fi}/d	z_{Gi}/d	f/L	g/L	z_{Fi}/L	z_{Gi}/L
1.000★-1	1.571★-1	6.414★1	−6.319★1	6.413★1	−6.318★1	4.083★1	−4.023★1	4.083★1	−4.022★1
2.000★-1	3.142★-1	1.640★1	−1.545★1	1.637★1	−1.542★1	1.044★1	−9.834	1.042★1	−9.814
3.000★-1	4.712★-1	7.572	−6.615	7.499	−6.546	4.820	−4.211	4.774	−4.167
4.000★-1	6.283★-1	4.494	−3.535	4.363	−3.414	2.861	−2.250	2.778	−2.173
5.000★-1	7.854★-1	3.085	−2.120	2.876	−1.935	1.964	−1.349	1.831	−1.232
6.000★-1	9.425★-1	2.338	−1.361	2.027	−1.101	1.489	−8.665★-1	1.290	−7.010★-1
7.000★-1	1.100	1.910	−9.131★-1	1.469	−5.690★-1	1.216	−5.813★-1	9.354★-1	−3.622★-1
8.000★-1	1.257	1.659	−6.309★-1	1.055	−1.950★-1	1.056	−4.016★-1	6.718★-1	−1.241★-1
9.000★-1	1.414	1.521	−4.449★-1	7.104★-1	8.900★-2	9.685★-1	−2.832★-1	4.522★-1	5.666★-2
1.000	1.571	1.467	−3.183★-1	3.906★-1	3.183★-1	9.340★-1	−2.026★-1	2.487★-1	2.026★-1
1.100	1.728	1.488	−2.302★-1	6.259★-2	5.128★-1	9.472★-1	−1.465★-1	3.984★-2	3.265★-1
1.200	1.885	1.592	−1.677★-1	−3.084★-1	6.842★-1	1.014	−1.068★-1	−1.963★-1	4.356★-1
1.300	2.042	1.813	−1.229★-1	−7.726★-1	8.397★-1	1.154	−7.823★-2	−4.918★-1	5.346★-1
1.400	2.199	2.236	−9.042★-2	−1.428	9.840★-1	1.423	−5.756★-2	−9.088★-1	6.264★-1
1.500	2.356	3.104	−6.675★-2	−2.527	1.120	1.976	−4.249★-2	−1.609	7.130★-1
1.600	2.513	5.419	−4.939★-2	−5.080	1.250	3.450	−3.144★-2	−3.234	7.958★-1
1.700	2.670	2.270★1	−3.661★-2	−2.261★1	1.375	1.445★1	−2.331★-2	−1.440★1	8.755★-1
1.800	2.827	−1.115★1	−2.718★-2	1.096★1	1.497	−7.098	−1.730★-2	6.976	9.530★-1
1.900	2.985	−4.816	−2.019★-2	4.311	1.616	−3.066	−1.285★-2	2.744	1.029

2.000	3.142	−3.310	−1.501⋆-2	2.441	1.732	−2.107	−9.557⋆-3	1.554	1.103
2.100	3.299	−2.723	−1.117⋆-2	1.416	1.847	−1.734	−7.109⋆-3	9.014⋆-1	1.176
2.200	3.456	−2.502	−8.310⋆-3	6.480⋆-1	1.960	−1.593	−5.290⋆-3	4.125⋆-1	1.248
2.300	3.613	−2.509	−6.185⋆-3	−6.295⋆-2	2.071	−1.597	−3.937⋆-3	−4.007⋆-2	1.319
2.400	3.770	−2.734	−4.604⋆-3	−8.448⋆-1	2.182	−1.740	−2.931⋆-3	−5.378⋆-1	1.389
2.500	3.927	−3.274	−3.427⋆-3	−1.862	2.291	−2.084	−2.182⋆-3	−1.185	1.459
2.600	4.084	−4.467	−2.551⋆-3	−3.492	2.400	−2.844	−1.624⋆-3	−2.223	1.528
2.700	4.241	−7.774	−1.899⋆-3	−7.221	2.508	−4.949	−1.209⋆-3	−4.597	1.597
2.800	4.398	−3.537⋆1	−1.413⋆-3	−3.525⋆1	2.615	−2.252⋆1	−8.998⋆-4	−2.244⋆-1	1.665
2.900	4.555	1.456⋆1	−1.052⋆-3	1.423⋆1	2.722	9.269	−6.696⋆-4	9.061	1.733
3.000	4.712	6.480	−7.826⋆-4	5.656	2.828	4.125	−4.982⋆-4	3.601	1.801
3.100	4.869	4.504	−5.822⋆-4	3.111	2.934	2.868	−3.706⋆-4	1.981	1.868
3.200	5.027	3.747	−4.330⋆-4	1.674	3.040	2.386	−2.757⋆-4	1.066	1.935
3.300	5.184	3.494	−3.220⋆-4	5.663⋆-1	3.145	2.225	−2.050⋆-4	3.605⋆-1	2.002
3.400	5.341	3.578	−2.394⋆-4	−4.931⋆-1	3.250	2.278	−1.524⋆-4	−3.139⋆-1	2.069
3.500	5.498	4.023	−1.780⋆-4	−1.714	3.354	2.561	−1.133⋆-4	−1.091	2.135
3.600	5.655	5.071	−1.323⋆-4	−3.428	3.458	3.228	−8.420⋆-5	−2.182	2.202
3.700	5.812	7.641	−9.827⋆-5	−6.611	3.562	4.865	−6.256⋆-5	−4.208	2.263
3.800	5.969	1.786⋆1	−7.300⋆-5	−1.742⋆1	3.666	1.137⋆1	−4.648⋆-5	−1.109⋆1	2.334
3.900	6.126	−4.904⋆1	−5.422⋆-5	4.887⋆1	3.770	−3.122⋆1	−3.452⋆-5	3.111⋆1	2.400
4.000	6.283	−1.093⋆1	−4.026⋆-5	1.012⋆1	3.873	−6.959	−2.563⋆-5	6.445	2.466

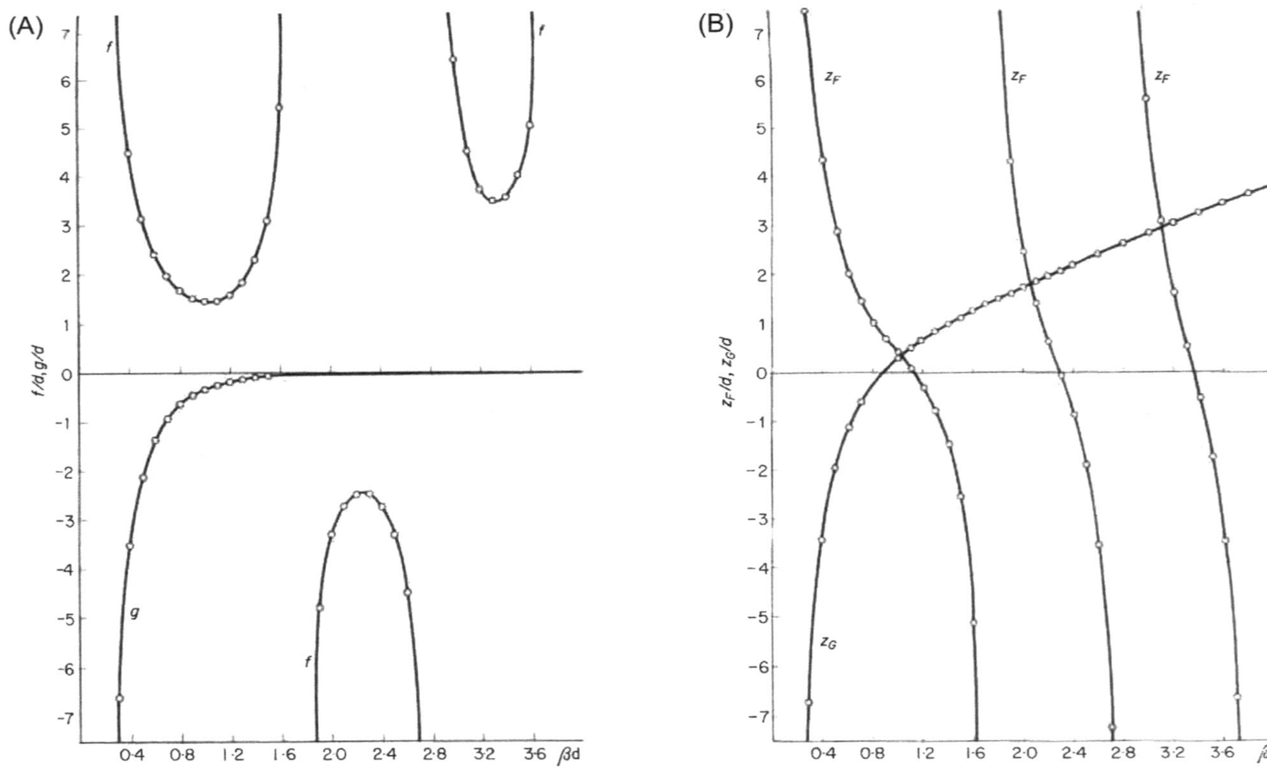

Fig. 3 The cardinal elements of a quadrupoles lens: bell-shaped model.

$$
\begin{pmatrix} (12)_{22} \\ (12)_{21} \\ (12)_{12} \\ (12)_{20} \\ (12)_{11} \\ (12)_{02} \\ (12)_{10} \\ (12)_{01} \\ (12)_{00} \end{pmatrix} =
$$

$$
\begin{pmatrix}
f^2g^2 & 0 & 0 & 0 & 0 & 0 & 0 & 0 & 0 \\
2f^2g\zeta_{Go} & f^2g & 0 & 0 & 0 & 0 & 0 & 0 & 0 \\
2fg^2\zeta_{Fo} & 0 & fg^2 & 0 & 0 & 0 & 0 & 0 & 0 \\
f^2\zeta_{Go}^2 & f^2\zeta_{Go} & 0 & f^2 & 0 & 0 & 0 & 0 & 0 \\
4fg\zeta_{Fo}\zeta_{Go} & 2fg\zeta_{Fo} & 2fg\zeta_{Go} & 0 & fg & 0 & 0 & 0 & 0 \\
g^2\zeta_{Fo}^2 & 0 & g^2\zeta_{Fo} & 0 & 0 & g^2 & 0 & 0 & 0 \\
2f\zeta_{Fo}\zeta_{Go}^2 & 2f\zeta_{Fo}\zeta_{Go} & f\zeta_{Go}^2 & 2f\zeta_{Fo} & f\zeta_{Go} & 0 & f & 0 & 0 \\
2g\zeta_{Fo}^2\zeta_{Go} & g\zeta_{Fo}^2 & 2g\zeta_{Fo}\zeta_{Go} & 0 & g\zeta_{Fo} & 2g\zeta_{Go} & 0 & g & 0 \\
\zeta_{Fo}^2\zeta_{Go}^2 & \zeta_{Fo}^2\zeta_{Go} & \zeta_{Fo}\zeta_{Go}^2 & \zeta_{Fo}^2 & \zeta_{Fo}\zeta_{Go} & \zeta_{Go}^2 & \zeta_{Fo} & \zeta_{Go} & 1
\end{pmatrix}
\begin{pmatrix} <12>_{22} \\ <12>_{21} \\ <12>_{12} \\ <12>_{20} \\ <12>_{11} \\ <12>_{02} \\ <12>_{10} \\ <12>_{01} \\ <12>_{00} \end{pmatrix} \tag{38}
$$

The coefficients are tabulated in Tables 12–19.

The asymptotic chromatic aberration coefficients are given by

$$
\begin{aligned}
\frac{C_{cx}}{d} &= -\frac{(n-1)}{8}\beta^2 d^2 \left\{ \frac{\sin 2\pi\omega_x - 2\pi\omega_x}{\omega_x^3}\zeta_x^2 + 2\frac{\cos 2\pi\omega_x - 1}{\omega_x^2}\zeta_x - \frac{\sin 2\pi\omega_x + 2\pi\omega_x}{\omega_x} \right\} \\
\frac{C_{cy}}{d} &= \frac{n-1}{8}\beta^2 d^2 \left\{ \frac{\sin 2\pi\omega_y - 2\pi\omega_y}{\omega_y^3}\zeta_y^2 + 2\frac{\cos 2\pi\omega_y - 1}{\omega_y^2}\zeta_y - \frac{\sin 2\pi\omega_y + 2\pi\omega_y}{\omega_y} \right\} \\
C_{Mx} &= -\frac{n-1}{8}\frac{\beta^2 d^2}{\omega_x^2}\left(-\frac{\sin 2\pi\omega_x - 2\pi\omega_x}{\omega_x}\zeta_x + 1 - \cos 2\pi\omega_x \right) \\
C_{My} &= \frac{n-1}{8}\frac{\beta^2 d^2}{\omega_y^2}\left(-\frac{\sin 2\pi\omega_y - 2\pi\omega_y}{\omega_y}\zeta_y + 1 - \cos 2\pi\omega_y \right)
\end{aligned} \tag{39}
$$

and their real counterparts by

Table 12 The bell-shaped model. The aperture aberration coefficients $\langle 0030 \rangle_r$ as a function of $\beta^2 d^2$.

Electrostatic

$\beta^2 d^2$	$\langle 30 \rangle_4$	$\langle 30 \rangle_3$	$\langle 30 \rangle_2$	$\langle 30 \rangle_1$	$\langle 30 \rangle_0$
2.000★-1	3.209★-2	2.203★-2	1.585★-1	5.112★-2	1.836★-1
4.000★-1	9.202★-2	4.805★-2	3.741★-1	1.726★-1	3.990★-1
6.000★-1	1.554★-1	7.867★-2	6.568★-1	2.842★-1	6.341★-1
8.000★-1	2.172★-1	1.150★-1	9.545★-1	3.364★-1	9.017★-1
1.000	2.755★-1	1.415★-1	1.225	3.364★-1	1.223
1.200	3.276★-1	1.437★-1	1.468	3.185★-1	1.610
1.400	3.712★-1	1.199★-1	1.709	3.089★-1	2.066
1.600	4.053★-1	8.056★-2	1.975	3.097★-1	2.585
1.800	4.309★-1	4.019★-2	2.277	3.056★-1	3.167
2.000	4.503★-1	9.952★-3	2.606	2.792★-1	3.814
2.200	4.657★-1	−5.912★-3	2.941	2.243★-1	4.534
2.400	4.792★-1	−9.466★-3	3.265	1.496★-1	5.337
2.600	4.918★-1	−6.202★-3	3.568	7.439★-2	6.226
2.800	5.039★-1	−1.817★-3	3.851	1.967★-2	7.199
3.200	5.262★-1	−1.688★-3	4.399	1.884★-2	9.358
3.400	5.359★-1	−5.553★-3	4.687	6.862★-2	1.052★1
3.600	5.448★-1	−9.154★-3	4.992	1.344★-1	1.173★1
3.800	5.529★-1	−1.017★-2	5.314	1.992★-1	1.298★1
4.000	5.606★-1	−7.259★-3	5.646	2.492★-1	1.427★1
4.200	5.680★-1	−4.530★-4	5.981	2.769★-1	1.562★1
4.400	5.755★-1	9.021★-3	6.309	2.818★-1	1.702★1
4.600	5.830★-1	1.920★-2	6.627	2.698★-1	1.848★1
4.800	5.906★-1	2.799★-2	6.932	2.497★-1	2.001★1
5.000	5.980★-1	3.371★-2	7.226	2.311★-1	2.161★1

Table 12 The bell-shaped model. The aperture aberration coefficients $\langle 0030 \rangle_{\Gamma}$ as a function of $\beta^2 d^2$.—cont'd

Magnetic

$\beta^2 d^2$	$\langle 30 \rangle_4$	$\langle 30 \rangle_3$	$\langle 30 \rangle_2$	$\langle 30 \rangle_1$	$\langle 30 \rangle_0$
2.000⋆-1	1.108⋆-2	2.684⋆-2	1.252⋆-1	5.747⋆-2	1.748⋆-1
4.000⋆-1	3.568⋆-2	7.170⋆-2	2.607⋆-1	2.100⋆-1	3.529⋆-1
6.000⋆-1	6.715⋆-2	1.294⋆-1	4.364⋆-1	3.739⋆-1	5.064⋆-1
8.000⋆-1	1.039⋆-1	1.921⋆-1	6.085⋆-1	4.837⋆-1	6.387⋆-1
1.000	1.430⋆-1	2.373⋆-1	7.380⋆-1	5.319⋆-1	7.686⋆-1
1.200	1.797⋆-1	2.467⋆-1	8.263⋆-1	5.452⋆-1	9.121⋆-1
1.400	2.093⋆-1	2.184⋆-1	9.044⋆-1	5.471⋆-1	1.076
1.600	2.296⋆-1	1.656⋆-1	1.004	5.400⋆-1	1.264
1.800	2.410⋆-1	1.069⋆-1	1.137	5.108⋆-1	1.479
2.000	2.455⋆-1	5.730⋆-2	1.296	4.462⋆-1	1.731
2.200	2.457⋆-1	2.410⋆-2	1.459	3.455⋆-1	2.033
2.400	2.441⋆-1	6.877⋆-3	1.607	2.245⋆-1	2.394
2.600	2.421⋆-1	7.262⋆-4	1.730	1.099⋆-1	2.818
2.800	2.405⋆-1	−1.790⋆-4	1.829	2.886⋆-2	3.301
3.200	2.385⋆-1	−2.396⋆-4	2.002	2.792⋆-2	4.384
3.400	2.380⋆-1	−1.253⋆-4	2.102	1.034⋆-1	4.949
3.600	2.378⋆-1	2.240⋆-3	2.220	2.078⋆-1	5.511
3.800	2.381⋆-1	8.610⋆-3	2.357	3.192⋆-1	6.064
4.000	2.390⋆-1	1.968⋆-2	2.504	4.187⋆-1	6.607
4.200	2.408⋆-1	3.473⋆-2	2.653	4.942⋆-1	7.145
4.400	2.434⋆-1	5.183⋆-2	2.794	5.420⋆-1	7.687
4.600	2.468⋆-1	6.836⋆-2	2.923	5.654⋆-1	8.240
4.800	2.508⋆-1	8.170⋆-2	3.037	5.723⋆-1	8.812
5.000	2.552⋆-1	8.983⋆-2	3.139	5.715⋆-1	9.405

Continued

Table 12 The bell-shaped model. The aperture aberration coefficients $\langle 0030 \rangle_r$ as a function of $\beta^2 d^2$.—cont'd

Achromatic

$\beta^2 d^2$	$\langle 30 \rangle_4$	$\langle 30 \rangle_3$	$\langle 30 \rangle_2$	$\langle 30 \rangle_1$	$\langle 30 \rangle_0$
2.000⋆-1	1.528⋆-2	2.587⋆-2	1.319⋆-1	5.620⋆-2	1.766⋆-1
4.000⋆-1	4.695⋆-2	6.697⋆-2	2.834⋆-1	2.025⋆-1	3.622⋆-1
6.000⋆-1	8.481⋆-2	1.192⋆-1	4.804⋆-1	3.560⋆-1	5.320⋆-1
8.000⋆-1	1.265⋆-1	1.767⋆-1	6.777⋆-1	4.542⋆-1	6.913⋆-1
1.000	1.695⋆-1	2.182⋆-1	8.355⋆-1	4.928⋆-1	8.594⋆-1
1.200	2.092⋆-1	2.261⋆-1	9.547⋆-1	4.999⋆-1	1.052
1.400	2.417⋆-1	1.987⋆-1	1.065	4.995⋆-1	1.274
1.600	2.648⋆-1	1.486⋆-1	1.198	4.940⋆-1	1.528
1.800	2.790⋆-1	9.354⋆-1	1.365	4.698⋆-1	1.817
2.000	2.864⋆-1	4.783⋆-2	1.558	4.128⋆-1	2.148
2.200	2.897⋆-1	1.810⋆-2	1.755	3.212⋆-1	2.533
2.400	2.911⋆-1	3.608⋆-3	1.939	2.095⋆-1	2.982
2.600	2.921⋆-1	−6.594⋆-4	2.097	1.028⋆-1	3.500
2.800	2.932⋆-1	−5.066⋆-4	2.233	2.702⋆-2	4.081
3.200	2.960⋆-1	−5.293⋆-4	2.481	2.611⋆-2	5.379
3.400	2.976⋆-1	−1.211⋆-3	2.619	9.647⋆-2	6.064
3.600	2.992⋆-1	−3.881⋆-5	2.775	1.931⋆-1	6.755
3.800	3.011⋆-1	4.855⋆-3	2.948	2.952⋆-1	7.447
4.000	3.033⋆-1	1.429⋆-2	3.132	3.848⋆-1	8.140
4.200	3.062⋆-1	2.769⋆-2	3.318	4.548⋆-1	8.840
4.400	3.098⋆-1	4.327⋆-2	3.497	4.900⋆-1	9.553
4.600	3.140⋆-1	5.852⋆-2	3.664	5.063⋆-1	1.029⋆1
4.800	3.188⋆-1	7.096⋆-2	3.816	5.078⋆-1	1.105⋆1
5.000	3.238⋆-1	7.861⋆-2	3.957	5.304⋆-1	1.185⋆1

Table 13 The bell-shaped model. The aperture aberration coefficients $\langle 0003\rangle_i$ as a function of $\beta^2 d^2$.

Electrostatic

$\beta^2 d^2$	$\langle 03\rangle_4$	$\langle 03\rangle_3$	$\langle 03\rangle_2$	$\langle 03\rangle_1$	$\langle 03\rangle_0$
2.000⋆-1	6.374⋆-2	1.105⋆-1	−1.748⋆-1	3.528⋆-2	−1.198⋆-1
4.000⋆-1	3.154⋆-1	8.837⋆-1	−1.725⋆-1	8.082⋆-2	−1.628⋆-1
6.000⋆-1	5.948⋆-1	3.288	1.269	3.582⋆-1	−1.387⋆-1
8.000⋆-1	−6.852⋆-1	6.620	6.360	2.243	1.061⋆-1
1.000	−8.145⋆10	4.072⋆10	1.600⋆1	6.889	1.047
1.200	−5.032⋆1	−8.456⋆1	−4.515⋆1	−5.910	7.624⋆-1
1.400	−1.838⋆2	−4.574⋆2	−4.189⋆2	−1.659⋆2	−2.381⋆1
1.600	−5.735⋆2	−1.767⋆3	−2.034⋆3	−1.036⋆3	−1.967⋆2
1.800	−1.611⋆3	−5.754⋆3	−7.695⋆3	−4.569⋆3	−1.016⋆3
2.000	−4.191⋆3	−1.675⋆4	−2.510⋆4	−1.670⋆4	−4.168⋆3
2.200	−1.026⋆4	−4.493⋆4	−7.379⋆4	−5.386⋆4	−1.474⋆4
2.400	−2.389⋆4	−1.130⋆5	−2.006⋆5	−1.582⋆5	−4.679⋆4
2.600	−5.337⋆4	−2.700⋆5	−5.123⋆5	−4.320⋆5	−1.366⋆5
2.800	−1.151⋆5	−6.176⋆5	−1.243⋆6	−1.112⋆6	−3.728⋆5
3.200	−4.892⋆5	−2.902⋆6	−6.457⋆6	−6.385⋆6	−2.368⋆6
3.400	−9.708⋆5	−6.016⋆6	−1.398⋆7	−1.444⋆7	−5.592⋆6
3.600	−1.885⋆6	−1.216⋆7	−2.940⋆7	−3.161⋆7	−1.274⋆7
3.800	−3.587⋆6	−2.401⋆7	−6.027⋆7	−6.723⋆7	−2.813⋆7
4.000	−6.705⋆6	−4.646⋆7	−1.207⋆8	−1.394⋆8	−6.035⋆7
4.200	−1.233⋆7	−8.820⋆7	−2.367⋆8	−2.822⋆8	−1.262⋆8
4.400	−2.231⋆7	−1.646⋆8	−4.552⋆8	−5.595⋆8	−2.579⋆8
4.600	−3.982⋆7	−3.022⋆8	−8.601⋆8	−1.088⋆9	−5.161⋆8
4.800	−7.014⋆7	−5.469⋆8	−1.599⋆9	−2.078⋆9	−1.013⋆9
5.000	−1.220⋆8	−9.762⋆8	−2.929⋆9	−3.905⋆9	−1.952⋆9

Continued

Table 13 The bell-shaped model. The aperture aberration coefficients $\langle 0003 \rangle_i$ as a function of $\beta^2 d^2$.—cont'd

Magnetic

$\beta^2 d^2$	$\langle 03 \rangle_4$	$\langle 03 \rangle_3$	$\langle 03 \rangle_2$	$\langle 03 \rangle_1$	$\langle 03 \rangle_0$
2.000★-1	1.025★-2	9.560★-2	−2.220★-1	2.513★-2	−1.254★-1
4.000★-1	−5.200★-2	6.556★-1	−4.285★-1	−1.732★-2	−1.888★-1
6.000★-1	−8.895★-1	1.767	2.006★-1	−7.927★-2	−2.300★-1
8.000★-1	−5.588	−6.031★-1	1.497	4.449★-1	−2.114★-1
1.000	−1.323★11	−1.018★10	−1.067★1	−1.722	−2.618★-1
1.200	−9.049★1	−1.813★2	−1.345★2	−4.360★1	−5.470
1.400	−2.887★2	−7.576★2	−7.434★2	−3.233★2	−5.285★1
1.600	−8.349★2	−2.627.3	−3.098★3	−1.623★3	−3.187★2
1.800	−2.236★3	−8.062★3	−1.090★4	−6.544★3	−1.474★3
2.000	−5.630★3	−2.261★4	−3.405★4	−2.279★4	−5.720★3
2.200	−1.346★4	−5.913★4	−9.738★4	−7.127★4	−1.956★4
2.400	−3.082★4	−1.461★5	−2.596★5	−2.050★5	−6.073★4
2.600	−6.796★4	−3.441★5	−6.534★5	−5.515★5	−1.745★5
2.800	−1.450★5	−7.785★5	−1.567★6	−1.403★6	−4.707★5
3.200	−6.068★5	−3.601★6	−8.013★6	−7.925★6	−2.939★6
3.400	−1.197★6	−7.420★6	−1.724★7	−1.781★7	−6.900★6
3.600	−2.313★6	−1.492★7	−3.609★7	−3.880★7	−1.564★7
3.800	−4.383★6	−2.934★7	−7.365★7	−8.216★7	−3.437★7
4.000	−8.162★6	−5.655★7	−1.469★8	−1.697★8	−7.347★7
4.200	−1.496★7	−1.070★8	−2.872★8	−3.425★8	−1.532★8
4.400	−2.699★7	−1.991★8	−5.507★8	−6.770★8	−3.121★8
4.600	−4.806★7	−3.647★8	−1.038★9	−1.313★9	−6.228★8
4.800	−8.445★7	−6.585★8	−1.925★9	−2.502★9	−1.220★9
5.000	−1.466★8	−1.173★9	−3.519★9	−4.692★9	−2.346.9

Table 13 The bell-shaped model. The aperture aberration coefficients $\langle 0003 \rangle_i$ as a function of $\beta^2 d^2$.—cont'd

Achromatic

$\beta^2 d^2$	$\langle 03 \rangle_4$	$\langle 03 \rangle_3$	$\langle 03 \rangle_2$	$\langle 03 \rangle_1$	$\langle 03 \rangle_0$
2.000⋆-1	2.095⋆-2	9.858⋆-2	−2.126⋆-1	2.176⋆-2	−1.243⋆-1
4.000⋆-1	2.148⋆-2	7.012⋆-1	−3.773⋆-1	2.309⋆-3	−1.836⋆-1
6.000⋆-1	−5.926⋆-1	2.071	4.143⋆-1	8.223⋆-3	−2.117⋆-1
8.000⋆-1	−4.607	8.415⋆-1	2.470	8.045⋆-1	−1.479⋆-1
1.000	−1.222⋆11	−1.074⋆1	−5.333	1.294⋆-10	4.851⋆-12
1.200	−8.246⋆1	−1.620⋆2	−1.166⋆2	−3.607⋆1	−4.223
1.400	−2.677⋆2	−6.975⋆2	−6.785⋆2	−2.918⋆2	−4.704⋆1
1.600	−7.826⋆2	−2.455⋆3	−2.885⋆3	−1.505⋆3	−2.943⋆2
1.800	−2.111⋆3	−7.600⋆3	−1.026⋆4	−6.149⋆3	−1.382⋆3
2.000	−5.342⋆3	−2.144⋆4	−3.226⋆4	−2.157⋆4	−5.409⋆3
2.200	−1.282⋆4	−5.629⋆4	−9.266⋆4	−6.779⋆4	−1.860⋆4
2.400	−2.944⋆4	−1.395⋆5	−2.478⋆5	−1.957⋆5	−5.764⋆4
2.600	−6.504.4	−3.293⋆5	−6.252⋆5	−5.276⋆5	−1.669⋆5
2.800	−1.390⋆5	−7.463⋆5	−1.502⋆6	−1.344⋆6	−4.511⋆5
3.200	−5.833⋆5	−3.461⋆6	−7.702⋆6	−7.617⋆6	−2.825⋆6
3.400	−1.152⋆6	−7.139⋆6	−1.659⋆7	−1.714⋆7	−6.638⋆6
3.600	−2.227⋆6	−1.437⋆7	−3.475⋆7	−3.736⋆7	−1.506⋆7
3.800	−4.224⋆6	−2.827⋆7	−7.097⋆7	−7.918⋆7	−3.312⋆7
4.000	−7.871⋆6	−5.453⋆7	−1.417⋆8	−1.636⋆8	−7.085⋆7
4.200	−1.443⋆7	−1.033⋆8	−2.771⋆8	−3.304⋆8	−1.478⋆8
4.400	−2.606⋆7	−1.922⋆8	−5.316⋆8	−6.535⋆8	−3.013⋆8
4.600	−4.641⋆7	−3.522⋆8	−1.002⋆9	−1.268⋆9	−6.015⋆8
4.800	−8.159⋆7	−6.362⋆8	−1.860⋆9	−2.418⋆9	−1.178⋆9
5.000	−1.417⋆8	−1.134⋆9	−3.401⋆9	−4.535⋆9	−2.267⋆9

Table 14 The bell-shaped model. The aperture aberration coefficients $\langle 0012 \rangle_{ij}$ as a function of $\beta^2 d^2$.

Electrostatic

$\beta^2 d^2$	$\langle 12 \rangle_{22}$	$\langle 12 \rangle_{21}$	$\langle 12 \rangle_{12}$	$\langle 12 \rangle_{20}$	$\langle 12 \rangle_{11}$	$\langle 12 \rangle_{02}$	$\langle 12 \rangle_{10}$	$\langle 12 \rangle_{01}$	$\langle 12 \rangle_{00}$
2.000⋆-1	−1.030⋆-2	−1.023⋆-1	−4.225⋆-2	1.920⋆-1	−1.429⋆-1	−2.813⋆-1	−1.262⋆-1	−9.777⋆-3	−1.212⋆-1
4.000⋆-1	−6.023⋆-2	−2.590⋆-1	−4.091⋆-1	1.874⋆-1	−2.447⋆-2	−8.661⋆-1	−2.625⋆-1	−3.425⋆-1	−1.948⋆-1
6.000⋆-1	−3.118⋆-1	−4.461⋆-1	−1.480	9.106⋆-2	−3.060⋆-1	−1.646	−2.966⋆-1	−1.378	−4.189⋆-1
8.000⋆-1	−1.075	−1.054	−3.349	−1.004⋆-1	−1.919	−2.182	−5.315⋆-1	−2.967	−1.007
1.000	−3.053	−3.089	−6.574	−6.493⋆-1	−5.042	−4.024	−1.394	−4.489	−1.843
1.200	−5.446	−6.924	−6.141	−2.092	−7.841	1.648⋆-1	−2.679	−3.104	−2.286
1.400	−9.235	−1.379⋆1	−2.962	−5.046	−5.481	3.590	−2.568	1.102	−1.476
1.600	−1.353⋆1	−2.316⋆1	7.228	−9.806	9.922	6.947	3.118	6.921	5.093⋆-1
1.800	−1.723⋆1	−3.309⋆1	2.724⋆1	−1.578⋆1	4.783⋆1	6.405	2.080⋆1	7.064	5.374⋆-1
2.000	−1.864⋆1	−3.961⋆1	5.823⋆1	−2.092⋆1	1.159⋆2	−4.735	5.749⋆1	−1.454⋆1	−1.044⋆1
2.200	−1.573⋆1	−3.686⋆1	9.807⋆1	−2.144⋆1	2.145⋆2	−3.607⋆1	1.172⋆2	−8.486⋆1	−5.032⋆1
2.400	−6.485	−1.777⋆1	1.396⋆2	−1.189⋆1	3.302⋆2	−9.926⋆1	1.951⋆2	−2.416⋆2	−1.475⋆2
2.600	1.055⋆1	2.422⋆1	1.691⋆2	1.382⋆1	4.278⋆2	−2.061⋆2	2.705⋆2	−5.289⋆2	−3.399⋆2
2.800	3.547⋆1	9.266⋆1	1.659⋆2	6.055⋆1	4.453⋆2	−3.651⋆2	2.986⋆2	−9.882⋆2	−6.693⋆2
3.200	9.778⋆1	2.875⋆2	−5.226⋆1	2.113⋆2	−1.550⋆2	−8.306⋆2	−1.150⋆2	−2.475⋆3	−1.844⋆3

3.400	1.214⋆2	3.736⋆2	−3.302⋆2	2.874⋆2	−1.023⋆3	−1.094⋆3	−7.926⋆2	−3.402⋆3	−2.645⋆3
3.600	1.244⋆2	3.986⋆2	−7.557⋆2	3.192⋆2	−2.437⋆3	−1.314⋆3	−1.965⋆3	−4.248⋆3	−3.436⋆3
3.800	9.017⋆1	2.990⋆2	−1.338⋆3	2.479⋆2	−4.476⋆3	−1.407⋆3	−3.745⋆3	−4.722⋆3	−3.963⋆3
4.000	−1.179	−6.954	−2.058⋆3	−8.430	−7.128⋆3	−1.267⋆3	−6.173⋆3	−4.403⋆3	−3.826⋆3
4.200	−1.710⋆2	−6.146⋆2	−2.860⋆3	−5.522⋆2	−1.023⋆4	−7.640⋆2	−9.152⋆3	−2.749⋆3	−2.473⋆3
4.400	−4.396⋆2	−1.624⋆3	−3.638⋆3	−1.500⋆3	−1.342⋆4	2.387⋆2	−1.237⋆4	8.638⋆2	7.805⋆2
4.600	−8.234⋆2	−3.128⋆3	−4.227⋆3	−2.970⋆3	−1.604⋆4	1.866⋆3	−1.522⋆4	7.065⋆3	6.685⋆3
4.800	−1.330⋆3	−5.188⋆3	−4.397⋆3	−5.060⋆3	−1.714⋆4	4.203⋆3	−1.671⋆4	1.637⋆4	1.593⋆4
5.000	−1.953⋆3	−7.817⋆3	−3.856⋆3	−7.820⋆3	−1.543⋆4	7.250⋆3	−1.543⋆4	2.898⋆4	2.896⋆4

Magnetic

$\beta^2 d^2$	$\langle 12\rangle_{22}$	$\langle 12\rangle_{21}$	$\langle 12\rangle_{12}$	$\langle 12\rangle_{20}$	$\langle 12\rangle_{11}$	$\langle 12\rangle_{02}$	$\langle 12\rangle_{10}$	$\langle 12\rangle_{01}$	$\langle 12\rangle_{00}$
2.000⋆-1	4.742⋆-2	−8.139⋆-2	−6.453⋆-2	2.012⋆-1	−1.052⋆-1	−2.594⋆-1	−1.192⋆-1	−1.499⋆-2	−1.109⋆-1
4.000⋆-1	1.580⋆-1	−1.086⋆-1	−5.761⋆-1	2.324⋆-1	2.621⋆-2	−6.980⋆-1	−2.258⋆-1	−3.415⋆-1	−1.650⋆-1
6.000⋆-1	1.433⋆-1	−3.517⋆-3	−1.964	2.294⋆-1	−4.807⋆-1	−9.846⋆-1	−2.569⋆-1	−1.145	−3.477⋆-1
8.000⋆-1	−3.154⋆-1	−1.406⋆-1	−4.214	2.187⋆-1	−2.656	−3.249⋆-1	−6.158⋆-1	−1.692	−7.177⋆-1
1.000	−1.381	−1.311	−8.574	−2.622⋆-2	−6.394	1.476	−1.742	−4.520⋆-2	−7.330⋆-1
1.200	−3.546	−4.070	−6.546	−9.333⋆-1	−9.001	8.316	−3.168	7.121	1.025
1.400	−6.011	−8.564	−1.487	−2.816	−4.272	1.767⋆1	−2.405	2.215⋆1	6.515
1.600	−7.883	−1.321⋆1	1.225⋆1	−5.302	1.717⋆1	2.916⋆1	5.726	4.502⋆1	1.697⋆1

Continued

Table 14 The bell-shaped model. The aperture aberration coefficients $\langle 0012 \rangle_{ij}$ as a function of $\beta^2 d^2$.—cont'd

Magnetic

$\beta^2 d^2$	$\langle 12 \rangle_{22}$	$\langle 12 \rangle_{21}$	$\langle 12 \rangle_{12}$	$\langle 12 \rangle_{20}$	$\langle 12 \rangle_{11}$	$\langle 12 \rangle_{02}$	$\langle 12 \rangle_{10}$	$\langle 12 \rangle_{01}$	$\langle 12 \rangle_{00}$
1.800	−7.383	−1.422⋆1	3.741⋆1	−6.605	6.562⋆1	3.913⋆1	2.859⋆1	6.976⋆1	3.071⋆1
2.000	−2.054	−5.101	7.433⋆1	−2.841	1.478⋆2	4.120⋆1	7.331⋆1	8.199⋆1	4.042⋆1
2.200	1.098⋆1	2.310⋆1	1.190⋆2	1.233⋆1	2.601⋆2	2.674⋆1	1.421⋆2	5.790⋆1	3.098⋆1
2.400	3.452⋆1	8.079⋆1	1.611⋆2	4.746⋆1	3.809⋆2	−1.355⋆1	2.251⋆2	−3.309⋆1	−2.054⋆1
2.600	7.063⋆1	1.778⋆2	1.828⋆2	1.121⋆2	4.625⋆2	−8.663⋆1	2.924⋆2	−2.206⋆2	−1.407⋆2
2.800	1.198⋆2	3.206⋆2	1.587⋆2	2.147⋆2	4.257⋆2	−1.926⋆2	2.855⋆2	−5.185⋆2	−3.494⋆2
3.200	2.458⋆2	7.284⋆2	−1.587⋆2	5.398⋆2	−4.708⋆2	−4.323⋆2	−3.492⋆2	−1.285⋆3	−9.554⋆2
3.400	3.083⋆2	9.544⋆2	−5.188⋆2	7.390⋆2	−1.607⋆3	−4.749⋆2	−1.245⋆3	−1.474⋆3	−1.145⋆3
3.600	3.544⋆2	1.142⋆3	−1.044⋆3	9.206⋆2	−3.366⋆3	−3.553⋆2	−2.714⋆3	−1.149⋆3	−9.296⋆2
3.800	3.680⋆2	1.231⋆3	−1.732⋆3	1.029⋆3	−5.795⋆3	5.298⋆1	−4.849⋆3	1.736⋆2	1.418⋆2
4.000	3.308⋆2	1.145⋆3	−2.544⋆3	9.914⋆2	−8.815⋆3	9.101⋆2	−7.634⋆3	3.149⋆3	2.723⋆3
4.200	2.254⋆2	8.056⋆2	−3.397⋆3	7.201⋆2	−1.215⋆4	2.403⋆3	−1.087⋆4	8.592⋆3	7.681⋆3
4.400	3.815⋆1	1.399⋆2	−4.144⋆3	1.285⋆2	−1.528⋆4	4.728⋆3	−1.409⋆4	1.743⋆4	1.607⋆4
4.600	−2.358⋆2	−8.956⋆2	−4.573⋆3	−8.502⋆2	−1.736⋆4	8.069⋆3	−1.647⋆4	3.062⋆4	2.904⋆4
4.800	−5.874⋆2	−2.291⋆3	−4.400⋆3	−2.234⋆3	−1.716⋆4	1.256⋆4	−1.672⋆4	4.897⋆4	4.772⋆4
5.000	−9.876⋆2	−3.951⋆3	−3.278⋆3	−3.952⋆3	−1.311⋆4	1.825⋆4	−1.311⋆4	7.299⋆4	7.299⋆4

Achromatic

$\beta^2 d^2$	$\langle 12\rangle_{22}$	$\langle 12\rangle_{21}$	$\langle 12\rangle_{12}$	$\langle 12\rangle_{20}$	$\langle 12\rangle_{11}$	$\langle 12\rangle_{02}$	$\langle 12\rangle_{10}$	$\langle 12\rangle_{01}$	$\langle 12\rangle_{00}$
2.000⋆-1	6.666⋆-2	−7.442⋆-2	−7.195⋆-2	2.043⋆-1	−9.262⋆-2	−2.521⋆-1	−1.169⋆-1	−1.673⋆-2	−1.075⋆-1
4.000⋆-1	2.308⋆-1	−5.847⋆-2	−6.317⋆-1	2.474⋆-1	4.310⋆-2	−6.420⋆-1	−2.136⋆-1	−3.412⋆-1	−1.551⋆-1
6.000⋆-1	2.950⋆-1	1.440⋆-1	−2.125	2.756⋆-1	−5.390⋆-1	−7.643⋆-1	−2.437⋆-1	−1.067	−3.240⋆-1
8.000⋆-1	−6.220⋆-2	1.639⋆-1	−4.503	3.250⋆-1	−2.902	2.940⋆-1	−6.439⋆-1	−1.268	−6.215⋆-1
1.000	−8.809⋆-1	−7.180⋆-1	−8.574	1.815⋆-1	−6.844	3.226	−1.858	1.436	−3.629⋆-1
1.200	−2.913	−3.119	−6.681	−5.471⋆-1	−9.387	1.103⋆1	−3.331	1.053⋆1	2.129
1.400	−4.937	−6.821	−9.957⋆-1	−2.073	−3.869	2.236⋆1	−2.351	2.917⋆1	9.178
1.600	−6.000	−9.897	1.393⋆1	−3.801	1.959⋆1	3.656⋆1	6.595	5.771⋆1	2.246⋆1
1.800	−4.103	−7.929	4.081⋆1	−3.547	7.155⋆1	5.004⋆1	3.118⋆1	9.066⋆1	4.077⋆1
2.000	3.474	6.402	7.969⋆1	3.187	1.584⋆2	5.651⋆1	7.858⋆1	1.142⋆2	5.737⋆1
2.200	1.988⋆1	4.308⋆1	1.259⋆2	2.359⋆1	2.753⋆2	4.768⋆1	1.584⋆2	1.055⋆2	5.808⋆1
2.400	4.819⋆1	1.136⋆2	1.682⋆2	6.725⋆1	3.978⋆2	1.501⋆1	2.351⋆2	3.640⋆1	2.178⋆1
2.600	9.066⋆1	2.290⋆2	1.874⋆2	1.449⋆2	4.740⋆2	−4.682⋆1	2.997⋆2	−1.178⋆2	−7.435⋆1
2.800	1.479⋆2	3.965⋆2	1.562⋆2	2.661⋆2	4.192⋆2	−1.350⋆2	2.812⋆2	−3.619⋆2	−2.428⋆2
3.200	2.951⋆2	8.753⋆2	−1.942⋆2	6.493⋆2	−5.761⋆2	−2.995⋆2	−4.273⋆2	−8.886⋆2	−6.593⋆2
3.400	3.706⋆2	1.148⋆3	−5.816⋆2	8.895⋆2	−1.802⋆3	−2.685⋆2	−1.396⋆3	−8.321⋆2	−6.449⋆2

Continued

Table 14 The bell-shaped model. The aperture aberration coefficients $\langle 0012 \rangle_{ij}$ as a function of $\beta^2 d^2$.—cont'd

Achromatic

$\beta^2 d^2$	$\langle 12 \rangle_{22}$	$\langle 12 \rangle_{21}$	$\langle 12 \rangle_{12}$	$\langle 12 \rangle_{20}$	$\langle 12 \rangle_{11}$	$\langle 12 \rangle_{02}$	$\langle 12 \rangle_{10}$	$\langle 12 \rangle_{01}$	$\langle 12 \rangle_{00}$
3.600	4.311⋆2	1.390⋆3	−1.140⋆3	1.121⋆3	−3.676⋆3	−3.586⋆1	−2.964⋆3	−1.160⋆2	−9.408⋆1
3.800	4.606⋆2	1.541⋆3	−1.863⋆3	1.290⋆3	−6.235⋆3	5.396⋆2	−5.217⋆3	1.806⋆3	1.510⋆3
4.000	4.415⋆2	1.529⋆3	−2.707⋆3	1.325⋆3	−9.377⋆3	1.636⋆3	−8.121⋆3	5.666⋆3	4.906⋆3
4.200	3.575⋆2	1.279⋆3	−3.576⋆3	1.144⋆3	−1.280⋆4	3.458⋆3	−1.145⋆4	1.237⋆4	1.107⋆4
4.400	1.974⋆2	7.280⋆2	−4.313⋆3	6.714⋆2	−1.591⋆4	6.224⋆3	−1.467⋆4	2.295⋆4	2.116⋆4
4.600	−3.995⋆1	−1.516⋆2	−4.689⋆3	−1.437⋆2	−1.779⋆4	1.014⋆4	−1.688⋆4	3.847⋆4	3.649⋆4
4.800	−3.400⋆2	−1.325⋆3	−4.401⋆3	−1.292⋆3	−1.716⋆4	1.535⋆4	−1.673⋆4	5.984⋆4	5.832⋆4
5.000	−6.657⋆2	−2.663⋆3	−3.085⋆3	−2.663⋆3	−1.234⋆4	2.192⋆4	−1.234⋆4	8.767⋆4	8.767⋆4

Table 15 The bell-shaped model. The aperture aberration coefficients $\langle 0021 \rangle_{ij}$ as a function of $\beta^2 d^2$.

Electrostatic

$\beta^2 d^2$	$\langle 21 \rangle_{22}$	$\langle 21 \rangle_{21}$	$\langle 21 \rangle_{12}$	$\langle 21 \rangle_{20}$	$\langle 21 \rangle_{11}$	$\langle 21 \rangle_{02}$	$\langle 21 \rangle_{10}$	$\langle 21 \rangle_{01}$	$\langle 21 \rangle_{00}$
2.000⋆-1	−3.989⋆-2	1.502⋆-2	−1.856⋆-1	8.689⋆-2	3.442⋆-1	−4.179⋆-1	7.580⋆-2	−2.566⋆-1	1.427⋆-1
4.000⋆-1	−2.638⋆-1	3.446⋆-3	−7.989⋆-1	1.166⋆-1	2.948⋆-1	−9.000⋆-1	2.677⋆-1	−1.133	9.572⋆-2
6.000⋆-1	−8.233⋆-1	−3.310⋆-1	−1.612	9.886⋆-2	−6.257⋆-1	−1.120	3.031⋆-1	−2.398	−3.145⋆-1
8.000⋆-1	−1.757	−1.371	−1.911	−1.283⋆-1	−2.016	−7.720⋆-1	−4.686⋆-2	−3.370	−1.007
1.000	−3.291	−3.427	−1.426	−7.400⋆-1	−2.165	−1.976	−4.646⋆-1	−3.813	−1.662
1.200	−3.675	−5.363	3.304	−1.755	1.654	8.923⋆-1	2.075⋆-1	−2.871	−2.101
1.400	−3.389	−5.320	1.022⋆1	−2.659	1.193⋆1	6.808⋆-2	3.680	−4.035	−3.002
1.600	−1.070	−3.725	1.906⋆1	−2.227	2.874⋆1	−4.494	1.110⋆1	−1.164⋆1	−6.617
1.800	4.179	5.350	2.665⋆1	1.478	4.704⋆1	−1.498⋆1	2.106⋆1	−3.210⋆1	−1.694⋆1
2.000	1.290⋆1	2.361⋆1	2.734⋆1	1.076⋆1	5.416⋆1	−3.242⋆1	2.715⋆1	−7.083⋆1	−3.856⋆1
2.200	2.495⋆1	5.241⋆1	1.318⋆1	2.753⋆1	2.846⋆1	−5.497⋆1	1.570⋆1	−1.272⋆2	−7.343⋆1
2.400	3.921⋆1	9.045⋆1	−2.505⋆1	5.219⋆1	−5.963⋆1	−7.573⋆1	−3.513⋆1	−1.867⋆2	−1.150⋆2
2.600	5.333⋆1	1.325⋆2	−9.596⋆1	8.235⋆1	−2.431⋆2	−8.051⋆1	−1.536⋆2	−2.120⋆2	−1.394⋆2
2.800	6.381⋆1	1.687⋆2	−2.046⋆2	1.116⋆2	−5.494⋆2	−4.591⋆1	−3.684⋆2	−1.324⋆2	−9.491⋆1
3.200	5.592⋆1	1.633⋆2	−5.155⋆2	1.192⋆2	−1.530⋆3	2.851⋆2	−1.134⋆3	8.349⋆2	6.112⋆2
3.400	2.903⋆1	8.724⋆1	−6.753⋆2	6.556⋆1	−2.093⋆3	6.743⋆2	−1.621⋆3	2.077⋆3	1.600⋆3
3.600	−1.605⋆1	−5.454⋆1	−7.807⋆2	−4.614⋆1	−2.518⋆3	1.280⋆3	−2.030⋆3	4.115⋆3	3.307⋆3
3.800	−7.684⋆1	−2.600⋆2	−7.642⋆2	−2.198⋆2	−2.558⋆3	2.145⋆3	−2.140⋆3	7.165⋆3	5.983⋆3

Continued

Table 15 The bell-shaped model. The aperture aberration coefficients $\langle 0021 \rangle_{ij}$ as a function of $\beta^2 d^2$.—cont'd

Electrostatic

$\beta^2 d^2$	$\langle 21 \rangle_{22}$	$\langle 21 \rangle_{21}$	$\langle 21 \rangle_{12}$	$\langle 21 \rangle_{20}$	$\langle 21 \rangle_{11}$	$\langle 21 \rangle_{02}$	$\langle 21 \rangle_{10}$	$\langle 21 \rangle_{01}$	$\langle 21 \rangle_{00}$
4.000	−1.447⋆2	−5.042⋆2	−5.394⋆2	−4.391⋆2	−1.869⋆3	3.294⋆3	−1.618⋆3	1.140⋆4	9.857⋆3
4.200	−2.027⋆2	−7.282⋆2	−6.897	−6.539⋆2	−2.497⋆1	4.718⋆3	−2.218⋆1	1.686⋆4	1.507⋆4
4.400	−2.236⋆2	−8.277⋆2	9.362⋆2	−7.659⋆2	3.452⋆3	6.362⋆3	3.183⋆3	2.345⋆4	2.160⋆4
4.600	−1.684⋆2	−6.423⋆2	2.380⋆3	−6.123⋆2	9.032⋆3	8.112⋆3	8.569⋆3	3.076⋆4	2.917⋆4
4.800	1.430⋆1	5.253⋆1	4.382⋆3	4.812⋆1	1.709⋆4	9.783⋆3	1.665⋆4	3.812⋆4	3.714⋆4
5.000	3.882⋆2	1.550⋆3	6.940⋆3	1.546⋆3	2.776⋆4	1.112⋆4	2.776⋆4	4.447⋆4	4.445⋆4

Magnetic

$\beta^2 d^2$	$\langle 21 \rangle_{22}$	$\langle 21 \rangle_{21}$	$\langle 21 \rangle_{12}$	$\langle 21 \rangle_{20}$	$\langle 21 \rangle_{11}$	$\langle 21 \rangle_{02}$	$\langle 21 \rangle_{10}$	$\langle 21 \rangle_{01}$	$\langle 21 \rangle_{00}$
2.000⋆-1	1.784⋆-2	3.593⋆-2	−2.079⋆-1	9.608⋆-2	3.819⋆-1	−3.960⋆-1	8.271⋆-2	−2.618⋆-1	1.530⋆-1
4.000⋆-1	−4.550⋆-2	1.538⋆-1	−9.658⋆-1	1.616⋆-1	3.455⋆-1	−7.320⋆-1	3.044⋆-1	−1.132	1.255⋆-1
6.000⋆-1	−3.681⋆-1	1.115⋆-1	−2.096	2.372⋆-1	−8.004⋆-1	−4.586⋆-1	3.428⋆-1	−2.165	−2.433⋆-1
8.000⋆-1	−9.974⋆-1	−4.572⋆-1	−2.776	1.988⋆-1	−2.754	1.085	−1.312⋆-1	−2.095	−7.184⋆-1
1.000	−1.619	−1.649	−3.426	−1.170⋆-1	−3.516	3.524	−8.130⋆-1	6.305⋆-1	−5.515⋆-1
1.200	−1.774	−2.509	2.899	−5.966⋆-1	4.942⋆-1	9.044	−2.814⋆-1	7.354	1.210
1.400	−1.656⋆-1	−1.090	1.169⋆1	−4.289⋆-1	1.314⋆1	1.414⋆1	3.843	1.702⋆1	4.988
1.600	4.581	6.225	2.408⋆1	2.277	3.598⋆1	1.772⋆1	1.370⋆1	2.646⋆1	9.848
1.800	1.402⋆1	2.422⋆1	3.683⋆1	1.065⋆1	6.483⋆1	1.775⋆1	2.885⋆1	3.060⋆1	1.324⋆1
2.000	2.949⋆1	5.812⋆1	4.343⋆1	2.884⋆1	8.604⋆1	1.352⋆1	4.297⋆1	2.569⋆1	1.231⋆1
2.200	5.167⋆1	1.124⋆2	3.408⋆1	6.130⋆1	7.402⋆1	7.839	4.058⋆1	1.560⋆1	7.866

2.400	8.022⋆1	1.890⋆2	−3.553	1.115⋆2	−8.898	9.971	−5.156	2.175⋆1	1.201⋆1
2.600	1.134⋆2	2.861⋆2	−8.223⋆1	1.807⋆2	−2.084⋆2	3.892⋆1	−1.317⋆2	9.632⋆1	5.985⋆1
2.800	1.481⋆2	3.966⋆2	−2.119⋆2	2.657⋆2	−5.689⋆2	1.266⋆2	−3.814⋆2	3.373⋆2	2.250⋆2
3.200	2.039⋆2	6.042⋆2	−6.220⋆2	4.477⋆2	−1.845⋆3	6.834⋆2	−1.368⋆3	2.024⋆3	1.499⋆3
3.400	2.159⋆2	6.681⋆2	−8.638⋆2	5.171⋆2	−2.677⋆3	1.294⋆3	−2.073⋆3	4.005⋆3	3.100⋆3
3.600	2.139⋆2	6.892⋆2	−1.069⋆3	5.552⋆2	−3.447⋆3	2.238⋆3	−2.779⋆3	7.214⋆3	5.814⋆3
3.800	2.010⋆2	6.718⋆2	−1.158⋆3	5.616⋆2	−3.877⋆3	3.605⋆3	−3.243⋆3	1.206⋆4	1.009⋆4
4.000	1.873⋆2	6.481⋆2	−1.026⋆3	5.607⋆2	−3.556⋆3	5.471⋆3	−3.079⋆3	1.895⋆4	1.641⋆4
4.200	1.936⋆2	6.920⋆2	−5.444⋆2	6.184⋆2	−1.948⋆3	7.885⋆3	−1.742⋆3	2.821⋆4	2.522⋆4
4.400	2.542⋆2	9.365⋆2	4.296⋆2	8.628⋆2	1.584⋆3	1.085⋆4	1.460⋆3	4.001⋆4	3.689⋆4
4.600	4.191⋆2	1.590⋆3	2.034⋆3	1.507⋆3	7.718⋆3	1.431⋆4	7.322⋆3	5.431⋆4	5.152⋆4
4.800	7.567⋆2	2.949⋆3	4.380⋆3	2.874⋆3	1.707⋆4	1.814⋆4	1.664⋆4	7.072⋆4	6.893⋆4
5.000	1.354⋆3	5.415⋆3	7.519⋆3	5.414⋆3	3.007⋆4	2.212⋆4	3.007⋆4	8.848⋆4	8.848⋆4

Achromatic

$\beta^2 d^2$	$\langle 21\rangle_{22}$	$\langle 21\rangle_{21}$	$\langle 21\rangle_{12}$	$\langle 21\rangle_{20}$	$\langle 21\rangle_{11}$	$\langle 21\rangle_{02}$	$\langle 21\rangle_{10}$	$\langle 21\rangle_{01}$	$\langle 21\rangle_{00}$
2.000⋆-1	3.708⋆-2	4.290⋆-2	−2.153⋆-1	9.914⋆-2	3.945⋆-1	−3.887⋆-1	8.502⋆-2	−2.635⋆-1	1.564⋆-1
4.000⋆-1	2.726⋆-2	2.040⋆-1	−1.021	1.766⋆-1	3.624⋆-1	−6.760⋆-1	3.166⋆-1	−1.132	1.354⋆-1
6.000⋆-1	−2.164⋆-1	2.590⋆-1	−2.257	2.834⋆-1	−8.586⋆-1	−2.383⋆-1	3.560⋆-1	−2.088	−2.195⋆-1

Continued

Table 15 The bell-shaped model. The aperture aberration coefficients $\langle 0021 \rangle_{ij}$ as a function of $\beta^2 d^2$.—cont'd

Achromatic

$\beta^2 d^2$	$\langle 21 \rangle_{22}$	$\langle 21 \rangle_{21}$	$\langle 21 \rangle_{12}$	$\langle 21 \rangle_{20}$	$\langle 21 \rangle_{11}$	$\langle 21 \rangle_{02}$	$\langle 21 \rangle_{10}$	$\langle 21 \rangle_{01}$	$\langle 21 \rangle_{00}$
8.000⋆-1	−7.443⋆-1	−1.528⋆-1	−3.065	3.052⋆-1	−3.000	1.704	−1.593⋆-1	−1.671	−6.221⋆-1
1.000	−1.119	−1.056	−3.426	9.074⋆-2	−3.966	5.274	−9.291⋆-1	2.112	−1.815⋆-1
1.200	−1.141	−1.558	2.765	−2.105⋆-1	1.076⋆-1	1.176⋆1	−4.443⋆-1	1.076⋆1	2.313
1.400	9.089⋆-1	6.533⋆-1	1.218⋆1	3.145⋆-1	1.354⋆1	1.884⋆1	3.898	2.404⋆1	7.652
1.600	6.464	9.542	2.576⋆1	3.778	3.840⋆1	2.512⋆1	1.457⋆1	3.916⋆1	1.534⋆1
1.800	1.730⋆1	3.051⋆1	4.022⋆1	1.371⋆1	7.076⋆1	2.866⋆1	3.144⋆1	5.150⋆1	2.330⋆1
2.000	3.501⋆1	6.963⋆1	4.880⋆1	3.487⋆1	9.667⋆1	2.883⋆1	4.824⋆1	5.787⋆1	2.926⋆1
2.200	6.057⋆1	1.324⋆2	4.104⋆1	7.255⋆1	8.921⋆1	2.878⋆1	4.887⋆1	6.319⋆1	3.496⋆1
2.400	9.389⋆1	2.219⋆2	3.612	1.313⋆2	8.012	3.854⋆1	4.835	9.125⋆1	5.434⋆1
2.600	1.334⋆2	3.373⋆2	−7.766⋆1	2.134⋆2	−1.969⋆2	7.872⋆1	−1.243⋆2	1.991⋆2	1.262⋆2
2.800	1.762⋆2	4.726⋆2	−2.143⋆2	3.171⋆2	−5.754⋆2	1.841⋆2	−3.858⋆2	4.939⋆2	3.316⋆2
3.200	2.533⋆2	7.511⋆2	−6.575⋆2	5.572⋆2	−1.951⋆3	8.162⋆2	−1.446⋆3	2.421⋆3	1.795⋆3

3.400	2.782⋆2	8.617⋆2	−9.267⋆2	6.676⋆2	−2.872⋆3	1.500⋆3	−2.224⋆3	4.647⋆3	3.600⋆3
3.600	2.906⋆2	9.371⋆2	−1.165⋆3	7.557⋆2	−3.757⋆3	2.558⋆3	−3.029⋆3	8.247⋆3	6.549⋆3
3.800	2.936⋆2	9.825⋆2	−1.290⋆3	8.221⋆2	−4.316⋆3	4.092⋆3	−3.611⋆3	1.369⋆4	1.146⋆4
4.000	2.980⋆2	1.032⋆3	−1.189⋆3	8.940⋆2	−4.118⋆3	6.196⋆3	−3.566⋆3	2.146⋆4	1.859⋆4
4.200	3.258⋆2	1.165⋆3	−7.235⋆2	1.043⋆3	−2.589⋆3	8.940⋆3	−2.316⋆3	3.199⋆4	2.861⋆4
4.400	4.134⋆2	1.525⋆3	2.608⋆2	1.406⋆3	9.610⋆2	1.235⋆4	8.859⋆2	4.554⋆4	4.198⋆4
4.600	6.150⋆2	2.334⋆3	1.919⋆3	2.214⋆3	7.280⋆3	1.638⋆4	6.906⋆3	6.217⋆4	5.898⋆4
4.800	1.004⋆3	3.915⋆3	4.379⋆3	3.816⋆3	1.707⋆4	2.093⋆4	1.664⋆4	8.159⋆4	7.953⋆4
5.000	1.676⋆3	6.703⋆3	7.711⋆3	6.703⋆3	3.084⋆4	2.579⋆4	3.084⋆4	1.032⋆5	1.032⋆5

Table 16 The bell-shaped model. The aperture aberration coefficients $(0030)_i$ as a function of $\beta^2 d^2$.

Electrostatic

$\beta^2 d^2$	$(30)_4 = (30)_0$	$(30)_3 = (30)_1$	$(30)_2$
2.000⋆-1	6.070	−2.207⋆1	3.220⋆1
4.000⋆-1	2.054	−6.400	9.200
6.000⋆-1	1.331	−3.191	4.746
8.000⋆-1	1.180	−1.843	3.252
1.000	1.276	−9.124⋆-1	2.824
1.200	1.594	1.352⋆-1	3.191
1.400	2.243	1.843	4.733
1.600	3.536	5.302	8.798
1.800	6.311	1.338⋆1	1.934⋆1
2.000	1.310⋆1	3.502⋆1	4.917⋆1
2.200	3.318⋆1	1.044⋆2	1.481⋆2
2.400	1.126⋆2	3.968⋆2	5.743⋆2
2.600	6.322⋆2	2.397⋆3	3.536⋆3
2.800	1.155⋆4	4.562⋆4	6.814⋆4
3.200	1.627⋆4	6.431⋆4	9.608⋆4
3.400	1.249⋆3	4.765⋆3	7.044⋆3
3.600	3.096⋆2	1.114⋆3	1.621⋆3
3.800	1.256⋆2	4.152⋆2	5.949⋆2
4.000	6.731⋆1	1.988⋆2	2.820⋆2
4.200	4.342⋆1	1.106⋆2	1.579⋆2
4.400	3.205⋆1	6.721⋆1	9.994⋆1
4.600	2.628⋆1	4.250⋆1	7.026⋆1
4.800	2.351⋆1	2.627⋆1	5.474⋆1
5.000	2.265⋆1	1.380⋆1	4.762⋆1

Table 16 The bell-shaped model. The aperture aberration coefficients $(0030)_i$ as a function of $\beta^2 d^2$.—cont'd

Magnetic

$\beta^2 d^2$	$(30)_4 = (30)_0$	$(30)_3 = (30)_1$	$(30)_2$
2.000⋆-1	2.095	−6.639	9.275
4.000⋆-1	7.966⋆-1	−1.937	2.742
6.000⋆-1	5.750⋆-1	−9.008⋆-1	1.533
8.000⋆-1	5.644⋆-1	−3.937⋆-1	1.211
1.000	6.626⋆-1	4.381⋆-2	1.272
1.200	8.744⋆-1	6.242⋆-1	1.710
1.400	1.265	1.621	2.792
1.600	2.003	3.613	5.276
1.800	3.529	8.113	1.128⋆1
2.000	7.140	1.974⋆1	2.743⋆1
2.200	1.750⋆1	5.576⋆1	7.892⋆1
2.400	5.736⋆1	2.028⋆2	2.936⋆2
2.600	3.112⋆2	1.181⋆3	1.742⋆3
2.800	5.512⋆3	2.177⋆4	3.252⋆4
3.200	7.376⋆3	2.915⋆4	4.355⋆4
3.400	5.548⋆2	2.116⋆3	3.127⋆3
3.600	1.352⋆2	4.855⋆2	7.068⋆2
3.800	5.407⋆1	1.780⋆2	2.553⋆2
4.000	2.870⋆1	8.394⋆1	1.195⋆2
4.200	1.840⋆1	4.596⋆1	6.622⋆1
4.400	1.355⋆1	2.745⋆1	4.166⋆1
4.600	1.113⋆1	1.694⋆1	2.925⋆1
4.800	9.985	1.005⋆1	2.289⋆1
5.000	9.664	4.737	2.012⋆1

Continued

Table 16 The bell-shaped model. The aperture aberration coefficients $(0030)_i$ as a function of $\beta^2 d^2$.—cont'd

Achromatic

$\beta^2 d^2$	$(30)_4 = (30)_0$	$(30)_3 = (30)_1$	$(30)_2$
2.000⋆-1	2.890	−9.726	1.386⋆1
4.000⋆-1	1.048	−2.829	4.034
6.000⋆-1	7.262⋆-1	−1.359	2.175
8.000⋆-1	6.875⋆-1	−6.836⋆-1	1.619
1.000	7.854⋆-1	−1.474⋆-1	1.582
1.200	1.018	5.264⋆-1	2.006
1.400	1.460	1.665	3.180
1.600	2.310	3.951	5.980
1.800	4.086	9.166	1.289⋆1
2.000	8.331	2.280⋆1	3.178⋆1
2.200	2.064⋆1	6.549⋆1	9.275⋆1
2.400	6.841⋆1	2.416⋆2	3.497⋆2
2.600	3.754⋆2	1.424⋆3	2.101⋆3
2.800	6.720⋆3	2.654⋆4	3.964⋆4
3.200	9.155⋆3	3.618⋆4	5.406⋆4
3.400	6.937⋆2	2.646⋆3	3.910⋆3
3.600	1.701⋆2	6.111⋆2	8.897⋆2
3.800	6.836⋆1	2.255⋆2	3.232⋆2
4.000	3.642⋆1	1.069⋆2	1.520⋆2
4.200	2.341⋆1	5.888⋆1	8.455⋆1
4.400	1.725⋆1	3.540⋆1	5.332⋆1
4.600	1.416⋆1	2.205⋆1	3.746⋆1
4.800	1.269⋆1	1.329⋆1	2.926⋆1
5.000	1.226⋆1	6.550	2.562⋆1

Table 17 The bell-shaped model. The aperture aberration coefficients $(0003)_i$ as a function of $\beta^2 d^2$.

Electrostatic

$\beta^2 d^2$	$(03)_4 = (03)_0$	$(03)_3 = (03)_1$	$(03)_2$
2.000★-1	3.629	−1.601★1	2.464★1
4.000★-1	6.345★-1	−3.421	5.353
6.000★-1	1.360★-1	−1.307	2.058
8.000★-1	−2.896★-2	−6.363★-1	9.972★-1
1.000	−8.361★8	−4.658★9	−8.957★9
1.200	−1.497★-1	−2.165★-1	3.261★-1
1.400	−1.795★-1	−1.406★-1	2.049★-1
1.600	−2.019★-1	−9.572★-2	1.340★-1
1.800	−2.200★-1	−6.762★-2	9.034★-2
2.000	−2.356★-1	−4.922★-2	6.243★-2
2.200	−2.494★-1	−3.671★-2	4.401★-2
2.400	−2.619★-1	−2.795★-2	3.155★-2
2.600	−2.736★-1	−2.165★-2	2.294★-2
2.800	−2.845★-1	−1.703★-2	1.688★-2
3.200	−3.046★-1	−1093★-2	9.438★-3
3.400	−3.141★-1	−8.895★-3	7.151★-3
3.600	−3.232★-1	−7.304★-3	5.461★-3
3.800	−3.320★-1	−6.045★-3	4.199★-3
4.000	−3.405★-1	−5.040★-3	3.250★-3
4.200	−3.488★-1	−4.229★-3	2.531★-3
4.400	−3.569★-1	−3.569★-3	1.978★-3
4.600	−3.647★-1	−3.028★-3	1.556★-3
4.800	−3.724★-1	−2.582★-3	1.231★-3
5.000	−3.799★-1	−2.212★-3	9.773★-4

Continued

Table 17 The bell-shaped model. The aperture aberration coefficients $(0003)_i$ as a function of $\beta^2 d^2$.—cont'd

Magnetic

$\beta^2 d^2$	$(03)_4 = (03)_0$	$(03)_3 = (03)_1$	$(03)_2$
2.000⋆-1	5.836⋆-1	−4.188	7.075
4.000⋆-1	−1.046⋆-1	−7.895⋆-1	1.554
6.000⋆-1	−2.033⋆-1	−2.553⋆-1	6.049⋆-1
8.000⋆-1	−2.362⋆-1	−9.975⋆-2	2.970⋆-1
1.000	−1.359⋆9	−5.106⋆9	−7.167⋆9
1.200	−2.691⋆-1	−1.613⋆-2	9.954⋆-2
1.400	−2.819⋆-1	−4.595⋆-3	6.324⋆-2
1.600	−2.939⋆-1	8.376⋆-4	4.177⋆-2
1.800	−3.054⋆-1	3.335⋆-3	2.843⋆-2
2.000	−3.165⋆-1	4.364⋆-3	1.981⋆-2
2.200	−3.274⋆-1	4.648⋆-3	1.407⋆-2
2.400	−3.380⋆-1	4.555⋆-3	1.016⋆-2
2.600	−3.483⋆-1	4.280⋆-3	7.436⋆-3
2.800	−3.584⋆-1	3.929⋆-3	5.506⋆-3
3.200	−3.779⋆-1	3.191⋆-3	3.111⋆-3
3.400	−3.873⋆-1	2.848⋆-3	2.368⋆-3
3.600	−3.966⋆-1	2.534⋆-3	1.816⋆-3
3.800	−4.056⋆-1	2.251⋆-3	1.403⋆-3
4.000	−4.145⋆-1	1.998⋆-3	1.089⋆-3
4.200	−4.232⋆-1	1.773⋆-3	8.513⋆-4
4.400	−4.318⋆-1	1.573⋆-3	6.654⋆-4
4.600	−4.401⋆-1	1.397⋆-3	5.237⋆-4
4.800	−4.484⋆-1	1.241⋆-3	4.178⋆-4
5.000	−4.565⋆-1	1.104⋆-3	3.252⋆-4

Table 17 The bell-shaped model. The aperture aberration coefficients $(0003)_i$ as a function of $\beta^2 d^2$.—cont'd

Achromatic

$\beta^2 d^2$	$(03)_4 = (03)_0$	$(03)_3 = (03)_1$	$(03)_2$
2.000⋆-1	1.193	−6.554	1.059⋆1
4.000⋆-1	4.322⋆-2	−1.316	2.314
6.000⋆-1	−1.355⋆-1	−4.656⋆-1	8.955⋆-1
8.000⋆-1	−1.948⋆-1	−2.071⋆-1	4.371⋆-1
1.000	−1.254⋆9	−5.017⋆9	−7.525⋆9
1.200	−2.453⋆-1	−5.621⋆-2	1.448⋆-1
1.400	−2.614⋆-1	−3.179⋆-2	9.156⋆-2
1.600	−2.755⋆-1	−1.847⋆-2	6.021⋆-2
1.800	−2.883⋆-1	−1.086⋆-2	4.081⋆-2
2.000	−3.003⋆-1	−6.352⋆-3	2.834⋆-2
2.200	−3.118⋆-1	−3.624⋆-3	2.006⋆-2
2.400	−3.228⋆-1	−1.946⋆-3	1.444⋆-2
2.600	−3.333⋆-1	−9.061⋆-4	1.054⋆-2
2.800	−3.436⋆-1	−2.622⋆-4	7.782⋆-3
3.200	−3.632⋆-1	3.669⋆-4	4.376⋆-3
3.400	−3.727⋆-1	4.997⋆-4	3.324⋆-3
3.600	−3.819⋆-1	5.668⋆-4	2.545⋆-3
3.800	−3.909⋆-1	5.918⋆-4	1.961⋆-3
4.000	−3.997⋆-1	5.903⋆-4	1.520⋆-3
4.200	−4.083⋆-1	5.723⋆-4	1.186⋆-3
4.400	−4.168⋆-1	5.447⋆-4	9.290⋆-4
4.600	−4.251⋆-1	5.119⋆-4	7.297⋆-4
4.800	−4.332⋆-1	4.767⋆-4	5.772⋆-4
5.000	−4.412⋆-1	4.411⋆-4	4.542⋆-4

Table 18 The bell-shaped model. The aperture aberration coefficients $(0012)_{ij}$ as a function of $\beta^2 d^2$.

Electrostatic

$\beta^2 d^2$	$(12)_{22} = (12)_{00}$	$(12)_{21} = (12)_{01}$	$(12)_{12} = (12)_{10}$	$(12)_{20} = (12)_{02}$	$(12)_{11}$
2.000⋆-1	−1.069	5.886	8.604⋆-1	−1.969	−7.556
4.000⋆-1	−4.036⋆-1	2.071	−5.841⋆-1	−4.544⋆-1	−1.494
6.000⋆-1	−4.363⋆-1	1.255	−6.230⋆-1	−1.696⋆-1	−3.496⋆-1
8.000⋆-1	−5.151⋆-1	9.438⋆-1	−5.596⋆-1	−6.423⋆-2	8.032⋆-2
1.000	−6.658⋆-1	7.849⋆-1	−6.226⋆-1	5.308⋆-2	−1.753⋆-2
1.200	−6.553⋆-1	7.357⋆-1	−4.285⋆-1	3.335⋆-2	4.928⋆-1
1.400	−7.093⋆-1	7.236⋆-1	−3.635⋆-1	7.301⋆-2	6.635⋆-1
1.600	−7.499⋆-1	7.576⋆-1	−2.858⋆-1	1.212⋆-1	8.652⋆-1
1.800	−7.703⋆-1	8.475⋆-1	−1.754⋆-1	1.907⋆-1	1.141
2.000	−7.537⋆-1	1.025	9.796⋆-3	3.055⋆-1	1.567
2.200	−6.548⋆-1	1.370	3.729⋆-1	5.225⋆-1	2.313
2.400	−3.292⋆-1	2.123	1.229	1.011	3.862
2.600	8.560⋆-1	4.237	3.916	2.478	8.123
2.800	8.443	1.529⋆1	1.988⋆1	1.083⋆1	3.026⋆1
3.200	1.357⋆1	1.311⋆1	2.736⋆1	1.313⋆1	2.593⋆1
3.400	3.335	3.112	7.663	3.652	5.952
3.600	1.228	1.336	3.735	1.820	2.411
3.800	4.133⋆-1	7.384⋆-1	2.262	1.157	1.224
4.000	−2.912⋆-3	4.720⋆-1	1.531	8.447⋆-1	6.961⋆-1
4.200	−2.514⋆-1	3.327⋆-1	1.102	6.751⋆-1	4.205⋆-1
4.400	−4.149⋆-1	2.521⋆-1	8.206⋆-1	5.763⋆-1	2.598⋆-1
4.600	−5.291⋆-1	2.022⋆-1	6.171⋆-1	5.179⋆-1	1.580⋆-1
4.800	−6.114⋆-1	1.700⋆-1	4.575⋆-1	4.855⋆-1	8.905⋆-2
5.000	−6.707⋆-1	1.488⋆-1	3.218⋆-1	4.721⋆-1	3.913⋆-2

Magnetic

$\beta^2 d^2$	$(12)_{22} = (12)_{00}$	$(12)_{21} = (12)_{01}$	$(12)_{12} = (12)_{10}$	$(12)_{20} = (12)_{02}$	$(12)_{11}$
2.000⋆-1	4.921	−6.231	−1.121⋆1	4.260	1.639⋆1
4.000⋆-1	1.059	−9.979⋆-1	−3.553	1.245	4.304
6.000⋆-1	2.005⋆-1	−1.550⋆-1	−1.876	7.012⋆-1	2.076

Table 18 The bell-shaped model. The aperture aberration coefficients $(0012)_{ij}$ as a function of $\beta^2 d^2$.—cont'd

Magnetic

$\beta^2 d^2$	$(12)_{22} = (12)_{00}$	$(12)_{21} = (12)_{01}$	$(12)_{12} = (12)_{10}$	$(12)_{20} = (12)_{02}$	$(12)_{11}$
8.000⋆-1	−1.511⋆-1	9.868⋆-2	−1.179	5.301⋆-1	1.308
1.000	−3.012⋆-1	2.957⋆-1	−1.114	5.405⋆-1	2.794⋆-1
1.200	−4.267⋆-1	2.536⋆-1	−4.853⋆-1	4.803⋆-1	8.044⋆-1
1.400	−4.617⋆-1	2.908⋆-1	−2.150⋆-1	5.298⋆-1	7.321⋆-1
1.600	−4.368⋆-1	3.283⋆-1	9.276⋆-2	6.314⋆-1	7.224⋆-1
1.800	−3.302⋆-1	3.782⋆-1	5.132⋆-1	8.117⋆-1	7.710⋆-1
2.000	−8.306⋆-2	4.566⋆-1	1.184	1.134	8.960⋆-1
2.200	4.569⋆-1	5.963⋆-1	2.424	1.758	1.155
2.400	1.752	8.870⋆-1	5.180	3.161	1.724
2.600	5.733	1.678	1.335⋆1	7.312	3.299
2.800	2.851⋆1	5.694	5.937⋆1	3.052⋆1	1.132⋆1
3.200	3.411⋆1	4.265	6.906⋆1	3.459⋆1	8.461
3.400	8.465	9.493⋆-1	1.819⋆1	9.378	1.829
3.600	3.499	3.848⋆-1	8.397	4.572	6.988⋆-1
3.800	1.687	2.028⋆-1	4.830	2.854	3.341⋆-1
4.000	8.169⋆-1	1.251⋆-1	3.103	2.049	1.781⋆-1
4.200	3.315⋆-1	8.628⋆-2	2.112	1.614	9.958⋆-2
4.400	3.600⋆-2	6.473⋆-2	1.470	1.361	5.538⋆-2
4.600	−1.515⋆-1	5.194⋆-2	1.011	1.209	2.830⋆-2
4.800	−2.701⋆-1	4.403⋆-2	6.519⋆-1	1.122	1.047⋆-2
5.000	−3.391⋆-1	3.907⋆-2	3.466⋆-1	1.082	−2.126⋆-3

Achromatic

$\beta^2 d^2$	$(12)_{22} = (12)_{00}$	$(12)_{21} = (12)_{01}$	$(12)_{12} = (12)_{10}$	$(12)_{20} = (12)_{02}$	$(12)_{11}$
2.000⋆-1	6.918	−1.027⋆1	−1.523⋆1	6.336	2.438⋆1
4.000⋆-1	1.547	−2.021	−4.543	1.812	6.237
6.000⋆-1	4.127⋆-1	−6.250⋆-1	−2.293	9.915⋆-1	2.884
8.000⋆-1	−2.981⋆-2	−1.830⋆-1	−1.385	7.282⋆-1	1.717
1.000	−1.921⋆-1	1.077⋆-1	−1.172	6.905⋆-1	5.899⋆-1
1.200	−3.504⋆-1	9.287⋆-2	−5.043⋆-1	6.292⋆-1	9.082⋆-1

Continued

Table 18 The bell-shaped model. The aperture aberration coefficients $(0012)_{ij}$ as a function of $\beta^2 d^2$.—cont'd

Achromatic

$\beta^2 d^2$	$(12)_{22} = (12)_{00}$	$(12)_{21} = (12)_{01}$	$(12)_{12} = (12)_{10}$	$(12)_{20} = (12)_{02}$	$(12)_{11}$
1.400	−3.792★-1	1.465★-1	−1.655★-1	6.821★-1	7.550★-1
1.600	−3.324★-1	1.852★-1	2.190★-1	8.015★-1	6.747★-1
1.800	−1.835★-1	2.218★-1	7.427★-1	1.019	6.477★-1
2.000	1.405★-1	2.672★-1	1.575	1.410	6.724★-1
2.200	8.275★-1	3.382★-1	3.108	2.170	7.693★-1
2.400	2.446	4.750★-1	6.497	3.877	1.012
2.600	7.359	8.255★-1	1.650★1	8.924	1.691
2.800	3.520★1	2.494	7.253★1	3.709★1	5.012
3.200	4.096★1	1.318	8.296★1	4.175★1	2.640
3.400	1.018★1	2.284★-1	2.170★1	1.129★1	4.542★-1
3.600	4.256	6.763★-2	9.951	5.490	1.279★-1
3.800	2.111	2.424★-2	5.686	3.419	3.760★-2
4.000	1.000	9.529★-3	3.627	2.450	5.489★-3
4.200	5.258★-1	4.138★-3	2.448	1.927	−7.375★-3
4.400	1.863★-1	2.270★-3	1.686	1.622	−1.276★-2
4.600	−2.567★-2	1.847★-3	1.142	1.440	−1.495★-2
4.800	−1.563★-1	2.035★-3	7.167★-1	1.335	−1.572★-2
5.000	−2.286★-1	2.480★-3	3.548★-1	1.285	−1.588★-2

Table 19 The bell-shaped model. The aperture aberration coefficients $(0021)_{ij}$ as a function of $\beta^2 d^2$.

Electrostatic

$\beta^2 d^2$	$(21)_{22} = (21)_{00}$	$(21)_{21} = (21)_{01}$	$(21)_{12} = (21)_{10}$	$(21)_{20} = (21)_{02}$	$(21)_{11}$
2.000★-1	−4.139	7.260	2.715	−1.969	−7.556
4.000★-1	−1.768	2.666	5.028★-1	−4.544★-1	−1.494
6.000★-1	−1.152	1.601	2.323★-1	−1.696★-1	−3.496★-1
8.000★-1	−8.420★-1	1.171	2.038★-1	−6.423★-2	8.032★-2
1.000	−7.178★-1	9.125★-1	1.702★-1	3.730★-2	1.010★-1
1.200	−4.421★-1	8.525★-1	3.142★-1	3.335★-2	4.928★-1

Table 19 The bell-shaped model. The aperture aberration coefficients $(0021)_{ij}$ as a function of $\beta^2 d^2$.—cont'd

Electrostatic

$\beta^2 d^2$	$(21)_{22} = (21)_{00}$	$(21)_{21} = (21)_{01}$	$(21)_{12} = (21)_{10}$	$(21)_{20} = (21)_{02}$	$(21)_{11}$
1.400	−2.603⋆-1	8.120⋆-1	4.204⋆-1	7.301⋆-2	6.635⋆-1
1.600	−5.926⋆-2	8.261⋆-1	5.735⋆-1	1.212⋆-1	8.652⋆-1
1.800	1.869⋆-1	9.016⋆-1	8.027⋆-1	1.907⋆-1	1.141
2.000	5.217⋆-1	1.068	1.171	3.055⋆-1	1.567
2.200	1.039	1.406	1.826	5.225⋆-1	2.313
2.400	1.990	2.152	3.186	1.011	3.862
2.600	4.329	4.261	6.910	2.478	8.123
2.800	1.519⋆1	1.531⋆1	2.603⋆1	1.083⋆1	3.026⋆1
3.200	7.761	1.312⋆1	2.073⋆1	1.313⋆1	2.593⋆1
3.400	7.972⋆-1	3.124	4.189	3.652	5.952
3.600	−1.585⋆-1	1.346	1.294	1.820	2.411
3.800	−3.522⋆-1	7.471⋆-1	3.209⋆-1	1.157	1.224
4.000	−3.573⋆-1	4.795⋆-1	−1.245⋆-1	8.447⋆-1	6.961⋆-1
4.200	−2.981⋆-1	3.392⋆-1	−3.761⋆-1	6.751⋆-1	4.205⋆-1
4.400	−2.110⋆-1	2.577⋆-1	−5.453⋆-1	5.763⋆-1	2.598⋆-1
4.600	−1.082⋆-1	2.071⋆-1	−6.785⋆-1	5.179⋆-1	1.580⋆-1
4.800	6.576⋆-3	1.743⋆-1	−7.984⋆-1	4.855⋆-1	8.905⋆-2
5.000	1.333⋆-1	1.526⋆-1	−9.185⋆-1	4.721⋆-1	3.913⋆-2

Magnetic

$\beta^2 d^2$	$(21)_{22} = (21)_{00}$	$(21)_{21} = (21)_{01}$	$(21)_{12} = (21)_{10}$	$(21)_{20} = (21)_{02}$	$(21)_{11}$
2.000⋆-1	1.851	−4.858	−9.355	4.260	1.639⋆1
4.000⋆-1	−3.049⋆-1	−4.024⋆-1	−2.466	1.245	4.304
6.000⋆-1	−5.151⋆-1	1.908⋆-1	−1.021	7.012⋆-1	2.076
8.000⋆-1	−4.780⋆-1	3.254⋆-1	−4.153⋆-1	5.301⋆-1	1.308
1.000	−3.531⋆-1	4.233⋆-1	−3.213⋆-1	5.247⋆-1	3.979⋆-1
1.200	−2.135⋆-1	3.703⋆-1	2.573⋆-1	4.803⋆-1	8.044⋆-1
1.400	−1.272⋆-2	3.791⋆-1	5.689⋆-1	5.298⋆-1	7.321⋆-1
1.600	2.538⋆-1	3.968⋆-1	9.521⋆-1	6.314⋆-1	7.224⋆-1

Continued

Table 19 The bell-shaped model. The aperture aberration coefficients $(0021)_{ij}$ as a function of $\beta^2 d^2$.—cont'd

Magnetic

$\beta^2 d^2$	$(21)_{22} = (21)_{00}$	$(21)_{21} = (21)_{01}$	$(21)_{12} = (21)_{10}$	$(21)_{20} = (21)_{02}$	$(21)_{11}$
1.800	6.270★-1	4.323★-1	1.491	8.117★-1	7.710★-1
2.000	1.192	4.999★-1	2.345	1.134	8.960★-1
2.200	2.150	6.314★-1	3.877	1.758	1.155
2.400	4.072	9.158★-1	7.138	3.161	1.724
2.600	9.206	1.702	1.635★1	7.312	3.299
2.800	3.526★1	5.713	6.552★1	3.052★1	1.132★1
3.200	2.830★1	4.279	6.243★1	3.459★1	8.461
3.400	5.928	9.612★-1	1.472★1	9.378	1.829
3.600	2.112	3.949★-1	5.955	4.572	6.988★-1
3.800	9.213★-1	2.115★-1	2.889	2.854	3.341★-1
4.000	4.625★-1	1.326★-1	1.448	2.049	1.781★-1
4.200	2.848★-1	9.276★-2	6.333★-1	1.614	9.958★-2
4.400	2.399★-1	7.035★-2	1.038★-1	1.361	5.538★-2
4.600	2.693★-1	5.683★-2	−2.850★-1	1.209	2.830★-2
4.800	3.479★-1	4.830★-2	−6.040★-1	1.122	1.047★-2
5.000	4.649★-1	4.280★-2	−8.938★-1	1.082	−2.126★-3

Achromatic

$\beta^2 d^2$	$(21)_{22} = (21)_{00}$	$(21)_{21} = (21)_{01}$	$(21)_{12} = (21)_{10}$	$(21)_{20} = (21)_{02}$	$(21)_{11}$
2.000★-1	3.848	−8.897	−1.338★1	6.336	2.438★1
4.000★-1	1.827★-1	−1.425	−3.456	1.812	6.237
6.000★-1	−3.028★-1	−2.793★-1	−1.438	9.915★-1	2.884
8.000★-1	−3.567★-1	4.368★-2	−6.216★-1	7.282★-1	1.717
1.000	−2.441★-1	2.353★-1	−3.794★-1	6.747★-1	7.084★-1
1.200	−1.373★-1	2.096★-1	2.384★-1	6.292★-1	9.082★-1
1.400	6.982★-2	2.349★-1	6.184★-1	6.821★-1	7.550★-1
1.600	3.582★-1	2.537★-1	1.078	8.015★-1	6.747★-1
1.800	7.737★-1	2.758★-1	1.721	1.019	6.477★-1
2.000	1.416	3.105★-1	2.736	1.410	6.724★-1

Table 19 The bell-shaped model. The aperture aberration coefficients $(0021)_{ij}$ as a function of $\beta^2 d^2$.—cont'd

Achromatic

$\beta^2 d^2$	$(21)_{22} = (21)_{00}$	$(21)_{21} = (21)_{01}$	$(21)_{12} = (21)_{10}$	$(21)_{20} = (21)_{02}$	$(21)_{11}$
2.200	2.521	3.733⋆-1	4.561	2.170	7.693⋆-1
2.400	4.766	5.037⋆-1	8.455	3.877	1.012
2.600	1.083⋆1	8.493⋆-1	1.949⋆1	8.924	1.691
2.800	4.195⋆1	2.514	7.868⋆1	3.789⋆1	5.012
3.200	3.515⋆1	1.332	7.633⋆1	4.175⋆1	2.640
3.400	7.638	2.403⋆-1	1.823⋆1	1.129⋆1	4.542⋆-1
3.600	2.869	7.780⋆-2	7.509	5.490	1.279⋆-1
3.800	1.346	3.296⋆-2	3.745	3.419	3.760⋆-2
4.000	7.358⋆-1	1.704⋆-2	1.972	2.450	5.489⋆-3
4.200	4.791⋆-1	1.062⋆-2	9.698⋆-1	1.927	−7.375⋆-3
4.400	3.902⋆-1	7.893⋆-3	3.202⋆-1	1.622	−1.276⋆-2
4.600	3.952⋆-1	6.738⋆-3	−1.538⋆-1	1.440	−1.495⋆-2
4.800	4.617⋆-1	6.303⋆-3	−5.392⋆-1	1.335	−1.572⋆-2
5.000	5.754⋆-1	6.215⋆-3	−8.855⋆-1	1.285	−1.588⋆-2

$$
\begin{aligned}
\frac{C_{cx}^*}{d} &= \frac{\pi}{4}(n-1)\frac{\beta^2 d^2}{\omega_x^3}\left(1+\zeta_x^2\right) \\
\frac{C_{cy}^*}{d} &= -\frac{\pi}{4}(n-1)\frac{\beta^2 d^2}{\omega_y^3}\left(1+\zeta_y^2\right) \\
C_{Mx}^* &= -\frac{\pi}{4}(n-1)\frac{\beta^2 d^2}{\omega_x^3}\zeta_x \\
C_{My}^* &= \frac{\pi}{4}(n-1)\frac{\beta^2 d^2}{\omega_y^3}\zeta_y.
\end{aligned}
\tag{40}
$$

Substituting for ζ_x and ζ_y in terms of m_x and m_y in the asymptotic expressions, we find

$$\frac{C_{cx}}{d}=-\frac{n-1}{8}\frac{\beta^2 d^2}{\sin^2\pi\omega_x}\left\{\left(\frac{\sin 2\pi\omega_x}{\omega_x}-2\pi\right)\left(m_x^2+1\right)+4m_x\left(\frac{\sin\pi\omega_x}{\omega_x}-\pi\cos\pi\omega_x\right)\right\}$$
$$\frac{C_{cy}}{d}=\frac{n-1}{8}\frac{\beta^2 d^2}{\sin^2\pi\omega_y}\left\{\left(\frac{\sin 2\pi\omega_y}{\omega_y}-2\pi\right)\left(m_y^2+1\right)+4m_y\left(\frac{\sin\pi\omega_y}{\omega_y}-\pi\cos\pi\omega_y\right)\right\}$$
$$C_{Mx}=\frac{n-1}{8}\frac{\beta^2 d^2 f}{\omega_x^2\, d}\left\{\left(\frac{\sin 2\pi\omega_x}{\omega_x}-2\pi\right)m_x+2\left(\frac{\sin\pi\omega_x}{\omega_x}-\pi\cos\pi\omega_x\right)\right\}$$
$$C_{My}=-\frac{n-1}{8}\frac{\beta^2 d^2 g}{\omega_y^2\, d}\left\{\left(\frac{\sin 2\pi\omega_y}{\omega_y}-2\pi\right)m_y+2\left(\frac{\sin\pi\omega_y}{\omega_y}-\pi\cos\pi\omega_y\right)\right\} \tag{41}$$

Table 20 lists $A_x\ldots D_y$ and $a_x\ldots b_y$, where

$$\frac{C_{cx}}{d}=(n-1)\left\{A_x\left(m_x^2+1\right)+B_x\right\};\qquad \frac{C_{cx}}{L}=(n-1)\left\{a_x\left(m_x^2+1\right)+b_x\right\}$$
$$\frac{C_{cy}}{d}=(n-1)\left\{A_y\left(m_y^2+1\right)+B_y\right\};\qquad \frac{C_{cy}}{L}=(n-1)\left\{a_y\left(m_y^2+1\right)+b_y\right\}$$
$$C_{Mx}=(n-1)(C_x m_x+D_x)$$
$$C_{My}=(n-1)\left(C_y m_y+D_y\right). \tag{42}$$

Formulae for the real aperture aberration coefficients are to be found in Dymnikov, Fishkova, and Yavor (1965c, 1965e); they are reproduced in QO.

1.3 The short asymmetric quadrupole: the bell-shaped model

We now use

$$k(z)=\frac{1}{\left\{1+(z/p)^2\right\}^2}\qquad z\le 0$$
$$k(z)=\frac{1}{\left\{1+(z/q)^2\right\}^2}\qquad z\ge 0. \tag{43}$$

The asymptotic cardinal elements (Hawkes, 1966a) are given by

$$\frac{1}{f}=-\frac{\cos\frac{\pi}{2}\theta_x\sin\frac{\pi}{2}\omega_x}{q\omega_x}-\frac{\sin\frac{\pi}{2}\theta_x\cos\frac{\pi}{2}\omega_x}{p\theta_x}$$
$$z_{Fi}=-f\left(\frac{q}{p}\cos\frac{\pi}{2}\theta_x\cos\frac{\pi}{2}\omega_x-\frac{\theta_x}{\omega_x}\sin\frac{\pi}{2}\theta_x\sin\frac{\pi}{2}\omega_x\right) \tag{44}$$
$$z_{Fo}=f\left(\frac{p}{q}\cos\frac{\pi}{2}\theta_x\cos\frac{\pi}{2}\omega_x-\frac{\omega_x}{\theta_x}\sin\frac{\pi}{2}\theta_x\sin\frac{\pi}{2}\omega_x\right)$$

Table 20 The bell-shaped model. The chromatic aberration coefficients $A_x \ldots D_y$ for a small range of values $\beta^2 d^2$ and the same quantities scaled with respect to L.

$\beta^2 d^2$	A_x	A_y	B_x	B_y	C_x	C_y	D_x	D_y
0.0	3.183⋆4	−3.183⋆4	−7.958⋆3	6.366⋆4	−5.000⋆-1	−5.000⋆-1	5.000⋆-1	5.000⋆-1
0.2	1.653	−1.644	−3.914⋆-1	3.145	−4.457⋆-1	−5.984⋆-1	4.221⋆-1	5.724⋆-1
0.4	9.301⋆-1	−8.935⋆-1	−1.836⋆-1	1.525	−4.279⋆-1	−7.501⋆-1	3.378⋆-1	6.403⋆-1
0.6	7.539⋆-1	−6.681⋆-1	−1.050⋆-1	9.741⋆-1	−4.407⋆-1	−9.661⋆-1	2.455⋆-1	7.043⋆-1
0.8	7.327⋆-1	−5.711⋆-1	−5.468⋆-2	6.937⋆-1	−4.799⋆-1	−1.259	1.433⋆-1	7.648⋆-1
1.0	7.964⋆-1	−5.236⋆-1	−1.044⋆-2	5.237⋆-1	−5.428⋆-1	−1.645	2.846⋆-2	8.224⋆-1
1.2	9.344⋆-1	−4.998⋆-1	3.820⋆-2	4.098⋆-1	−6.291⋆-1	−2.140	−1.029⋆-1	8.775⋆-1
1.4	1.161	−4.887⋆-1	1.005⋆-1	3.288⋆-1	−7.408⋆-1	−2.765	−2.564⋆-1	9.302⋆-1
1.6	1.519	−4.851⋆-1	1.896⋆-1	2.687⋆-1	−8.839⋆-1	−3.542	−4.412⋆-1	9.808⋆-1
1.8	2.095	−4.861⋆-1	3.288⋆-1	2.226⋆-1	−1.071	−4.497	−6.723⋆-1	1.030
2.0	3.082	−4.901⋆-1	5.670⋆-1	1.865⋆-1	−1.327	−5.660	−9.767⋆-1	1.077
$\beta^2 d^2$	a_x	a_y	b_x	b_y	C_x	C_y	D_x	D_y
0.0	2.026⋆4	−2.026⋆4	−5.066⋆3	4.053⋆4	−5.000⋆-1	−5.000⋆-1	5.000⋆-1	5.000⋆-1
0.2	1.052	−1.047	−2.492⋆-1	2.002	−4.457⋆-1	−5.984⋆-1	4.221⋆-1	5.724⋆-1
0.4	5.921⋆-1	−5.688⋆-1	−1.169⋆-1	9.710⋆-1	−4.279⋆-1	−7.501⋆-1	3.378⋆-1	6.403⋆-1
0.6	4.800⋆-1	−4.253⋆-1	−6.683⋆-2	6.201⋆-1	−4.407⋆-1	−9.661⋆-1	2.455⋆-1	7.043⋆-1

Continued

Table 20 The bell-shaped model. The chromatic aberration coefficients $A_x \ldots D_y$ for a small range of values $\beta^2 d^2$ and the same quantities scaled with respect to *L*.—cont'd

$\beta^2 d^2$	a_x	a_y	b_x	b_y	C_x	C_y	D_x	D_y
0.8	4.664⋆-1	−3.636⋆-1	−3.481⋆-2	4.416⋆-1	−4.799⋆-1	−1.259	1.433⋆-1	7.648⋆-1
1.0	5.070⋆-1	−3.333⋆-1	−6.646⋆-3	3.334⋆-1	−5.428⋆-1	−1.645	2.846⋆-2	8.224⋆-1
1.2	5.949⋆-1	−3.182⋆-1	2.432⋆-2	2.609⋆-1	−6.291⋆-1	−2.140	−1.029⋆-1	8.775⋆-1
1.4	7.394⋆-1	−3.111⋆-1	6.398⋆-2	2.093⋆-1	−7.488⋆-1	−2.765	−2.564⋆-1	9.302⋆-1
1.6	9.670⋆-1	−3.088⋆-1	1.207⋆-1	1.710⋆-1	−8.839⋆-1	−3.542	−4.412⋆-1	9.808⋆-1
1.8	1.334	−3.095⋆-1	2.093⋆-1	1.417⋆-1	−1.071	−4.497	−6.723⋆-1	1.030
2.0	1.962	−3.120⋆-1	3.610⋆-1	1.187⋆-1	−1.327	−5.660	−9.767⋆-1	1.077

with similar expressions in the $y-z$ plane; we have written

$$\begin{aligned}\omega_x^2 &= 1+\beta^2p^2 \quad \theta_x^2 = 1+\beta^2q^2\\ \omega_y^2 &= 1-\beta^2p^2 \quad \theta_y^2 = 1-\beta^2q^2.\end{aligned} \tag{45}$$

When βp and βq are small, the focal length tends to

$$f \simeq 1/\frac{\pi}{4}\beta^2(p+q) \tag{46}$$

and since the equivalent length is now

$$L = \frac{\pi}{4}(p+q) \tag{47}$$

$$\frac{f}{L} \simeq \frac{1}{\beta^2L^2} \tag{48}$$

as we should expect.

We have calculated the aberrations by treating the asymmetric lens as two separate adjoining lenses; this is convenient since a change of variable makes the formulae for the second half-lens identical with those of the first. The dependence upon m_x and m_y was obtained with the aid of Eqs. (37) and (38) by evaluating the coefficients for five object positions [for (0030) and (0003)] or for nine pairs of object positions [for (0012) and (0021)] and solving the resulting sets of simultaneous equations for the coefficients $(0030)_i$, $(0003)_i$, $(0012)_{ij}$ and $(0021)_{ij}$.

The cardinal elements are listed in Tables 21 and 22 for $p/q = \frac{1}{4}, \frac{1}{2}$ and 2,4 (normalized to $p+q=2$) and the coefficients $(00rs)_i$ and $(00rs)_{ij}$ are given in Tables 23–26 for the same values of the asymmetry parameter.

1.4 The intermediate quadrupole: The modified rectangular model

We now have

$$\begin{aligned}k(z) &= 1 && |z| \leqslant L/2\\ k(z) &= \frac{1}{\left\{1+[(z-L/2)/d]^2\right\}^2} && z \geqslant L/2\\ k(z) &= \frac{1}{\left\{1+[(z+L/2)/d]^2\right\}^2} && z \leqslant -L/2\end{aligned} \tag{49}$$

Table 21 The asymmetric bell-shaped model. Cardinal elements for $p/q = \frac{1}{4}$ and $\frac{1}{2}$ $(p + q = 2)$.

$p/q = \frac{1}{4}$

β^2	βp	βq	f	z_{Fi}	z_{Fo}	g	z_{Gi}	z_{Go}
2.000⋆-1	1.789⋆-1	7.155⋆-1	3.781	4.147	−2.913	−2.701	−1.581	3.376
4.000⋆-1	2.530⋆-1	1.012	2.266	2.182	−1.275	−1.154	2.880⋆-1	1.753
6.000⋆-1	3.098⋆-1	1.239	1.830	1.226	−6.895⋆-1	−6.608⋆-1	1.075	1.193
8.000⋆-1	3.578⋆-1	1.431	1.684	4.667⋆-1	−3.586⋆-1	−4.281⋆-1	1.580	9.019⋆-1
1.000	4.000⋆-1	1.600	1.680	−2.816⋆-1	−1.185⋆-1	−2.972⋆-1	1.964	7.186⋆-1
1.200	4.382⋆-1	1.753	1.782	−1.115	9.082⋆-2	−2.159⋆-1	2.282	5.902⋆-1
1.400	4.733⋆-1	1.893	1.997	−2.133	3.045⋆-1	−1.619⋆-1	2.558	4.936⋆-1
1.600	5.060⋆-1	2.024	2.373	−3.492	5.580⋆-1	−1.244⋆-1	2.806	4.174⋆-1
1.800	5.367⋆-1	2.147	3.035	−5.522	9.108⋆-1	−9.739⋆-2	3.033	3.550⋆-1
2.000	5.657⋆-1	2.263	4.338	−9.113	1.513	−7.744⋆-2	3.244	3.025⋆-1
2.200	5.933⋆-1	2.373	7.756	−1.791⋆1	2.965	−6.237⋆-2	3.441	3.573⋆-1
2.400	6.197⋆-1	2.479	3.528⋆1	−8.612⋆1	1.419⋆1	−5.078⋆-2	3.627	2.178⋆-1
2.600	6.450⋆-1	2.580	−1.464⋆1	3.654⋆1	−6.004	−4.174⋆-2	3.804	1.827⋆-1
2.800	6.693⋆-1	2.677	−6.299	1.560⋆1	−2.565	−3.459⋆-2	3.973	1.512⋆-1
3.000	6.928⋆-1	2.771	−4.146	9.898	−1.637	−2.887⋆-2	4.135	1.226⋆-1
3.200	7.155⋆-1	2.862	−3.180	7.103	−1.191	−2.425⋆-2	4.290	9.641⋆-2
3.400	7.376⋆-1	2.950	−2.649	5.355	−9.204⋆-1	−2.049⋆-2	4.441	7.231⋆-2
3.600	7.589⋆-1	3.036	−2.325	4.091	−7.322⋆-1	−1.740⋆-2	4.586	4.996⋆-2
3.800	7.797⋆-1	3.119	−2.120	3.084	−5.891⋆-1	−1.485⋆-2	4.727	2.912⋆-2
4.000	8.000⋆-1	3.200	−1.989	2.220	−4.728⋆-1	−1.273⋆-2	4.863	9.592⋆-3

$p/q = \frac{1}{2}$

β^2	βp	βq	f	z_{Fi}	z_{Fo}	g	z_{Gi}	z_{Go}
2.000⋆-1	2.981⋆-1	5.963⋆-1	3.731	3.885	−3.193	−2.733	−2.064	3.057
4.000⋆-1	4.216⋆-1	8.433⋆-1	2.202	2.054	−1.534	−1.179	−2.883⋆-1	1.414
6.000⋆-1	5.164⋆-1	1.033	1.747	1.257	−9.279⋆-1	−6.819⋆-1	4.141⋆-1	8.361⋆-1
8.000⋆-1	5.963⋆-1	1.193	1.574	6.895⋆-1	−5.766⋆-1	−4.455⋆-1	8.405⋆-1	5.264⋆-1
1.000	6.667⋆-1	1.333	1.530	1.817⋆-1	−3.175⋆-1	−3.117⋆-1	1.152	3.256⋆-1
1.200	7.303⋆-1	1.461	1.570	−3.360⋆1	−9.267⋆-2	−2.280⋆ 1	1.402	1.801⋆-1
1.400	7.888⋆-1	1.578	1.687	−9.125⋆-1	1.285⋆-1	−1.721⋆-1	1.614	6.703⋆-2
1.600	8.433⋆-1	1.687	1.892	−1.603	3.785⋆-1	−1.330⋆-1	1.802	−2.533⋆-2
1.800	8.944⋆-1	1.789	2.222	−2.492	6.633⋆-1	−1.047⋆-1	1.971	−1.035⋆-1
2.000	9.428⋆-1	1.886	2.760	−3.742	1.058	−8.370⋆-2	2.127	−1.713⋆-1
2.200	9.888⋆-1	1.978	3.709	−5.731	1.671	−6.774⋆-2	2.272	−2.315⋆-1
2.400	1.033	2.066	5.701	−9.615	2.851	−5.541⋆-2	2.407	−2.856⋆-1
2.600	1.075	2.150	1.211⋆1	−2.154⋆1	6.450	−4.574⋆-2	2.536	−3.350⋆-1
2.800	1.116	2.231	−1.533⋆2	2.797⋆2	−8.424⋆1	−3.805⋆-2	2.658	−3.805⋆-1
3.000	1.155	2.309	−1.109⋆1	2.021⋆1	−6.113	−3.188⋆-2	2.775	−4.228⋆-1
3.200	1.193	2.385	−5.990	1.064⋆1	−3.231	−2.688⋆-2	2.886	−4.624⋆-1
3.400	1.229	2.459	−4.240	7.173	−2.186	−2.279⋆-2	2.994	−4.996⋆-1
3.600	1.265	2.530	−3.376	5.303	−1.624	−1.942⋆-2	3.098	−5.349⋆-1
3.800	1.300	2.599	−2.876	4.081	−1.259	−1.662⋆-2	3.198	−5.684⋆-1
4.000	1.333	2.667	−2.564	3.179	−9.916⋆-1	−1.429⋆-2	3.296	−6.004⋆-1

Table 22 The asymmetric bell-shaped model. Cardinal elements for $p/q = 2$ and 4 ($p + q = 2$).

$p/q = 4$

β^2	βp	βq	f	z_{Fi}	z_{Fo}	g	z_{Gi}	z_{Go}
2.000⋆-1	7.155⋆-1	1.789⋆-1	3.781	2.913	−4.147	−2.701	−3.376	1.581
4.000⋆-1	1.012	2.530⋆-1	2.266	1.275	−2.182	−1.154	−1.753	−2.880⋆-1
6.000⋆-1	1.239	3.098⋆-1	1.830	6.895⋆-1	−1.226	−6.608⋆-1	−1.193	−1.075
8.000⋆-1	1.431	3.578⋆-1	1.684	3.586⋆-1	−4.667⋆-1	−4.281⋆-1	−9.019⋆-1	−1.588
1.000	1.600	4.000⋆-1	1.680	1.185⋆-1	2.816⋆-1	−2.972⋆-1	−7.186⋆-1	−1.964
1.200	1.753	4.382⋆-1	1.782	−9.082⋆-2	1.115	−2.159⋆-1	−5.902⋆-1	−2.282
1.400	1.893	4.733⋆-1	1.997	−3.045⋆-1	2.133	−1.619⋆-1	−4.936⋆-1	−2.558
1.600	2.024	5.060⋆-1	2.373	−5.580⋆-1	3.492	−1.244⋆-1	−4.174⋆-1	−2.806
1.800	2.147	5.367⋆-1	3.035	−9.108⋆-1	5.522	−9.739⋆-2	−3.550⋆-1	−3.033
2.000	2.263	5.657⋆-1	4.338	−1.513	9.113	−7.744⋆-2	−3.025⋆-1	−3.244
2.200	2.373	5.933⋆-1	7.756	−2.965	1.791⋆1	−6.237⋆-2	−2.573⋆-1	−3.441
2.400	2.479	6.197⋆-1	3.528⋆1	−1.419⋆1	8.612⋆1	−5.078⋆-2	−2.178⋆-1	−3.627
2.600	2.580	6.450⋆-1	−1.464⋆1	6.004	−3.654⋆1	−4.174⋆-2	−1.827⋆-1	−3.804
2.800	2.677	6.693⋆-1	−6.299	2.565	−1.560⋆1	−3.459⋆-2	−1.512⋆-1	−3.973
3.000	2.771	6.928⋆-1	−4.146	1.637	−9.898	−2.887⋆-2	−1.226⋆-1	−4.135
3.200	2.862	7.155⋆-1	−3.180	1.191	−7.103	−2.425⋆-2	−9.641⋆-2	−4.290

3.4 00	2.950	7.376⋆-1	−2.649	9.204⋆-1	−5.355	−2.049⋆-2	−7.231⋆-2	−4.441
3.600	3.036	7.589⋆-1	−2.325	7.322⋆-1	−4.091	−1.740⋆-2	−4.996⋆-2	−4.586
3.800	3.119	7.797⋆-1	−2.120	5.891⋆-1	−3.084	−1.485⋆-2	−2.912⋆-2	−4.727
4.000	3.200	8.000⋆-1	−1.989	4.728⋆-1	−2.220	−1.273⋆-2	−9.593⋆-3	−4.863
				$p/q = 2$				
β^2	βp	βq	f	z_{Fi}	z_{Fo}	g	z_{Gi}	z_{Go}
2.000⋆-1	5.963⋆-1	2.981⋆-1	3.731	3.193	−3.885	−2.733	−3.057	2.064
4.000⋆-1	8.433⋆-1	4.216⋆-1	2.202	1.534	−2.054	−1.179	−1.414	2.883⋆-1
6.000⋆-1	1.033	5.164⋆-1	1.747	9.279⋆-1	−1.257	−6.819⋆-1	−8.361⋆-1	−4.141⋆-1
8.000⋆-1	1.193	5.963⋆-1	1.574	5.766⋆-1	−6.895⋆-1	−4.455⋆-1	−5.264⋆-1	−8.405⋆-1
1.000	1.333	6.667⋆-1	1.530	3.175⋆-1	−1.817⋆-1	−3.117⋆-1	−3.256⋆-1	−1.152
1.200	1.461	7.303⋆-1	1.570	9.267⋆-2	3.360⋆-1	−2.280⋆-1	−1.801⋆-1	−1.402
1.400	1.578	7.888⋆-1	1.687	−1.285⋆-1	9.125⋆-1	−1.721⋆-1	−6.703⋆-2	−1.614
1.600	1.687	8.433⋆-1	1.892	−3.705⋆-1	1.603	−1.330⋆-1	2.533⋆-2	−1.802
1.800	1.789	8.944⋆-1	2.222	−6.633⋆-1	2.492	−1.047⋆-1	1.035⋆-1	−1.971
2.000	1.886	9.428⋆-1	2.760	−1.058	3.742	−8.370⋆-2	1.713⋆-1	−2.127

Continued

Table 22 The asymmetric bell-shaped model. Cardinal elements for $p/q = 2$ and 4 $(p + q = 2)$.—cont'd

$p/q = 2$								
β^2	βp	βq	f	z_{Fi}	z_{Fo}	g	z_{Gi}	z_{Go}
2.200	1.978	9.888⋆-1	3.709	−1.671	5.731	−6.774⋆-2	2.315⋆-1	−2.272
2.400	2.066	1.033	5.701	−2.851	9.615	−5.541⋆-2	2.856⋆-1	−2.407
2.600	2.150	1.075	1.211⋆1	−6.450	2.154⋆1	−4.574⋆-2	3.350⋆-1	−2.536
2.800	2.231	1.116	−1.533⋆2	8.424⋆1	−2.797⋆2	−3.805⋆-2	3.805⋆-1	−2.658
3.000	2.309	1.155	−1.109⋆1	6.113	−2.021⋆1	−3.180⋆-2	4.228⋆-1	−2.775
3.200	2.385	1.193	−5.990	3.231	−1.064⋆1	−2.688⋆-2	4.624⋆-1	−2.886
3.400	2.459	1.229	−4.240	2.186	−7.173	−2.279⋆-2	4.996⋆-1	−2.994
3.600	2.530	1.265	−3.376	1.624	−5.303	−1.942⋆-2	5.349⋆-1	−3.098
3.800	2.599	1.300	−2.876	1.259	−4.081	−1.662⋆-2	5.684⋆-1	−3.198
4.000	2.667	1.333	−2.564	9.916⋆-1	−3.179	−1.429⋆-2	6.004⋆-1	−3.296

Table 23 The asymmetric bell-shaped model. The aperture aberration coefficients $(0030)_i$. For $p/q = \frac{1}{4}, \frac{1}{2}, 2, 4$ $(p + q = 2)$.

Electrostatic $p/q = \frac{1}{4}$

β^2	$(30)_4$	$(30)_3$	$(30)_2$	$(30)_1$	$(30)_0$
2.000⋆-1	7.748	−2.498⋆1	3.242⋆1	−1.982⋆1	4.943
4.000⋆-1	3.505	−8.178	9.278	−5.091	1.370
6.000⋆-1	3.123	−4.495	4.805	−2.184	7.323⋆-1
8.000⋆-1	3.806	−2.574	3.424	−9.696⋆-1	5.533⋆-1
1.000	5.639	−3.576⋆-1	3.462	−6.562⋆-2	5.427⋆-1
1.200	9.784	4.183	5.378	1.166	6.696⋆-1
1.400	1.994⋆1	1.640⋆1	1.212⋆1	3.813	1.036
1.600	4.932⋆1	5.587⋆1	3.547⋆1	1.146⋆1	2.059
1.800	1.591⋆2	2.170⋆2	1.331⋆2	4.075⋆1	5.626
2.000	7.857⋆2	1.192⋆3	7.297⋆2	2.111⋆2	2.467⋆1
2.200	9.354⋆3	1.499⋆4	9.185⋆3	2.551⋆3	2.712⋆2
2.400	4.605⋆6	7.487⋆6	4.569⋆6	1.240⋆6	1.264⋆5
2.600	1.550⋆5	2.479⋆5	1.494⋆5	4.021⋆4	4.081⋆3
2.800	5.976⋆3	9.141⋆3	5.399⋆3	1.455⋆3	1.512⋆2
3.000	1.248⋆3	1.780⋆3	1.025⋆3	2.776⋆2	3.022⋆1

Electrostatic $p/q = \frac{1}{2}$

β^2	$(30)_4$	$(30)_3$	$(30)_2$	$(30)_1$	$(30)_0$
2.000⋆-1	6.919	−2.360⋆1	3.227⋆1	−2.075⋆1	5.389
4.000⋆-1	2.728	−7.323	9.225	−5.624	1.610
6.000⋆-1	2.092	−3.888	4.763	−2.596	9.151⋆-1
8.000⋆-1	2.190	−2.335	3.294	−1.350	7.179⋆-1
1.000	2.771	−1.043	2.969	−5.065⋆-1	7.043⋆-1
1.200	4.026	8.376⋆-1	3.651	4.382⋆-1	8.255⋆-1
1.400	6.614	4.665	6.137	2.024	1.132
1.600	1.237⋆1	1.397⋆1	1.315⋆1	5.465	1.814
1.800	2.701⋆1	4.008⋆1	3.386⋆1	1.448⋆1	3.470
2.000	7.232⋆1	1.278⋆2	1.050⋆2	4.340⋆1	8.310

Continued

Table 23 The asymmetric bell-shaped model. The aperture aberration coefficients $(0030)_i$. For $p/q = \frac{1}{4}, \frac{1}{2}, 2, 4$ $(p + q = 2)$.—cont'd

Electrostatic $p/q = \frac{1}{2}$

β^2	$(30)_4$	$(30)_3$	$(30)_2$	$(30)_1$	$(30)_0$
2.200	2.616⋆2	5.171⋆2	4.246⋆2	1.676⋆2	2.748⋆1
2.400	1.601⋆3	3.387⋆3	2.793⋆3	1.061⋆3	1.572⋆2
2.600	3.540⋆4	7.734⋆4	6.389⋆4	2.365⋆4	3.311⋆3
2.800	9.822⋆8	2.155⋆9	1.772⋆9	6.481⋆8	8.887⋆7
3.000	2.878⋆4	6.183⋆4	5.034⋆4	1.839⋆4	2.546⋆3

Electrostatic $p/q = 4$

β^2	$(30)_4$	$(30)_3$	$(30)_2$	$(30)_1$	$(30)_0$
2.000⋆-1	4.943	−1.982⋆1	3.242⋆1	−2.498⋆1	7.748
4.000⋆-1	1. 370	−5.091	9.278	−8.178	3.505
6.000⋆-1	7.323⋆-1	−2.184	4.805	−4.495	3.123
8.000⋆-1	5.533⋆-1	−9.696⋆-1	3.424	−2.574	3.806
1.000	5.427⋆-1	−6.562⋆-2	3.462	−3.576⋆-1	5.639
1.200	6.696⋆-1	1.166	5.378	4.183	9.784
1.400	1.036	3.813	1.212⋆1	1.640⋆1	1.994⋆1
1.600	2.059	1.146⋆1	3.547⋆1	5.587⋆1	4.932⋆1
1.800	5.626	4.075⋆1	1.331⋆2	2.170⋆2	1.591⋆2
2.000	2.467⋆1	2.111⋆2	7.297⋆2	1.192⋆3	7.857⋆2
2.200	2.712⋆2	2.551⋆3	9.185⋆3	1.499⋆4	9.354⋆3
2.400	1.264⋆5	1.240⋆6	4.569⋆6	7.487⋆6	4.605⋆6
2.600	4.081⋆3	4.021⋆4	1.494⋆5	2.479⋆5	1.550⋆5
2.800	1.512⋆2	1.455⋆3	5.399⋆3	9.141⋆3	5.976⋆3
3.000	3.022⋆1	2.776⋆2	1.025⋆3	1.780⋆3	1.248⋆3

Electrostatic $p/q = 2$

β^2	$(30)_4$	$(30)_3$	$(30)_2$	$(30)_1$	$(30)_0$
2.000⋆-1	5.389	−2.075⋆1	3.227⋆1	−2.360⋆1	6.919
4.000⋆-1	1.610	−5.624	9.225	−7.323	2.728
6.000⋆-1	9.151⋆-1	−2.596	4.763	−3.888	2.092
8.000⋆-1	7.179⋆-1	−1.350	3.294	−2.335	2.190

Table 23 The asymmetric bell-shaped model. The aperture aberration coefficients $(0030)_i$. For $p/q = \frac{1}{4}, \frac{1}{2}, 2, 4$ $(p + q = 2)$.—cont'd

Electrostatic $p/q = 2$

β^2	$(30)_4$	$(30)_3$	$(30)_2$	$(30)_1$	$(30)_0$
1.000	7.043⋆-1	−5.065⋆-1	2.969	−1.043	2.771
1.200	8.255⋆-1	4.382⋆-1	3.651	8.376⋆-1	4.026
1.400	1.132	2.024	6.137	4.665	6.614
1.600	1.814	5.465	1.315⋆1	1.397⋆1	1.237⋆1
1.800	3.470	1.448⋆1	3.386⋆1	4.008⋆1	2.701⋆1
2.000	8.310	4.340⋆1	1.050⋆2	1.278⋆2	7.232⋆1
2.200	2.748⋆1	1.676⋆2	4.246⋆2	5.171⋆2	2.616⋆2
2.400	1.572⋆2	1.061⋆3	2.793⋆3	3.387⋆3	1.601⋆3
2.600	3.311⋆3	2.365⋆4	6.389⋆4	7.734⋆4	3.540⋆4
2.800	8.887⋆7	6.481⋆8	1.772⋆9	2.155⋆9	9.822⋆8
3.000	2.546⋆3	1.839⋆4	5.034⋆4	6.183⋆4	2.878⋆4

Magnetic $p/q = \frac{1}{4}$

β^2	$(30)_4$	$(30)_3$	$(30)_2$	$(30)_1$	$(30)_0$
2.000⋆-1	2.621	−7.401	9.355	−6.042	1.761
4.000⋆-1	1.355	−2.322	2.800	−1.622	5.572⋆-1
6.000⋆-1	1.377	−1.049	1.624	−6.884⋆-1	3.218⋆-1
8.000⋆-1	1.843	−1.738⋆-1	1.423	−2.364⋆-1	2.546⋆-1
1.000	2.862	1.066	1.823	1.844⋆-1	2.588⋆-1
1.200	5.018	3.656	3.221	8.515⋆-1	3.314⋆-1
1.400	1.009⋆1	1.028⋆1	7.225	2.325	5.303⋆-1
1.600	2.430⋆1	3.054⋆1	1.984⋆1	6.450	1.074
1.800	7.597⋆1	1.094⋆2	6.927⋆1	2.152⋆1	2.912
2.000	3.638⋆2	5.686⋆2	3.564⋆2	1.049⋆2	1.233⋆1
2.200	4.218⋆3	6.853⋆3	4.256⋆3	1.196⋆3	1.284⋆2
2.400	2.031⋆6	3.311⋆6	2.026⋆6	5.517⋆5	5.636⋆4
2.600	6.714⋆4	1.067⋆5	6.388⋆4	1.709⋆4	1.724⋆3
2.800	2.550⋆3	3.845⋆3	2.239⋆3	5.957⋆2	6.124⋆1
3.000	5.262⋆2	7.338⋆2	4.142⋆2	1.104⋆2	1.191⋆1

Continued

Table 23 The asymmetric bell-shaped model. The aperture aberration coefficients $(0030)_i$. For $p/q = \frac{1}{4}, \frac{1}{2}, 2, 4$ $(p + q = 2)$.—cont'd

Magnetic $p/q = \frac{1}{2}$

β^2	$(30)_4$	$(30)_3$	$(30)_2$	$(30)_1$	$(30)_0$
2.000⋆-1	2.358	−7.039	9.300	−6.290	1.892
4.000⋆-1	1.051	−2.142	2.760	−1.754	6.375⋆-1
6.000⋆-1	9.146⋆-1	−1.002	1.559	−7.861⋆-1	3.955⋆-1
8.000⋆-1	1.064	−3.642⋆-1	1.268	−3.309⋆-1	3.341⋆-1
1.000	1.444	3.155⋆-1	1.409	4.797⋆-2	3.496⋆-1
1.200	2.171	1.418	2.056	5.500⋆-1	4.311⋆-1
1.400	3.592	3.648	3.700	1.447	6.110⋆-1
1.600	6.632	8.843	7.825	3.382	9.929⋆-1
1.800	1.412⋆1	2.277⋆1	1.920⋆1	8.296	1.889
2.000	3.667⋆1	6.767⋆1	5.637⋆1	2.348⋆1	4.429
2.200	1.285⋆2	2.602⋆2	2.166⋆2	8.628⋆1	1.416⋆1
2.400	7.643⋆2	1.636⋆3	1.363⋆3	5.218⋆2	7.776⋆1
2.600	1.647⋆4	3.617⋆4	3.002⋆4	1.116⋆4	1.568⋆3
2.800	4.471⋆8	9.804⋆8	8.062⋆8	2.947⋆8	4.039⋆7
3.000	1.286⋆4	2.749⋆4	2.227⋆4	8.096⋆3	1.115⋆3

Magnetic $p/q = 4$

β^2	$(30)_4$	$(30)_3$	$(30)_2$	$(30)_1$	$(30)_0$
2.000⋆-1	1.761	−6.042	9.355	−7.401	2.621
4.000⋆-1	5.572⋆-1	−1.622	2.800	−2.322	1.355
6.000⋆-1	3.218⋆-1	−6.884⋆-1	1.624	−1.049	1.377
8.000⋆-1	2.546⋆-1	−2.364⋆-1	1.423	−1.738⋆-1	1.843
1.000	2.588⋆-1	1.844⋆-1	1.823	1.066	2.862
1.200	3.314⋆-1	8.515⋆-1	3.221	3.656	5.018
1.400	5.303⋆-1	2.325	7.225	1.028⋆1	1.009⋆1
1.600	1.074	6.450	1.984⋆1	3.054⋆1	2.430⋆1

Table 23 The asymmetric bell-shaped model. The aperture aberration coefficients $(0030)_i$. For $p/q = \frac{1}{4}, \frac{1}{2}, 2, 4$ $(p + q = 2)$.—cont'd

Magnetic $p/q = 4$

β^2	$(30)_4$	$(30)_3$	$(30)_2$	$(30)_1$	$(30)_0$
1.800	2.912	2.152⋆1	6.927⋆1	1.094⋆2	7.597⋆1
2.000	1.233⋆1	1.049⋆2	3.564⋆2	5.686⋆2	3.638⋆2
2.200	1.284⋆2	1.196⋆3	4.256⋆3	6.853⋆3	4.218⋆3
2.400	5.636⋆4	5.517⋆5	2.026⋆6	3.311⋆6	2.031⋆6
2.600	1.724⋆3	1.709⋆4	6.388⋆4	1.067⋆5	6.714⋆4
2.800	6.124⋆1	5.957⋆2	2.239⋆3	3.845⋆3	2.550⋆3
3.000	1.191⋆1	1.104⋆2	4.142⋆2	7.338⋆2	5.262⋆2

Magnetic $p/q = 2$

β^2	$(30)_4$	$(30)_3$	$(30)_2$	$(30)_1$	$(30)_0$
2.000⋆-1	1.892	−6.290	9.300	−7.039	2.358
4.000⋆-1	6.375⋆-1	−1.754	2.760	−2.142	1.051
6.000⋆-1	3.955⋆-1	−7.861⋆-1	1.559	−1.002	9.146⋆-1
8.000⋆-1	3.341⋆-1	−3.309⋆-1	1.268	−3.642⋆-1	1.064
1.000	3.496⋆-1	4.797⋆-2	1.409	3.155⋆-1	1.444
1.200	4.311⋆-1	5.500⋆-1	2.056	1.418	2.171
1.400	6.110⋆-1	1.447	3.700	3.648	3.592
1.600	9.929⋆-1	3.382	7.825	8.843	6.632
1.800	1.889	8.296	1.920⋆1	2.277⋆1	1.412⋆1
2.000	4.429	2.348⋆1	5.637⋆1	6.767⋆1	3.667⋆1
2.200	1.416⋆1	8.628⋆1	2.166⋆2	2.602⋆2	1.285⋆2
2.400	7.776⋆1	5.218⋆2	1.363⋆3	1.636⋆3	7.643⋆2
2.600	1.568⋆3	1.116⋆4	3.002⋆4	3.617⋆4	1.647⋆4
2.800	4.039⋆7	2.947⋆8	8.062⋆8	9.804⋆8	4.471⋆8
3.000	1.115⋆3	8.096⋆3	2.227⋆4	2.749⋆4	1.286⋆4

Continued

Table 23 The asymmetric bell-shaped model. The aperture aberration coefficients $(0030)_i$. For $p/q = \frac{1}{4}, \frac{1}{2}, 2, 4$ $(p + q = 2)$.—cont'd

Achromatic $p/q = \frac{1}{4}$					
$\beta^2 d^2$	**(30)$_4$**	**(30)$_3$**	**(30)$_2$**	**(30)$_1$**	**(30)$_0$**
2.000⋆-1	3.646	−1.092⋆1	1.397⋆1	−8.797	2.397
4.000⋆-1	1.785	−3.493	4.095	−2.316	7.197⋆-1
6.000⋆-1	1.727	−1.738	2.260	−9.875⋆-1	4.039⋆-1
8.000⋆-1	2.236	−6.537⋆-1	1.823	−3.831⋆-1	3.143⋆-1
1.000	3.418	7.815⋆-1	2.151	1.344⋆-1	3.155⋆-1
1.200	5.971	3.762	3.652	9.145⋆-1	3.990⋆-1
1.400	1.206⋆1	1.150⋆1	8.204	2.622	6.314⋆-1
1.600	2.930⋆1	3.561⋆1	2.297⋆1	7.452	1.271
1.800	9.261⋆1	1.309⋆2	8.203⋆1	2.537⋆1	3.455
2.000	4.482⋆2	6.934⋆2	4.311⋆2	1.261⋆2	1.480⋆1
2.200	5.245⋆3	8.479⋆3	5.242⋆3	1.467⋆3	1.569⋆2
2.400	2.546⋆6	4.147⋆6	2.535⋆6	6.894⋆5	7.037⋆4
2.600	8.472⋆4	1.349⋆5	8.098⋆4	2.172⋆4	2.195⋆3
2.800	3.235⋆3	4.904⋆3	2.871⋆3	7.675⋆2	7.923⋆1
3.000	6.706⋆2	9.432⋆2	5.363⋆2	1.438⋆2	1.557⋆1
Achromatic $p/q = \frac{1}{2}$					
β^2	**(30)$_4$**	**(30)$_3$**	**(30)$_2$**	**(30)$_1$**	**(30)$_0$**
2.000⋆-1	3.270	−1.035⋆1	1.389⋆1	−9.182	2.591
4.000⋆-1	1.387	−3.178	4.053	−2.528	8.320⋆-1
6.000⋆-1	1.150	−1.579	2.200	−1.148	4.994⋆-1
8.000⋆-1	1.290	−7.584⋆-1	1.673	−5.348⋆-1	4.109⋆-1
1.000	1.709	4.379⋆-2	1.721	−6.292⋆-2	4.205⋆-1
1.200	2.542	1.302	2.375	5.277⋆-1	5.100⋆-1
1.400	4.196	3.852	4.188	1.562	7.152⋆-1

Table 23 The asymmetric bell-shaped model. The aperture aberration coefficients $(0030)_i$. For $p/q = \frac{1}{4}, \frac{1}{2}, 2, 4$ $(p + q = 2)$.—cont'd

Achromatic $p/q = \frac{1}{2}$

β^2	$(30)_4$	$(30)_3$	$(30)_2$	$(30)_1$	$(30)_0$
1.600	7.780	9.868	8.890	3.799	1.157
1.800	1.670⋆1	2.623⋆1	2.213⋆1	9.532	2.205
2.000	4.380⋆1	7.970⋆1	6.610⋆1	2.746⋆1	5.205
2.200	1.551⋆2	3.116⋆2	2.582⋆2	1.025⋆2	1.683⋆1
2.400	9.317⋆2	1.986⋆3	1.649⋆3	6.296⋆2	9.365⋆1
2.600	2.026⋆4	4.440⋆4	3.679⋆4	1.366⋆4	1.917⋆3
2.800	5.541⋆8	1.215⋆9	9.995⋆8	3.654⋆8	5.009⋆7
3.000	1.604⋆4	3.436⋆4	2.788⋆4	1.016⋆4	1.401⋆3

Achromatic $p/q = 4$

β^2	$(30)_4$	$(30)_3$	$(30)_2$	$(30)_1$	$(30)_0$
2.000⋆-1	2.397	−8.797	1.397⋆1	−1.092⋆1	3.646
4.000⋆-1	7.197⋆-1	−2.316	4.095	−3.493	1.785
6.000⋆-1	4.039⋆-1	−9.875⋆-1	2.260	−1.738	1.727
8.000⋆-1	3.143⋆-1	−3.831⋆-1	1.823	−6.537⋆-1	2.236
1.000	3.155⋆-1	1.344⋆-1	2.151	7.815⋆-1	3.418
1.200	3.990⋆-1	9.145⋆-1	3.652	3.762	5.971
1.400	6.314⋆-1	2.622	8.204	1.150⋆1	1.206⋆1
1.600	1.271	7.452	2.297⋆1	3.561⋆1	2.930⋆1
1.800	3.455	2.537⋆1	8.203⋆1	1.309⋆2	9.261⋆1
2.000	1.480⋆1	1.261⋆2	4.311⋆2	6.934⋆2	4.482⋆2
2.200	1.569⋆2	1.467⋆3	5.242⋆3	8.203⋆1	5.245⋆3
2.400	7.037⋆4	6.894⋆5	2.535⋆6	4.147⋆6	2.546⋆6
2.600	2.195⋆3	2.172⋆4	8.098⋆4	1.349⋆5	8.472⋆4
2.800	7.923⋆1	7.675⋆2	2.871⋆3	4.904⋆3	3.235⋆3
3.000	1.557⋆1	1.438⋆2	5.363⋆2	9.432⋆2	6.706⋆2

Continued

Table 23 The asymmetric bell-shaped model. The aperture aberration coefficients $(0030)_i$. For $p/q = \frac{1}{4}, \frac{1}{2}, 2, 4$ $(p+q=2)$.—cont'd

Achromatic $p/q = 2$

β^2	$(30)_4$	$(30)_3$	$(30)_2$	$(30)_1$	$(30)_0$
2.000⋆-1	2.591	−9.182	1.389⋆1	−1.035⋆1	3.270
4.000⋆-1	8.320⋆-1	−2.528	4.053	−3.178	1.387
6.000⋆-1	4.994⋆-1	−1.148	2.200	−1.579	1.150
8.000⋆-1	4.109⋆-1	−5.348⋆-1	1.673	−7.584⋆-1	1.290
1.000	4.205⋆-1	−6.292⋆-2	1.721	4.379⋆-2	1.709
1.200	5.100⋆-1	5.277⋆-1	2.375	1.302	2.542
1.400	7.152⋆-1	1.562	4.188	3.852	4.196
1.600	1.157	3.799	8.890	9.868	7.780
1.800	2.205	9.532	2.213⋆1	2.623⋆1	1.670⋆1
2.000	5.205	2.746⋆1	6.610⋆1	7.970⋆1	4.380⋆1
2.200	1.683⋆1	1.025⋆2	2.582⋆2	3.116⋆2	1.551⋆2
2.400	9.365⋆1	6.296⋆2	1.649⋆3	1.986⋆3	9.317⋆2
2.600	1.917⋆3	1.366⋆4	3.679⋆4	4.440⋆4	2.026⋆4
2.800	5.009⋆7	3.654⋆8	9.995⋆8	1.215⋆9	5.541⋆8
3.000	1.401⋆3	1.016⋆4	2.788⋆4	3.436⋆4	1.604⋆4

Notice that the $(30)_i$ for $p/q = r$ is equal to $(30)_{4-i}$ for $q/p = r$.

Table 24 The Asymmetric bell-shaped model. The aperture aberration coefficients $(0003)_i$ for $p/q = \frac{1}{4}, \frac{1}{2}, 2, 4$ $(p+q=2)$.

Electrostatic $p/q = \frac{1}{4}$

β^2	$(03)_4$	$(03)_3$	$(03)_2$	$(03)_1$	$(03)_0$
2.000⋆-1	2.752	−1.424⋆1	2.435⋆1	−1.766⋆1	4.606
4.000⋆-1	1.992⋆-1	−2.734	5.212	−4.118	1.140
6.000⋆-1	−1.988⋆-1	−9.517⋆-1	1.970	−1.702	5.012⋆-1
8.000⋆-1	−3.380⋆-1	−4.284⋆-1	9.379⋆-1	−8.885⋆-1	2.759⋆-1
1.000	−4.137⋆-1	−2.243⋆-1	5.063⋆-1	−5.265⋆-1	1.693⋆-1
1.200	−4.671⋆-1	−1.300⋆-1	2.946⋆-1	−3.310⋆-1	9.811⋆-2
1.400	−5.103⋆-1	−8.143⋆-2	1.838⋆-1	−2.399⋆-1	9.450⋆-2

Table 24 The Asymmetric bell-shaped model. The aperture aberration coefficients $(0003)_i$ for $p/q = \frac{1}{4}, \frac{1}{2}, 2, 4$ $(p+q=2)$.—cont'd

Electrostatic $p/q = \frac{1}{4}$

β^2	$(03)_4$	$(03)_3$	$(03)_2$	$(03)_1$	$(03)_0$
1.600	−5.479⋆-1	−5.388⋆-2	1.137⋆-1	−1.205⋆-1	−5.332⋆-2
1.800	−5.821⋆-1	−3.158⋆-2	−7.941⋆-2	1.358	−3.537
2.000	−6.136⋆-1	−2.765⋆-2	1.364⋆-1	−1.582	4.619
2.200	−6.433⋆-1	−3.244⋆-2	4.632⋆-1	−5.646	1.660⋆1
2.400	−6.718⋆-1	1.562⋆-2	−1.596	2.630⋆1	−8.553⋆1
2.600	−7.018⋆-1	6.239⋆-1	−3.489⋆1	6.279⋆2	−2.132⋆3
2.800	−7.124⋆-1	−2.458	1.576⋆2	−3.312⋆3	1.166⋆4
3.000	−7.626⋆-1	3.763	−3.091⋆2	7.892⋆3	−2.822⋆4

Electrostatic $p/q = \frac{1}{2}$

β^2	$(03)_4$	$(03)_3$	$(03)_2$	$(03)_1$	$(03)_0$
2.000⋆-1	3.131	−1.504⋆1	2.455⋆1	−1.695⋆1	4.158
4.000⋆-1	3.877⋆-1	−3.035	5.309	−3.811	9.030⋆-1
6.000⋆-1	−4.908⋆-2	−1.103	2.030	−1.524	3.298⋆-1
8.000⋆-1	−1.951⋆-1	−5.144⋆-1	9.786⋆-1	−7.727⋆-1	1.349⋆-1
1.000	−2.677⋆-1	−2.770⋆-1	5.349⋆-1	−4.465⋆-1	4.521⋆-2
1.200	−3.142⋆-1	−1.642⋆-1	3.166⋆-1	−2.807⋆-1	−4.309⋆-3
1.400	−3.492⋆-1	−1.043⋆-1	1.977⋆-1	−1.864⋆-1	−3.574⋆-2
1.600	−3.781⋆-1	−6.982⋆-2	1.282⋆-1	−1.281⋆-1	−5.673⋆-2
1.800	−4.035⋆-1	−4.854⋆-2	8.205⋆-2	−7.472⋆-2	−5.015⋆-2
2.000	−4.265⋆-1	−3.534⋆-2	6.134⋆-2	−7.760⋆-2	−1.121⋆-1
2.200	−4.478⋆-1	−2.692⋆-2	5.674⋆-2	−1.069⋆-1	−4.894⋆-1
2.400	−4.678⋆-1	−1.932⋆-2	9.782⋆-3	4.379⋆-2	9.426⋆-1
2.600	−4.868⋆-1	−1.923⋆-2	1.447⋆-1	−3.887⋆-1	−1.093⋆1
2.800	−5.050⋆-1	−1.651⋆-2	1.463⋆-1	−4.545⋆-2	−1.807⋆1
3.000	−5.223⋆-1	−3.405⋆-2	8.531⋆-1	1.848	−1.849⋆2

Electrostatic $p/q = 4$

β^2	$(03)_4$	$(03)_3$	$(03)_2$	$(03)_1$	$(03)_0$
2.000⋆-1	4.606	−1.766⋆1	2.435⋆1	−1.424⋆1	2.752
4.000⋆-1	1.140	−4.118	5.212	−2.734	1.992⋆-1
6.000⋆-1	5.012⋆-1	−1.702	1.970	−9.517⋆-1	−1.908⋆-1

Continued

Table 24 The Asymmetric bell-shaped model. The aperture aberration coefficients $(0003)_i$ for $p/q = \frac{1}{4}, \frac{1}{2}, 2, 4$ $(p+q=2)$.—cont'd

Electrostatic $p/q = 4$

β^2	$(03)_4$	$(03)_3$	$(03)_2$	$(03)_1$	$(03)_0$
8.000⋆-1	2.759⋆-1	−8.885⋆-1	9.379⋆-1	−4.284⋆-1	−3.380⋆-1
1.000	1.698⋆-1	−5.269⋆-1	5.064⋆-1	−2.243⋆-1	−4.137⋆-1
1.200	1.105⋆-1	−3.387⋆-1	2.960⋆-1	−1.301⋆-1	−4.671⋆-1
1.400	7.353⋆-2	−2.302⋆-1	1.828⋆-1	−8.142⋆-2	−5.103⋆-1
1.600	4.845⋆-2	−1.629⋆-1	1.176⋆-1	−5.400⋆-2	−5.479⋆-1
1.800	3.038⋆-2	−1.190⋆-1	7.808⋆-2	−3.750⋆-2	−5.820⋆-1
2.000	1.675⋆-2	−8.904⋆-2	5.315⋆-2	−2.702⋆-2	−6.136⋆-1
2.200	6.053⋆-3	−6.798⋆-2	3.693⋆-2	−2.007⋆-2	−6.434⋆-1
2.400	−2.595⋆-3	−5.277⋆-2	2.611⋆-2	−1.527⋆-2	−6.715⋆-1
2.600	−9.766⋆-3	−4.154⋆-2	1.874⋆-2	−1.185⋆-2	−6.981⋆-1
2.800	−1.584⋆-2	−3.309⋆-2	1.362⋆-2	−9.410⋆-3	−7.247⋆-1
3.000	−2.108⋆-2	−2.664⋆-2	1.001⋆-2	−7.659⋆-3	−7.512⋆-1

Electrostatic $p/q = 2$

β^2	$(03)_4$	$(03)_3$	$(03)_2$	$(03)_1$	$(03)_0$
2.000⋆-1	4.158	−1.695⋆1	2.455⋆1	−1.504⋆1	3.131
4.000⋆-1	9.030⋆-1	−3.811	5.309	−3.035	3.877⋆-1
6.000⋆-1	3.298⋆-1	−1.524	2.030	−1.103	−4.908⋆-2
8.000⋆-1	1.349⋆-1	−7.727⋆-1	9.786⋆-1	−5.144⋆-1	−1.951⋆-1
1.000	4.522⋆-2	−4.465⋆-1	5.349⋆-1	−2.770⋆-1	−2.677⋆-1
1.200	−4.407⋆-3	−2.805⋆-1	3.165⋆-1	−1.642⋆-1	−3.142⋆-1
1.400	−3.558⋆-2	−1.868⋆-1	1.978⋆-1	−1.043⋆-1	−3.492⋆-1
1.600	−5.709⋆-2	−1.299⋆-1	1.287⋆-1	−6.986⋆-2	−3.781⋆-1
1.800	−7.304⋆-2	−9.336⋆-2	8.639⋆-2	−4.879⋆-2	−4.035⋆-1
2.000	−8.555⋆-2	−6.890⋆-2	5.942⋆-2	−3.525⋆-2	−4.265⋆-1
2.200	−9.579⋆-2	−5.195⋆-2	4.171⋆-2	−2.619⋆-2	−4.478⋆-1
2.400	−1.045⋆-1	−3.989⋆-2	2.978⋆-2	−1.991⋆-2	−4.679⋆-1

Table 24 The Asymmetric bell-shaped model. The aperture aberration coefficients $(0003)_i$ for $p/q = \frac{1}{4}, \frac{1}{2}, 2, 4$ $(p + q = 2)$.—cont'd

Electrostatic $p/q = 2$

β^2	$(03)_4$	$(03)_3$	$(03)_2$	$(03)_1$	$(03)_0$
2.600	−1.121⋆-1	−3.109⋆-2	2.156⋆-2	−1.544⋆-2	−4.870⋆-1
2.800	−1.188⋆-1	−2.456⋆-2	1.582⋆-2	−1.217⋆-2	−5.048⋆-1
3.000	−1.249⋆-1	−1.962⋆-2	1.173⋆-2	−9.735⋆-3	−5.228⋆-1

Magnetic $p/q = \frac{1}{4}$

β^2	$(03)_4$	$(03)_3$	$(03)_2$	$(03)_1$	$(03)_0$
2.000⋆-1	2.550⋆-1	−3.658	6.996	−4.686	8.985⋆-1
4.000⋆-1	−3.498⋆-1	−5.799⋆-1	1.514	−1.011	9.479⋆-2
6.000⋆-1	−4.552⋆-1	−1.461⋆-1	5.798⋆-1	−3.861⋆-1	−2.705⋆-2
8.000⋆1	−5.110⋆-1	−3.600⋆-2	2.796⋆-1	−1.861⋆-1	−6.184⋆-2
1.000	−5.559⋆-1	−1.610⋆-3	1.528⋆-1	−1.018⋆-1	−7.514⋆-2
1.200	−5.966⋆-1	9.574⋆-3	8.983⋆-2	−5.745⋆-2	−8.570⋆-2
1.400	−6.348⋆-1	1.253⋆-2	5.678⋆-2	−4.173⋆-2	−7.559⋆-2
1.600	−6.710⋆-1	1.258⋆-2	3.508⋆-2	−8.204⋆-3	−1.252⋆-1
1.800	−7.056⋆-1	1.375⋆-2	−3.671⋆-2	5.577⋆-1	−1.474
2.000	−7.386⋆-1	9.837⋆-3	4.909⋆-2	−5.904⋆-1	1.697
2.200	−7.704⋆-1	3.931⋆-3	1.767⋆-1	−2.166	6.329
2.400	−8.066⋆-1	7.614⋆-1	−3.295⋆1	4.995⋆2	−1.594⋆3
2.600	−8.318⋆-1	2.514⋆-1	−1.345⋆1	2.420⋆2	−8.215⋆2
2.800	−8.546⋆-1	−9.364⋆-1	6.061⋆1	−1.274⋆3	4.487⋆3
3.000	−8.920⋆-1	1.453	−1.187⋆2	3.031⋆3	−1.084⋆4

Magnetic $p/q = \frac{1}{2}$

β^2	$(03)_4$	$(03)_3$	$(03)_2$	$(03)_1$	$(03)_0$
2.000⋆-1	4.033⋆-1	−3.896	7.051	−4.471	7.597⋆-1
4.000⋆-1	−2.338⋆-1	−6.699⋆-1	1.541	−9.133⋆-1	1.030⋆-2
6.000⋆-1	−3.324⋆-1	−1.909⋆-1	5.971⋆-1	−3.269⋆-1	−9.802⋆-2
8.000⋆-1	−3.749⋆-1	−6.086⋆-2	2.916⋆-1	−1.461⋆-1	−1.290⋆-1
1.000	−4.057⋆-1	−1.624⋆-2	1.614⋆-1	−7.314⋆-2	−1.420⋆-1
1.200	−4.327⋆-1	6.203⋆-4	9.665⋆-2	−3.908⋆-2	−1.492⋆-1

Continued

Table 24 The Asymmetric bell-shaped model. The aperture aberration coefficients $(0003)_i$ for $p/q = \frac{1}{4}, \frac{1}{2}, 2, 4$ $(p+q=2)$.—cont'd

Magnetic $p/q = \frac{1}{2}$

β^2	$(03)_4$	$(03)_3$	$(03)_2$	$(03)_1$	$(03)_0$
1.400	−4.578⋆-1	6.982⋆-3	6.100⋆-2	−2.130⋆-2	−1.546⋆-1
1.600	−4.817⋆-1	9.033⋆-3	3.993⋆-2	−1.120⋆-2	−1.587⋆-1
1.800	−5.046⋆-1	9.325⋆-3	2.547⋆-2	8.740⋆-4	−1.539⋆-1
2.000	−5.266⋆-1	8.641⋆-3	1.959⋆-2	−6.651⋆-3	−1.771⋆-1
2.200	−5.479⋆-1	7.556⋆-3	1.920⋆-2	−2.279⋆-2	−3.242⋆-1
2.400	−5.684⋆-1	7.180⋆-3	1.790⋆-3	3.238⋆-2	2.338⋆-1
2.600	−5.883⋆-1	4.619⋆-3	5.485⋆-2	−1.386⋆-1	−4.385
2.800	−6.076⋆-1	3.631⋆-3	5.580⋆-2	−7.276⋆-3	−7.148
3.000	−6.264⋆-1	−4.798⋆-3	3.297⋆-1	7.246⋆-1	−7.177⋆1

Magnetic $p/q = 4$

β^2	$(03)_4$	$(03)_3$	$(03)_2$	$(03)_1$	$(03)_0$
2.000⋆-1	8.985⋆-1	−4.686	6.996	−3.658	2.550⋆-1
4.000⋆-1	9.479⋆-2	−1.011	1.514	−5.799⋆-1	−3.498⋆-1
6.000⋆-1	−2.705⋆-2	−3.861⋆-1	5.798⋆-1	−1.461⋆-1	−4.552⋆-1
8.000⋆-1	−6.184⋆-2	−1.861⋆-1	2.796⋆-1	−3.600⋆-2	−5.110⋆-1
1.000	−7.496⋆-2	−1.019⋆-1	1.529⋆-1	−1.615⋆-3	−5.559⋆-1
1.200	−8.082⋆-2	−6.048⋆-2	9.038⋆-2	9.537⋆-3	−5.966⋆-1
1.400	−8.379⋆-2	−3.793⋆-2	5.640⋆-2	1.254⋆-2	−6.348⋆-1
1.600	−8.549⋆-2	−2.476⋆-2	3.663⋆-2	1.253⋆-2	−6.710⋆-1
1.800	−8.661⋆-2	−1.665⋆-2	2.452⋆-2	1.146⋆-2	−7.055⋆-1
2.000	−8.744⋆-2	−1.145⋆-2	1.682⋆-2	1.008⋆-2	−7.387⋆-1
2.200	−8.815⋆-2	−8.016⋆-3	1.177⋆-2	8.713⋆-3	−7.705⋆-1
2.400	−8.880⋆-2	−5.686⋆-3	8.379⋆-3	7.465⋆-3	−8.018⋆-1
2.600	−8.943⋆-2	−4.072⋆-3	6.049⋆-3	6.416⋆-3	−8.303⋆-1
2.800	−9.006⋆-2	−2.936⋆-3	4.421⋆-3	5.463⋆-3	−8.589⋆-1
3.000	−9.070⋆-2	−2.125⋆-3	3.266⋆-3	4.856⋆-3	−8.755⋆-1

Magnetic $p/q = 2$

β^2	$(03)_4$	$(03)_3$	$(03)_2$	$(03)_1$	$(03)_0$
2.000⋆-1	7.597⋆-1	−4.471	7.051	−3.896	4.033⋆-1
4.000⋆-1	1.030⋆-2	−9.133⋆-1	1.541	−6.699⋆-1	−2.338⋆-1
6.000⋆-1	−9.802⋆-2	−3.269⋆-1	5.971⋆-1	−1.909⋆-1	−3.324⋆-1

Table 24 The Asymmetric bell-shaped model. The aperture aberration coefficients $(0003)_i$ for $p/q = \frac{1}{4}, \frac{1}{2}, 2, 4$ $(p + q = 2)$.—cont'd

Magnetic $p/q = 2$

β^2	$(03)_4$	$(03)_3$	$(03)_2$	$(03)_1$	$(03)_0$
8.000★-1	−1.290★-1	−1.461★-1	2.916★1	−6.086★-2	−3.749★-1
1.000	−1.420★-1	−7.314★-2	1.614★-1	−1.624★-2	−4.057★-1
1.200	−1.493★-1	−3.902★-2	9.664★-2	6.222★-4	−4.327★-1
1.400	−1.545★-1	−2.148★-2	6.106★-2	6.978★-3	−4.578★-1
1.600	−1.589★-1	−1.190★-2	4.012★-2	9.018★-3	−4.817★-1
1.800	−1.629★-1	−6.439★-3	2.717★-2	9.227★-3	−5.046★-1
2.000	−1.667★-1	−3.244★-3	1.884★-2	8.673★-3	−5.266★-1
2.200	−1.704★-1	−1.347★-3	1.333★-2	7.842★-3	−5.479★-1
2.400	−1.741★-1	−2.177★-4	9.579★-3	6.951★-3	−5.684★-1
2.600	−1.778★-1	4.458★-4	6.981★-3	6.066★-3	−5.888★-1
2.800	−1.814★-1	8.226★-4	5.150★-3	5.301★-3	−6.080★-1
3.000	−1.849★-1	1.021★-3	3.839★-3	4.601★-3	−6.273★-1

Achromatic $p/q = \frac{1}{4}$

β^2	$(03)_4$	$(03)_3$	$(03)_2$	$(03)_1$	$(03)_0$
2.000★-1	7.543★-1	−5.775	1.047★1	−7.280	1.640
4.000★-1	−2.400★-1	−1.011	2.254	−1.633	3.039★-1
6.000★-1	−4.039★-1	−3.072★-1	8.578★-1	−6.493★-1	7.860★-2
8.000★-1	−4.764★-1	−1.145★-1	4.113★-1	−3.266★-1	5.704★-3
1.000	−5.275★-1	−4.614★-2	2.235★-1	−1.867★-1	−2.624★-2
1.200	−5.707★-1	−1.835★-2	1.308★-1	−1.122★-1	−4.894★-2
1.400	−6.099★-1	−6.258★-3	8.218★-2	−8.136★-2	−4.157★-2
1.600	−6.464★-1	−7.157★-4	5.079★-2	−3.066★-2	−1.108★-1
1.800	−6.809★-1	4.687★-3	−4.525★-2	7.178★-1	−1.886
2.000	−7.136★-1	2.339★-3	6.655★-2	−7.888★-1	2.281
2.200	−7.450★-1	−3.343★-3	2.340★-1	−2.862	8.383
2.400	−7.753★-1	1.865★-2	−8.135★-1	1.339★1	−4.359★1

Continued

Table 24 The Asymmetric bell-shaped model. The aperture aberration coefficients $(0003)_i$ for $p/q = \frac{1}{4}, \frac{1}{2}, 2, 4$ $(p + q = 2)$.—cont'd

β^2	$(03)_4$	$(03)_3$	$(03)_2$	$(03)_1$	$(03)_0$
			Achromatic $p/q=\frac{1}{4}$		
2.600	−8.058⋆-1	3.259⋆-1	−1.773⋆1	3.192⋆2	−1.084⋆3
2.800	−8.261⋆-1	−1.241	8.000⋆1	−1.682⋆3	5.923⋆3
3.000	−8.661⋆-1	1.915	−1.568⋆2	4.004⋆3	−1.432⋆4
			Achromatic $p/q=\frac{1}{2}$		
β^2	$(03)_4$	$(03)_3$	$(03)_2$	$(03)_1$	$(03)_0$
2.000⋆-1	9.487⋆-1	−6.125	1.055⋆1	−6.967	1.439
4.000⋆-1	−1.095⋆-1	−1.143	2.295	−1.493	1.888⋆-1
6.000⋆-1	−2.757⋆-1	−3.732⋆-1	8.838⋆-1	−5.663⋆-1	−1.245⋆-2
9.000⋆-1	−3.390⋆-1	−1.516⋆-1	4.290⋆-1	−2.714⋆-1	−7.624⋆-2
1.000	−3.781⋆-1	−6.839⋆-2	2.361⋆-1	−1.478⋆-1	−1.045⋆-1
1.200	−4.090⋆-1	−3.234⋆-2	1.406⋆-1	−8.740⋆-2	−1.203⋆-1
1.400	−4.361⋆-1	−1.527⋆-2	8.834⋆-2	−5.432⋆-2	−1.308⋆-1
1.600	−4.609⋆-1	−6.738⋆-3	5.760⋆-2	−3.458⋆-2	−1.383⋆-1
1.800	−4.843⋆-1	−2.249⋆-3	3.679⋆-2	−1.424⋆-2	−1.332⋆-1
2.000	−5.066⋆-1	−1.546⋆-4	2.794⋆-2	−2.084⋆-2	−1.641⋆-1
2.200	−5.278⋆-1	6.598⋆-4	2.670⋆-2	−3.960⋆2	−3.572⋆-1
2.400	−5.483⋆-1	1.879⋆-3	3.388⋆-3	3.467⋆-2	3.756⋆-1
2.600	−5.680⋆-1	−1.511⋆-4	7.282⋆-2	−1.886⋆-1	−5.694
2.800	−5.871⋆-1	−3.968⋆-4	7.391⋆-2	−1.491⋆-2	−9.334
3.000	−6.055⋆1	−1.065⋆-2	4.344⋆-1	9.493⋆-1	−9.440⋆-1
			Achromatic $p/q = 4$		
β^2	$(03)_4$	$(03)_3$	$(03)_2$	$(03)_1$	$(03)_0$
2.000⋆-1	1.640	−7.280	1.047⋆1	−5.775	7.543⋆-1
4.000⋆-1	3.039⋆-1	−1.633	2.254	−1.011	−2.400⋆-1
6.000⋆-1	7.860⋆-2	−6.493⋆-1	8.578⋆-1	−3.072⋆-1	−4.039⋆-1
8.000⋆-1	5.706⋆-3	−3.266⋆-1	4.113⋆-1	−1.145⋆-1	−4.764⋆-1
1.000	−2.602⋆-2	−1.869⋆-1	2.236⋆-1	−4.615⋆-2	−5.275⋆-1
1.200	−4.254⋆-2	−1.161⋆-1	1.315⋆-1	−1.840⋆-2	−5.707⋆-1

Table 24 The Asymmetric bell-shaped model. The aperture aberration coefficients $(0003)_i$ for $p/q = \frac{1}{4}, \frac{1}{2}, 2, 4$ $(p + q = 2)$.—cont'd

β^2	$(03)_4$	$(03)_3$	$(03)_2$	$(03)_1$	$(03)_0$
Achromatic $p/q = 4$					
1.400	−5.232⋆-2	−7.639⋆-2	8.169⋆-2	−6.256⋆-3	−6.099⋆-1
1.600	−5.870⋆-2	−5.239⋆-2	5.283⋆-2	−7.764⋆-4	−6.464⋆-1
1.800	−6.321⋆-2	−3.711⋆-2	3.523⋆-2	1.663⋆-3	−6.808⋆-1
2.000	−6.661⋆-2	−2.697⋆-2	2.409⋆-2	2.664⋆-3	−7.136⋆-1
2.200	−6.931⋆-2	−2.001⋆-2	1.681⋆-2	2.965⋆-3	−7.449⋆-1
2.400	−7.156⋆-2	−1.510⋆-2	1.193⋆-2	2.950⋆-3	−7.747⋆-1
2.600	−7.350⋆-2	−1.157⋆-2	8.587⋆-3	2.761⋆-3	−8.039⋆-1
2.800	−7.521⋆-2	−8.967⋆-3	6.263⋆-3	2.609⋆-3	−8.261⋆-1
3.000	−7.677⋆-2	−7.028⋆-3	4.614⋆-3	1.991⋆-3	−8.682⋆-1
Achromatic $p/q = 2$					
β^2	$(03)_4$	$(03)_3$	$(03)_2$	$(03)_1$	$(03)_0$
2.000⋆-1	1.439	−6.967	1.055⋆1	−6.125	9.487⋆-1
4.000⋆-1	1.888⋆-1	−1.493	2.295	−1.143	−1.095⋆-1
6.000⋆-1	−1.245⋆-2	−5.663⋆-1	8.838⋆-1	−3.732⋆-1	−2.757⋆-1
8.000⋆-1	−7.624⋆-2	−2.714⋆-1	4.290⋆-1	−1.516⋆-1	−3.390⋆-1
1.000	−1.045⋆-1	−1.478⋆-1	2.361⋆-1	−6.839⋆-2	−3.781⋆-1
1.200	−1.203⋆-1	−8.732⋆-2	1.406⋆-1	−3.234⋆-2	−4.090⋆-1
1.400	−1.307⋆-1	−5.455⋆-2	8.841⋆-2	−1.528⋆-2	−4.361⋆-1
1.600	−1.385⋆-1	−3.550⋆-2	5.784⋆-2	−6.757⋆-3	−4.609⋆-1
1.800	−1.449⋆-1	−2.382⋆-2	3.901⋆-2	−2.377⋆-3	−4.843⋆-1
2.000	−1.505⋆-1	−1.637⋆-2	2.696⋆-2	−1.116⋆-4	−5.066⋆-1
2.200	−1.555⋆-1	−1.147⋆-2	1.900⋆-2	1.037⋆-3	−5.278⋆-1
2.400	−1.602⋆-1	−8.152⋆-3	1.362⋆-2	1.576⋆-3	−5.484⋆-1
2.600	−1.646⋆-1	−5.862⋆-3	9.898⋆-3	1.783⋆-3	−5.681⋆-1
2.800	−1.689⋆-1	−4.254⋆-3	7.283⋆-3	1.811⋆-3	−5.873⋆-1
3.000	−1.729⋆-1	−3.107⋆-3	5.417⋆-3	1.724⋆-3	−6.067⋆-1

Notice that $(03)_i$ for $p/q = r$ is equal to $(03)_{4-i}$ for $q/p = r$; this rule is not obeyed for some high values of β^2 in these tables, and erratically varying coefficients cannot then be relied upon.

Table 25 The asymmetric bell-shaped model. The aperture aberration coefficients $(0012)_{ij}$ for $p/q = \frac{1}{4}, \frac{1}{2}, 2, 4$ $(p + q = 2)$.

Electrostatic $p/q=\frac{1}{4}$

β^2	$(12)_{22}$	$(12)_{21}$	$(12)_{12}$	$(12)_{20}$	$(12)_{11}$	$(12)_{02}$	$(12)_{10}$	$(12)_{01}$	$(12)_{00}$
2.000⋆-1	−1.484	6.659	6.748⋆-1	−2.475	−7.462	−1.559	1.050	5.236	−8.399⋆-1
4.000⋆-1	−8.967⋆-1	2.645	−5.456⋆-1	−7.427⋆-1	−1.390	−2.650⋆-1	−6.498⋆-1	1.654	−1.458⋆-1
6.000⋆-1	−9.903⋆-1	1.799	−5.006⋆-1	−3.921⋆-1	−2.246⋆-1	−5.006⋆-2	−8.032⋆-1	9.193⋆-1	−1.559⋆-1
8.000⋆-1	−1.110	1.517	−3.775⋆-1	−2.604⋆-1	2.397⋆-1	2.382⋆-2	−8.296⋆-1	6.492⋆-1	−2.193⋆-1
1.000	−1.199	1.453	−2.453⋆-1	−1.952⋆-1	5.322⋆-1	6.543⋆-2	−8.550⋆-1	5.303⋆-1	−2.880⋆-1
1.200	−1.235	1.535	−9.105⋆-2	−1.556⋆-1	8.032⋆-1	1.030⋆-1	−9.038⋆-1	4.848⋆-1	−3.593⋆-1
1.400	−1.180	1.780	1.240⋆-1	−1.244⋆-1	1.142	1.523⋆-1	−9.881⋆-1	4.925⋆-1	−4.380⋆-1
1.600	−9.420⋆-1	2.294	4.863⋆-1	−8.651⋆-2	1.671	2.344⋆-1	−1.124	5.619⋆-1	−5.335⋆-1
1.800	−2.546⋆-1	3.397	1.232	−1.142⋆-2	2.673	4.013⋆-1	−1.341	7.444⋆-1	−6.637⋆-1
2.000	1.898	6.254	3.231	2.150⋆-1	5.130	8.386⋆-1	−1.691	1.240	−8.694⋆-1
2.200	1.210⋆1	1.798⋆1	1.201⋆1	1.373	1.493⋆1	2.704	−2.136	3.261	−1.264
2.400	3.272⋆2	3.343⋆2	2.711⋆2	4.134⋆1	2.737⋆2	5.594⋆1	1.921⋆1	5.614⋆1	7.972⋆-1
2.600	6.033⋆1	5.176⋆1	4.857⋆1	9.132	4.074⋆1	9.575	1.315⋆1	8.143	3.620
2.800	1.026⋆1	8.632	9.012	2.037	6.376	1.757	4.027	1.287	1.067
3.000	3.430	3.375	3.783	1.017	2.280	7.520⋆-1	2.320	4.813⋆-1	5.257⋆-1

Electrostatic $p/q=\frac{1}{2}$

β^2	$(12)_{22}$	$(12)_{21}$	$(12)_{12}$	$(12)_{20}$	$(12)_{11}$	$(12)_{02}$	$(12)_{10}$	$(12)_{01}$	$(12)_{00}$
2.000⋆-1	−1.279	6.303	7.593⋆-1	−2.236	−7.527	−1.731	9.624⋆-1	5.507	−9.164⋆-1
4.000⋆-1	−6.510⋆-1	2.372	−5.584⋆-1	−5.981⋆-1	−1.462	−3.405⋆-1	−6.187⋆-1	1.818	−2.291⋆-1
6.000⋆-1	−7.140⋆-1	1.530	−5.497⋆-1	−2.734⋆-1	−3.115⋆-1	−9.607⋆-2	−7.145⋆-1	1.044	−2.435⋆-1
8.000⋆-1	−8.156⋆-1	1.221	−4.542⋆-1	−1.490⋆-1	1.279⋆-1	−9.934⋆-3	−6.921⋆-1	7.505⋆-1	−3.082⋆-1
1.000	−9.039⋆-1	1.099	−3.522⋆-1	−8.160⋆-2	3.792⋆-1	3.587⋆-2	−6.639⋆-1	6.139⋆-1	−3.739⋆-1
1.200	−9.687⋆-1	1.076	−2.531⋆-1	−3.293⋆-2	5.798⋆-1	7.116⋆-2	−6.447⋆-1	5.506⋆-1	−4.360⋆-1
1.400	−1.001	1.134	−1.330⋆-1	1.396⋆-2	7.908⋆-1	1.089⋆-1	−6.345⋆-1	5.340⋆-1	−4.956⋆-1
1.600	−9.820⋆-1	1.286	3.254⋆-2	7.305⋆-2	1.063	1.601⋆-1	−6.287⋆-1	5.594⋆-1	−5.547⋆-1
1.800	−8.680⋆-1	1.585	2.997⋆-1	1.659⋆-1	1.475	2.424⋆-1	−6.163⋆-1	6.392⋆-1	−6.147⋆-1
2.000	−5.451⋆-1	2.170	8.048⋆-1	3.416⋆-1	2.193	3.961⋆-1	−5.659⋆-1	8.155⋆-1	−6.741⋆-1
2.200	3.444⋆-1	3.465	1.963	7.496⋆-1	3.697	7.411⋆-1	−3.661⋆-1	1.219	−7.175⋆-1
2.400	3.376	7.223	5.570	2.040	7.944	1.788	5.360⋆-1	2.393	−6.380⋆-1
2.600	2.325⋆1	2.873⋆1	2.810⋆1	1.025⋆1	3.187⋆1	8.150	7.906	9.025	9.106⋆-1
2.800	4.388⋆3	4.068⋆3	4.801⋆3	1.790⋆3	4.457⋆3	1.313⋆3	2.003⋆3	1.221⋆3	5.597⋆2
3.000	2.315⋆1	1.882⋆1	2.576⋆1	1.003⋆1	1.995⋆1	6.870	1.354⋆1	5.442	4.025

Continued

Table 25 The asymmetric bell-shaped model. The aperture aberration coefficients $(0012)_{ij}$ for $p/q = \frac{1}{4}, \frac{1}{2}, 2, 4$ $(p + q = 2)$.—cont'd

Electrostatic $p/q = 4$

β^2	$(12)_{22}$	$(12)_{21}$	$(12)_{12}$	$(12)_{20}$	$(12)_{11}$	$(12)_{02}$	$(12)_{10}$	$(12)_{01}$	$(12)_{00}$
2.000⋆-1	−8.399⋆-1	5.236	1.050	−1.559	−7.462	−2.475	6.748⋆-1	6.659	−1.484
4.000⋆-1	−1.458⋆-1	1.654	−6.498⋆-1	−2.650⋆-1	−1.390	−7.427⋆-1	−5.456⋆-1	2.645	−8.967⋆-1
6.000⋆-1	−1.559⋆-1	9.193⋆-1	−8.032⋆-1	−5.006⋆-2	−2.246⋆-1	−3.921⋆-1	−5.006⋆-1	1.799	−9.903⋆-1
8.000⋆-1	−2.193⋆-1	6.492⋆-1	−8.296⋆-1	2.382⋆-2	2.397⋆-1	−2.604⋆-1	−3.775⋆-1	1.517	−1.110
1.000	−2.880⋆-1	5.303⋆-1	−8.550⋆-1	6.543⋆-2	5.322⋆-1	−1.952⋆-1	−2.453⋆-1	1.453	−1.199
1.200	−3.593⋆-1	4.848⋆-1	−9.038⋆-1	1.030⋆-1	8.032⋆-1	−1.556⋆-1	−9.105⋆-2	1.535	−1.235
1.400	−4.380⋆-1	4.925⋆-1	−9.881⋆-1	1.523⋆-1	1.142	−1.244⋆-1	1.240⋆-1	1.780	−1.180
1.600	−5.335⋆-1	5.619⋆-1	−1.124	2.344⋆-1	1.671	−8.650⋆-2	4.863⋆-1	2.294	−9.420⋆-1
1.800	−6.637⋆-1	7.444⋆-1	−1.341	4.013⋆-1	2.673	−1.140⋆-2	1.232	3.397	−2.546⋆-1
2.000	−8.695⋆-1	1.240	−1.691	8.386⋆-1	5.130	2.149⋆-1	3.231	6.254	1.898
2.200	−1.264	3.261	−2.135	2.704	1.493⋆1	1.374	1.201⋆1	1.798⋆1	1.210⋆1
2.400	7.924⋆-1	5.614⋆1	1.919⋆1	5.594⋆1	2.737⋆2	4.131⋆1	2.711⋆2	3.343⋆2	3.272⋆2
2.600	3.620	8.143	1.315⋆1	9.575	4.074⋆1	9.135	4.857⋆1	5.176⋆1	6.033⋆1
2.800	1.067	1.287	4.026	1.757	6.376	2.035	9.013	8.633	1.026⋆1
3.000	5.258⋆-1	4.813⋆-1	2.321	7.519⋆-1	2.279	1.018	3.782	3.375	3.430

Electrostatic $p/q = 2$

β^2	$(12)_{22}$	$(12)_{21}$	$(12)_{12}$	$(12)_{20}$	$(12)_{11}$	$(12)_{02}$	$(12)_{10}$	$(12)_{01}$	$(12)_{00}$
2.000⋆-1	−9.164⋆-1	5.507	9.624⋆-1	−1.731	−7.527	−2.236	7.593⋆-1	6.303	−1.279
4.000⋆-1	−2.291⋆-1	1.818	−6.187⋆-1	−3.405⋆-1	−1.462	−5.981⋆-1	−5.584⋆-1	2.372	−6.510⋆-1
6.000⋆-1	−2.435⋆-1	1.044	−7.145⋆-1	−9.607⋆-2	−3.115⋆-1	−2.734⋆-1	−5.497⋆-1	1.530	−7.140⋆-1
8.000⋆-1	−3.082⋆-1	7.505⋆-1	−6.921⋆-1	−9.934⋆-3	1.279⋆-1	−1.490⋆-1	−4.542⋆-1	1.221	−8.156⋆-1
1.000	−3.739⋆-1	6.139⋆-1	−6.639⋆-1	3.587⋆-2	3.792⋆-1	−8.160⋆2	−3.552⋆-1	1.099	−9.039⋆-1
1.200	−4.360⋆-1	5.506⋆-1	−6.447⋆-1	7.116⋆-2	5.798⋆-1	−3.293⋆-2	−2.531⋆-1	1.076	−9.687⋆-1
1.400	−4.956⋆-1	5.340⋆-1	−6.345⋆-1	1.089⋆-1	7.908⋆-1	1.396⋆-2	−1.330⋆-1	1.134	−1.001
1.600	−5.547⋆-1	5.594⋆-1	−6.287⋆-1	1.601⋆-1	1.063	7.305⋆-2	3.254⋆-2	1.286	−9.820⋆-1
1.800	−6.147⋆-1	6.392⋆-1	−6.163⋆-1	2.426⋆-1	1.475	1.659⋆-1	2.997⋆-1	1.585	−8.680⋆-1
2.000	−6.741⋆-1	8.155⋆-1	−5.659⋆-1	3.961⋆-1	2.193	3.416⋆-1	8.048⋆-1	2.170	−5.451⋆-1
2.200	−7.175⋆-1	1.219	−3.661⋆-1	7.411⋆-1	3.697	7.496⋆-1	1.963	3.465	3.444⋆-1
2.400	−6.380⋆-1	2.393	5.359⋆-1	1.788	7.944	2.040	5.570	7.223	3.376
2.600	9.106⋆-1	9.025	7.906	8.150	3.187⋆1	1.025⋆1	2.810⋆1	2.873⋆1	2.325⋆1
2.800	5.598⋆2	1.221⋆3	2.003⋆3	1.313⋆3	4.458⋆3	1.790⋆3	4.801⋆3	4.068⋆3	4.388⋆3
3.000	4.025	5.442	1.354⋆1	6.869	1.995⋆1	1.003⋆1	2.576⋆1	1.882⋆1	2.315⋆1

Magnetic $p/q=\frac{1}{4}$

β^2	$(12)_{22}$	$(12)_{21}$	$(12)_{12}$	$(12)_{20}$	$(12)_{11}$	$(12)_{02}$	$(12)_{10}$	$(12)_{01}$	$(12)_{00}$
2.000⋆-1	4.549	−6.969	−1.013⋆1	5.344	1.640⋆1	3.469	−1.238⋆1	−5.536	5.072
4.000⋆-1	6.317⋆-1	−1.296	−2.892	1.928	4.316	8.822⋆-1	−4.288	−7.452⋆-1	1.235

Continued

Table 25 The asymmetric bell-shaped model. The aperture aberration coefficients $(0012)_{ij}$ for $p/q = \frac{1}{4}, \frac{1}{2}, 2, 4$ $(p + q = 2)$.—cont'd

Magnetic $p/q = \frac{1}{4}$

β^2	$(12)_{22}$	$(12)_{21}$	$(12)_{12}$	$(12)_{20}$	$(12)_{11}$	$(12)_{02}$	$(12)_{10}$	$(12)_{01}$	$(12)_{00}$
6.000⋆-1	−2.460⋆-1	−3.086⋆-1	−1.319	1.302	2.090	4.692⋆-1	−2.480	−4.445⋆-2	3.921⋆-1
8.000⋆-1	−5.705⋆-1	2.139⋆-2	−6.269⋆-1	1.154	1.329	3.576⋆-1	−1.724	1.411⋆-1	4.724⋆-2
1.000	−6.565⋆-1	1.786⋆-1	−1.625⋆-1	1.202	1.002	3.378⋆-1	−1.286	2.042⋆-1	−1.370⋆-1
1.200	−5.467⋆-1	2.843⋆-1	2.821⋆-1	1.411	8.590⋆-1	3.691⋆-1	−9.453⋆-1	2.321⋆-1	−2.478⋆-1
1.400	−1.732⋆-1	3.887⋆-1	8.535⋆-1	1.841	8.250⋆-1	4.541⋆-1	−5.725⋆-1	2.535⋆-1	−3.104⋆-1
1.600	6.877⋆-1	5.344⋆-1	1.780	2.686	8.908⋆-1	6.289⋆-1	2.007⋆-3	2.839⋆-1	−3.183⋆-1
1.800	2.697	8.017⋆-1	3.633	4.521	1.110	1.008	1.194	3.438⋆-1	−2.124⋆-1
2.000	8.316	1.453	8.469	9.465	1.714	2.013	4.545	4.898⋆-1	2.751⋆-1
2.200	3.313⋆1	4.046	2.913⋆1	3.091⋆1	4.091	6.288	2.014⋆1	1.031	3.026
2.400	7.579⋆2	7.242⋆1	6.208⋆2	6.514⋆2	6.197⋆1	1.270⋆2	5.090⋆2	1.326⋆1	9.917⋆1
2.600	1.315⋆2	1.080⋆1	1.062⋆2	1.139⋆2	7.580	2.134⋆1	9.946⋆1	1.337	2.148⋆1
2.800	2.260⋆1	1.745	1.894⋆1	2.139⋆1	9.515⋆-1	3.859	1.990⋆1	1.384⋆-1	4.407
3.000	8.344	6.669⋆-1	7.663	9.375	2.579⋆-1	1.632	8.972	3.384⋆-2	1.933

Magnetic $p/q = \frac{1}{2}$

β^2	$(12)_{22}$	$(12)_{21}$	$(12)_{12}$	$(12)_{20}$	$(12)_{11}$	$(12)_{02}$	$(12)_{10}$	$(12)_{01}$	$(12)_{00}$
2.000⋆-1	4.735	−6.627	−1.059⋆1	4.821	1.640⋆1	3.789	−1.186⋆1	−5.849	5.040
4.000⋆-1	8.446⋆-1	−1.149	−3.173	1.574	4.309	1.014	−3.960	−8.609⋆-1	1.196
6.000⋆-1	−2.793-2	−2.245⋆-1	−1.559	9.702⋆-1	2.082	5.423⋆-1	−2.213	−9.743⋆-2	3.491⋆-1

8.000⋆-1	−3.763⋆-1	7.160⋆-2	−8.752⋆-1	7.885⋆-1	1.317	4.018⋆-1	−1.490	1.174⋆-1	2.245⋆-3
1.000	−5.307⋆-1	2.032⋆-1	−4.589⋆-1	7.529⋆-1	9.811⋆-1	3.614⋆-1	−1.082	1.962⋆-1	−1.818⋆-1
1.200	−5.662⋆-1	2.820⋆-1	−1.197⋆-1	8.040⋆-1	8.235⋆-1	3.704⋆-1	−7.888⋆-1	2.324⋆-1	−2.911⋆-1
1.400	−4.875⋆-1	3.483⋆-1	2.384⋆-1	9.397⋆-1	7.622⋆-1	4.196⋆-1	−5.189⋆-1	2.565⋆-1	−3.543⋆-1
1.600	−2.529⋆-1	4.252⋆-1	7.076⋆-1	1.195	7.723⋆-1	5.197⋆-1	−1.980⋆-1	2.824⋆-1	−3.762⋆-1
1.800	2.529⋆-1	5.382⋆-1	1.440	1.661	8.597⋆-1	7.060⋆-1	2.850⋆-1	3.215⋆-1	−3.420⋆-1
2.000	1.330	7.357⋆-1	2.784	2.578	1.070	1.072	1.178	3.923⋆-1	−1.970⋆-1
2.200	3.899	1.152	5.774	4.672	1.554	1.904	3.228	5.415⋆-1	2.433⋆-1
2.400	1.192⋆1	2.326	1.480⋆1	1.105⋆-1	2.936	4.418	9.669	9.515⋆-1	1.814
2.600	6.12611	8.912	6.950⋆1	4.985⋆1	1.051⋆1	1.956⋆1	5.040⋆1	3.126	1.245⋆1
2.800	1.024⋆4	1.213⋆3	1.123⋆4	7.987⋆3	1.323⋆3	3.079⋆3	8.809⋆3	3.607⋆2	2.429⋆3
3.000	5.190⋆1	5.405	5.756⋆1	4.170⋆1	5.353	1.581⋆1	4.812⋆1	1.355	1.361⋆1

Magnetic $p/q = 4$

β^2	$(12)_{22}$	$(12)_{21}$	$(12)_{12}$	$(12)_{20}$	$(12)_{11}$	$(12)_{02}$	$(12)_{10}$	$(12)_{01}$	$(12)_{00}$
2.000⋆-1	5.072	−5.536	−1.238⋆1	3.469	1.640⋆1	5.344	−1.013⋆1	−6.969	4.549
4.000⋆-1	1.235	−7.452⋆-1	−4.288	8.822⋆-1	4.314	1.928	−2.892	−1.296	6.317⋆-1
6.000⋆-1	3.921⋆-1	−4.445⋆-2	−2.480	4.692⋆-1	2.090	1.302	−1.319	−3.086⋆-1	−2.460⋆-1

Continued

Table 25 The asymmetric bell-shaped model. The aperture aberration coefficients $(0012)_{ij}$ for $p/q = \frac{1}{4}, \frac{1}{2}, 2, 4$ $(p + q = 2)$.—cont'd

Magnetic $p/q = 4$

β^2	$(12)_{22}$	$(12)_{21}$	$(12)_{12}$	$(12)_{20}$	$(12)_{11}$	$(12)_{02}$	$(12)_{10}$	$(12)_{01}$	$(12)_{00}$
8.000⋆-1	4.724⋆-2	1.411⋆-1	−1.724	3.576⋆-1	1.329	1.154	−6.269⋆-1	2.139⋆-2	−5.705⋆-1
1.000	−1.370⋆-1	2.042⋆-1	−1.286	3.378⋆-1	1.002	1.202	−1.625⋆-1	1.786⋆-1	−6.565⋆-1
1.200	−2.478⋆-1	2.321⋆-1	−9.453⋆-1	3.691⋆-1	8.590⋆-1	1.411	2.821⋆-1	2.843⋆-1	−5.467⋆-1
1.400	−3.104⋆-1	2.535⋆-1	−5.725⋆-1	4.541⋆-1	8.250⋆-1	1.841	8.535⋆-1	3.887⋆-1	−1.732⋆-1
1.600	−3.183⋆1	2.839⋆-1	2.012⋆-3	6.289⋆-1	8.908⋆-1	2.687	1.780	5.344⋆-1	6.877⋆-1
1.800	−2.124⋆-1	3.438⋆-1	1.194	1.008	1.110	4.521	3.633	8.017⋆-1	2.697
2.000	2.751⋆-1	4.898⋆-1	4.545	2.013	1.714	9.465	8.469	1.453	8.316
2.200	3.026	1.031	2.014⋆1	6.288	4.091	3.091⋆1	2.913⋆1	4.046	3.313⋆1
2.400	9.917⋆1	1.326⋆1	5.090⋆2	1.270⋆2	6.197⋆1	6.514⋆2	6.208⋆2	7.242⋆1	7.579⋆2
2.600	2.148⋆1	1.337	9.946⋆1	2.134⋆1	7.579	1.139⋆2	1.062⋆2	1.080⋆1	1.315⋆2
2.800	4.407	1.384⋆-1	1.990⋆1	3.859	9.516⋆-1	2.139⋆1	1.894⋆1	1.745	2.260⋆1
3.000	1.933	3.384⋆-2	8.972	1.632	2.578⋆-1	9.375	7.663	6.669⋆-1	8.344

Magnetic $p/q = 2$

β^2	$(12)_{22}$	$(12)_{21}$	$(12)_{12}$	$(12)_{20}$	$(12)_{11}$	$(12)_{02}$	$(12)_{10}$	$(12)_{01}$	$(12)_{00}$
2.000⋆-1	5.040	−5.849	−1.186⋆1	3.789	1.640⋆1	4.821	−1.059⋆1	−6.627	4.735
4.000⋆-1	1.196	−8.609⋆-1	−3.960	1.014	4.309	1.574	−3.173	−1.149	8.446⋆-1
6.000⋆-1	3.491⋆-1	−9.743⋆-2	−2.213	5.423⋆-1	2.082	9.702⋆-1	−1.559	−2.245⋆-1	−2.793⋆-2
8.000⋆-1	2.245⋆-3	1.174⋆-1	−1.490	4.018⋆-1	1.317	7.885⋆-1	−8.752⋆-1	7.160⋆-2	−3.763⋆-1

1.000	−1.818⋆-1	1.962⋆-1	−1.082	3.614⋆-1	9.811⋆-1	7.529⋆-1	−4.589⋆-1	2.032⋆-1	−5.307⋆-1
1.200	−2.911⋆-1	2.324⋆-1	−7.888⋆-1	3.784⋆-1	8.235⋆-1	8.040⋆-1	−1.197⋆-1	2.820⋆-1	−5.662⋆-1
1.400	−3.543⋆-1	2.585⋆-1	−5.189⋆-1	4.196⋆-1	7.622⋆-1	9.397⋆-1	2.384⋆-1	3.483⋆-1	−4.875⋆-1
1.600	−3.762⋆-1	2.826⋆-1	−1.980⋆1	5.197⋆-1	7.723⋆-1	1.195	7.076⋆-1	4.252⋆-1	−2.529⋆-1
1.800	−3.420⋆-1	3.215⋆-1	2.850⋆-1	7.060⋆-1	8.597⋆-1	1.661	1.440	5.382⋆-1	2.529⋆-1
2.000	−1.970⋆-1	3.923⋆-1	1.178	1.072	1.070	2.578	2.784	7.357⋆-1	1.330
2.200	2.433⋆-1	5.415⋆-1	3.228	1.904	1.554	4.672	5.774	1.152	3.899
2.400	1.814	9.515⋆-1	9.669	4.418	2.936	1.105⋆1	1.480⋆1	2.326	1.192⋆1
2.600	1.245⋆1	3.126	5.040⋆1	1.956⋆1	1.051⋆1	4.985⋆1	6.950⋆1	8.912	6.126⋆1
2.800	2.429⋆3	3.607⋆2	8.809⋆3	3.079⋆3	1.323⋆3	7.987⋆3	1.123⋆4	1.213⋆3	1.024⋆4
3.000	1.361⋆1	1.355	4.812⋆1	1.581⋆1	5.353	4.170⋆1	5.756⋆1	5.405	5.190⋆1

Achromatic $p/q=\frac{1}{4}$

β^2	$(12)_{22}$	$(12)_{21}$	$(12)_{12}$	$(12)_{20}$	$(12)_{11}$	$(12)_{02}$	$(12)_{10}$	$(12)_{01}$	$(12)_{00}$
2.000⋆-1	6.560	−1.151⋆1	−1.373⋆1	7.950	2.435⋆1	5.146	−1.685⋆1	−9.127	7.042
4.000⋆-1	1.141	−2.610	−3.674	2.818	6.215	1.265	−5.501	−1.545	1.695
6.000⋆-1	2.161⋆3	−1.011	−1.592	1.867	2.861	6.423⋆-1	−3.039	−3.657⋆-1	5.748⋆-1

Continued

Table 25 The asymmetric bell-shaped model. The aperture aberration coefficients $(0012)_{ij}$ for $p/q = \frac{1}{4}, \frac{1}{2}, 2, 4$ $(p + q = 2)$.—cont'd

Achromatic $p/q=\frac{1}{4}$

β^2	$(12)_{22}$	$(12)_{21}$	$(12)_{12}$	$(12)_{20}$	$(12)_{11}$	$(12)_{02}$	$(12)_{10}$	$(12)_{01}$	$(12)_{00}$
8.000⋆-1	−3.905⋆-1	−4.772⋆-1	−7.100⋆-1	1.626	1.692	4.689⋆-1	−2.022	−2.829⋆-2	1.361⋆-1
1.000	−4.755⋆-1	−2.463⋆-1	−1.349⋆-1	1.668	1.159	4.285⋆-1	−1.430	9.552⋆-2	−8.665⋆-2
1.200	−3.172⋆-1	−1.327⋆-1	4.065⋆-1	1.933	8.776⋆-1	4.578⋆ = 1	−9.591⋆-1	1.479⋆-1	−2.106⋆-1
1.400	1.623⋆-1	−7.527⋆-2	1.097	2.496	7.192⋆-1	5.547⋆-1	−4.339⋆-1	1.738⋆-1	−2.678⋆-1
1.600	1.231	−5.222⋆-2	2.211	3.611	6.308⋆-1	7.604⋆-1	3.773⋆-1	1.913⋆-1	−2.466⋆-1
1.800	3.680	−6.334⋆-2	4.434	6.031	5.887⋆-1	1.210	2.039	2.102⋆-1	−6.200⋆-2
2.000	1.046⋆1	−1.478⋆-1	1.022⋆1	1.255⋆1	5.753⋆-1	2.405	6.624	2.399⋆-1	6.566⋆-1
2.200	4.014⋆1	−5.983⋆-1	3.484⋆1	4.075⋆1	4.790⋆-1	7.483	2.756⋆1	2.882⋆-1	4.456
2.400	9.014⋆2	−1.487⋆1	7.374⋆2	8.548⋆2	−8.597	1.507⋆2	6.722⋆2	−1.036	1.320⋆2
2.600	1.552⋆2	−2.853	1.254⋆2	1.489⋆2	−3.476	2.527⋆1	1.282⋆2	−9.313⋆-1	2.743⋆1
2.800	2.672⋆1	−5.520⋆-1	2.225⋆1	2.784⋆1	−8.567⋆-1	4.559	2.519 × 1	−2.443⋆-1	5.520
3.000	9.982	−2.358⋆-1	8.956	1.216⋆1	−4.160⋆-1	1.925	1.119⋆1	−1.153⋆-1	2.402

Achromatic $p/q=\frac{1}{2}$

β^2	$(12)_{22}$	$(12)_{21}$	$(12)_{12}$	$(12)_{20}$	$(12)_{11}$	$(12)_{02}$	$(12)_{10}$	$(12)_{01}$	$(12)_{00}$
2.000⋆-1	6.739	−1.094⋆1	−1.438⋆1	7.173	2.437⋆1	5.629	−1.613⋆1	−9.634	7.025
4.000⋆-1	1.343	−2.322	−4.044	2.298	6.233	1.465	−5.074	−1.754	1.671
6.000⋆-1	2.008⋆-1	−8.094⋆-1	−1.895	1.385	2.879	7.551⋆-1	−2.713	−4.778⋆-1	5.466⋆-1

8.000⋆-1	−2.299⋆-1	−3.116⋆-1	−1.015	1.101	1.713	5.391⋆-1	−1.756	−9.370⋆-2	1.057⋆-1
1.000	−4.063⋆-1	−9.517⋆-2	−4.935⋆-1	1.031	1.182	4.699⋆-1	−1.221	5.691⋆-2	−1.178⋆-1
1.200	−4.321⋆-1	1.731⋆-2	−7.520⋆-2	1.083	9.047⋆-1	4.701⋆-1	−8.368⋆-1	1.264⋆-1	−2.429⋆-1
1.400	−3.163⋆-1	8.624⋆-2	3.623⋆-1	1.248	7.527⋆-1	5.231⋆-1	−4.804⋆-1	1.640⋆-1	−3.072⋆-1
1.600	−9.813⋆-3	1.381⋆-1	9.326⋆-1	1.569	6.753⋆-1	6.396⋆-1	−5.438⋆-2	1.900⋆-1	−3.167⋆-1
1.800	6.265⋆-1	1.893⋆-1	1.820	2.160	6.546⋆-1	8.606⋆-1	5.854⋆-1	2.156⋆-1	−2.511⋆-1
2.000	1.955	2.576⋆-1	3.444	3.323	6.953⋆-1	1.297	1.759	2.512⋆-1	−3.791⋆-2
2.200	5.084	3.803⋆-1	7.044	5.979	8.393⋆-1	2.291	4.426	3.156⋆-1	5.636⋆-1
2.400	1.476⋆1	6.939⋆-1	1.788⋆1	1.406⋆1	1.267	5.295	1.271⋆1	4.710⋆-1	2.631
2.600	7.393⋆1	2.306	8.330⋆1	6.305⋆1	3.388	2.336⋆1	6.456⋆1	1.160	1.629⋆1
2.800	1.220⋆4	2.616⋆2	1.338⋆4	1.005⋆4	2.782⋆2	3.667⋆3	1.108⋆4	7.390⋆1	3.852⋆3
3.000	6.148⋆1	9.340⋆-1	6.816⋆1	5.225⋆1	4.863⋆-1	1.879⋆1	5.965⋆1	−7.164⋆-3	1.680⋆1

Achromatic $p/q = 4$

β^2	$(12)_{22}$	$(12)_{21}$	$(12)_{12}$	$(12)_{20}$	$(12)_{11}$	$(12)_{02}$	$(12)_{10}$	$(12)_{01}$	$(12)_{00}$
2.000⋆-1	7.042	−9.127	−1.685⋆1	5.146	2.435⋆1	7.950	−1.373⋆1	−1.151⋆1	6.560
4.000⋆-1	1.695	−1.545	−5.501	1.265	6.215	2.818	−3.674	−2.610	1.141
6.000⋆-1	5.748⋆-1	−3.657⋆-1	−3.039	6.423⋆-1	2.861	1.867	−1.592	−1.011	2.161⋆-3

Continued

Table 25 The asymmetric bell-shaped model. The aperture aberration coefficients $(0012)_{ij}$ for $p/q = \frac{1}{4}, \frac{1}{2}, 2, 4$ $(p + q = 2)$.—cont'd

Achromatic $p/q = 4$

β^2	$(12)_{22}$	$(12)_{21}$	$(12)_{12}$	$(12)_{20}$	$(12)_{11}$	$(12)_{02}$	$(12)_{10}$	$(12)_{01}$	$(12)_{00}$
8.000⋆-1	1.361⋆-1	−2.829⋆-2	−2.022	4.689⋆-1	1.692	1.626	−7.100⋆-1	−4.772⋆-1	−3.905⋆-1
1.000	−8.665⋆-2	9.552⋆-2	−1.430	4.285⋆-1	1.159	1.668	−1.349⋆-1	−2.463⋆-1	−4.755⋆-1
1.200	−2.106⋆-1	1.479⋆-1	−9.591⋆-1	4.578⋆-1	8.776⋆-1	1.933	4.065⋆-1	−1.327⋆-1	−3.172⋆-1
1.400	−2.678⋆-1	1.730⋆-1	−4.339⋆-1	5.547⋆-1	7.192⋆-1	2.496	1.097	−7.527⋆-2	1.623⋆-1
1.600	−2.466⋆-1	1.913⋆-1	3.773⋆-1	7.604⋆-1	6.308⋆-1	3.611	2.211	−5.222⋆-2	1.231
1.800	−6.200⋆-2	2.102⋆-1	2.039	1.210	5.887⋆-1	6.031	4.434	−6.335⋆-2	3.680
2.000	6.566⋆-1	2.399⋆-1	6.624	2.405	5.753⋆-1	1.255⋆1	1.022⋆1	−1.478⋆-1	1.046⋆1
2.200	4.456	2.882⋆-1	2.756⋆1	7.483	4.789⋆-1	4.075⋆1	3.484⋆1	−5.983⋆1	4.014⋆1
2.400	1.320⋆2	−1.036	6.722⋆2	1.507⋆2	−8.597	8.548⋆2	7.374⋆2	−1.487⋆1	9.014⋆2
2.600	2.743⋆1	−9.313⋆-1	1.282⋆2	2.527⋆1	−3.475	1.489⋆2	1.254⋆2	−2.852	1.552⋆2
2.800	5.520	−2.443⋆-1	2.519⋆1	4.559	−8.567⋆-1	2.784⋆1	2.225⋆1	−5.509⋆1	2.672⋆1
3.000	2.402	−1.253⋆-1	1.119⋆1	1.925	−4.160⋆-1	1.216⋆1	8.956	−2.358⋆-1	9.982

Achromatic $p/q = 2$

β^2	$(12)_{22}$	$(12)_{21}$	$(12)_{12}$	$(12)_{20}$	$(12)_{11}$	$(12)_{02}$	$(12)_{10}$	$(12)_{01}$	$(12)_{00}$
2.000⋆-1	7.025	−9.634	−1.613⋆1	5.629	2.437⋆1	7.173	−1.438⋆1	−1.094⋆1	6.739
4.000⋆-1	1.671	−1.754	−5.074	1.465	6.233	2.298	−4.044	−2.322	1.343
6.000⋆-1	5.466⋆-1	−4.778⋆-1	−2.713	7.551⋆-1	2.879	1.385	−1.895	−8.094⋆-1	2.008⋆-1
8.000⋆-1	1.057⋆-1	−9.370⋆-2	−1.756	5.391⋆-1	1.713	1.101	−1.015	−3.116⋆-1	−2.299⋆-1

1.000	−1.178⋆-1	5.691⋆-2	−1.221	4.699⋆1	1.182	1.031	−4.935⋆-1	−9.517⋆-2	−4.063⋆-1
1.200	−2.429⋆-1	1.264⋆-1	−8.368⋆-1	4.701⋆-1	9.047⋆-1	1.083	−7.520⋆-2	1.731⋆-2	−4.321⋆-1
1.400	−3.072⋆-1	1.640⋆-1	−4.804⋆-1	5.231⋆-1	7.527⋆-1	1.248	3.623⋆-1	8.624⋆-2	−3.163⋆-1
1.600	−3.167⋆-1	1.900⋆-1	−5.438⋆-2	6.396⋆-1	6.753⋆-1	1.569	9.326⋆-1	1.381⋆-1	−9.813⋆-3
1.800	−2.511⋆-1	2.156⋆-1	5.854⋆-1	8.686⋆-1	6.546⋆-1	2.160	1.820	1.893⋆-1	6.265⋆-1
2.000	−3.791⋆-2	2.512⋆-1	1.759	1.297	6.953⋆-1	3.323	3.444	2.576⋆-1	1.955
2.200	5.636⋆-1	3.156⋆-1	4.426	2.291	8.393⋆-1	5.979	7.044	3.803⋆-1	5.084
2.400	2.631	4.710⋆-1	1.271⋆1	5.295	1.267	1.406⋆1	1.788⋆1	6.939⋆-1	1.476⋆1
2.600	1.629⋆1	1.160	6.456⋆1	2.336⋆1	3.388	6.305⋆1	8.330⋆1	2.306	7.393⋆1
2.800	3.052⋆3	7.390⋆1	1.188⋆4	3.667⋆3	2.782⋆2	1.005⋆4	1.338⋆4	2.616⋆2	1.220⋆4
3.000	1.680⋆1	−7.165⋆-3	5.965⋆1	1.879⋆1	4.863⋆-1	5.225⋆1	6.816⋆1	9.340⋆-1	6.148⋆1

Notice that the $(12)_{ij}$ for $p/q = r$ are related to those for $q/p = r$ as follows: $(12)_{22} \rightarrow (12)_{00}$, $(12)_{21} \rightarrow (12)_{01}$, $(12)_{12} \rightarrow (12)_{10}$, $(12)_{20} \rightarrow (12)_{02}$, $(12)_{11} \rightarrow (12)_{11}$.

Table 26 The asymmetric bell-shaped model. the aperture aberration coefficients $(0021)_{ij}$ $p/q = \frac{1}{4}, \frac{1}{2}, 2, 4$ $(p + q = 2)$.

Electrostatic $p/q = \frac{1}{4}$

β^2	$(12)_{22}$	$(12)_{21}$	$(12)_{12}$	$(12)_{20}$	$(12)_{11}$	$(12)_{02}$	$(12)_{10}$	$(12)_{01}$	$(12)_{00}$
2.000⋆-1	−4.349	8.009	2.565	−2.475	−7.462	−1.559	2.941	6.586	−3.985
4.000⋆-1	−1.844	3.222	5.872⋆-1	−7.427⋆-1	−1.390	−2.650⋆-1	4.830⋆-1	2.230	−1.660
6.000⋆-1	−1.066	2.129	4.142⋆-1	−3.921⋆-1	−2.246⋆-1	−5.006⋆-2	1.117⋆-1	1.250	−1.097
8.000⋆-1	−5.537⋆-1	1.731	4.645⋆-1	−2.604⋆-1	2.397⋆-1	2.382⋆-2	1.247⋆-2	8.632⋆-1	−8.496⋆-1
1.000	−7.676⋆-2	1.602	5.947⋆-1	−1.952⋆-1	5.322⋆-1	6.543⋆-2	−1.502⋆-2	6.789⋆-1	−7.066⋆-1
1.200	4.634⋆-1	1.643	7.998⋆-1	−1.556⋆-1	8.032⋆-1	1.030⋆-1	−1.299⋆-2	5.928⋆-I	−6.090⋆-4
1.400	1.166	1.861	1.122	−1.244⋆-1	1.142	1.523⋆-1	1.033⋆-2	5.734⋆-1	−5.326⋆-1
1.600	2.207	2.356	1.673	−8.651⋆-2	1.671	2.344⋆-1	6.261⋆-2	6.240⋆-1	−4.632⋆-1
1.800	4.023	3.446	2.749	−1.142⋆-2	2.673	4.013⋆-1	1.767⋆-1	7. 931⋆-1	−3.858⋆-1
2.000	8.076	6.293	5.400	2.150⋆-1	5.130	8.386⋆-1	4.780⋆-1	1.278	−2.644⋆-1
2.200	2.278⋆1	1.801⋆1	1.589⋆1	1.373	1.493⋆1	2.704	1.742	3.292	8.952⋆-2
2.400	3.721⋆2	3.343⋆2	2.888⋆2	4.134⋆1	2.737⋆2	5.594⋆1	3.685⋆1	5.617⋆1	7.785

2.600	4.396⋆1	5.178⋆1	4.125⋆1	9.132	4.074⋆1	9.575	5.833	8.164	5.267⋆-1
2.800	4.443	8.650	5.863	2.037	6.376	1.757	8.771⋆-1	1.304	−2.909⋆-1
3.000	5.484⋆-1	3.390	1.710	1.017	2.280	7.520⋆-1	2.474⋆-1	4.958⋆-1	−3.542⋆-1
				Electrostatic $p/q=\frac{1}{2}$					
β^2	$(12)_{22}$	$(12)_{21}$	$(12)_{12}$	$(12)_{20}$	$(12)_{11}$	$(12)_{02}$	$(12)_{10}$	$(12)_{01}$	$(12)_{00}$
2.000⋆-1	−4.254	7.670	2.625	−2.236	−7.527	−1.731	2.828	6.873	−4.041
4.000⋆-1	−1.822	2.961	5.425⋆-1	−5.981⋆-1	−1.462	−3.405⋆-1	4.822⋆-1	2.408	−1.703
6.000⋆-1	−1.135	1.871	3.236⋆-1	−2.734⋆-1	−3.115⋆-1	−9.607⋆-2	1.588⋆-1	1.385	−1.125
8.000⋆-1	−7.401⋆-1	1.444	3.326⋆-1	1.490⋆-1	1.279⋆-1	−9.934⋆-3	9.470⋆-2	9.733⋆-1	−8.597⋆-1
1.000	−4.189⋆-1	1.254	4.095⋆-1	−8.160⋆-2	3.792⋆-1	3.587⋆-2	1.009⋆-1	7.698⋆-1	−6.954⋆-1
1.200	−9.984⋆-2	1.190	5.320⋆-1	−3.293⋆-2	5.798⋆-1	7.116⋆-2	1.405⋆-1	6.646⋆-1	−5.724⋆-1
1.400	2.625⋆-1	1.220	7.105⋆-1	1.396⋆-2	7.908⋆-1	1.089⋆-1	2.091⋆-1	6.200⋆-1	−4.649⋆-1
1.600	7.203⋆-1	1.353	9.786⋆-1	7.305⋆-2	1.063	1.601⋆-1	3.174⋆-1	6.259⋆-1	−3.568⋆-1
1.800	1.363	1.637	1.411	1.659⋆-1	1.475	2.424⋆-1	4.948⋆-1	6.916⋆-1	−2.313⋆-1
2.000	2.389	2.212	2.185	3.416⋆-1	2.193	3.961⋆-1	8.142⋆-1	8.573⋆-1	−5.919⋆-2
2.200	4.346	3.499	3.818	7.496⋆-1	3.697	7.411⋆-1	1.489	1.253	2.340⋆-1

Continued

Table 26 The asymmetric bell-shaped model. the aperture aberration coefficients $(0021)_{ij}$ $p/q = \frac{1}{4}, \frac{1}{2}, 2, 4$ $(p + q = 2)$.—cont'd

Electrostatic $p/q = \frac{1}{2}$

β^2	$(12)_{22}$	$(12)_{21}$	$(12)_{12}$	$(12)_{20}$	$(12)_{11}$	$(12)_{02}$	$(12)_{10}$	$(12)_{01}$	$(12)_{00}$
2.400	9.388	7.251	8.420	2.040	7.944	1.788	3.386	2.421	9.306⋆-1
2.600	3.529⋆1	2.875⋆1	3.416⋆1	1.025⋆1	3.187⋆1	8.150	1.396⋆1	9.048	4.303
2.800	4.249⋆3	4.068⋆3	4.725⋆3	1.790⋆3	4.457⋆3	1.313⋆3	1.926⋆3	1.221⋆3	5.178⋆2
3.000	1.444⋆1	1.884⋆1	2.022⋆1	1.003⋆1	1.995⋆1	6.870	7.990	5.458	1.180

Electrostatic $p/q = 4$

β^2	$(12)_{22}$	$(12)_{21}$	$(12)_{12}$	$(12)_{20}$	$(12)_{11}$	$(12)_{02}$	$(12)_{10}$	$(12)_{01}$	$(12)_{00}$
2.000⋆-1	−3.985	6.586	2.941	−1.559	−7.462	−2.475	2.565	8.009	−4.349
4.000⋆-1	−1.660	2.230	4.830⋆-1	−2.650⋆-1	−1.390	−7.427⋆-1	5.872⋆-1	3.222	−1.844
6.000⋆-1	−1.097	1.250	1.117⋆-1	−5.006⋆-2	−2.246⋆-1	−3.921⋆-1	4.142⋆-1	2.129	−1.066
8.000⋆-1	−8.496⋆-1	8.632⋆-1	1.247⋆-2	2.382⋆-2	2.397⋆-1	−2.604⋆-1	4.645⋆-1	1.731	−5.537⋆-1
1.000	−7.066⋆-1	6.789⋆-1	−1.502⋆-2	6.543⋆-2	5.322⋆-1	−1.952⋆-1	5.947⋆-1	1.602	−7.676⋆-2
1.200	−6.090⋆-1	5.928⋆-1	−1.299⋆-2	1.030⋆-1	8.032⋆-1	−1.556⋆-1	7.998⋆-1	1.643	4.634⋆-1
1.400	−5.326⋆-1	5.734⋆-1	1.033⋆-2	1.523⋆-1	1.142	−1.244⋆-1	1.122	1.861	1.166
1.600	−4.632⋆-1	6.240⋆-1	6.262⋆-2	2.344⋆-1	1.671	−8.650⋆-2	1.673	2.356	2.207
1.800	−3.858⋆-1	7.931⋆-1	1.767⋆-1	4.013⋆-1	2.673	−1.140⋆-2	2.749	3.446	4.023
2.000	−2.644⋆-1	1.278	4.779⋆-1	8.386⋆-1	5.130	2.149⋆-1	5.400	6.293	8.076
2.200	8.974⋆-2	3.292	1.743	2.704	1.493⋆1	1.374	1.589⋆1	1.801⋆1	2.278⋆1

2.400	7.780	5.617⋆1	3.683⋆1	5.594⋆1	2.737⋆2	4.131⋆1	2.888⋆2	3.343⋆2	3.721⋆2
2.600	5.269⋆-1	8.164	5.834	9.575	4.074⋆1	9.135	4.126⋆1	5.178⋆1	4.396⋆1
2.800	−2.911⋆-1	1.304	8.759⋆-1	1.757	6.376	2.035	5.863	8.650	4.443
3.000	−3.541⋆-1	4.957⋆-1	2.484⋆-1	7.520⋆-1	2.279	1.010	1.710	3.389	5.483⋆-1

Electrostatic $p/q = 2$

β^2	$(12)_{22}$	$(12)_{21}$	$(12)_{12}$	$(12)_{20}$	$(12)_{11}$	$(12)_{02}$	$(12)_{10}$	$(12)_{01}$	$(12)_{00}$
2.000⋆-1	−4.041	6.873	2.828	−1.731	−7.527	−2.236	2.625	7.670	−4.254
4.000⋆-1	−1.703	2.408	4.822⋆-1	−3.405⋆-1	−1.462	−5.981⋆-1	5.425⋆-1	2.961	−1.822
6.000⋆-1	−1.125	1.385	1.588⋆-1	−9.607⋆-2	−3.115⋆-1	−2.734⋆-1	3.236⋆-1	1.871	−1.135
8.000⋆-1	−8.597⋆-1	9.733⋆-1	9.470⋆-1	−9.934⋆-3	1.279⋆-1	−1.0490⋆-1	3.326⋆-1	1.444	−7.401⋆-1
1.000	−6.954⋆-1	7.698⋆-1	1.009⋆-1	3.587⋆-2	3.792⋆-1	−8.160⋆-2	4.095⋆-1	1.254	−4.189⋆-1
1.200	−5.724⋆-1	6.646⋆-1	1.405⋆-1	7.116⋆-2	5.798⋆-1	−3.293⋆-2	5.320⋆-1	1.190	−9.984⋆-1
1.400	−4.649⋆-1	6.200⋆-1	2.091⋆-1	1.089⋆-1	7.908⋆-1	1.396⋆-2	7.105⋆-1	1.220	2.625⋆-1
1.600	−3.568⋆-1	6.259⋆-1	3.174⋆-1	1.601⋆-1	1.063	7.305⋆-2	9.786⋆-1	1.353	7.203⋆-1
1.800	−2.313⋆-1	6.916⋆-1	4.948⋆-1	2.424⋆-1	1.475	1.659⋆-1	1.411	1.637	1.363
2.000	−5.919⋆-1	8.573⋆-1	8.142⋆-1	3.961⋆-1	2.193	3.416⋆-1	2.155	2.212	2.389
2.200	2.340⋆-1	1.253	1.489	7.411⋆-1	3.697	7.496⋆-1	3.810	3.499	4.346
2.400	9.305⋆-1	2.421	3.386	1.788	7.944	2.040	8.420	7.251	9.388

Continued

Table 26 The asymmetric bell-shaped model. the aperture aberration coefficients $(0021)_{ij}$ $p/q = \frac{1}{4}, \frac{1}{2}, 2, 4$ $(p + q = 2)$.—cont'd

Electrostatic $p/q = 2$

β^2	$(12)_{22}$	$(12)_{21}$	$(12)_{12}$	$(12)_{20}$	$(12)_{11}$	$(12)_{02}$	$(12)_{10}$	$(12)_{01}$	$(12)_{00}$
2.600	4.303	9.048	1.396⋆1	8.150	3.187⋆1	1.025⋆-1	3.416⋆1	2.875⋆1	3.529⋆1
2.800	5.178⋆2	1.221⋆-3	1.926⋆3	1.313⋆3	4.458⋆3	1.790⋆-1	4.725⋆3	4.068⋆3	4.249⋆3
3.000	1.180	5.458	7.991	6.869	1.995⋆1	1.003⋆-1	2.021⋆1	1.884⋆1	1.444⋆1

Magnetic $p/q = \frac{1}{4}$

β^2	$(12)_{22}$	$(12)_{21}$	$(12)_{12}$	$(12)_{20}$	$(12)_{11}$	$(12)_{02}$	$(12)_{10}$	$(12)_{01}$	$(12)_{00}$
2.000⋆-1	1.684	−5.618	−8.238	5.344	1.640⋆1	3.469	−1.049.1	−4.186	1.927
4.000⋆-1	−3.152⋆-1	−7.195⋆-1	−1.759	1.928	4.314	8.822⋆-1	−3.155	−1.684⋆-1	−2.793⋆-1
6.000⋆-1	−3.212⋆-1	2.178⋆-2	−4.040⋆-1	1.302	2.090	4.692⋆-1	−1.565	2.859⋆-1	−5.494⋆-1
8.000⋆-1	−1.379⋆-2	2.354⋆-1	2.151⋆-1	1.154	1.329	3.576⋆-1	−8.819⋆-1	3.551⋆-1	−5.830⋆-1
1.000	4.662⋆-1	3.272⋆-1	6.775⋆-1	1.202	1.002	3.378⋆-1	−4.462⋆-1	3.528⋆-1	−5.555⋆-1
1.200	1.152	3.922⋆-1	1.173	1.411	8.590⋆-1	3.691⋆-1	−5.446⋆-2	3.401⋆-1	−4.975⋆-1
1.400	2.172	4.696⋆-1	1.852	1.841	8.250⋆-1	4.541⋆-1	4.260⋆-1	3.344⋆-1	−4.049⋆-1
1.600	3.837	5.966⋆-1	2.966	2.686	8.908⋆-1	6.289⋆-1	1.189	3.461⋆-1	−2.480⋆-1
1.800	6.974	8.504⋆-1	5.151	4.521	1.110	1.008	2.711	3.925⋆-1	6.546⋆-2
2.000	1.449⋆1	1.491	1.064⋆1	9.465	1.714	2.013	6.714	5.285⋆-1	8.801⋆-1
2.200	4.381⋆1	4.077	3.301⋆1	3.091⋆1	4.091	6.288	2.402⋆1	1.063	4.380
2.400	8.028⋆2	7.245⋆1	6.385⋆2	6.514⋆2	6.197⋆1	1.270⋆2	5.266⋆2	1.328⋆1	1.062⋆2

2.600	1.151⋆2	1.082⋆1	9.890⋆1	1.139⋆2	7.580	2.134⋆1	9.214⋆1	1.358	1.839⋆1
2.800	1.679⋆1	1.762	1.579⋆1	2.139⋆1	9.515⋆-1	3.859	1.675⋆1	1.557⋆-1	3.049
3.000	5.463	6.813⋆-1	5.590	9.375	2.579⋆-1	1.632	6.899	4.828⋆-2	1.053

Magnetic $p/q=\frac{1}{2}$

β^2	$(12)_{22}$	$(12)_{21}$	$(12)_{12}$	$(12)_{20}$	$(12)_{11}$	$(12)_{02}$	$(12)_{10}$	$(12)_{01}$	$(12)_{00}$
2.000⋆-1	1.760	−5.261	−8.727	4.821	1.640⋆1	3.789	−9.990	−4.482	1.915
4.000⋆-1	−3.268⋆-1	−5.590⋆-1	−2.072	1.574	4.309	1.014	−2.859	−2.713⋆-1	−2.781⋆-1
6.000⋆-1	−4.494⋆-1	1.164⋆-1	−6.857⋆-1	9.702⋆-1	2.082	5.423⋆-1	−1.340	2.435⋆-1	−5.329⋆-1
8.000⋆-1	−3.008⋆-1	2.943⋆-1	−8.836⋆-2	7.885⋆-1	1.317	4.018⋆-1	−7.030⋆-1	3.401⋆-1	−5.492⋆-1
1.000	−4.576⋆-2	3.591⋆-1	3.059⋆-1	7.529⋆-1	9.811⋆-1	3.614⋆-1	−3.171⋆-1	3.520⋆-1	−5.034⋆-1
1.200	3.026⋆-1	3.960⋆-1	6.654⋆-1	8.040⋆-1	8.235⋆-1	3.704⋆-1	−3.655⋆-3	3.464⋆-1	−4.275⋆-1
1.400	7.759⋆-1	4.343⋆-1	1.082	9.397⋆-1	7.622⋆-1	4.196⋆-1	3.246⋆-1	3.425⋆-1	−3.235⋆-1
1.600	1.449	4.917⋆-1	1.654	1.195	7.723⋆-1	5.197⋆-1	7.481⋆-1	3.489⋆-1	−1.783⋆-1
1.800	2.484	5.905⋆-1	2.551	1.661	8.597⋆-1	7.060⋆-1	1.396	3.739⋆-1	4.138⋆-2
2.000	4.264	7.776⋆-1	4.164	2.578	1.070	1.072	2.558	4.341⋆-1	4.179⋆-1
2.200	7.901	1.185	7.628	4.672	1.554	1.904	5.082	5.753⋆-1	1.195

Continued

Table 26 The asymmetric bell-shaped model. the aperture aberration coefficients $(0021)_{ij}$ $p/q=\frac{1}{4}, \frac{1}{2}, 2, 4$ $(p+q=2)$.—cont'd

Magnetic $p/q=\frac{1}{2}$									
β^2	$(12)_{22}$	$(12)_{21}$	$(12)_{12}$	$(12)_{20}$	$(12)_{11}$	$(12)_{02}$	$(12)_{10}$	$(12)_{01}$	$(12)_{00}$
2.400	1.793⋆1	2.354	1.765⋆1	1.105.1	2.936	4.418	1.252⋆1	9.792⋆-1	3.382
2.600	7.330⋆1	8.935	7.555⋆1	4.985⋆1	1.051⋆1	1.956.1	5.645⋆1	3.149	1.584⋆1
2.800	1.011⋆4	1.213⋆3	1.116⋆4	7.987⋆3	1.323⋆3	3.079.3	8.732⋆3	3.607⋆2	2.387⋆3
3.000	4.318⋆1	5.421	5.201⋆1	4.170⋆1	5.353	1.581⋆1	4.258⋆1	1.371	1.076⋆1

Magnetic $p/q=4$									
β^2	$(12)_{22}$	$(12)_{21}$	$(12)_{12}$	$(12)_{20}$	$(12)_{11}$	$(12)_{02}$	$(12)_{10}$	$(12)_{01}$	$(12)_{00}$
2.000⋆-1	1.927	−4.186	−1.049⋆1	3.469	1.640⋆1	5.344	−8.238	−5.618	1.684
4.000⋆-1	−2.793⋆-1	−1.684⋆-1	−3.155	8.822⋆-1	4.314	1.928	−1.759	−7.195⋆-1	−3.152⋆-1
6.000⋆-1	−5.494⋆-1	2.859⋆-1	−1.565	4.692⋆-1	2.090	1.302	−4.040⋆-1	2.178⋆-2	−3.212⋆-1
8.000⋆-1	−5.830⋆-1	3.551⋆-1	−8.019⋆-1	3.576⋆-1	1.329	1.154	2.151⋆-1	2.354⋆-1	−1.379⋆-2
1.000	−5.555⋆-1	3.528⋆-1	−4.462⋆-1	3.378⋆-1	1.002	1.202	6.775⋆-1	3.272⋆-1	4.662⋆-1
1.200	−4.975⋆-1	3.401⋆-1	−5.446⋆-2	3.691⋆-1	8.590⋆-1	1.411	1.173	3.922⋆-1	1.152
1.400	−4.049⋆-1	3.344⋆-1	4.260⋆-1	4.541⋆-1	8.250⋆-1	1.841	1.852	4.696⋆-1	2.172
1.600	−2.400⋆-1	3.461⋆-1	1.189	6.289⋆-1	8.908⋆-1	2.687	2.966	5.966⋆-1	3.837
1.800	6.546⋆-2	3.925⋆-1	2.711	1.008	1.110	4.521	5.151	8.504⋆-1	6.974
2.000	8.801⋆-1	5.285⋆-1	6.714	2.013	1.714	9.465	1.064⋆1	1.491	1.499⋆1
2.200	4.380	1.063	2.402⋆1	6.288	4.091	3.091⋆1	3.301⋆1	4.077	4.381⋆1

2.400	1.062⋆2	1.328⋆1	5.266⋆2	1.270⋆2	6.197⋆1	6.514⋆2	6.385⋆2	7.245⋆1	8.028⋆2
2.600	1.839⋆1	1.358	9.214⋆1	2.134⋆1	7.579	1.139⋆2	9.890⋆1	1.082⋆1	1.151⋆2
2.800	3.049	1.557⋆-1	1.675⋆1	3.859	9.516⋆-1	2.139⋆1	1.579⋆1	1.762	1.679⋆1
3.000	1.053	4.827⋆-2	6.900	1.632	2.578⋆-1	9.375	5.590	6.813⋆-1	5.462

Magnetic $p/q = 2$

β^2	$(12)_{22}$	$(12)_{21}$	$(12)_{12}$	$(12)_{20}$	$(12)_{11}$	$(12)_{02}$	$(12)_{10}$	$(12)_{01}$	$(12)_{00}$
2.000⋆-1	1.915	−4.482	−9.990	3.789	1.640⋆1	4.821	−8.727	−5.261	1.760
4.000⋆-1	−2.781⋆-1	−2.713⋆-1	−2.859	1.014	4.309	1.574	−2.072	−5.590⋆-1	−3.268⋆-1
6.000⋆-1	−5.329⋆-1	2.435⋆-1	−1.340	5.423⋆-1	2.082	9.702⋆-1	−6.857⋆-1	1.164⋆-1	−4.494⋆-1
8.000⋆-1	−5.492⋆-1	3.401⋆-1	−7.030⋆-1	4.018⋆-1	1.317	7.885⋆-1	8.836⋆-2	2.943⋆-1	−3.008⋆-1
1.000	−5.034⋆-1	3.520⋆-1	−3.171⋆-1	3.614⋆-1	9.811⋆-1	7.529⋆-1	3.059⋆-1	3.591⋆-1	−4.576⋆-2
1.200	−4.275⋆-1	3.464⋆-1	−3.655⋆-3	3.704⋆-1	8.235⋆-1	8.040⋆-1	6.654⋆-1	3.960⋆-1	3.026⋆-1
1.400	−3.235⋆-1	3.425⋆-1	3.246⋆-1	4.196⋆-1	7.622⋆-1	9.397⋆-1	1.082	4.343⋆-1	7.759⋆-1
1.600	−1.783⋆-1	3.489⋆-1	7.481⋆-1	5.197⋆-1	7.723⋆-1	1.195	1.654	4.917⋆-1	1.449
1.800	4.138⋆-2	3.739⋆-1	1.396	7.060⋆-1	8.597⋆-1	1.661	2.551	5.905⋆-1	2.484
2.000	4.179⋆-1	4.341⋆-1	2.558	1.072	1.070	2.578	4.164	7.776⋆-1	4.264
2.200	1.195	5.753⋆-1	5.082	1.904	1.554	4.672	7.628	1.185	7.901

Continued

Table 26 The asymmetric bell-shaped model. the aperture aberration coefficients $(0021)_{ij}$ $p/q = \frac{1}{4}, \frac{1}{2}, 2, 4$ $(p + q = 2)$.—cont'd

Magnetic $p/q = 2$

β^2	$(12)_{22}$	$(12)_{21}$	$(12)_{12}$	$(12)_{20}$	$(12)_{11}$	$(12)_{02}$	$(12)_{10}$	$(12)_{01}$	$(12)_{00}$
2.400	3.382	9.792⋆-1	1.252⋆1	4.418	2.936	1.105⋆1	1.765⋆1	2.354	1.793⋆1
2.600	1.584⋆1	3.149	5.645⋆1	1.956⋆1	1.051⋆1	4.985⋆1	7.555⋆1	8.935	7.330⋆1
2.800	2.387⋆3	3.607⋆2	8.732⋆3	3.079⋆3	1.323⋆3	7.987⋆3	1.116⋆4	1.213⋆3	1.011⋆4
3.000	1.076⋆1	1.371	4.258⋆1	1.581⋆1	5.353	4.170⋆1	5.291⋆1	5.421	4.318⋆1

Achromatic $p/q=\frac{1}{4}$

β^2	$(12)_{22}$	$(12)_{21}$	$(12)_{12}$	$(12)_{20}$	$(12)_{11}$	$(12)_{02}$	$(12)_{10}$	$(12)_{01}$	$(12)_{00}$
2.000⋆-1	3.695	−1.016⋆1	−1.184⋆1	7.950	2.435⋆1	5.146	−1.496⋆1	−7.777	3.898
4.000⋆-1	1.943⋆-1	−2.033	−2.541	2.818	6.215	1.265	−4.368	−9.681⋆-1	1.809⋆-1
6.000⋆-1	−7.309⋆-2	−6.807⋆-1	−6.757⋆-1	1.867	2.861	6.423⋆-1	−2.124	−3.531⋆-2	−3.667⋆-1
8.000⋆-1	1.662⋆-1	−2.632⋆-1	1.320⋆-1	1.626	1.692	4.689⋆-1	−1.180	1.858⋆-1	−4.941⋆-1
1.000	6.472⋆-1	−9.764⋆-2	7.051⋆-1	1.668	1.159	4.285⋆-1	−5.900⋆-1	2.441⋆-1	−5.052⋆-1
1.200	1.381	−2.476⋆-2	1.297	1.933	8.776⋆-1	4.578⋆-1	−6.828⋆-2	2.559⋆-1	−4.603⋆-1
1.400	2.508	5.683⋆-3	2.095	2.496	7.192⋆-1	5.547⋆-1	5.645⋆-1	2.548⋆-1	−3.624⋆-1
1.600	4.380	9.976⋆-3	3.398	3.611	6.308⋆-1	7.604⋆-1	1.564	2.535⋆-1	−1.763⋆-1
1.800	7.958	−1.465⋆-2	5.951	6.031	5.887⋆-1	1.210	3.556	2.589⋆-1	2.159⋆-1
2.000	1.663⋆1	−1.091⋆-1	1.238⋆1	1.255⋆1	8.753⋆-1	2.405	8.792	2.786⋆-1	1.262
2.200	5.082⋆1	−5.671⋆-1	3.871⋆1	4.075⋆1	4.790⋆-1	7.483	3.144⋆1	3.194⋆-1	5.810

2.400	9.463⋆2	−1.485⋆1	7.550⋆2	8.548⋆2	−8.597	1.507⋆2	6.899⋆2	−1.011	1.390⋆2
2.600	1.389⋆2	−2.832	1.181⋆2	1.489⋆2	−3.476	2.527⋆1	1.209⋆2	−9.105⋆-1	2.434⋆1
2.800	2.090⋆1	−5.337⋆-1	1.910⋆1	2.784⋆1	−8.567⋆-1	4.539	2.204⋆1	−2.270⋆-1	4.162
3.000	7.101	−2.214⋆-1	6.883	1.216⋆1	−4.160⋆-1	1.925	9.117	−1.009⋆-1	1.522

Achromatic $p/q=\frac{1}{2}$

β^2	$(12)_{22}$	$(12)_{21}$	$(12)_{12}$	$(12)_{20}$	$(12)_{11}$	$(12)_{02}$	$(12)_{10}$	$(12)_{01}$	$(12)_{00}$
2.000⋆-1	3.768	−9.571	−1.251⋆1	7.173	2.437⋆1	5.629	−1.426⋆1	−8.268	3.900
4.000⋆-1	1.717⋆-1	−1.732	−2.943	2.298	6.233	1.465	−3.973	−1.164	1.969⋆-1
6.000⋆-1	−2.207⋆-1	−4.685⋆-1	−1.022	1.385	2.879	7.551⋆-1	−1.839	−1.369⋆-1	−3.353⋆-1
8.000⋆-1	−1.544⋆-1	−8.885⋆-2	−2.287⋆-1	1.101	1.713	5.391⋆-1	−9.689⋆-1	1.291⋆-1	−4.458⋆-1
1.000	7.862⋆-2	6.068⋆-2	2.713⋆-1	1.031	1.182	4.499⋆-1	−4.564⋆-1	2.128⋆-1	−4.393⋆-1
1.200	4.368⋆-1	1.313⋆-1	7.099⋆-1	1.083	9.047⋆-1	4.701⋆-1	−5.169⋆-2	2.404⋆-1	−3.793⋆-1
1.400	9.471⋆-1	1.723⋆-1	1.206	1.248	7.527⋆-1	5.231⋆-1	3.631⋆-1	2.500⋆-1	−2.764⋆-1
1.600	1.693	2.046⋆-1	1.879	1.569	6.753⋆-1	6.396⋆-1	8.917⋆-1	2.565⋆-1	−1.188⋆-1
1.800	2.858	2.417⋆-1	2.931	2.160	6.546⋆-1	8.606⋆-1	1.697	2.680⋆-1	1.323⋆-1
2.000	4.889	2.995⋆-1	4.824	3.323	6.953⋆-1	1.297	3.140	2.930⋆-1	5.770⋆-1
2.200	9.086	4.142⋆-1	8.899	5.979	8.393⋆-1	2.291	6.280	3.495⋆-1	1.515

Continued

Table 26 The asymmetric bell-shaped model. the aperture aberration coefficients $(0021)_{ij}$ $p/q = \frac{1}{4}, \frac{1}{2}, 2, 4$ $(p + q = 2)$.—cont'd

Achromatic $p/q = \frac{1}{2}$

β^2	$(12)_{22}$	$(12)_{21}$	$(12)_{12}$	$(12)_{20}$	$(12)_{11}$	$(12)_{02}$	$(12)_{10}$	$(12)_{01}$	$(12)_{00}$
2.400	2.077⋆1	7.216⋆-1	2.073⋆1	1.406⋆1	1.267	5.295	1.556⋆1	4.987⋆-1	4.200
2.600	8.596⋆1	2.329	8.935⋆1	6.305⋆1	3.388	2.336⋆1	7.062⋆1	1.183	1.969⋆1
2.800	1.206⋆4	2.616⋆2	1.330⋆-4	1.005⋆4	2.782⋆2	3.667⋆3	1.100⋆4	7.392⋆1	3.010⋆3
3.900	5.276⋆1	9.499⋆-1	6.261⋆1	5.225⋆1	4.863⋆-1	1.879⋆1	5.410⋆1	8.777⋆-3	1.396⋆1

Achromatic $p/q = 4$

β^2	$(12)_{22}$	$(12)_{21}$	$(12)_{12}$	$(12)_{20}$	$(12)_{11}$	$(12)_{02}$	$(12)_{10}$	$(12)_{01}$	$(12)_{00}$
2.000⋆-1	3.898	−7.777	−1.496⋆1	5.146	2.435⋆1	7.950	−1.184⋆1	−1.016⋆1	3.695
4.000⋆-1	1.809⋆-1	−9.681⋆-1	−4.368	1.265	6.215	2.818	−2.541	−2.033	1.943⋆-1
6.000⋆-1	−3.667⋆-1	−3.531⋆-2	−2.124	6.423⋆-1	2.861	1.867	−6.767⋆-1	−6.807⋆-1	−7.309⋆-2
8.000⋆-1	−4.941⋆-1	1.858⋆-1	−1.180	4.689⋆-1	1.692	1.626	1.320⋆-1	−2.632⋆-1	1.662⋆-1
1.000	−5.052⋆-1	2.441⋆-1	−5.900⋆-1	4.285⋆-1	1.159	1.668	7.051⋆-1	−9.764⋆-2	6.472⋆-1
1.200	−4.603⋆-1	2.559⋆-1	−6.828⋆-2	4.578⋆-1	8.776⋆-1	1.933	1.297	−2.476⋆-2	1.381
1.400	−3.624⋆-1	2.548⋆-1	5.645⋆-1	5.547⋆-1	7.192⋆-1	2.496	2.095	5.683⋆-3	2.508
1.600	−1.763⋆-1	2.535⋆-1	1.564	7.604⋆-1	6.308⋆-1	3.611	3.398	9.975⋆-3	4.300
1.800	2.159⋆-1	2.589⋆-1	3.556	1.210	5.887⋆-1	6.031	5.951	−1.465⋆-2	7.958
2.000	1.262	2.786⋆-1	8.792	2.405	5.753⋆-1	1.255⋆1	1.238⋆1	−1.091⋆-1	1.663⋆1
2.200	5.810	3.194⋆-1	3.144⋆1	7.483	4.789⋆-1	4.075⋆1	3.871⋆1	−5.671⋆-1	5.082⋆1
2.400	1.390⋆2	−1.011	6.899⋆2	1.507⋆2	−8.597	8.548⋆2	7.550⋆2	−1.485⋆1	9.463⋆2

2.600	2.434⋆1	−9.105⋆-1	1.209⋆2	2.527⋆1	−3.475	1.489⋆2	1.181⋆2	−2.832	1.389⋆2
2.800	4.162	−2.270⋆-1	2.204⋆1	4.559	−8.567⋆-1	2.784⋆1	1.910⋆1	−5.336⋆-1	2.090⋆1
3.000	1.522	−1.009⋆-1	9.117	1.925	−4.160⋆-1	1.216⋆1	6.883	−2.213⋆-1	7.100

Achromatic $p/q = 2$

β^2	$(12)_{22}$	$(12)_{21}$	$(12)_{12}$	$(12)_{20}$	$(12)_{11}$	$(12)_{02}$	$(12)_{10}$	$(12)_{01}$	$(12)_{00}$
2.000⋆-1	3.900	−8.268	−1.426⋆1	5.629	2.437⋆1	7.173	−1.251⋆1	−9.571	3.765
4.000⋆-1	1.969⋆-1	−1.164	−3.973	1.465	6.233	2.298	−2.943	−1.732	1.717⋆-1
6.000⋆-1	−3.353⋆-1	−1.369⋆-1	−1.839	7.551⋆-1	2.879	1.365	−1.022	−4.685⋆-1	−2.207⋆-1
8.000⋆-1	−4.458⋆-1	1.291⋆-1	−9.689⋆-1	5.391⋆-1	1.713	1.101	−2.287⋆-1	−8.885⋆-2	−1.544⋆-1
1.000	−4.393⋆-1	2.128⋆-1	−4.564⋆-1	4.699⋆-1	1.182	1.031	2.713⋆-1	6.068⋆-1	7.862⋆-1
1.200	−3.793⋆-1	2.404⋆-1	−5.169⋆-2	4.701⋆-1	9.047⋆-1	1.083	7.099⋆-1	1.313⋆-1	4.368⋆-1
1.400	−2.764⋆-1	2.500⋆-1	3.631⋆-1	5.231⋆-1	7.527⋆-1	1.248	1.206	1.723⋆-1	9.471⋆-1
1.600	−1.188⋆-1	2.565⋆-1	8.917⋆-1	6.396⋆-1	6.753⋆-1	1.569	1.879	2.046⋆-1	1.693
1.800	1.323⋆-1	2.680⋆-1	1.697	8.606⋆-1	6.546⋆-1	2.160	2.931	2.417⋆-1	2.858
2.000	5.770⋆-1	2.930⋆-1	3.140	1.297	6.953⋆-1	5.323	4.824	2.995⋆-1	4.889
2.200	1.515	3.495⋆-1	6.280	2.291	8.393⋆-1	5.979	8.899	4.142⋆-1	9.086
2.400	4.200	4.987⋆-1	1.556⋆1	5.295	1.267	1.406⋆1	2.073⋆-1	7.216⋆-1	2.077⋆1
2.600	1.969⋆1	1.183	7.062⋆1	2.336⋆1	3.388	6.305⋆1	8.935⋆-1	2.329	8.596⋆1
2.800	3.010⋆3	7.982⋆-1	1.100⋆6	3.667⋆3	2.782⋆2	1.005⋆1	1.330⋆4	2.616⋆2	1.206⋆1
3.000	1.396⋆1	8.776⋆-3	5.410⋆1	1.879⋆1	4.863⋆-1	5.225⋆1	6.261⋆1	9.499⋆-1	5.276⋆1

Notice that the $(21)_{ij}$ for $p/q = r$ are related to those for $q/p = r$ as follows: $(21)_{22} \rightarrow (21)_{00}$, $(21)_{21} \rightarrow (21)_{01}$, $(21)_{12} \rightarrow (21)_{10}$, $(21)_{20} \rightarrow (21)_{02}$, $(21)_{11} \rightarrow (21)_{11}$.

and the asymptotic cardinal elements are given by

$$\frac{d}{f} = -\cos\theta\frac{\sin\pi\omega_x}{\omega_x} + \frac{d}{L}\theta\sin\theta\left(\frac{\sin^2\frac{\pi}{2}\omega_x}{\omega_x^2} + \frac{\cos^2\frac{\pi}{2}\omega_x}{1-\omega_x^2}\right)$$

$$\frac{d}{g} = -\cosh\theta\frac{\sin\pi\omega_y}{\omega_y} - \frac{d}{L}\theta\sinh\theta\left(\frac{\sin^2\frac{\pi}{2}\omega_y}{\omega_y^2} + \frac{\cos^2\frac{\pi}{2}\omega_y}{1-\omega_y^2}\right)$$

$$\frac{z_{Fi}}{d} = -\frac{z_{Fo}}{d} = \frac{L}{2d} - \frac{f}{d}\left(\cos\theta\cos\pi\omega_x + \frac{1-2\omega_x^2}{2\omega_x\left(\omega_x^2-1\right)^{1/2}}\sin\theta\sin\pi\omega_x\right)$$

$$\frac{z_{Gi}}{d} = -\frac{z_{Go}}{d} = \frac{L}{2d} - \frac{g}{d}\left(\cosh\theta\cos\pi\omega_y + \frac{1-2\omega_y^2}{2\omega_y\left(1-\omega_y^2\right)^{1/2}}\sinh\theta\sin\pi\omega_y\right). \tag{50}$$

Very similar formulae were first published by Bernard (1955); they are given in the above form in QO. Kanaya and Matsuura (1967) and Kanaya et al. (1968a) have derived them afresh, together with formulae for the real cardinal elements. (Graphs illustrating the behavior of both sets of cardinal elements were sent by Kanaya and Matsuura to the conference on Electron Optics, Instrumentation and Quantitative Electron Microscopy, held at St. Andrews in 1967.) The asymptotic cardinal elements are listed in Table 27 and their behavior is illustrated in Figs. 4A and B. Formulae for the chromatic and aperture aberrations and graphs illustrating their behavior are to be found in Kanaya, Okayama, and Matsuura (1968); Kanaya, Okayama, and Mishina (1968b, 1968c) and Kanaya and Okayama (1969). The behavior of the chromatic aberration coefficients is illustrated in Fig. 4C–H (in the case of high magnification in both principal sections).

1.5 The short achromatic quadrupole

There are a number of features of combined electromagnetic quadrupoles that are more conveniently treated together than scattered over the preceding sections; in particular, we discuss the behavior of combined lenses in which the electrostatic and magnetic potentials are not exactly the same and the aperture aberrations of achromatic lenses.

The consequences of a difference between the electrostatic and magnetic functions $k(z)$, which we now write $k_E(z)$ and $k_M(z)$, were first analyzed by Shpak and Yavor (1964, 1965). Using the rectangular model, they derived

Table 27 The rectangular model terminated by half-bells. Cardinal elements as a function of β; $d = 1$.

***L* = 0.25**

β	βL	f	g	z_F	z_G
0.1	0.025	5.542⋆1	−5.443⋆1	5.541.1	−5.442⋆1
0.2	0.050	1.423⋆1	−1.335⋆1	1.420.1	−1.321⋆1
0.3	0.075	6.621	−5.633	6.536	−5.553
0.4	0.100	3.973	−2.981	3.818	−2.840
0.5	0.125	2.768	−1.766	2.517	−1.551
0.6	0.150	2.138	−1.118	1.761	−8.183⋆-1
0.7	0.175	1.788	−7.379⋆-1	1.249	−3.439⋆-1
0.8	0.200	1.601	−5.008⋆-1	8.496⋆-1	−4.900⋆-3
0.9	0.225	1.524	−3.465⋆-1	4.934⋆-1	2.569⋆-1
1.0	0.250	1.543	−2.429⋆-1	1.320⋆-1	4.720⋆-1
1.1	0.275	1.669	−1.720⋆-1	−2.810⋆-1	6.571⋆-1
1.2	0.300	1.955	−1.226⋆-1	−8.232⋆-1	8.222⋆-1
1.3	0.325	2.552	−8.783⋆-2	−1.663	9.736⋆-1
1.4	0.350	4.019	−6.316⋆-2	−3.369	1.115
1.5	0.375	1.064⋆1	−4.555⋆-2	−1.024⋆1	1.249
1.6	0.400	−1.658⋆1	−3.291⋆-2	1.672⋆1	1.378
1.7	0.425	−4.960	−2.382⋆-2	4.806	1.502
1.8	0.450	−3.137	−1.726⋆-2	2.647	1.623
1.9	0.475	−2.482	−1.251⋆-2	1.591	1.742
2.0	0.500	−2.231	−9.079⋆-3	8.418⋆-1	1.858

***L* = 0.50**

β	βL	f	g	z_F	z_G
0.1	0.050	4.881⋆-1	−4.778⋆1	4.880⋆1	−4.777⋆1
0.2	0.100	1.260⋆1	−1.157⋆1	1.256⋆1	−1.153⋆1
0.3	0.150	5.912	−4.881	5.810	−4.787
0.4	0.200	3.593	−2.555	3.406	−2.391
0.5	0.250	2.546	−1.493	2.243	−1.245

Continued

Table 27 The rectangular model terminated by half-bells. Cardinal elements as a function of β; $d = 1$.—cont'd

$L = 0.50$

β	βL	f	g	z_F	z_G
0.6	0.300	2.010	−9.305⋆-1	1.551	−5.855⋆-1
0.7	0.350	1.731	−6.035⋆-1	1.061	−1.527⋆-1
0.8	0.400	1.608	−4.019⋆-1	6.541⋆-1	1.615⋆-1
0.9	0.450	1.609	−2.724⋆-1	2.569⋆1	4.082⋆-1
1.0	0.500	1.742	−1.869⋆-1	−1.968⋆-1	6.138⋆-1
1.1	0.550	2.079	−1.294⋆-1	−8.109⋆-1	7.929⋆-1
1.2	0.600	2.861	−9.016⋆-2	−1.848	9.542⋆-1
1.3	0.650	5.234	−6.310⋆-2	−4.468	1.103
1.4	0.700	4.952⋆1	−4.431⋆-2	−4.901⋆1	1.243
1.5	0.750	−6.791	−3.120⋆-2	7.029	1.376
1.6	0.800	3.382	−2.200⋆-2	3.315	1.504
1.7	0.850	−2.431	−1.554⋆-2	2.009	1.628
1.8	0.900	−2.065	−1.099⋆-2	1.210	1.749
1.9	0.950	−1.965	−7.771⋆-3	5.552⋆-1	1.867
2.0	1.000	−2.064	−5.501⋆-3	−1.082⋆-1	1.983

$L = 1.00$

β	βL	f	g	z_F	z_G
0.1	0.100	3.947⋆1	−3.834⋆1	3.946⋆1	−3.832⋆1
0.2	0.200	1.031⋆1	−9.178	1.025⋆1	−9.119
0.3	0.300	4.940	−3.800	4.795	−3.669
0.4	0.400	3.095	−1.939	2.825	−1.716
0.5	0.500	2.288	−1.099	1.838	−7.665⋆-1
0.6	0.600	1.913	−6.613⋆-1	1.206	−2.064⋆-1
0.7	0.700	1.780	−4.126⋆-1	6.982⋆-1	1.724⋆-1
0.8	0.800	1.842	−2.635⋆-1	1.813⋆-1	4.560⋆-1
0.9	0.900	2.163	−1.709⋆-1	−4.933⋆-1	6.848⋆-1
1.0	1.000	3.065	−1.121⋆-1	−1.698	8.798⋆-1

Table 27 The rectangular model terminated by half-bells. Cardinal elements as a function of β; $d = 1$.—cont'd

$L = 1.00$

β	βL	f	g	z_F	z_G
1.1	1.100	6.760	−7.404⋆-2	−5.666	1.053
1.2	1.200	−1.953⋆1	−4.918⋆-2	2.036⋆1	1.210
1.3	1.300	−4.043	−3.200⋆-2	4.560	1.357
1.4	1.400	−2.401	−2.193⋆-2	2.621	1.495
1.5	1.500	−1.860	−1.470⋆-2	1.703	1.627
1.6	1.600	−1.676	−9.869⋆-3	1.046	1.755
1.7	1.700	−1.703	−6.633⋆-3	4.298⋆-1	1.879
1.8	1.800	−1.951	−4.462⋆-3	−2.899⋆-1	1.999
1.9	1.900	−2.602	−3.003⋆-3	−1.359	2.117
2.0	2.000	−4.515	−2.023⋆-3	−3.676	2.233

$L = 2.00$

β	βL	f	g	z_F	z_G
0.1	0.200	2.871⋆1	−2.732⋆1	2.868⋆1	−2.729⋆1
0.2	0.400	7.747	−6.348	7.625	−6.237
0.3	0.600	3.927	−2.506	3.634	−2.269
0.4	0.800	2.683	−1.202	2.108	−8.092⋆-1
0.5	1.000	2.257	−6.330⋆-1	1.213	−6.696⋆-2
0.6	1.200	2.297	−3.508⋆-1	4.091⋆-1	3.953⋆-1
0.7	1.400	2.976	−2.004⋆-1	−7.820⋆-1	7.254⋆-1
0.8	1.600	5.503	−1.167⋆-1	−4.703	9.839⋆-1
0.9	1.800	−1.211⋆1	−6.887⋆-2	1.357⋆1	1.200
1.0	2.000	−3.056	−4.099⋆-2	4.159	1.380
1.1	2.200	−1.859	−2.456⋆-2	2.616	1.557
1.2	2.400	−1.503	−1.478⋆-2	1.816	1.712
1.3	2.600	−1.462	−8.925⋆-3	1.166	1.858
1.4	2.800	−1.698	−5.404⋆-3	4.254⋆-1	1.996
1.5	3.000	−2.511	−3.278⋆-3	−8.148⋆-1	2.128

Continued

Table 27 The rectangular model terminated by half-bells. Cardinal elements as a function of β; $d = 1$.—cont'd

$L = 2.00$

β	βL	f	g	z_F	z_G
1.6	3.200	−6.680	−1.992⋆-3	−5.389	2.255
1.7	3.400	9.326	−1.211⋆-3	1.019⋆-1	2.379
1.8	3.600	3.100	−7.374⋆-4	3.477	2.499
1.9	3.800	2.192	−4.492⋆-4	1.950	2.617
2.0	4.000	2.056	−2.737⋆-4	9.333⋆-1	2.733

$L = 4.00$

β	βL	f	g	z_F	z_G
0.1	0.400	1.898⋆-1	−1.699⋆1	1.890⋆1	−1.691⋆1
0.2	0.800	5.655	−3.625	5.265	−3.317
0.3	1.200	3.468	−1.258	2.405	−6.459⋆-1
0.4	1.600	3.370	−5.148⋆-1	6.678⋆-1	4.245⋆-1
0.5	2.000	6.465	−2.271⋆-1	−3.366	1.026
0.6	2.400	−8.914	−1.043⋆-1	1.139⋆1	1.433
0.7	2.800	−2.443	−4.910⋆-2	4.377	1.741
0.8	3.200	−1.635	−2.349⋆-2	2.956	1.991
0.9	3.600	−1.614	−1.136⋆-2	1.992	2.203
1.0	4.000	−2.454	−5.543⋆-2	5.281⋆-1	2.389
1.1	4.400	−2.565⋆1	−2.720⋆-2	−2.317⋆1	2.557
1.2	4.800	2.834	−1.340⋆-2	4.818	2.713
1.3	5.200	1.612	−6.628⋆-4	3.001	2.858
1.4	5.600	1.510	−3.286⋆-4	2.001	2.996
1.5	6.000	2.145	−1.632⋆-4	6.774⋆-1	3.128
1.6	6.400	8.744	−8.119⋆-5	−6.501	3.255
1.7	6.800	−3.978	−4.043⋆-5	5.618	3.379
1.8	7.200	−2.046	−2.015⋆-5	2.921	3.499
1.9	7.600	−1.946	−1.005⋆-5	1.592	3.617
2.0	8.000	−2.996	−5.013⋆-6	−3.360⋆-1	3.733

Table 27 The rectangular model terminated by halfells. Cardinal elements as a function of β; $d = 1$.—cont'd

$L = 8.00$

β	βL	f	g	z_F	z_G
0.1	0.800	1.228⋆1	−8.975	1.185⋆1	−8.620
0.2	1.600	5.393	−1.485	2.955	−3.462⋆-1
0.3	2.400	1.301⋆1	−3.666⋆-1	−7.782	1.570
0.4	3.200	−4.169	−1.028⋆-1	7.952	2.478
0.5	4.000	−2.208	−3.061⋆-2	4.533	3.040
0.6	4.800	−3.814	−9.449⋆-3	1.382	3.437
0.7	5.600	4.037	−2.984⋆-3	8.295	3.742
0.8	6.400	1.617	−9.573⋆-4	4.859	3.991
0.9	7.200	2.201	−3.105⋆-4	2.954	4.203
1.0	8.000	−6.714	−1.015⋆-4	1.105⋆1	4.389
1.1	8.800	−1.543	−3.339⋆-5	4.996	4.557
1.2	9.600	−1.766	−1.103⋆-5	3.351	4.713
1.3	10.400	3.350⋆1	−3.656⋆-6	3.778⋆1	4.856
1.4	11.200	1.724	−1.215⋆-6	5.103	4.996
1.5	12.000	1.744	−4.046⋆-7	3.354	5.128
1.6	12.800	2.347⋆1	−1.349⋆-7	−1.932⋆1	5.255
1.7	13.600	−2.078	−4.503⋆-8	5.216	5.379
1.8	14.400	−1.980	−1.504⋆-8	3.138	5.499
1.9	15.200	−1.975⋆1	−5.029⋆-9	−1.576⋆1	5.617
2.0	16.000	2.505	−1.682⋆-9	5.279	5.733

the achromatic condition in the converging plane when the electrostatic and magnetic effective lengths are slightly different but the components have a common center and when the lengths are the same but the centers do not exactly coincide. This work is recapitulated in QO and we merely summarize their results here: in the first situation, we denote the common length by L and the extent to which the electrostatic length exceeds L by 2λ.

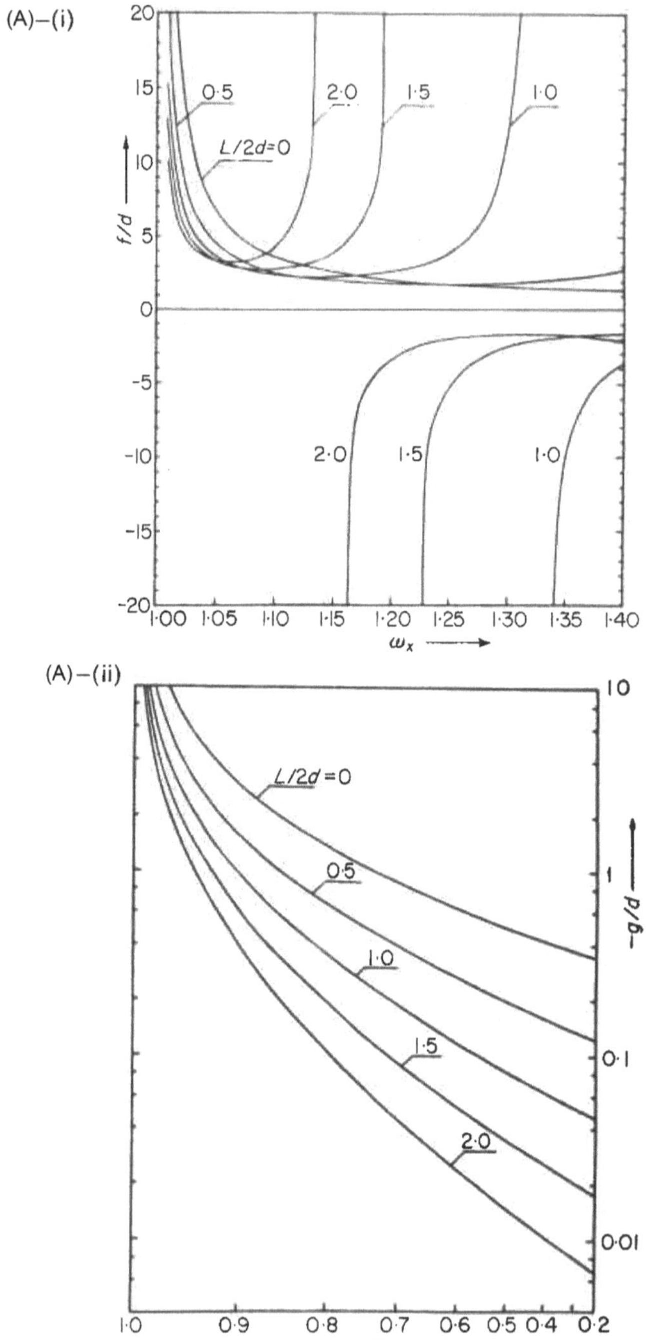

Fig. 4 (A,B) The asymptotic cardinal elements of a quadrupole lens: rectangular model terminated by half-bells. (After Kanaya et al., 1968a). (C–H) The corresponding chromatic aberration coefficients.

$$\Delta x = (C_{Fx}\overline{\alpha} + C_{mx}x_o)\frac{\Delta\Phi_0}{\Phi_0}, \qquad \Delta y = (C_{Fy}\overline{\gamma} + C_{my}y_o)\frac{\Delta\Phi_0}{\Phi_0}$$

$$C_{Fx} = x_\gamma C_{02x} - x_\alpha C_{11x}; \qquad C_{Fy} = -y_\gamma C_{02y} + y_\alpha C_{11y}$$

$$C_{mx} = x_\gamma C_{11x} - x_\alpha C_{20x}; \qquad C_{mx} = -y_\gamma C_{11y} + y_\alpha C_{20y}$$

(After Kanaya, Okayama, & Mishina, 1968b, q.v. for further details, especially of the notation.)

(Continued)

(B)–(i)

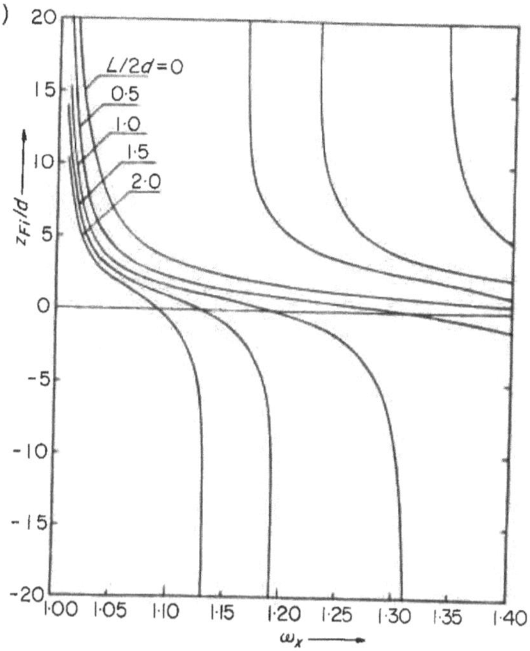

(B)–(ii)

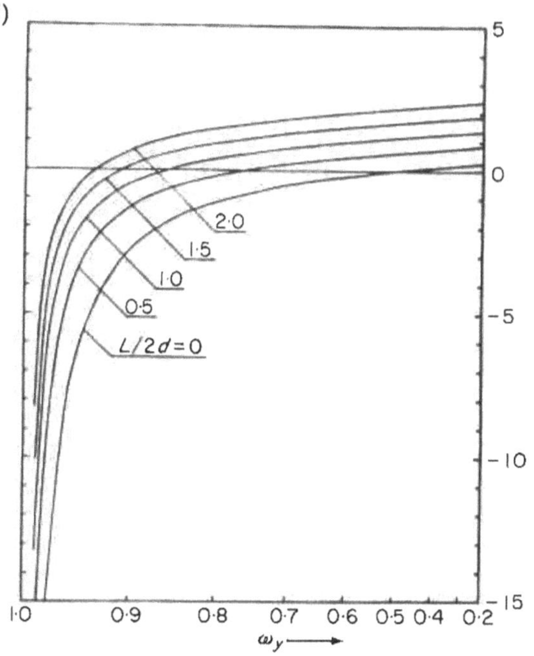

Fig. 4—Cont'd

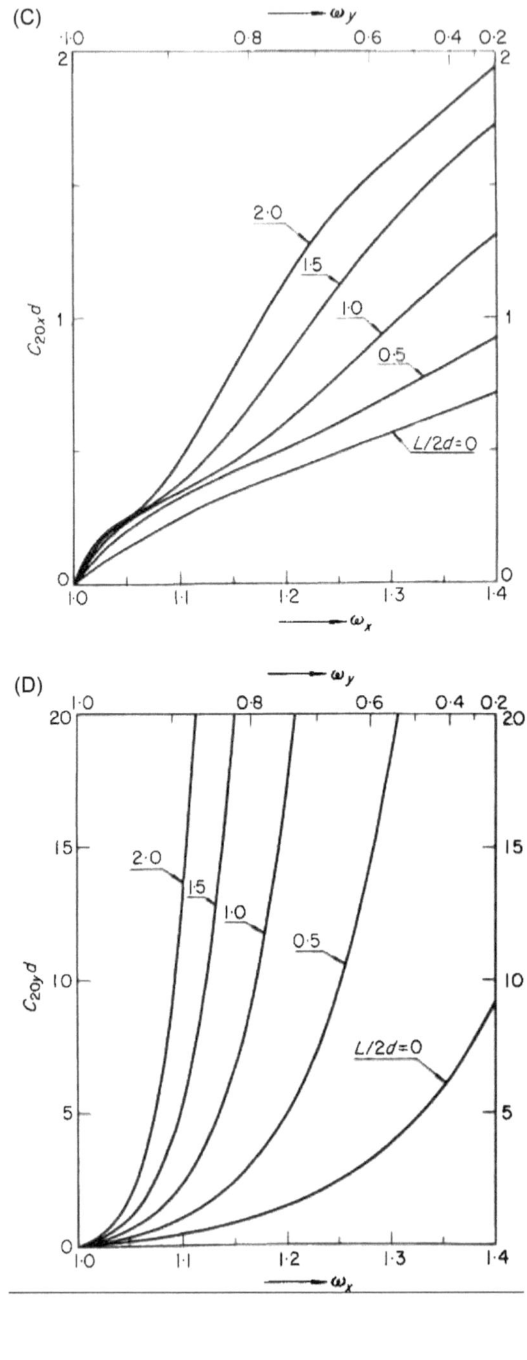
(C)
ωy
1·0 0·8 0·6 0·4 0·2
2.0
1·5
1·0
0·5
L/2d=0
$C_{20x}d$
1·0 1·1 1·2 1·3 1·4
ωx
(D)
ωy
1·0 0·8 0·6 0·4 0·2
2·0
1·5
1·0
0·5
L/2d=0
$C_{20y}d$
1·0 1·1 1·2 1·3 1·4
ωx

Fig. 4—Cont'd

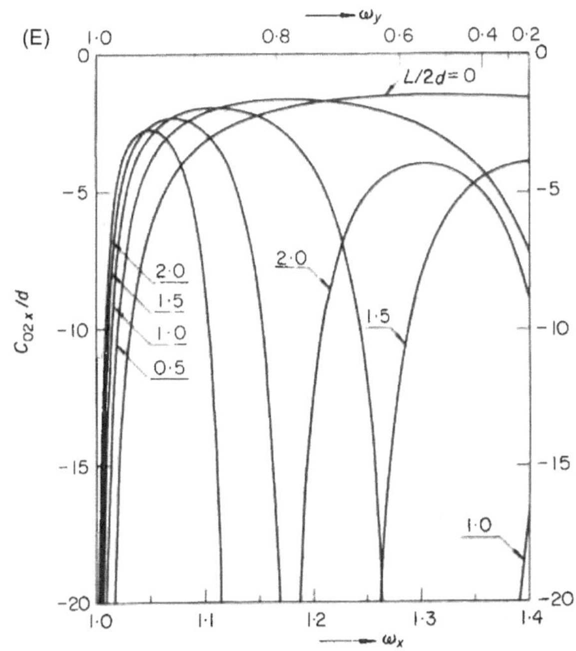
(E)
ω_y
1·0
0·8
0·6
0·4
0·2
0
-5
-10
-15
-20
$L/2d = 0$
2·0
1·5
1·0
0·5
C_{02x}/d
1·1
1·2
1·3
1·4
ω_x

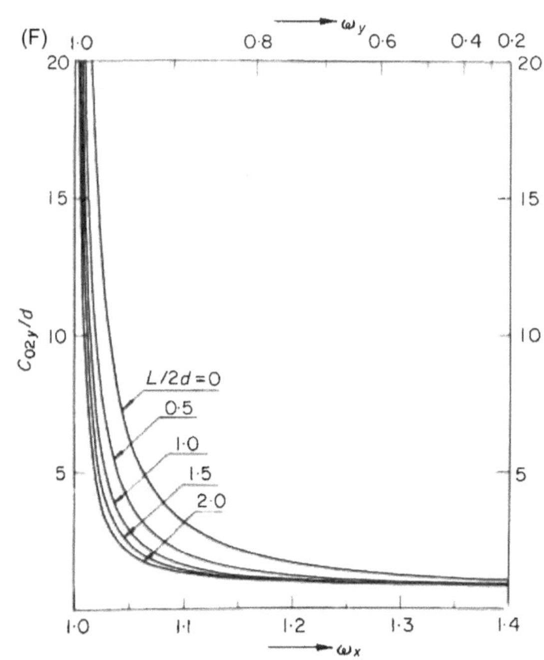
(F)
ω_y
1·0
0·8
0·6
0·4
0·2
20
15
10
5
C_{02y}/d
$L/2d = 0$
0·5
1·0
1·5
2·0
1·1
1·2
1·3
1·4
ω_x

Fig. 4—Cont'd

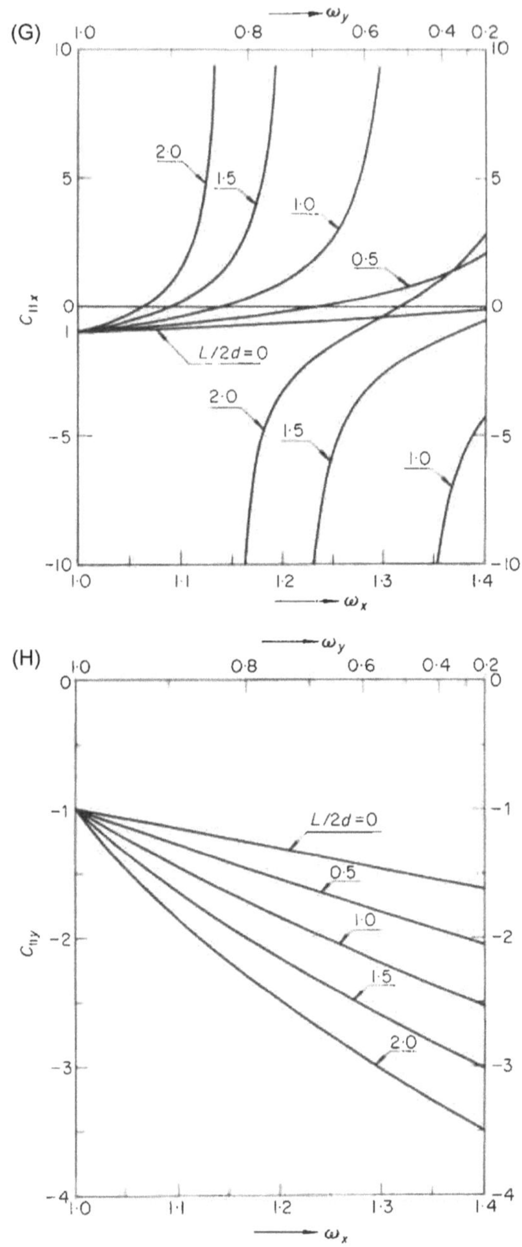

Fig. 4—Cont'd

For small values of λ, the axial chromatic aberration in the x–z plane vanishes if

$$\frac{\beta_E^2}{\beta_M^2} = \frac{1}{2}\left(1 - \frac{\lambda}{L} f_1\right) \tag{51}$$

in which f_1 is a complicated function of the object distance and βL; the chromatic aberration of magnification in the x–z plane vanishes if

$$\frac{\beta_E^2}{\beta_M^2} = \frac{1}{2}\left(1 - \frac{\lambda}{L} f_2\right) \tag{52}$$

where f_2 is another complicated function. For each value of the object distance, Eqs. (51) and (52) have a common solution only for isolated values of βL. The case in which the centers do not coincide yields similar results. The theory is experimentally verified in Shpak and Yavor (1965).

Similar situations have been analyzed by Hardy (1967), who has used the bell-shaped model as well as the rectangular one. The aberrations are considered at the line foci, and are compared with the predictions of the weak lens approximation. The latter gives $C_{cx} = 0$ when $2\beta_E^2 L_E - \beta_M^2 L_M = 0$. When the centers of the rectangular distributions $k_E(z)$ and $k_M(z)$ coincide but $L_M > L_E$, the value of $n^* = 2\beta_E^2 L_E / \beta_M^2 L_M$ for which $C_{cx} = 0$ remains quite close to 1; Fig. 5A shows how n^* varies with the position of the focus for various values of L_M/L_E and Fig. 5B illustrates the behavior of C_{cy}. Hardy also discusses the magnitude of the secondary chromatic aberration coefficient in the x–z plane when $C_{cx} = 0$. The bell-shaped model (Fig. 6A) leads to very similar results (Fig. 6B).

We now turn to Hardy's work on the aperture aberrations of achromatic quadrupoles. We have seen (Eq. 149 in Definitions, for example) that each of the aperture aberration coefficients may be written in the form $C_2 n^2 + C_1 n + C_0$, C_1 differing from C_2 only by a constant factor. Hardy plots (30) and (12) as a function of n for a particular position of z_{Fi} (Fig. 7) as Dymnikov et al. (1965c) have done, and shows that both are quite small in absolute value in the vicinity of the achromatic working point. He then considers, theoretically and experimentally, the consequences of having $k_E(z) \neq k_M(z)$; the general trend is maintained but the minimum of (30) and the maximum of (12), regarded as functions of n, shift.

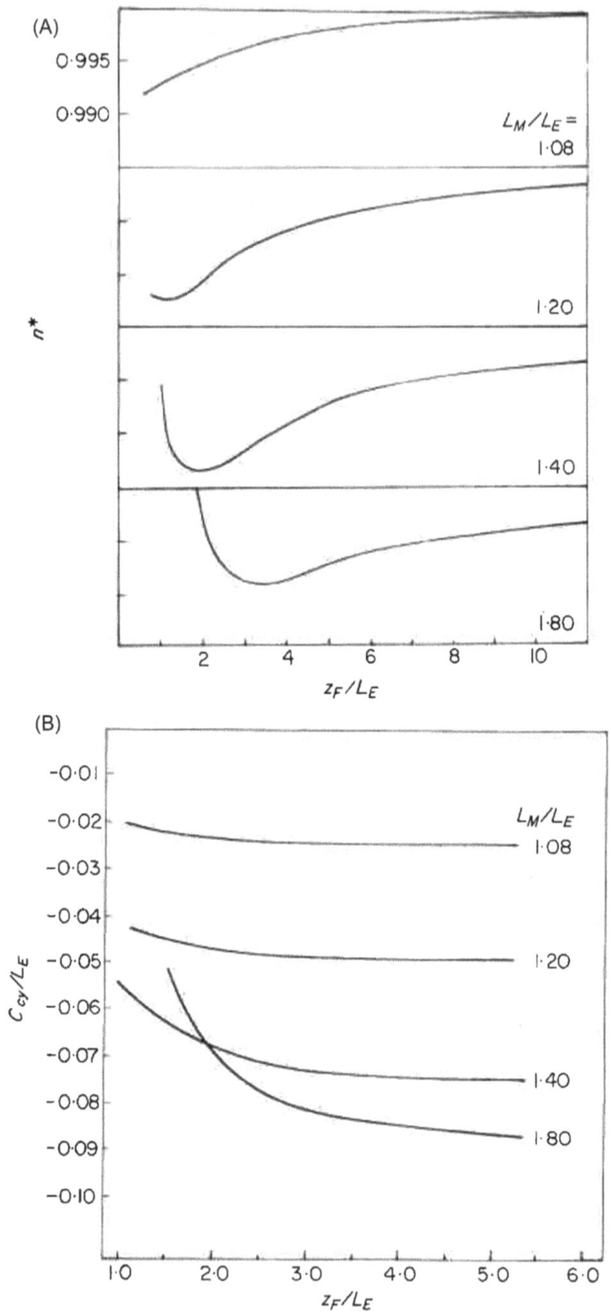

Fig. 5 The combined electromagnetic quadrupole in which the magnetic and electrostatic functions, k_M and k_E differ. (A) Variation of the parameter n^* with z_F and the ratio of the lengths (rectangular model). (B) Magnitude of the primary chromatic aberration coefficient at the virtual line image when the real image is achromatic. (After Hardy, 1967.).

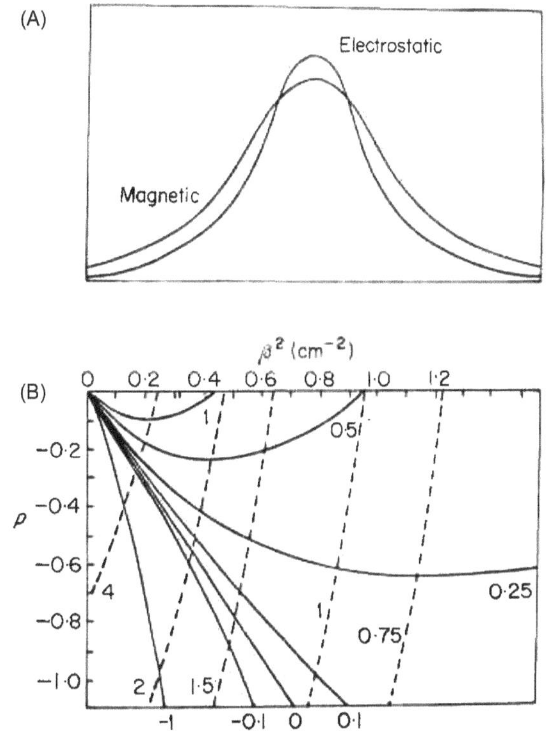

Fig. 6 (A) The functions $k_E(z)$ and $k_M(z)$ in an imperfect combined lens (bell-shaped model). (B) Contours of equal focal distance (dashed curve) and primary chromatic aberration (full curve). (After Hardy, 1967.) L is here the equivalent length, $L = \pi d/2$; $p = -\beta_E^2$.

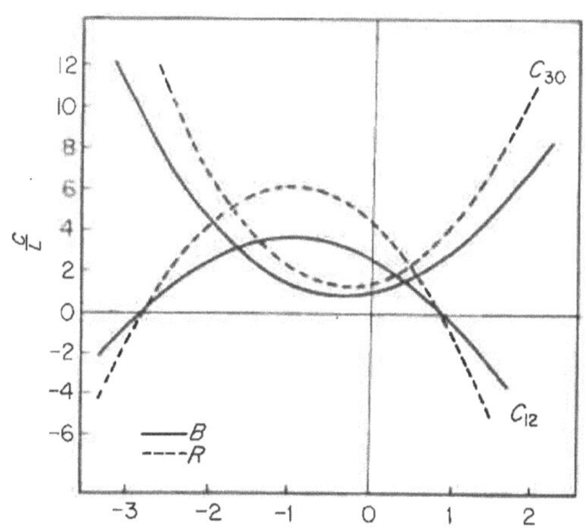

Fig. 7 The aperture aberration coefficients of an ideal combined lens for a fixed value of z_F/L. (After Hardy, 1967).

2. The quadrupole doublet

The cardinal elements of quadrupole doublets have been calculated by several authors, using both the rectangular and bell-shaped models. If the asymptotic Dušek transfer matrices of the i-th quadrupole ($i = 1, 2$) are (cf. Eq. 39 in Definitions).

$$\mathbf{T}_x = \begin{pmatrix} \alpha_{xi} & \beta_{xi} \\ \gamma_{xi} & \delta_{xi} \end{pmatrix}, \quad \mathbf{T}_y = \begin{pmatrix} \alpha_{yi} & \beta_{yi} \\ \gamma_{yi} & \delta_{yi} \end{pmatrix} \tag{53}$$

and the lens centers are D apart, then the cardinal elements of the doublet will be given by

$$\begin{aligned} f &= -1/(\alpha_{x1}\gamma_{x2} + \gamma_{x1}\delta_{x2} + D\gamma_{x1}\gamma_{x2}) \\ z_{Fo} &= -f(\beta_{x1}\gamma_{x2} + \delta_{x1}\delta_{x2} + D\delta_{x1}\gamma_{x2}) \\ z_{Fi} &= f(\alpha_{x1}\alpha_{x2} + \gamma_{x1}\beta_{x2} + D\gamma_{x1}\alpha_{x2}) \end{aligned} \tag{54}$$

with similar expressions for g, z_{Go} and z_{Gi}. The position of the object focus is measured with respect to the center of the first quadrupole and that of the image focus with respect to the center of the second quadrupole. If the doublet is antisymmetric, so that

$$\begin{aligned} \alpha_{x1} &= \delta_{y2} & \alpha_{y1} &= \delta_{x2} \\ \delta_{x1} &= \alpha_{y2} & \delta_{y1} &= \alpha_{x2} \\ \gamma_{x1} &= \gamma_{y2} & \gamma_{y1} &= \gamma_{x2} \\ \beta_{x1} &= \beta_{y2} & \beta_{y1} &= \beta_{x2} \end{aligned} \tag{55}$$

then

$$\begin{aligned} f &= -1/\left(\alpha_{x1}\gamma_{y1} + \gamma_{x1}\alpha_{y1} + D\gamma_{x1}\gamma_{y1}\right) = g \\ z_{Fo} &= -f\left(\beta_{x1}\gamma_{y1} + \delta_{x1}\alpha_{y1} + D\delta_{x1}\gamma_{y1}\right) = -z_{Gi} \\ z_{Fi} &= f\left(\alpha_{x1}\delta_{y1} + \gamma_{x1}\beta_{y1} + D\gamma_{x1}\delta_{y1}\right) = -z_{Go}. \end{aligned} \tag{56}$$

When the doublet is not antisymmetric, it is not possible to make it regular for a continuous range of values of the three parameters available (the two excitations and the lens separation) since three conditions must be satisfied: $f = g$, $z_{Fo} = z_{Go}$ and $z_{Fi} = z_{Gi}$. Except in the degenerate case of antisymmetry, the combination will be regular only for isolated sets of values. It is, on the other hand, relatively easy to make the doublet stigmatic but not orthomorphic for given separation and object distance. Graphs corresponding to this pseudo-stigmatic imagery have been obtained by Enge (1959), King (1962), Baker (1963) and Junior and Antony-Spies (1967), using the rectangular model; tables are given in Blewett (1959) and Dymnikov, Ovsyannikova, and Yavor (1963). Septier and van Acker (1960a, 1960b, 1961) plot theoretical and experimental curves, showing the dependence of the cardinal elements in the two principal sections on the various parameters. Numerous other papers in which doublet properties are described are listed in *QO*.

When the system is antisymmetric, only one condition need be satisfied for regularity while two parameters are free: the common excitation (β^2) and the separation (D). We may therefore set

$$D = f\left(\beta^2\right). \tag{57}$$

This function has been established by Dhuicq (1961) for the rectangular model and by Glaser (1955) and Dušek (1959) for the symmetric bell-shaped model. We have also calculated it for the asymmetric bell-shaped model.

The aperture aberrations of a range of pseudo-stigmatic doublets have been calculated by Dymnikov, Fishkova, and Yavor (1967), using the thin-lens approximation, and by Fishkova et al. (1968a), using the rectangular model. We have calculated them for a range of antisymmetric doublets using the symmetric and asymmetric bell-shaped models.

Most of the information about doublets is in the form of graphs, with the exception of Blewett's tables Blewett (1959) and of Tables 28–31. Blewett lists f, g, z_P and z_Q for a range of doublets, using the rectangular model in the range $0 \cdot 1 \leq \theta_1,\ \theta_2 \leq 1 \cdot 6$; both quadrupoles are of length L, and the distance between the end-faces is $L/2$, L, or $2L$. These figures are not reproduced here. Table 28 (Dymnikov et al., 1963) contains first-order data for pseudo-stigmatic doublets (rectangular model). Tables 29–31 list the computed values of the cardinal elements and aperture aberrations of a regular antisymmetric doublet (bell-shaped model).

Table 28 The pseudo-stigmatic doublet (rectangular model).

			L_2/L_1														
			1					2					4				
s/L_1	z_0/L_1	z_i/L_1	θ_1	θ_2	M_x	M_y	M_xM_y	θ_1	θ_2	M_x	M_y	M_xM_y	θ_1	θ_2	M_x	M_y	M_xM_y
0	1	1	1.41	1.41	−4.0	−0.25	1.0	1.28	1.79	−6.1	−0.27	1.6	1.16	2.19	−10.5	−0.30	3.1
0	1	2	1.35	1.22	−5.2	−0.43	2.2	1.24	1.56	−7.2	−0.41	2.9	1.11	1.96	−12.0	−0.40	4.8
0	1	4	1.30	1.09	−7.6	−0.82	6.2	1.20	1.39	−9.8	−0.70	6.9	1.10	1.72	−14.0	−0.59	8.4
0	0.5	1	1.62	1.47	−6.2	−0.29	1.8	1.49	1.85	−9.4	−0.30	2.8	1.37	2.25	−16.0	−0.32	5.1
0	0	1	2.18	1.55	−10.4	−0.33	3.5	2.02	1.94	−15.8	−0.33	5.2	1.87	2.33	−27.0	−0.34	9.2
0.5	1	1	1.23	1.23	−4.6	−0.22	1.0	1.17	1.62	−6.7	−0.25	1.7	1.10	2.03	−11.1	−0.29	3.3
1	1	1	1.15	1.15	−5.3	−0.19	1.0	1.11	1.53	−7.5	−0.23	1.7	1.06	1.94	−12.0	−0.27	3.3
0	1	1	1.69	3.82	−61	0.22	−14	1.53	4.21	−71	0.24	−17	1.36	4.70	−107	0.26	−27
0	1	1	3.82	1.69	4.5	−0.016	−0.074	3.68	2.06	7.7	−0.015	−0.12	3.58	2.42	15	−0.014	−0.21
0	1	1	4.16	4.16	55	0.018	1.0	3.92	4.59	80	0.017	1.4	3.73	5.05	142	0.015	2.2

Note: z_0 denotes the object distance, measured from the beginning of the first quadrupole, and z_i denotes the image distance, measured from the end of the second quadrupole; s is the distance from the end of the first quadrupole to the beginning of the second.

Table 29 The regular antisymmetric doublet of asymmetric or symmetric quadrupoles: the bell-shaped model. Cardinal elements and spacing as a function OF β^2. (Negative values of the spacing, t are included to show the behavior of the other quantities more clearly. $p + q = 2$).

$p/q = \frac{1}{4}$

β^2	f	z_{F0}	t
2.000⋆-1	−9.175	9.931	1.453
4.000⋆-1	−2.081	2.812	1.214
6.000⋆-1	−7.820⋆-1	1.476	7.552⋆-1
8.000⋆-1	−3.425⋆-1	9.890⋆-1	−5.787⋆-2
1.000	−1.529⋆-1	7.456⋆-1	−1.584
1.200	−6.141⋆-2	5.976⋆-1	−5.097
1.400	−1.543⋆-2	4.949⋆-1	−2.052⋆1
1.600	7.390⋆-3	4.170⋆-1	3.926⋆1
1.800	1.785⋆-2	3.544⋆-1	1.047⋆1
2.000	2.161⋆-2	3.021⋆-1	9.675
2.200	2.178⋆-2	2.571⋆-1	7.745
2.400	2.012⋆-2	2.177⋆-1	6.572
2.600	1.764⋆-2	1.827⋆-1	5.706
2.800	1.491⋆-2	1.512⋆-1	4.967
3.000	1.225⋆-2	1.226⋆-1	4.266

$p/q = \frac{1}{2}$

β^2	f	z_{F0}	t
2.000⋆-1	−9.875	1.029⋆1	7.888⋆-1
4.000⋆-1	−2.214	2.600	5.936⋆-1
6.000⋆-1	−8.125⋆-1	1.153	2.053⋆-1
8.000⋆-1	−3.395⋆-1	6.225⋆-1	−5.350⋆-1
1.000	−1.367⋆-1	3.535⋆-1	−2.153
1.200	−4.046⋆-2	1.860⋆-1	−7.782
1.400	6.336⋆-3	6.638⋆-2	4.652⋆1
1.600	2.796⋆-2	−2.730⋆-2	9.200

Continued

Table 29 The regular antisymmetric doublet of asymmetric or symmetric quadrupoles: the bell-shaped model. Cardinal elements and spacing as a function of β^2. (Negative values of the spacing, t are included to show the behavior of the other quantities more clearly. $p + q = 2$).—cont'd

$p/q = \frac{1}{2}$

β^2	f	z_{FO}	t
1.800	3.621⋆-2	−1.052⋆-1	5.906
2.000	3.733⋆-2	−1.724⋆-1	4.575
2.200	3.476⋆-2	−2.321⋆-1	3.769
2.400	3.049⋆-2	−2.859⋆-1	3.151
2.600	2.563⋆-2	−3.351⋆-1	2.593
2.800	2.081⋆-2	−3.805⋆-1	2.020
3.000	1.636⋆-2	−4.227⋆-1	1.372

$p/q = 1$

β^2	f	z_{FO}	t
2.000⋆-1	−1.007⋆1	1.006⋆1	−6.514⋆-2
4.000⋆-1	−2.136	2.075	−2.932⋆-1
6.000⋆-1	−6.915⋆-1	5.590⋆-1	−8.385⋆-1
8.000⋆-1	−2.129⋆-1	−1.161⋆-2	−2.447
1.000	−1.647⋆-2	−3.147⋆-1	−2.765⋆1
1.200	6.873⋆-2	−5.153⋆-1	5.629
1.400	1.025⋆-1	−6.683⋆-1	3.118
1.600	1.108⋆-1	−7.954⋆-1	2.317
1.800	1.061⋆-1	−9.068⋆-1	1.881
2.000	9.524⋆-2	−1.007	1.568
2.200	8.191⋆-2	−1.100	1.299
2.400	6.819⋆-2	−1.186	1.035
2.600	5.519⋆-2	−1.266	7.487⋆-1
2.800	4.348⋆-2	−1.342	4.151⋆-1
3.000	3.327⋆-2	−1.415	−1.154

Table 29 The regular antisymmetric doublet of asymmetric or symmetric quadrupoles: the bell-shaped model. Cardinal elements and spacing as a function of β^2. (Negative values of the spacing, t are included to show the behavior of the other quantities more clearly. $p+q=2$).—cont'd

$p/q = 2$

β^2	f	z_{F0}	t
2.000★-1	−9.401	8.950	−9.482★-1
4.000★-1	−1.760	1.231	−1.356
6.000★-1	−3.883★-1	−2.625★-1	−2.976
8.000★-1	4.648★-2	−8.537★-1	1.513★1
1.000	2.062★-1	−1.194	2.304
1.200	2.577★-1	−1.439	1.302
1.400	2.605★-1	−1.641	9.191★-1
1.600	2.405★-1	−1.819	7.013★-1
1.800	2.108★-1	−1.981	5.443★-1
2.000	1.781★-1	−2.132	4.099★-1
2.200	1.462★-1	−2.274	2.788★-1
2.400	1.169★-1	−2.409	1.371★-1
2.600	9.097★-2	−2.536	−2.862★-2
2.800	6.875★-2	−2.658	−2.368★-1
3.000	5.012★-2	−2.774	−5.176★-1

$p/q = 4$

β^2	f	z_{F0}	t
2.000★-1	−8.357	7.551	−1.685
4.000★-1	−1.302	3.747★-1	−2.486
6.000★-1	−6.250★-2	−1.053	−1.985★1
8.000★-1	3.026★-1	−1.657	1.839
1.000	4.101★-1	−2.036	6.174★-1
1.200	4.177★-1	−2.332	2.397★-1
1.400	3.829★-1	−2.589	4.620★-2
1.600	3.310★-1	−2.824	−8.374★-2
1.800	2.748★-1	−3.042	−1.903★-1

Continued

Table 29 The regular antisymmetric doublet of asymmetric or symmetric quadrupoles: the bell-shaped model. Cardinal elements and spacing as a function of β^2. (Negative values of the spacing, t are included to show the behavior of the other quantities more clearly. $p + q = 2$).—cont'd

$p/q = 4$

β^2	f	z_{FO}	t
2.000	2.207★-1	−3.247	−2.928★-1
2.200	1.717★-1	−3.442	−4.051★-1
2.400	1.292★-1	−3.627	−5.416★-1
2.600	9.334★-2	−3.803	−7.248★-1
2.800	6.386★-2	−3.972	−9.978★-1
3.000	4.013★-2	−4.134	−1.467

Fig. 8A–D (Enge, 1959) show the relation between θ_1^2, θ_2^2, M_x, M_y and the image distance, b/l, for two geometries: $s = 0$ and $s = l$ where s is the gap between rectangles and l is the same as our L, the effective length. Fig. 9A–C show similar information plotted in a different way (King, 1962). Figs. 10A–D and 11A–D are a selection of Baker's curves (Baker, 1963). Fig. 12A–D (Junior & Antony-Spies, 1967) show the domains in the $\theta_1 - \theta_2$ plane in which various types of imagery are possible, for four geometries and Fig. 13 (Junior & Antony-Spies, 1967) shows curves of constant object distance in the $\theta_1 - \theta_2$ plane for a particular geometry; Fig. 14A–D (AntonySpies and Junior, 1967) show contours of various parameters in the $\theta_1 - \theta_2$ plane for the same geometry as Fig. 13. Fig. 15A (Septier, 1958a, 1958b) and 15B (Dhuicq, 1961) show the cardinal elements of an antisymmetric doublet and the separation for which the doublet is regular, using the rectangular model. Fig. 16A and B show the corresponding quantities for the bell-shaped model, symmetric (Dušek, 1959) and unsymmetric.

The aberrations of the pseudo-stigmatic doublet have been studied by Fishkova et al. (1968a); Fig. 17A–C show their results for the magnetic case, Fig. 18A–C for the electrostatic case and Fig. 19A–C for the achromatic situation. (The thin lens approximation had already been discussed in Dymnikov et al., 1967.)

Fuller details of all these figures are given in their captions.

Table 30 The regular antisymmetric doublet of asymmetric or symmetric quadrupoles: the bell-shaped model. The aperture aberration coefficients $(0030)_i$ and $(0003)_i$.

Electrostatic

	β^2	t	$(30)_4 = (03)_0$	$(30)_3 = (03)_1$	$(30)_2 = (03)_2$	$(30)_1 = (03)_3$	$(30)_0 = (03)_4$
$p/q = \frac{1}{4}$	2.000★-1	1.453	7.586★2	−2.999★3	4.451★3	−2.942★-3	7.309★2
	4.000★-1	1.214	1.407★1	−5.155★1	7.289★1	−4.867★-1	1.359★1
	6.000★-1	7.552★-1	1.760	−5.354	6.919	−5.719	3.163
	1.600	3.926★1	4.794★-2	9.639★-3	6.709★-4	2.428★-2	1.492
	1.800	1.407★1	2.969★-2	2.166★-2	4.184★-3	5.861★-2	1.471
	2.000	9.675	1.630★-2	2.474★-2	6.550★-3	7.083★-2	1.452
	2.200	7.745	5.863★-3	2.370★-2	7.080★-3	7.065★-2	1.433
	2.400	6.572	−2.625★-3	2.092★-2	6.396★-3	6.397★-2	1.417
	2.600	5.706	−9.715★-3	1.760★-2	5.196★-3	5.455★-2	1.406
	2.800	4.967	−1.576★-2	1.433★-2	3.928★-3	4.467★-2	1.402
	3.000	4.266	−2.100★-2	1.132★-2	2.712★-3	3.548★-2	1.403
$p/q = \frac{1}{2}$	2.000★-1	7.886★-1	9.934★2	−3.931★3	5.833★-3	−3.848★3	9.524★2
	4.000★-1	5.939★-1	1.638★1	−6.086★1	8.527★-1	−5.435★1	1.382★1
	6.000★-1	2.053★-1	1.685	−5.509	6.857	−4.665	2.125
	1.400	4.652★1	−3.628★-2	6.824★-3	3.466★-4	5.976★-3	1.065

Continued

Table 30 The regular antisymmetric doublet of asymmetric or symmetric quadrupoles: the bell-shaped model. The aperture aberration coefficients $(0030)_i$ and $(0003)_i$.—cont'd

Electrostatic

	β^2	t	$(30)_4 = (03)_0$	$(30)_3 = (03)_1$	$(30)_2 = (03)_2$	$(30)_1 = (03)_3$	$(30)_0 = (03)_4$
	1.600	9.200	−5.898⋆-2	2.653⋆-2	7.193⋆-3	2.050⋆-2	1.140
	1.800	5.906	−7.453⋆-2	3.134⋆-2	1.306⋆-2	1.893⋆-2	1.235
	2.000	4.575	−8.646⋆-2	3.001⋆-2	1.516⋆-2	9.539⋆-3	1.350
	2.200	3.769	−9.627⋆-2	2.625⋆-2	1.439⋆-2	−3.367⋆-3	1.495
	2.400	3.151	−1.047⋆-1	2.176⋆-2	1.209⋆-2	−1.633⋆-2	1.676
	2.600	2.593	−1.121⋆-1	1.736⋆-2	9.294⋆-3	−2.682⋆-2	1.901
	2.800	2.020	−1.188⋆-1	1.342⋆-2	6.642⋆-3	−3.361⋆-2	2.167
	3.000	1.372	−1.249⋆-1	1.007⋆-2	4.423⋆-3	−3.655⋆-2	2.473
$p/q = 1$	1.200	5.629	−1.590⋆-1	5.509⋆-2	3.383⋆-2	−8.030⋆-2	1.838
	1.400	3.118	−1.878⋆-1	6.624⋆-2	8.402⋆-2	−1.949⋆-1	2.641
	1.600	2.317	−2.074⋆-1	6.294⋆-2	1.130⋆-1	−3.091⋆-1	3.624
	1.800	1.881	−2.234⋆-1	5.566⋆-2	1.210⋆-1	−4.226⋆-1	4.819
	2.000	1.568	−2.375⋆-1	4.729⋆-2	1.135⋆-1	−5.289⋆-1	6.285
	2.200	1.299	−2.504⋆-1	3.892⋆-2	9.657⋆-2	−6.133⋆-1	8.072
	2.400	1.035	−2.624⋆-1	3.113⋆-2	7.571⋆-2	−6.614⋆-1	1.020⋆1
	2.600	7.487⋆-1	−2.737⋆-1	2.421⋆-2	5.522⋆-2	−6.671⋆-1	1.263⋆1
	2.800	4.151⋆-1	−2.845⋆-1	1.823⋆-2	3.765⋆-2	−6.332⋆-1	1.534⋆1

$p/q = 2$	8.000⋆-1	1.513⋆1	−2.095⋆-1	4.780⋆-2	1.320⋆-2	−8.983⋆-2	2.471
	1.000	2.304	−2.962⋆-1	9.786⋆-2	2.631⋆-1	−7.477⋆-1	5.051
	1.200	1.302	−3.333⋆-1	7.929⋆-2	4.559⋆-1	−1.315	8.453
	1.400	9.191⋆-1	−3.606⋆-1	6.864⋆-2	5.414⋆-1	−1.700	1.235⋆1
	1.600	7.013⋆-1	−3.847⋆-1	6.366⋆-2	5.493⋆-1	−1.968	1.658⋆1
	1.800	5.443⋆-1	−4.071⋆-1	5.930⋆-2	5.032⋆-1	−2.183	2.120⋆1
	2.000	4.099⋆-1	−4.284⋆-1	5.340⋆-2	4.229⋆-1	−2.342	2.637⋆1
	2.200	2.788⋆-1	−4.487⋆-1	4.575⋆-2	3.292⋆-1	−2.410	3.219⋆1
	2.400	1.371⋆-1	−4.682⋆-1	3.733⋆-2	2.365⋆-1	−2.356	3.864⋆1
$p/q = 4$	8.000⋆-1	1.839	−3.890⋆-1	1.062⋆-1	4.865⋆-1	−1.490	7.885
	1.000	6.174⋆-1	−4.434⋆-1	4.015⋆-2	9.369⋆-1	−2.955	1.548⋆1
	1.200	2.397⋆-1	−4.833⋆-1	2.847⋆-2	1.103	−3.675	2.432⋆1
	1.400	4.620⋆-2	−5.193⋆-1	3.974⋆-2	1.102	−3.788	3.321⋆1

Magnetic

	β^2	t	$(30)_4 = (03)_0$	$(30)_3 = (03)_1$	$(30)_2 = (03)_2$	$(30)_1 = (03)_3$	$(30)_0 = (03)_4$
$p/q = \frac{1}{4}$	2.000⋆-1	1.453	2.055⋆2	−8.036⋆2	1.172⋆3	−7.557⋆2	1.821⋆2
	4.000⋆-1	1.214	3.465	−1.331⋆1	1.844⋆1	−1.205⋆1	3.755
	6.000⋆-1	7.552⋆-1	2.860⋆-1	−1.397	2.106	−2.321	1.853

Continued

Table 30 The regular antisymmetric doublet of asymmetric or symmetric quadrupoles: the bell-shaped model. The aperture aberration coefficients $(0030)_i$ and $(0003)_i$.—cont'd

Magnetic

	β^2	t	$(30)_4 = (03)_0$	$(30)_3 = (03)_1$	$(30)_2 = (03)_2$	$(30)_1 = (03)_3$	$(30)_0 = (03)_4$
	1.600	3.926⋆1	−8.557⋆-2	1.458⋆-3	2.742⋆-4	1.369⋆-2	7.546⋆-1
	1.800	1.407⋆1	−8.670⋆-2	3.010⋆-3	1.671⋆-3	3.110⋆-2	7.063⋆-1
	2.000	9.675	8.750⋆-2	3.169⋆-3	2.565⋆-3	3.570⋆-2	6.700⋆-1
	2.200	7.745	−8.817⋆-2	2.797⋆-3	2.725⋆-3	3.402⋆-2	6.418⋆-1
	2.400	6.572	−8.880⋆-2	2.264⋆-3	2.422⋆-3	2.955⋆-2	6.222⋆-1
	2.600	5.706	−8.942⋆-2	1.740⋆-3	1.937⋆-3	2.434⋆-2	6.118⋆-1
	2.800	4.967	−9.005⋆-2	1.290⋆-3	1.444⋆-3	1.946⋆-2	6.099⋆-1
	3.000	4.266	−9.069⋆-2	9.076⋆-4	9.848⋆-4	1.531⋆-2	6.139⋆-1
$p/q = \frac{1}{2}$	2.000⋆-1	7.888⋆-1	2.698⋆2	−1.057⋆3	1.543⋆3	−9.950⋆-2	2.394⋆2
	4.000⋆-1	5.936⋆-1	4.006	−1.542⋆1	2.064⋆1	−1.208⋆-1	3.055
	6.000⋆-1	2.053⋆-1	2.298⋆-1	−1.378	1.862	−1.533	1.151
	1.400	4.652⋆-1	−1.546⋆-1	7.742⋆-4	1.297⋆-4	4.464⋆-3	5.626⋆-1
	1.600	9.200	−1.590⋆-1	2.263⋆-3	2.638⋆-3	1.523⋆-2	5.575⋆-1
	1.800	5.906	−1.630⋆-1	1.935⋆-3	4.725⋆-3	1.478⋆-2	5.699⋆-1
	2.000	4.575	−1.668⋆-1	1.225⋆-3	5.457⋆-3	9.415⋆-3	5.967⋆-1

	2.200	3.769	−1.705⋆-1	5.597⋆-4	5.177⋆-3	1.920⋆-3	6.436⋆-1
	2.400	3.151	−1.741⋆-1	5.629⋆-5	4.343⋆-3	−5.452⋆-3	7.1600⋆-1
	2.600	2.593	−1.778⋆-1	−2.741⋆-4	3.324⋆-3	−1.105⋆-2	8.169⋆-1
	2.800	2.020	−1.814⋆-1	−4.543⋆-4	2.363⋆-3	−1.423⋆-2	9.404⋆-1
	3.000	1.372	−1.849⋆-1	−5.230⋆-4	1.566⋆-3	−1.520⋆-2	1.080
$p/q = 1$	1.200	5.629	−2.697⋆-1	2.059⋆-3	1.158⋆-2	−1.263⋆-2	8.274⋆-1
	1.400	3.118	−2.819⋆-1	−2.284⋆-3	2.935⋆-2	−4.817⋆-2	1.059
	1.600	2.317	−2.936⋆-1	−5.555⋆-3	4.081⋆-2	−9.381⋆-2	1.352
	1.800	1.881	−3.051⋆-1	−7.054⋆-3	4.563⋆-2	−1.534⋆-1	1.751
	2.000	1.568	−3.163⋆-1	−7.321⋆-3	4.449⋆-2	−2.218⋆-1	2.327
	2.200	1.299	−3.272⋆-1	−6.907⋆-3	3.878⋆-2	−2.832⋆-1	3.138
	2.400	1.035	3.379⋆-1	−6.178⋆-3	3.067⋆-2	−3.219⋆-1	4.191
	2.600	7.487⋆-1	−3.482⋆-1	−5.329⋆-3	2.228⋆-2	−3.307⋆-1	5.443
	2.800	4.151⋆-1	−3.584⋆-1	−4.455⋆-3	1.501⋆-2	−3.119⋆-1	6.821
$p/q = 2$	8.000⋆-1	1.513⋆1	−3.765⋆-1	4.582⋆-3	4.242⋆-3	−1.876⋆-2	1.117
	1.000	2.304	−4.051⋆-1	−1.664⋆-2	9.150⋆-2	−2.116⋆-1	1.878
	1.200	1.302	−4.299⋆-1	−3.620⋆-2	1.632⋆-1	−3.950⋆-1	2.861
	1.400	9.191⋆-1	−4.551⋆-1	−4.162⋆-2	1.983⋆-1	−5.338⋆-1	3.925

Continued

Table 30 The regular antisymmetric doublet of asymmetric or symmetric quadrupoles: the bell-shaped model. The aperture aberration coefficients $(0030)_i$ and $(0003)_i$.—cont'd

Magnetic

	β^2	t	$(30)_4 = (03)_0$	$(30)_3 = (03)_1$	$(30)_2 = (03)_2$	$(30)_1 = (03)_3$	$(30)_0 = (03)_4$
	1.600	7.013⋆-1	−4.797⋆-1	−3.915⋆-2	2.107⋆-1	−6.924⋆-1	5.145
	1.800	5.443⋆-1	−5.033⋆-1	−3.338⋆-2	2.022⋆-1	−8.970⋆-1	6.777
	2.000	4.099⋆-1	−5.259⋆-1	−2.698⋆-2	1.753⋆-1	−1.102	9.081
	2.200	2.788⋆-1	−5.475⋆-1	−2.128⋆-2	1.374⋆-1	−1.236	1.215⋆1
	2.400	1.371⋆-1	−5.683⋆-1	−1.660⋆-2	9.816⋆-2	−1.255	1.586⋆1
$p/q = 4$	8.000⋆-1	1.839	−5.093⋆-1	−3.737⋆-2	1.704⋆-1	−4.395⋆-1	2.808
	1.000	6.174⋆-1	−5.480⋆-1	−8.768⋆-2	3.372⋆-1	−9.144⋆-1	5.043
	1.200	2.397⋆-1	−5.895⋆-1	−9.519⋆-2	3.939⋆-1	−1.087	7.346
	1.400	4.620⋆-2	−6.300⋆-1	−8.428⋆-2	4.036⋆-1	−1.130	9.403

Achromatic

	β^2	t	$(30)_4 = (03)_0$	$(30)_3 = (03)_1$	$(30)_2 = (03)_2$	$(30)_1 = (03)_3$	$(30)_0 = (03)_4$
$p/q = \frac{1}{4}$	2.000⋆-1	1.453	3.161⋆2	−1.243⋆3	1.828⋆3	−1.193⋆3	2.918⋆2
	4.000⋆-1	1.214	5.587	−2.096⋆1	2.935⋆1	−1.937⋆1	5.723
	6.000⋆-1	7.552⋆-1	5.809⋆-1	−2.188	3.069	−3.000	2.115
	1.600	3.926⋆1	−5.887⋆-2	3.094⋆-3	3.535⋆-4	1.581⋆-2	9.021⋆-1
	1.800	1.407⋆1	−6.343⋆-2	6.741⋆-3	3.174⋆-3	3.661⋆-2	8.593⋆-1
	2.000	9.675	−6.674⋆-2	7.483⋆-3	3.362⋆-3	4.273⋆-2	8.263⋆-1

	2.200	7.745	−6.936⋆-2	6.978⋆-3	3.596⋆-3	4.134⋆-2	8.000⋆-1
	2.400	6.572	−7.157⋆-2	5.996⋆-3	3.217⋆-3	3.643⋆-2	7.811⋆-1
	2.600	5.706	−7.348⋆-2	4.912⋆-3	2.589⋆-3	3.038⋆-2	7.707⋆-1
	2.800	4.967	−7.519⋆-2	3.899⋆-3	1.941⋆-3	2.450⋆-2	7.683⋆-1
	3.000	4.266	−7.675⋆-2	2.991⋆-3	1.330⋆-3	1.935⋆-2	7.717⋆-1
$p/q = \frac{1}{2}$	2.000⋆-1	7.888⋆-1	4.145⋆2	−1.632⋆3	2.401⋆3	−1.566⋆3	3.820⋆2
	4.000⋆-1	5.936⋆-1	6.481	−2.451⋆1	3.357⋆1	−2.053⋆1	5.209
	6.000⋆-1	2.053⋆-1	5.208⋆-1	−2.204	2.861	−2.159	1.345
	1.400	4.652⋆-1	−1.309⋆-1	1.984⋆-3	1.731⋆-4	4.767⋆-3	6.632⋆-1
	1.600	9.200	−1.390⋆-1	7.116⋆-3	3.549⋆-3	1.628⋆-2	6.740⋆-1
	1.800	5.906	−1.453⋆-1	7.815⋆-3	6.393⋆-3	1.561⋆-2	7.029⋆-1
	2.000	4.575	−1.507⋆-1	6.983⋆-3	7.398⋆-3	9.440⋆-3	7.475⋆-1
	2.200	3.769	−1.556⋆-1	5.697⋆-3	7.020⋆-3	8.624⋆-4	8.138⋆-1
	2.400	3.151	−1.602⋆-1	4.396⋆-3	5.894⋆-3	−7.627⋆-3	9.086⋆-1
	2.600	2.593	−1.646⋆-1	3.252⋆-3	4.518⋆-3	−1.421⋆-2	1.034
	2.800	2.020	−1.689⋆-1	2.321⋆-3	3.219⋆-3	−1.811⋆-2	1.186
	3.000	1.372	−1.729⋆-1	1.596⋆-3	2.137⋆-3	−1.947⋆-2	1.359

Continued

Table 30 The regular antisymmetric doublet of asymmetric or symmetric quadrupoles: the bell-shaped model. The aperture aberration coefficients $(0030)_i$ and $(0003)_i$.—cont'd

Achromatic

	β^2	t	$(30)_4 = (03)_0$	$(30)_3 = (03)_1$	$(30)_2 = (03)_2$	$(30)_1 = (03)_3$	$(30)_0 = (03)_4$
$p/q = 1$	1.200	5.629	−2.475⋆-1	1.267⋆-2	1.603⋆-2	−2.617⋆-2	1.030
	1.400	3.118	−2.631⋆-1	1.142⋆-2	4.028⋆-2	−7.752⋆-2	1.376
	1.600	2.317	−2.764⋆-1	8.144⋆-3	5.525⋆-2	−1.369⋆-1	1.807
	1.800	1.881	−2.887⋆-1	5.488⋆-3	6.071⋆-2	2.072⋆-1	2.364
	2.000	1.568	−3.005⋆-1	3.601⋆-3	5.830⋆-2	−2.832⋆-1	3.119
	2.200	1.299	−3.118⋆-1	2.259⋆-3	5.034⋆-2	−3.492⋆-1	4.125
	2.400	1.035	−3.228⋆-1	1.283⋆-3	3.968⋆-2	−3.898⋆-1	5.392
	2.600	7.487⋆-1	−3.333⋆-1	5.798⋆-4	2.887⋆-2	−3.979⋆-1	6.881
	2.800	4.151⋆-1	−3.436⋆-1	1.002⋆-4	1.954⋆-2	−3.761⋆-1	8.524
$p/q = 2$	8.000⋆-1	1.513⋆1	−3.431⋆-1	1.323⋆-2	6.034⋆-3	−3.297⋆-2	1.388
	1.000	2.304	−3.833⋆-1	6.257⋆-3	1.258⋆-1	−3.188⋆-1	2.512
	1.200	1.302	−4.106⋆-1	−1.311⋆-2	2.210⋆-1	−5.970⋆-1	3.979
	1.400	9.191⋆-1	−4.362⋆-1	−1.957⋆-2	2.669⋆-1	−7.671⋆-1	5.610
	1.600	7.013⋆-1	−4.607⋆-1	−1.859⋆-2	2.784⋆-1	−9.476⋆-1	7.432

	1.800	5.443⋆-1	−4.841⋆-1	−1.484⋆-2	2.624⋆-1	−1.154	9.662
	2.000	4.099⋆-1	−5.064⋆-1	−1.090⋆-2	2.248⋆-1	−1.350	1.254⋆1
	2.200	2.788⋆-1	−5.277⋆-1	−7.876⋆-3	1.175⋆-1	−1.471	1.616⋆1
	2.400	1.371⋆-1	−5.482⋆-1	−5.814⋆-3	1.258⋆-1	−1.476	2.042⋆1
$p/q = 4$	8.000⋆-1	1.839	−4.852⋆-1	−8.663⋆-3	2.336⋆-1	−6.496⋆-1	3.824
	1.000	6.174⋆-1	−5.271⋆-1	−6.211⋆-2	4.571⋆-1	−1.322	7.130
	1.200	2.397⋆-1	−5.682⋆-1	−7.046⋆-2	5.358⋆-1	−1.605	1.074⋆1
	1.400	4.620⋆-2	−6.078⋆-1	−5.947⋆-2	5.432⋆-1	−1.662	1.417⋆1

Table 31 The regular antisymmetric doublet of asymmetric or symmetric quadrupoles: the bell-shaped model. The aperture aberration coefficients $(0012)_{ij}$ and $(0021)_{ij}$.

Electrostatic $p/q = \frac{1}{4}$

β^2	$(12)_{22} = (12)_{00}$	$(12)_{21} = (21)_{10}$	$(12)_{12} = (21)_{01}$	$(12)_{20} = (21)_{20}$	$(12)_{11} = (21)_{11}$	$(12)_{02} = (21)_{02}$	$(12)_{10} = (21)_{21}$	$(12)_{01} = (21)_{12}$	$(21)_{22} = (21)_{00}$
2.000⋆-1	−5.557⋆2	9.732⋆2	1.223⋆3	−4.411.2	−2.155⋆3	−6.859⋆2	9.778⋆2	1.219⋆3
4.000⋆-1	−1.943⋆1	2.188⋆1	4.697⋆1	−7.749	−6.043⋆1	−3.050⋆1	2.292⋆1	4.593⋆1
6.000⋆-1	−3.655	1.965	7.461	−5.185⋆-1	−6.616	−4.316	2.356	7.070
1.600	−4.656⋆-2	2.392⋆-4	−3.172⋆-2	−1.537⋆-6	−5.182⋆-4	1.218⋆-3	−3.456⋆-3	−2.803⋆-2
1.800	−4.161⋆-1	1.092⋆-3	−7.659⋆-2	6.086⋆-7	−3.093⋆-3	6.701⋆-3	−7.835⋆-3	−6.766⋆-2
2.000	−3.794⋆-1	1.780⋆-3	−9.303⋆-2	1.223⋆-5	−4.573⋆-3	3.547⋆-3	−9.025⋆-3	−8.223⋆-2
2.200	−3.505⋆-1	2.169⋆-3	−9.494⋆-2	2.246⋆-5	−4.652⋆-3	−9.787⋆-3	−8.719⋆-3	−8.405⋆-2
2.400	−3.270⋆-1	2.298⋆-3	−8.964⋆-2	2.699⋆-5	−3.974⋆-3	−2.504⋆-2	−7.760⋆-3	−7.958⋆-2
2.600	−3.074⋆-1	2.238⋆-3	−8.076⋆-2	2.635⋆-5	−3.068⋆-3	−3.344⋆-2	−6.582⋆-3	−7.194⋆-2
2.800	−2.907⋆-1	2.055⋆-3	−7.020⋆-2	2.258⋆-5	−2.208⋆-3	−3.114⋆-2	−5.402⋆-3	−6.275⋆-2
3.000	−2.764⋆-1	1.803⋆-3	−5.912⋆-2	1.762⋆-4	−1.501⋆-3	−1.991⋆-2	−4.323⋆-3	−5.299⋆-2

Electrostatic $p/q = \frac{1}{2}$

β^2	$(12)_{22} = (12)_{00}$	$(12)_{21} = (21)_{10}$	$(12)_{12} = (21)_{01}$	$(12)_{20} = (21)_{20}$	$(12)_{11} = (21)_{11}$	$(12)_{02} = (21)_{02}$	$(12)_{10} = (21)_{21}$	$(12)_{01} = (21)_{12}$	$(21)_{22} = (21)_{00}$
2.000⋆-1	−6.864⋆2	1.224⋆3	1.497⋆3	−5.622⋆2	−2.676⋆3	−8.294⋆2	1.299⋆3	1.492
4.000⋆-1	−2.120⋆1	2.495⋆1	5.132⋆1	−9.166	−6.764⋆1	−3.325⋆1	2.605⋆1	5.022⋆1
6.000⋆-1	−3.538	1.933	7.327	−5.475⋆-1	−6.538	−4.384	2.339	6.921

1.400	−4.703⋆-1	7.906⋆-4	−2.184⋆-2	5.183⋆-7	−1.954⋆-4	2.649⋆-4	−2.377⋆-3	−1.867⋆-2
1.600	−4.135⋆-1	4.288⋆-3	−9.678⋆-2	4.040⋆-5	−3.263⋆-3	4.241⋆-3	−9.692⋆-3	−8.280⋆-2
1.800	−3.765⋆-1	6.144⋆-3	−1.278⋆-1	1.021⋆-4	−4.319⋆-3	−5.671⋆-3	−1.196⋆-2	−1.097⋆-1
2.000	−3.501⋆-1	6.755⋆-3	−1.376⋆-1	1.358⋆-4	−3.104⋆-3	−3.464⋆-2	−1.191⋆-2	−1.189⋆-1
2.200	−3.299⋆-1	6.594⋆-3	−1.375⋆-1	1.374⋆-4	−1.275⋆-3	−6.563⋆-2	−1.079⋆-2	−1.201⋆-1
2.400	−3.139⋆-1	5.988⋆-3	−1.318⋆-1	1.190⋆-4	1.283⋆-4	−7.820⋆-2	−9.259⋆-3	−1.165⋆-1
2.600	−3.007⋆-1	5.159⋆-3	−1.218⋆-1	9.245⋆-5	8.513⋆-4	−6.372⋆-2	−7.659⋆-3	−1.090⋆-1
2.800	−2.896⋆-1	4.255⋆-3	−1.084⋆-1	6.594⋆-5	1.043⋆-3	−2.859⋆-2	−6.151⋆-3	−9.800⋆-2
3.000	−2.801⋆-1	3.374⋆-3	−9.276⋆-2	4.352⋆-5	9.363⋆-4	1.178⋆-2	−4.805⋆-3	−8.458⋆-2

Electrostatic $p/q = 1$

β^2	$(12)_{22} = (12)_{00}$	$(12)_{21} = (21)_{10}$	$(12)_{12} = (21)_{01}$	$(12)_{20} = (21)_{20}$	$(12)_{11} = (21)_{11}$	$(12)_{02} = (21)_{02}$	$(12)_{10} = (21)_{21}$	$(12)_{01} = (21)_{12} = (21)_{22} = (21)_{00}$
1.200	−4.959⋆-1	1.324⋆-2	−2.275⋆-1	2.952⋆-4	−1.362⋆-3	4.584⋆-4	−2.112⋆-2	−1.931⋆-1
1.400	−4.360⋆-1	1.940⋆-2	−3.765⋆-1	1.037⋆-3	1.138⋆-2	−7.181⋆-2	−3.187⋆-2	−3.253⋆-1
1.600	−4.044⋆-1	2.033⋆-2	−4.604⋆-1	1.417⋆-3	3.262⋆-2	−3.399⋆-1	−3.506⋆-2	−4.050⋆-1
1.800	−3.855⋆-1	1.925⋆-2	−5.239⋆-1	1.396⋆-3	5.107⋆-2	−7.674⋆-	−3.380⋆-2	−4.709⋆-1
2.000	−3.727⋆-1	1.737⋆-2	−5.857⋆-1	1.171⋆-3	6.003⋆-2	−1.172	−3.025⋆-2	−5.380⋆-1

Continued

Table 31 The regular antisymmetric doublet of asymmetric or symmetric quadrupoles: the bell-shaped model. The aperture aberration coefficients $(0012)_{ij}$ and $(0021)_{ij}$.—cont'd

Electrostatic $p/q = 1$

β^2	$(12)_{22} = (12)_{00}$	$(12)_{21} = (21)_{10}$	$(12)_{12} = (21)_{01}$	$(12)_{20} = (21)_{20}$	$(12)_{11} = (21)_{11}$	$(12)_{02} = (21)_{02}$	$(12)_{10} = (21)_{21}$	$(12)_{01} = (21)_{12} = (21)_{22} = (21)_{00}$
2.200	−3.632⋆-1	1.508⋆-2	−6.372⋆-1	8.932⋆-4	5.842⋆-2	−1.357	−2.587⋆-2	−5.963⋆-1
2.400	−3.552⋆-1	1.260⋆-2	−6.607⋆-1	6.363⋆-4	4.937⋆-2	−1.231	−2.149⋆-2	−6.266⋆-1
2.600	−3.475⋆-1	1.014⋆-2	−6.450⋆-1	4.277⋆-4	3.731⋆-2	−8.431⋆-1	−1.746⋆-2	−6.174⋆-1
2.800	−3.397⋆-1	7.852⋆-3	−5.919⋆-1	2.717⋆-4	2.570⋆-2	−3.404⋆-1	−1.389⋆-2	−5.702⋆-1

Electrostatic $p/q = 2$

β^2	$(12)_{22} = (12)_{00}$	$(12)_{21} = (21)_{10}$	$(12)_{12} = (21)_{01}$	$(12)_{20} = (21)_{20}$	$(12)_{11} = (21)_{11}$	$(12)_{02} = (21)_{02}$	$(12)_{10} = (21)_{21}$	$(12)_{01} = (21)_{12} = (21)_{22} = (21)_{00}$
8.000⋆-1	−7.325⋆-1	9.762⋆-3	−1.495⋆-1	2.621⋆-5	3.847⋆-4	−2.704⋆-3	−1.348⋆-2	−1.263⋆-1
1.000	−5.046⋆-1	3.353⋆-2	−8.160⋆-1	4.342⋆-3	7.518⋆-2	−1.133⋆-1	−6.959⋆-2	−7.129⋆1
1.200	−4.230⋆-1	2.998⋆-2	−1.150	8.476⋆-3	1.865⋆-1	−1.003	−9.889⋆-2	−1.822
1.400	−4.003⋆-1	2.513⋆-2	−1.240	8.892⋆-3	2.676⋆-1	−3.502	−1.051⋆-1	−1.110
1.600	−3.945⋆-1	2.250⋆2	−1.331	7.368⋆-3	3.220⋆-1	−7.427	−9.775⋆-2	−1.211
1.800	−3.930⋆-1	2.124⋆-2	−1.570	5.429⋆-3	3.407⋆-1	−1.155⋆1	−8.415⋆-2	−1.465
2.000	−3.939⋆-1	1.996⋆-2	−1.908	3.736⋆-3	3.156⋆-1	−1.434⋆1	−6.911⋆-2	−1.819
2.200	−3.956⋆-1	1.786⋆-2	−2.190	2.453⋆-3	2.570⋆-1	−1.479⋆1	−5.524⋆-2	−2.117
2.400	−3.954⋆-1	1.498⋆-2	−2.289	1.544⋆-3	1.862⋆-1	−1.286⋆1	−4.345⋆-2	−2.230

Electrostatic $p/q = 4$

β^2	$(12)_{22} = (12)_{00}$	$(12)_{21} = (21)_{10}$	$(12)_{12} = (21)_{01}$	$(12)_{20} = (21)_{20}$	$(12)_{11} = (21)_{11}$	$(12)_{02} = (21)_{02}$	$(12)_{10} = (21)_{21}$	$(12)_{01} = (21)_{12}$	$(21)_{22} = (21)_{00}$
8.000⋆-1	−5.380⋆-1	3.978⋆-2	−1.279	9.916⋆-3	1.790⋆-1	−2.385⋆-1	−1.115⋆-1	−1.128
1.000	−3.687⋆-1	1.726⋆-2	−1.985	2.186⋆-2	4.572⋆-1	−2.063	−1.878⋆-1	−1.780
1.200	−3.636⋆-1	8.619⋆-3	−1.852	2.229⋆-2	5.611⋆-1	−8.099	−2.003⋆-1	−1.643
1.400	−3.883⋆-1	9.035⋆-3	−1.600	1.740⋆-2	6.219⋆-1	−1.882⋆-1	−1.824⋆-1	−1.409

Magnetic $p/q=\frac{1}{4}$

β^2	$(12)_{22} = (12)_{00}$	$(12)_{21} = (21)_{10}$	$(12)_{12} = (21)_{01}$	$(12)_{20} = (21)_{20}$	$(12)_{11} = (21)_{11}$	$(12)_{02} = (21)_{02}$	$(12)_{10} = (21)_{21}$	$(12)_{01} = (21)_{12}$	$(21)_{22} = (21)_{00}$
2.000⋆-1	3.396⋆2	−8.179⋆2	−5.434⋆2	4.590⋆2	1.394⋆3	1.902⋆2	−8.133⋆2	−5.480⋆2
4.000⋆-1	−5.604	−3.535	2.234⋆1	5.039	−1.293⋆1	−1.821⋆1	−2.495	2.130⋆1
6.000⋆-1	−1.916	−1.878⋆-1	4.903	4.792⋆-1	−3.095	−2.654	2.032⋆-1	4.512
1.600	−3.082⋆-1	2.715⋆-3	−1.770⋆-2	5.214⋆-5	−3.096⋆-1	3.501⋆-3	−9.801⋆-4	−1.400⋆-2
1.800	−2.783⋆-1	6.222⋆-3	−4.001⋆-2	3.132⋆-4	−1.807⋆-3	2.452⋆-2	−2.704⋆-3	−3.108⋆-2
2.000	−2.555⋆-1	7.143⋆-3	−4.542⋆-2	4.702⋆-4	−2.570⋆-3	3.667⋆-2	−3.662⋆-3	−3.461⋆-2
2.200	−2.371⋆-1	6.831⋆-3	−4.397⋆-2	4.875⋆-4	−2.507⋆-3	3.177⋆-2	−4.057⋆-3	−3.308⋆-2
2.400	−2.217⋆-1	5.994⋆-3	−4.027⋆-2	4.236⋆-4	−2.075⋆-3	1.907⋆-2	−4.064⋆-3	−3.821⋆-2

Continued

Table 31 The regular antisymmetric doublet of asymmetric or symmetric quadrupoles: the bell-shaped model. The aperture aberration coefficients $(0012)_{ij}$ and $(0021)_{ij}$.—cont'd

Magnetic $p/q=\frac{1}{4}$

β^2	$(12)_{22}=(12)_{00}$	$(12)_{21}=(21)_{10}$	$(12)_{12}=(21)_{01}$	$(12)_{20}=(21)_{20}$	$(12)_{11}=(21)_{11}$	$(12)_{02}=(21)_{02}$	$(12)_{10}=(21)_{21}$	$(12)_{01}=(21)_{12}=(21)_{22}=(21)_{00}$
2.600	−2.086⋆-1	4.993⋆-3	−3.580⋆-2	3.308⋆-4	−1.582⋆-3	1.052⋆-2	−3.027⋆-3	−2.698⋆-2
2.800	−1.971⋆-1	4.010⋆-3	−3.093⋆-2	2.397⋆-4	−1.149⋆-3	1.216⋆-2	−3.447⋆-3	−2.347⋆-2
3.000	−1.869⋆-1	3.128⋆-3	−2.584⋆-2	1.638⋆-4	−8.004⋆-4	2.276⋆-2	−2.999⋆-3	−1.972⋆-2

Magnetic $p/q=\frac{1}{2}$

β^2	$(12)_{22}=(12)_{00}$	$(12)_{21}=(21)_{10}$	$(12)_{12}=(21)_{01}$	$(12)_{20}=(21)_{20}$	$(12)_{11}=(21)_{11}$	$(12)_{02}=(21)_{02}$	$(12)_{10}=(21)_{21}$	$(12)_{01}=(21)_{12}=(21)_{22}=(21)_{00}$
2.000⋆-1	4.656⋆2	−1.094⋆3	−7.762⋆2	6.069⋆2	1.907⋆3	2.950⋆2	−1.089⋆3	−7.812⋆2
4.000⋆-1	−7.042	−2.086	2.620⋆1	5.451	−1.670⋆1	−2.113⋆1	−1.779	2.509⋆1
6.000⋆-1	−2.302	1.840⋆-1	5.529	4.085⋆-1	−3.910	−3.220	5.902⋆-1	5.122
1.400	−3.330⋆-1	9.693⋆-4	−1.170⋆-2	2.655⋆-5	−2.042⋆-4	1.072⋆-3	−2.199⋆-3	−8.536⋆-3
1.600	−2.797⋆-1	3.740⋆-3	−4.471⋆-2	5.247⋆-4	−3.771⋆-3	2.633⋆-2	−1.024⋆-2	−3.073⋆-2
1.800	−2.403⋆-1	4.101⋆-3	−4.981⋆-2	8.873⋆-4	−5.675⋆-3	4.296⋆-2	−1.401⋆-2	−3.171⋆-2
2.000	−2.087⋆-1	3.512⋆-3	−4.627⋆-2	9.467⋆-4	−5.083⋆-3	3.147⋆-2	−1.515⋆-2	−2.760⋆-2
2.200	−1.821⋆-1	2.667⋆-3	−4.256⋆-2	8.228⋆-4	−3.527⋆-3	9.850⋆-2	−1.472⋆-2	−2.518⋆-2
2.400	−1.591⋆-1	1.856⋆-3	−4.000⋆-2	6.333⋆-4	−2.111⋆-3	4.260⋆-2	−1.339⋆-2	−2.475⋆-2

2.600	−1.387⋆-1	1.182⋆-3	−3.719⋆-2	4.473⋆-4	−1.169⋆-3	2.654⋆-2	−1.164⋆-2	−2.437⋆-2
2.800	−1.202⋆-1	6.704⋆-4	−3.304⋆-2	2.944⋆-4	−6.296⋆-4	6.889⋆-2	−9.736⋆-3	−2.264⋆-2
3.000	−1.034⋆-1	3.116⋆-4	−2.763⋆-2	1.814⋆-4	−3.339⋆-4	1.118⋆-1	−7.868⋆-3	−1.945⋆-2
Magnetic $p/q = 1$								
β^2	$(12)_{22} = (12)_{00}$	$(12)_{21} = (21)_{10}$	$(12)_{12} = (21)_{01}$	$(12)_{20} = (21)_{20}$	$(12)_{11} = (21)_{11}$	$(12)_{02} = (21)_{02}$	$(12)_{10} = (21)_{21}$	$(12)_{01} = (21)_{12} = (21)_{22} = (21)_{00}$
1.200	−2.392⋆-1	−1.538⋆-5	−6.022⋆-2	2.116⋆-3	−1.710⋆-2	8.421⋆-2	−3.438⋆-2	−2.585⋆-2
1.400	−1.589⋆-1	−2.155⋆-3	−5.536⋆-2	4.712⋆-3	−3.055⋆-2	2.338⋆-1	−5.343⋆-2	−4.086⋆-3
1.600	−9.959⋆-2	−4.906⋆-3	−2.748⋆-2	5.439⋆-3	−2.318⋆-2	1.950⋆-1	−6.030⋆-2	2.791⋆-2
1.800	−5.094⋆-2	−6.710⋆-3	−2.446⋆-2	4.907⋆-3	−5.600⋆-3	−5.611⋆-2	−5.976⋆-2	2.859⋆-2
2.000	−9.830⋆-3	−7.353⋆-3	−6.304⋆-2	3.884⋆-3	1.005⋆-2	−3.271⋆-1	−5.497⋆-2	−1.542⋆-2
2.200	2.526⋆-2	−7.205⋆-3	−1.252⋆-1	2.825⋆-3	1.779⋆-2	−3.888⋆-1	−4.816⋆-2	−8.420⋆-2
2.400	5.589⋆-2	−6.649⋆-3	−1.804⋆-1	1.928⋆-3	1.808⋆-2	−1.351⋆-1	−4.074⋆-2	−1.463⋆-1
2.600	8.364⋆-2	−5.909⋆-3	−2.078⋆-1	1.245⋆-3	1.429⋆-2	3.655⋆-1	−3.350⋆-2	−1.802⋆-1
2.800	1.096⋆-1	−5.084⋆-3	−2.037⋆-1	7.626⋆-4	9.579⋆-3	9.229⋆-1	−2.682⋆-2	−1.820⋆-1
Magnetic $p/q = 2$								
β^2	$(12)_{22} = (12)_{00}$	$(12)_{21} = (21)_{10}$	$(12)_{12} = (21)_{01}$	$(12)_{20} = (21)_{20}$	$(12)_{11} = (21)_{11}$	$(12)_{02} = (21)_{02}$	$(12)_{10} = (21)_{21}$	$(12)_{01} = (21)_{12} = (21)_{22} = (21)_{00}$
8.000⋆-1	−2.817⋆-1	−4.026⋆-3	−2.252⋆-2	7.127⋆-4	−7.444⋆-3	1.960⋆-2	−2.727⋆-2	7.215⋆-4
1.000	−9.620⋆-2	−1.478⋆-2	−9.216⋆-2	1.426⋆-2	−9.849⋆-2	9.330⋆-1	−1.179⋆-1	1.095⋆-2
1.200	5.591⋆-3	−2.705⋆-2	−2.034⋆-2	2.199⋆-2	−1.135⋆-1	2.026	−1.559⋆-1	1.085⋆-1
1.400	8.330⋆-2	−3.611⋆-2	2.337⋆-1	2.182⋆-2	−6.653⋆-2	1.422	−1.663⋆-1	3.640⋆-1

Continued

Table 31 The regular antisymmetric doublet of asymmetric or symmetric quadrupoles: the bell-shaped model. The aperture aberration coefficients $(0012)_{ij}$ and $(0021)_{ij}$.—cont'd

Magnetic $p/q = 2$

β^2	$(12)_{22} = (12)_{00}$	$(12)_{21} = (21)_{10}$	$(12)_{12} = (21)_{01}$	$(12)_{20} = (21)_{20}$	$(12)_{11} = (21)_{11}$	$(12)_{02} = (21)_{02}$	$(12)_{10} = (21)_{21}$	$(12)_{01} = (21)_{12}$	$(21)_{22} = (21)_{00}$
1.600	1.566⋆-1	−3.920⋆-2	3.975⋆-1	1.794⋆-2	2.036⋆-2	−1.370	−1.595⋆-1	5.177⋆-1
1.800	2.242⋆-1	−3.701⋆-2	2.584⋆-1	1.327⋆-2	1.015⋆-1	−5.181	−1.424⋆-1	3.638⋆-1
2.000	2.804⋆-1	−3.203⋆-2	−1.370⋆-1	9.156⋆-3	1.396⋆-1	−8.129	−1.211⋆-1	−4.792⋆-2
2.200	3.252⋆-1	−2.653⋆-2	−5.803⋆-1	5.987⋆-3	1.316⋆-1	−8.816	−9.963⋆-2	−5.072⋆-1
2.400	3.635⋆-1	−2.159⋆-2	−8.798⋆-1	3.731⋆-3	9.776⋆-2	−7.009	−8.002⋆-2	−8.213⋆-1

Magnetic $p/q = 4$

β^2	$(12)_{22} = (12)_{00}$	$(12)_{21} = (21)_{10}$	$(12)_{12} = (21)_{01}$	$(12)_{20} = (21)_{20}$	$(12)_{11} = (21)_{11}$	$(12)_{02} = (21)_{02}$	$(12)_{10} = (21)_{21}$	$(12)_{01} = (21)_{12}$	$(21)_{22} = (21)_{00}$
8.000⋆-1	−3.308⋆-2	−2.941⋆-2	−1.659⋆-1	2.541⋆-2	−1.642⋆-1	1.940	−1.807⋆-1	−1.466⋆-2
1.000	1.105⋆-1	−5.593⋆-2	−4.048⋆-1	4.590⋆-2	−1.867⋆-1	6.619	−2.610⋆-1	−1.997⋆-1
1.200	1.950⋆-1	−7.328⋆-2	3.199⋆-1	4.571⋆-2	−1.757⋆-1	7.362	−2.821⋆-1	5.287⋆-1
1.400	2.905⋆-1	−7.971⋆-2	1.339	3.656⋆-2	−5.011⋆-1	1.508⋆-1	−2.711⋆-1	1.530

Achromatic $p/q=\frac{1}{4}$

β^2	$(12)_{22} = (12)_{00}$	$(12)_{21} = (21)_{10}$	$(12)_{12} = (21)_{01}$	$(12)_{20} = (21)_{20}$	$(12)_{11} = (21)_{11}$	$(12)_{02} = (21)_{02}$	$(12)_{10} = (21)_{21}$	$(12)_{01} = (21)_{12}$	$(21)_{22} = (21)_{00}$
2.000⋆-1	6.380⋆2	−1.415⋆3	−1.132⋆3	7.591⋆2	2.577⋆3	4.823⋆2	−1.410⋆3	−1.137⋆3
4.000⋆-1	−9.944⋆-1	−1.201⋆1	1.413⋆1	9.302	2.909	−1.411⋆1	−1.097⋆1	1.308⋆1
6.000⋆-1	−1.337	−9.053⋆-1	4.050	8.117⋆-1	−1.922	−2.100	−5.143⋆-1	3.659

1.600	−2.558⋆-1	3.540⋆-3	−1.303⋆-2	7.003⋆-5	−2.401⋆-4	4.262⋆-3	−1.549⋆-4	−9.330⋆-3
1.800	−2.324⋆-1	7.933⋆-3	−2.782⋆-2	4.174⋆-4	−1.378⋆-3	3.046⋆-2	−9.939⋆-4	−1.889⋆-2
2.000	−2.142⋆-1	8.930⋆-3	−2.955⋆-2	6.229⋆-4	−1.903⋆-3	4.771⋆-2	−1.875⋆-3	−1.874⋆-2
2.200	−1.993⋆-1	8.386⋆-3	−2.698⋆-2	6.425⋆-4	−1.792⋆-3	4.562⋆-2	−2.592⋆-3	−1.609⋆-2
2.400	−1.866⋆-1	7.226⋆-3	−2.381⋆-2	5.558⋆-4	−1.442⋆-3	3.376⋆-2	−2.833⋆-3	−1.375⋆-2
2.600	−1.756⋆-1	5.911⋆-3	−2.082⋆-2	4.323⋆-4	−1.087⋆-3	2.517⋆-2	−2.908⋆-3	−1.200⋆-2
2.800	−1.659⋆-1	4.661⋆-3	−1.784⋆-2	3.121⋆-4	−7.957⋆-4	2.659⋆-2	−2.796⋆-3	−1.038⋆-2
3.000	−1.571⋆-1	3.569⋆-3	−1.475⋆-2	2.125⋆-4	−5.669⋆-4	3.699⋆-2	−2.558⋆-3	−8.622⋆-3

Achromatic $p/q=\frac{1}{2}$

β^2	$(12)_{22}=(12)_{00}$	$(12)_{21}=(21)_{10}$	$(12)_{12}=(21)_{01}$	$(12)_{20}=(21)_{20}$	$(12)_{11}=(21)_{11}$	$(12)_{02}=(21)_{02}$	$(12)_{10}=(21)_{21}$	$(12)_{01}=(21)_{12}=(21)_{22}=(21)_{00}$
2.000⋆-1	8.496⋆2	−1.867⋆3	−1.534⋆3	9.966⋆3	3.435⋆3	6.698⋆2	−1.862⋆3	−1.539⋆3
4.000⋆-1	−2.322	−1.216⋆1	1.783⋆1	1.032⋆1	2.822⋆-1	−1.709⋆1	−1.106⋆1	1.672⋆1
6.000⋆-1	−1.890	−3.990⋆-1	4.929	7.271⋆-1	−3.033	−2.832	7.297⋆-3	4.523
1.400	−2.873⋆-1	1.029⋆-3	−8.325⋆-3	3.523⋆-5	−2.071⋆-4	1.341⋆-3	−2.139⋆-3	−5.157⋆-3
1.600	−2.351⋆-1	3.558⋆-3	−2.736⋆-2	6.862⋆-4	−3.941⋆-3	3.370⋆-2	−1.042⋆-2	−1.338⋆-2
1.800	−1.949⋆-1	3.420⋆-3	−2.382⋆-2	1.149⋆-3	−6.127⋆-3	5.917⋆-2	−1.469⋆-2	−5.713⋆-3
2.000	−1.616⋆-1	2.431⋆-3	−1.583⋆-2	1.217⋆-3	−5.742⋆-3	5.351⋆2	−1.623⋆-2	2.837⋆-3
2.200	−1.328⋆-1	1.359⋆-3	−1.093⋆-2	1.051⋆-3	−4.277⋆-3	3.501⋆-2	−1.602⋆-2	6.455⋆-3

Continued

Table 31 The regular antisymmetric doublet of asymmetric or symmetric quadrupoles: the bell-shaped model. The aperture aberration coefficients $(0012)_{ij}$ and $(0021)_{ij}$.—cont'd

Achromatic $p/q = \frac{1}{2}$

β^2	$(12)_{22} = (12)_{00}$	$(12)_{21} = (21)_{10}$	$(12)_{12} = (21)_{01}$	$(12)_{20} = (21)_{20}$	$(12)_{11} = (21)_{11}$	$(12)_{02} = (21)_{02}$	$(12)_{10} = (21)_{21}$	$(12)_{01} = (21)_{12} = (21)_{22} = (21)_{00}$
2.400	−1.075⋆-1	4.789⋆-4	−9.411⋆-3	8.047⋆-3	−2.857⋆-3	3.175⋆-2	−1.477⋆-2	5.836⋆-3
2.600	−8.465⋆-1	1.434⋆-4	−8.984⋆-3	5.655⋆-4	−1.843⋆-3	5.662⋆-2	−1.296⋆-2	3.833⋆-3
2.800	−6.378⋆-2	−5.246⋆-4	−7.922⋆-3	3.785⋆-4	−1.187⋆-3	1.014⋆-1	−1.093⋆-2	2.485⋆-3
3.000	−4.443⋆-2	−7.092⋆-4	−5.919⋆-3	2.274⋆-4	−7.573⋆-4	1.452⋆-1	−8.889⋆-3	2.260⋆-3

Achromatic $p/q = 1$

β^2	$(12)_{22} = (12)_{00}$	$(12)_{21} = (21)_{10}$	$(12)_{12} = (21)_{01}$	$(12)_{20} = (21)_{20}$	$(12)_{11} = (21)_{11}$	$(12)_{02} = (21)_{02}$	$(12)_{10} = (21)_{21}$	$(12)_{01} = (21)_{12} = (21)_{22} = (21)_{00}$
1.200	−1.536⋆-1	−4.434⋆-3	−4.470⋆-3	2.723⋆-3	−2.234⋆-2	1.121⋆-1	−3.880⋆-2	2.990⋆-2
1.400	−6.657⋆-2	−9.341⋆-3	5.170⋆-2	5.937⋆-3	−4.453⋆-2	3.356⋆-1	−6.062⋆-2	1.030⋆-1
1.600	2.029⋆-3	−1.332⋆-2	1.160⋆-1	6.779⋆-3	−4.178-2	3.733⋆-1	−6.871⋆-2	1.722⋆-1
1.800	6.058⋆-2	−1.536⋆-2	1.420⋆-1	6.078⋆-3	−2.449⋆-2	1.810⋆-1	−6.841⋆-2	1.951⋆-1
2.000	1.111⋆-1	−1.559⋆-2	1.112⋆-1	4.788⋆-3	−6.610⋆-2	−4.563⋆-2	−6.321⋆-2	1.588⋆-1
2.200	1.547⋆-1	−1.463⋆-2	4.554⋆-2	3.469⋆-3	4.254⋆-3	−6.615⋆-2	−5.559⋆-2	8.650⋆-2
2.400	1.929⋆-1	−1.307⋆-2	−2.025⋆-2	2.358⋆-3	7.653⋆-3	2.303⋆-1	−4.716⋆-2	1.384⋆-2
2.600	2.274⋆-1	1.126⋆-2	−6.204⋆-2	1.518⋆-3	6.614⋆-3	7.683⋆-1	−3.885⋆-2	−3.444⋆-2
2.800	2.594⋆-1	−9.396⋆-3	−7.429⋆-2	9.263⋆-4	4.206⋆-3	1.344	−3.113⋆-2	−5.255⋆-2

Achromatic $p/q = 2$

β^2	$(12)_{22} = (12)_{00}$	$(12)_{21} = (21)_{10}$	$(12)_{12} = (21)_{01}$	$(12)_{20} = (21)_{20}$	$(12)_{11} = (21)_{11}$	$(12)_{02} = (21)_{02}$	$(12)_{10} = (21)_{21}$	$(12)_{01} = (21)_{12} = (21)_{22} = (21)_{00}$
8.000⋆-1	−1.315⋆-1	−8.622⋆-3	−1.982⋆-2	9.415⋆-4	−1.005⋆-2	−2.703⋆-2	−3.186⋆-2	4.306⋆-2
1.000	3.994⋆-2	−3.089⋆-2	1.491⋆-1	1.756⋆-2	−1.564⋆-1	1.282	−1.340⋆-1	2.522⋆-1
1.200	1.485⋆-1	−4.606⋆-2	3.564⋆-1	2.650⋆-2	−2.134⋆-1	3.036	−1.749⋆-1	4.853⋆-1
1.400	2.445⋆-1	−5.652⋆-2	7.250⋆-1	2.613⋆-2	−1.779⋆-1	3.064	−1.868⋆-1	8.552⋆-1
1.600	3.403⋆-1	−5.977⋆-2	9.736⋆-1	2.147⋆-2	−8.019⋆-2	6.497⋆-1	−1.800⋆-1	1.094
1.800	4.299⋆-1	−5.643⋆-2	8.680⋆-1	1.589⋆-2	2.178⋆-2	−3.057	−1.618⋆-1	9.734⋆-1
2.000	5.051⋆-1	−4.936⋆-2	4.533⋆-1	1.096⋆-2	8.097⋆-2	−6.059	−1.384⋆-1	5.424⋆-1
2.200	5.655⋆-1	−4.133⋆-2	−4.383⋆-2	7.165⋆-3	8.979⋆-2	−6.824	−1.44⋆-1	2.928⋆-2
2.400	6.164⋆-1	−3.378⋆-2	−4.101⋆-1	4.459⋆-3	6.824⋆-2	−5.058	−9.221⋆-2	−3.517⋆-1

Achromatic $p/q = 4$

β^2	$(12)_{22} = (12)_{00}$	$(12)_{21} = (21)_{10}$	$(12)_{12} = (21)_{01}$	$(12)_{20} = (21)_{20}$	$(12)_{11} = (21)_{11}$	$(12)_{02} = (21)_{02}$	$(12)_{10} = (21)_{21}$	$(12)_{01} = (21)_{12} = (21)_{22} = (21)_{00}$
8.000⋆-1	1.352⋆-1	−5.248⋆-2	2.051⋆-1	3.081⋆-2	−2.786⋆-1	2.666	−2.038⋆-1	3.564⋆-1
1.000	2.703⋆-1	−8.033⋆-2	1.218⋆-1	5.392⋆-2	−4.013⋆-1	9.513	−2.854⋆-1	3.269⋆-1
1.200	3.813⋆-1	−1.006⋆-1	1.044	5.351⋆-2	−4.213⋆-1	1.252⋆1	−3.094⋆-1	1.253
1.400	5.168⋆-1	−1.093⋆-1	2.318	4.295⋆-2	−2.741⋆-1	6.473	−3.007⋆-1	2.510

(A)

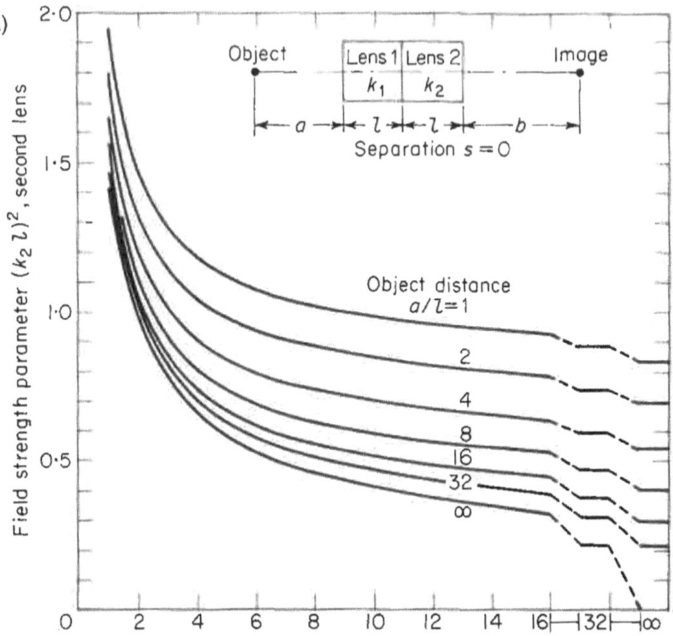

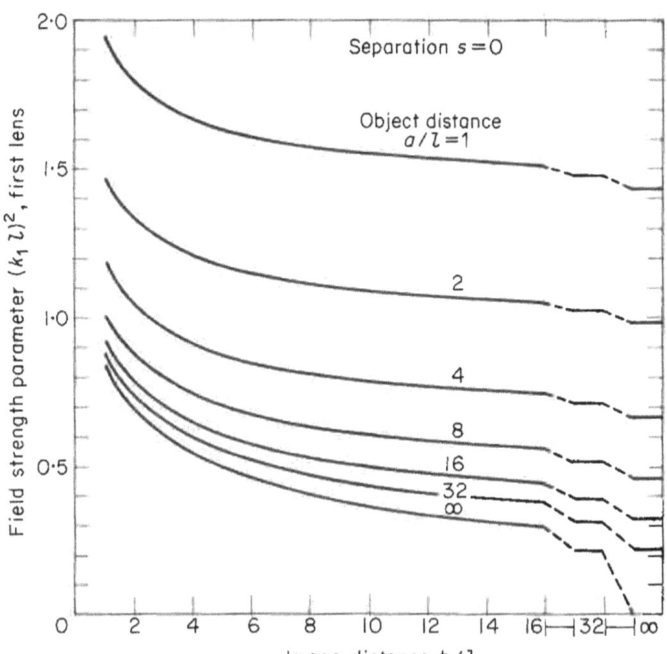

Fig. 8 The behavior of the quadrupole doublet (rectangular model). (A) and (C) lenses in contact: (B) and (D) lens separation equal to lens length. (After Enge, 1959. With the permission of AIP Publishing.)

(Continued)

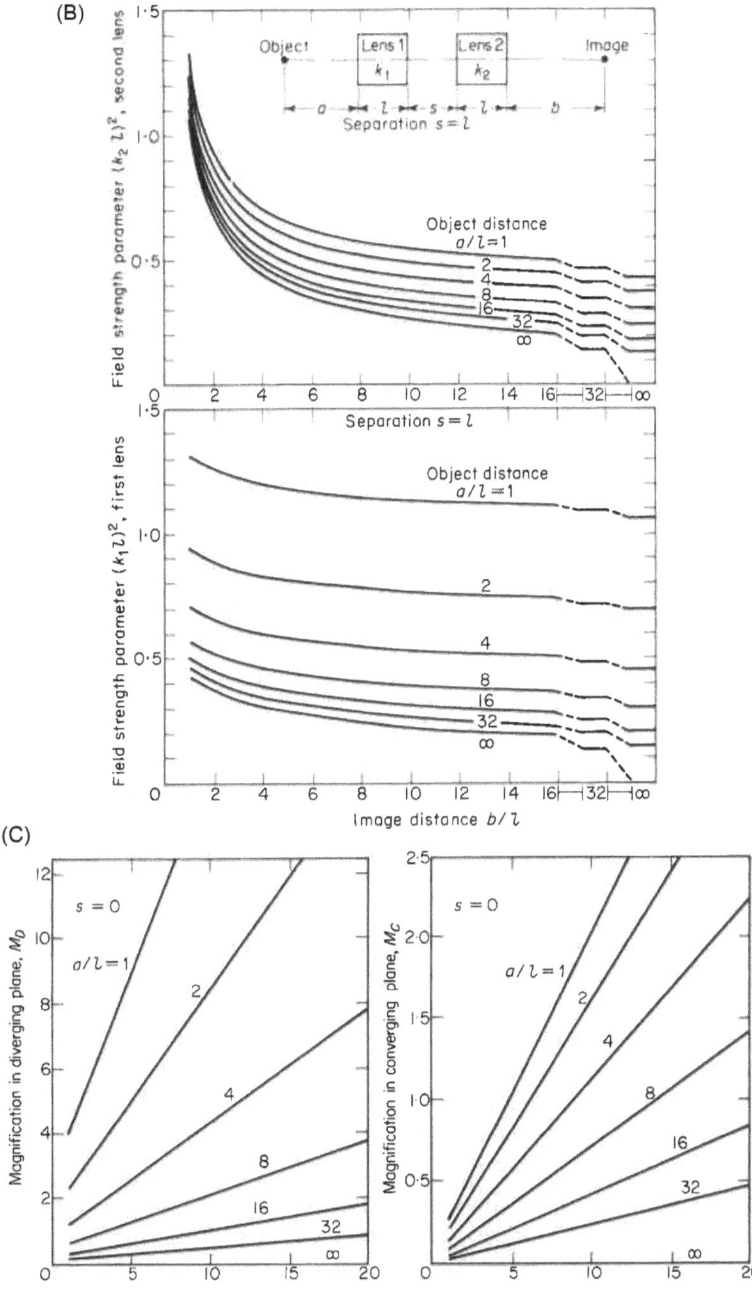
(B)
Object
Lens 1
k_1
Lens 2
k_2
Image
a
l
s
l
b
Separation s = l
Object distance
a/l = 1
2
4
8
16
32
∞
Field strength parameter $(k_2 l)^2$, second lens
0
0·5
1·0
1·5
2
4
6
8
10
12
14
16
32
∞
Separation s = l
Object distance
a/l = 1
2
4
8
16
32
∞
Field strength parameter $(k_1 l)^2$, first lens
Image distance b/l
(C)
s = 0
a/l = 1
2
4
8
16
32
∞
Magnification in diverging plane, M_D
0
2
4
6
8
10
12
5
10
15
20
s = 0
a/l = 1
2
4
8
16
32
∞
Magnification in converging plane, M_C
0·5
1·0
1·5
2·0
2·5
Image distance, b/l

Fig. 8—Cont'd

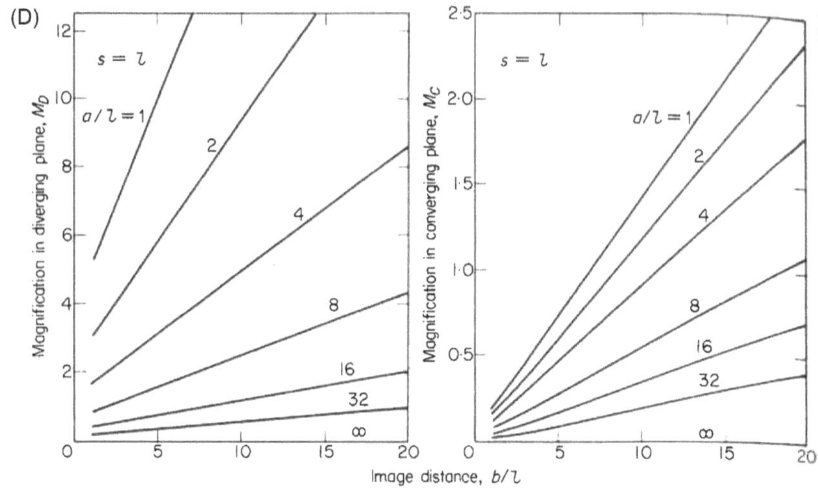

Fig. 8—Cont'd

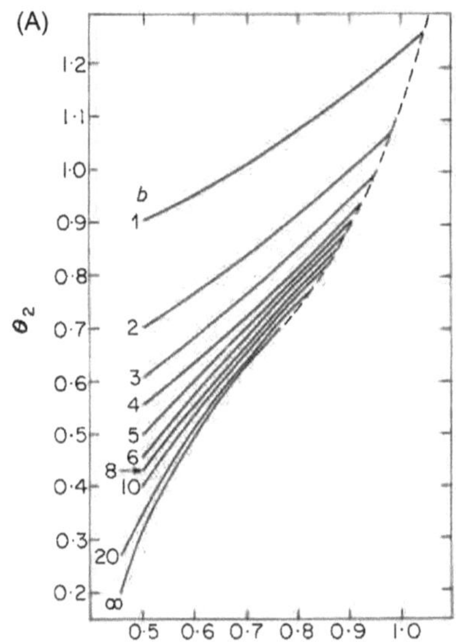

Fig. 9 The behavior of a quadrupole doublet (rectangular model); the lenses are both of unit length and the object distance (to the beginning of the first quadrupole) is set at 5 in this example. (A) θ_2 against θ_1 for various values of the image distance, b, measured from the end-face of the second quadrupole. (B) The corresponding center spacing, X. (C) Angular magnification in the principal sections. (After King, 1962.)

(Continued)

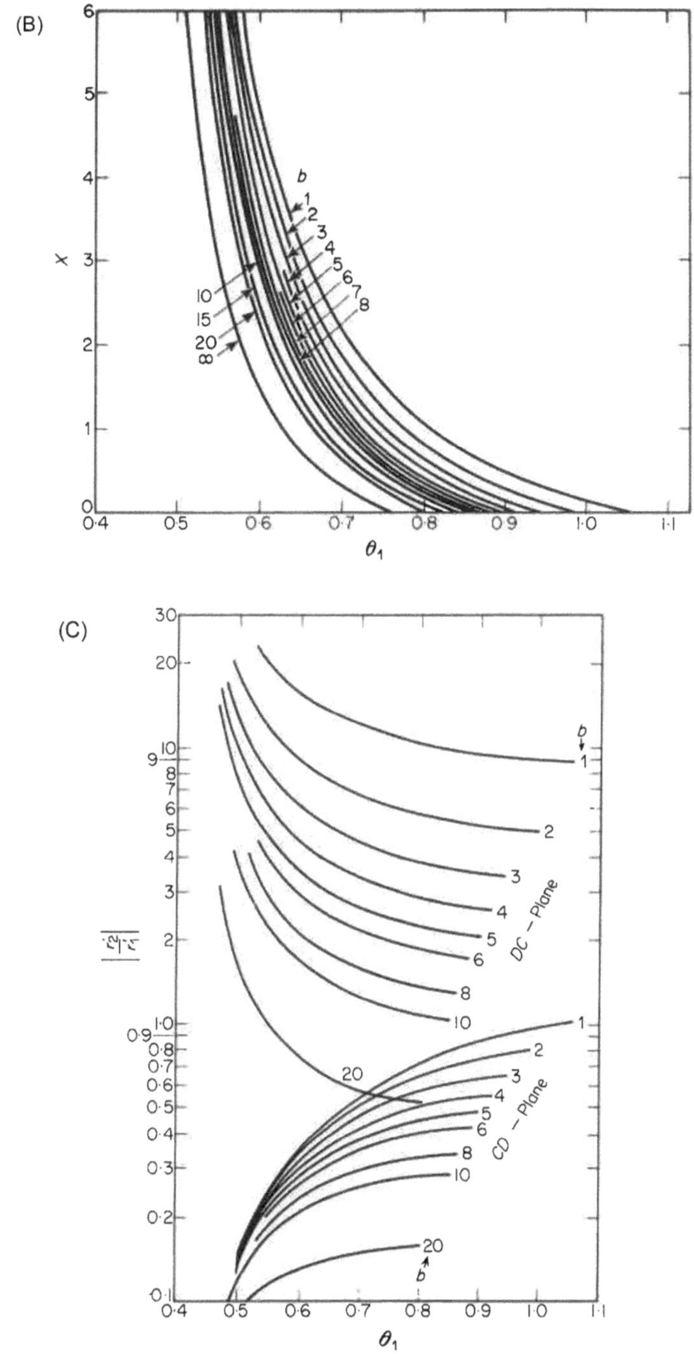

Fig. 9—Cont'd

(A)

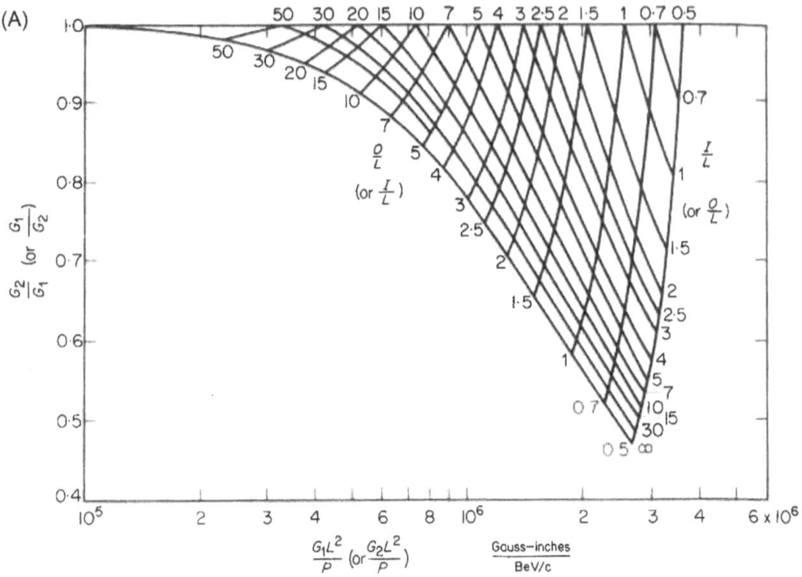

(B)

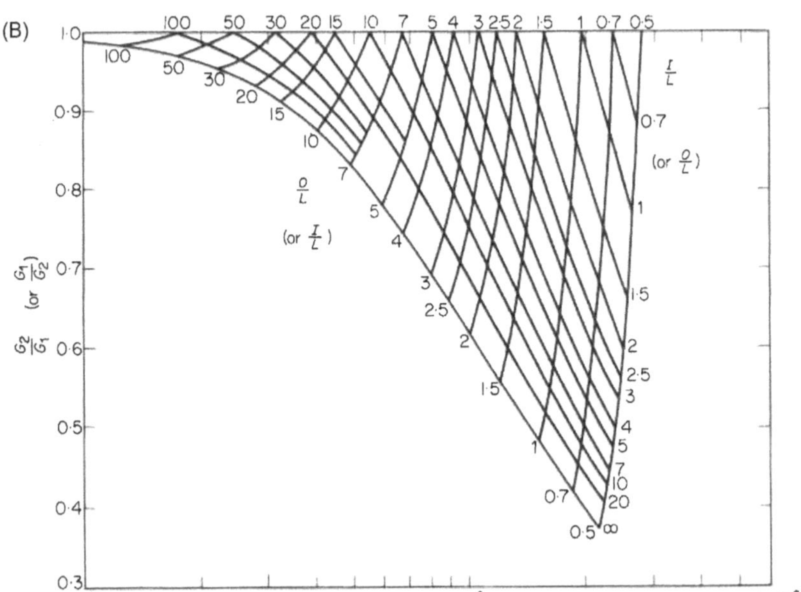

Fig. 10 The properties of a quadrupole doublet (rectangular model) for various separations (s). The object and image distances, measured from the end-faces of the first and second lenses, respectively, are denoted by O and I. Notice that $\beta^2 = 7 \cdot 62 \times 10^{-7} G/P$ in gauss $(\text{BeV}/c)^{-1}$. (a) $s = 0$; (b) $s = 0.5L$; (c) $s = L$ (d) $s = 2L$. (After Baker, 1963.)

(Continued)

(C)

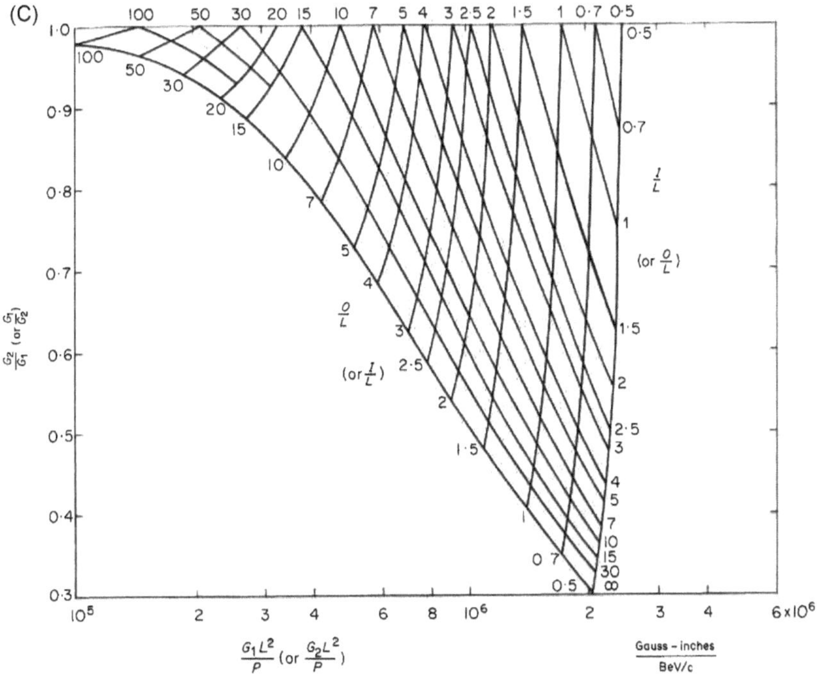

(D)

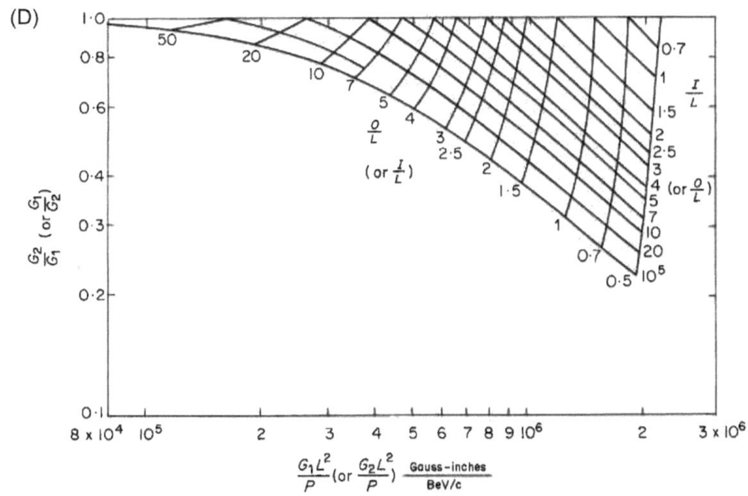

Fig. 10—Cont'd

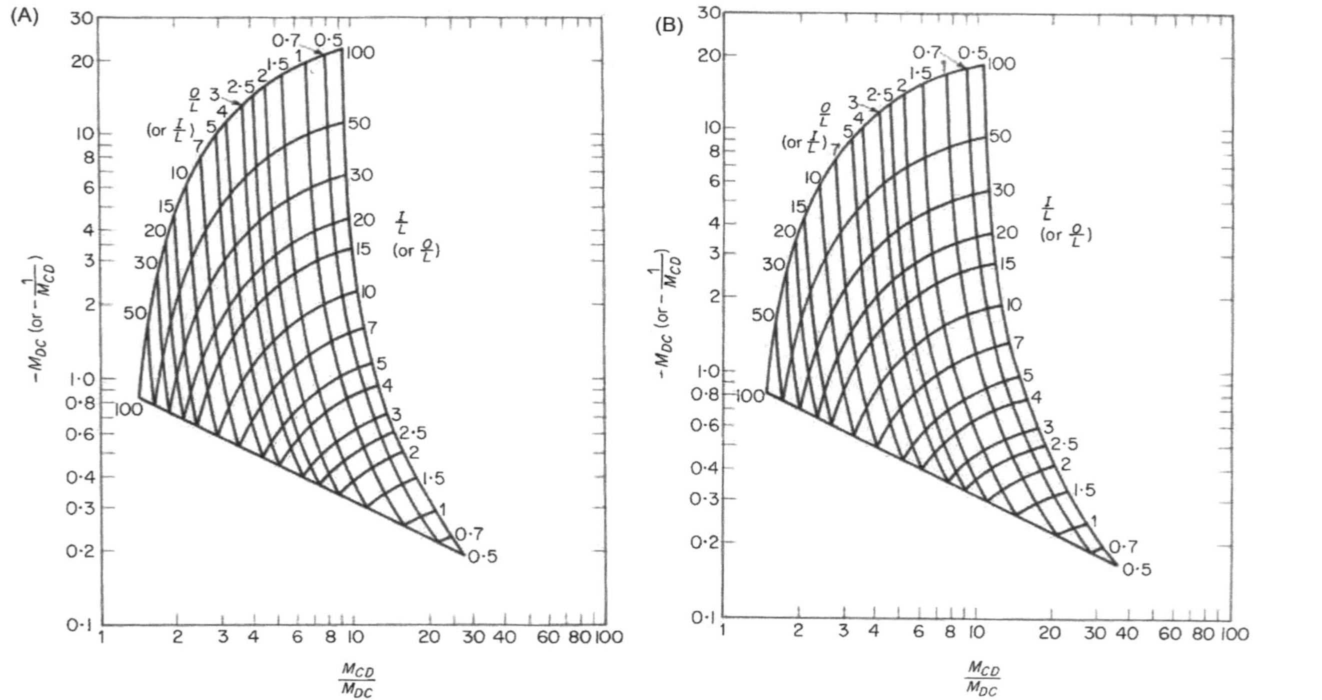

Fig. 11 The properties of a quadrupole doublet; magnifications [cf. Fig. 10A–D]. (A) $s = 0$; (B) $s = 0.5\,L$; (C) $s = L$; (D) $s = 2\,L$, (After Baker, 1963.)

(Continued)

(C)

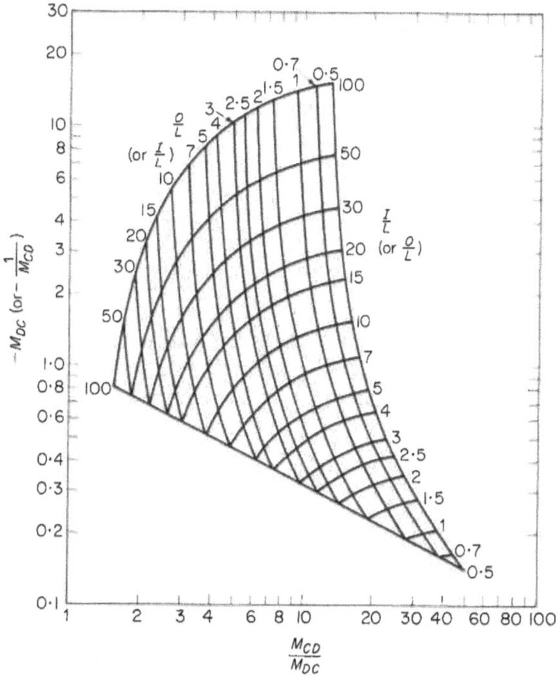

(D)

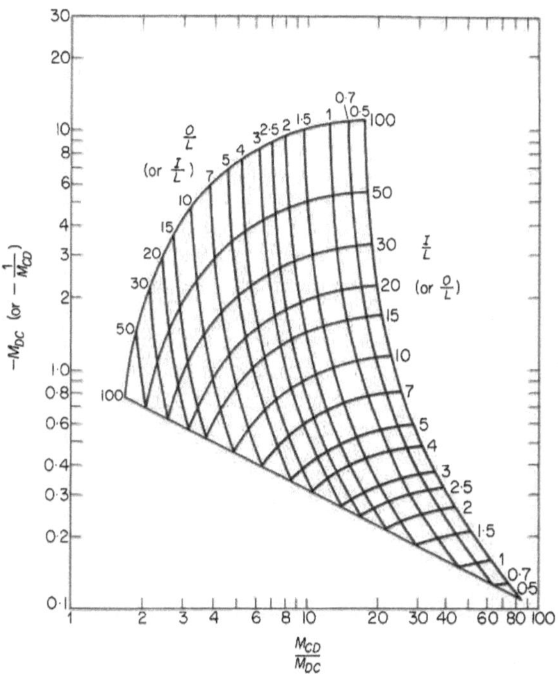

Fig. 11—Cont'd

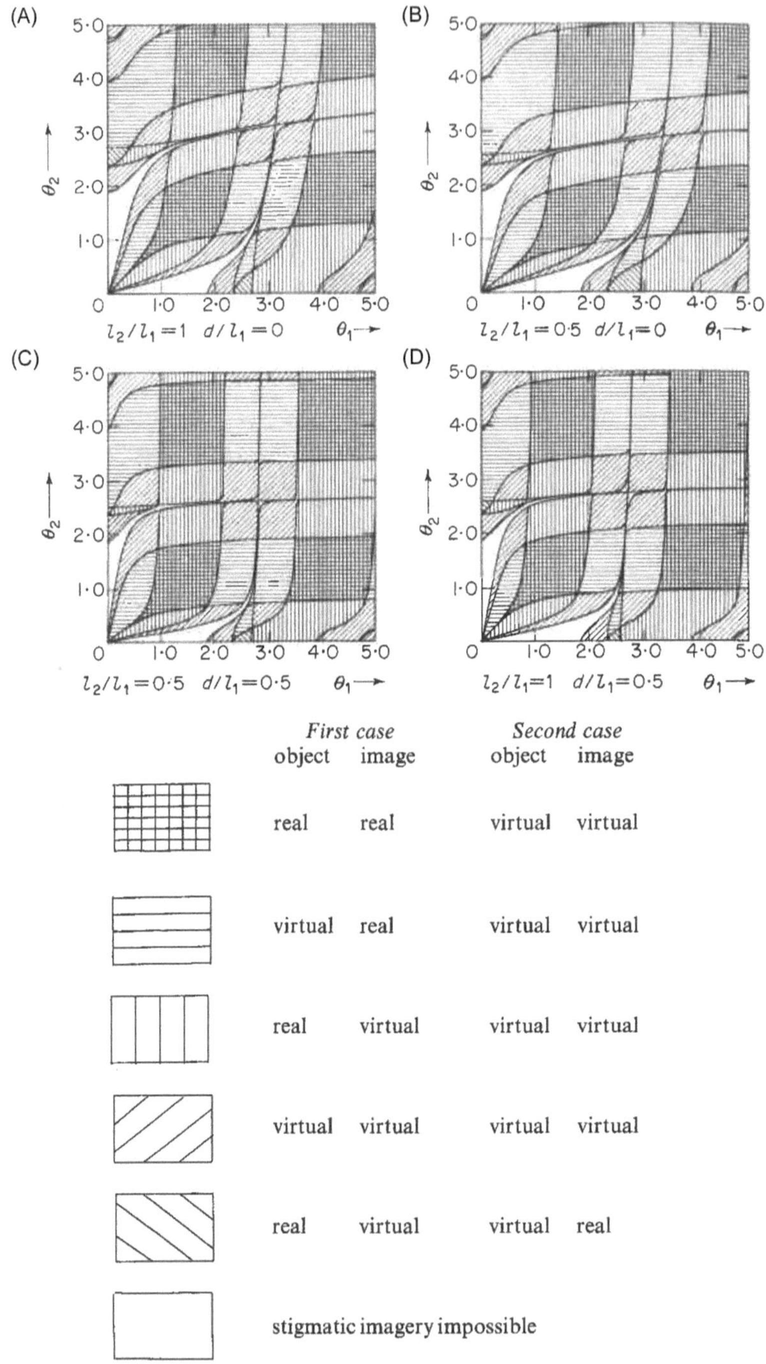

Fig. 12 Types of pseudo-stigmatic imagery produced by a doublet (rectangular model). The lens separation (between end-faces) is denoted by d. (A) $l_2/l_1(=L_2/L_1) = 1, d/l_1 = 0$; (B) $l_2/l_1 = 0 \cdot 5, d/l_1 = 0$; (C) $l_2/l_1 = 0 \cdot 5, d/l_1 = 0 \cdot 5$; (D) $l_2/l_1 = 1, d/l_1 = 0 \cdot 5$. (After Junior & Antony-Spies, 1967.)

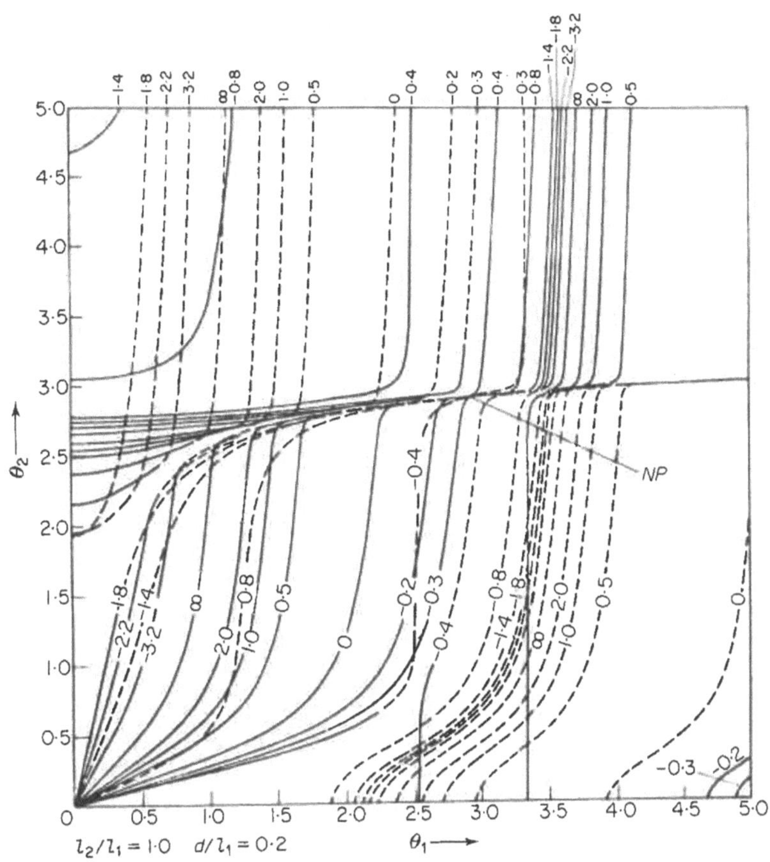

Fig. 13 Contours of equal object distance for a doublet (rectangular model) with $l_1/l_2 = L_1/L_2 = 1$, $d/l_1 = 0 \cdot 2$ (cf. Fig. 12). At NP, the doublet is regular; $kl = \theta$. Full curves correspond to one case and dotted curves to the other [there are always two pairs of stigmatic object-image positions or none (Carathéodory, 1937); cf. Lenz, 1958]. (After Junior & Antony-Spies, 1967.)

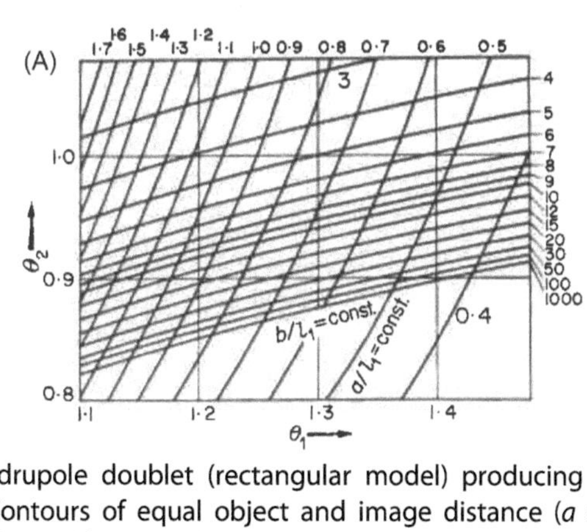

Fig. 14 The quadrupole doublet (rectangular model) producing a real image of a real object. (A) Contours of equal object and image distance (a and b respectively, measured from the endfaces) in the $k_2l_2 - k_1l_1 (= \theta_2 - \theta_1)$ plane. $l_2/l_1 = 1$, $d_1/l_1 = 0 \cdot 2$, $0 \cdot 4 \leq a/l_1 \leq 1 \cdot 7$, $3 \leq b/l \leq 1000$.

(Continued)

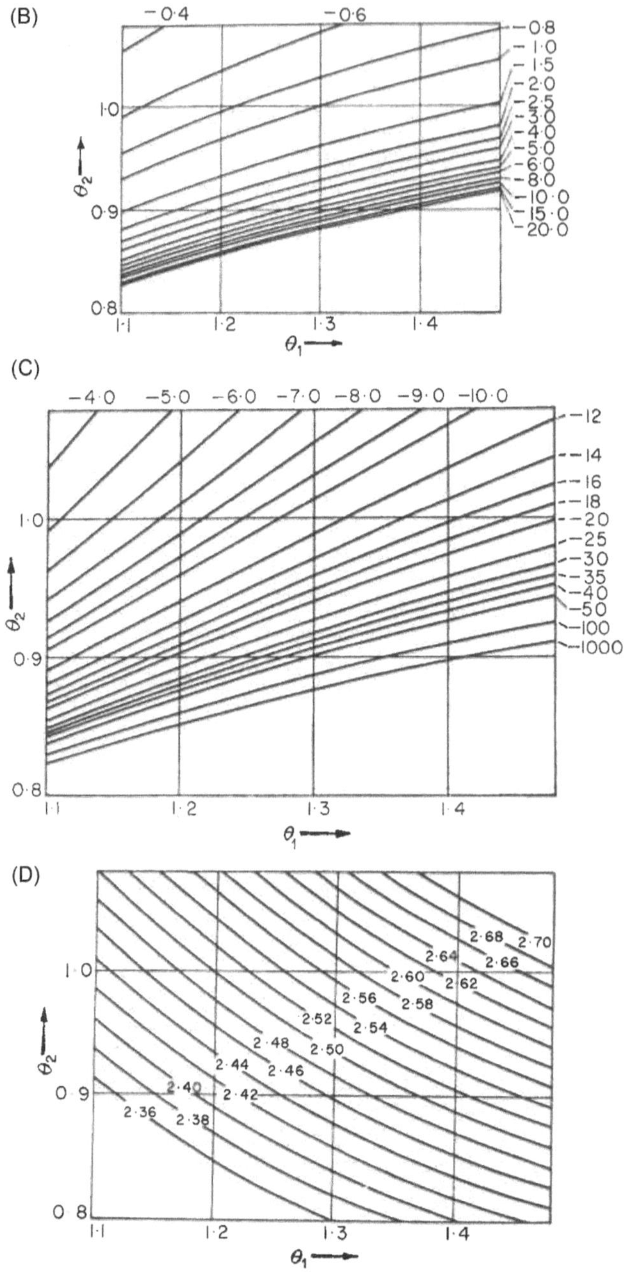

Fig. 14—Cont'd (B) Contours of equal $V_y (= M_y)$ for $-0 \cdot 4 \geq V_y \geq -20$. (C) Contours of equal $V_x (= M_x)$ for $-1000 \leq V_x \leq -4 \cdot 0$. (D) Contours of equal anamorphism; the parameters labeling the contours are the strengths of a third quadrupole ($k_3 l_3 = \theta_3$) that will cancel this distortion and produce a stigmatic and orthomorphic (though not regular) final image; $2 \cdot 36 \leq k_3 l_3 \leq 2 \cdot 70$. (After Antony-Spies & Junior, 1967.)

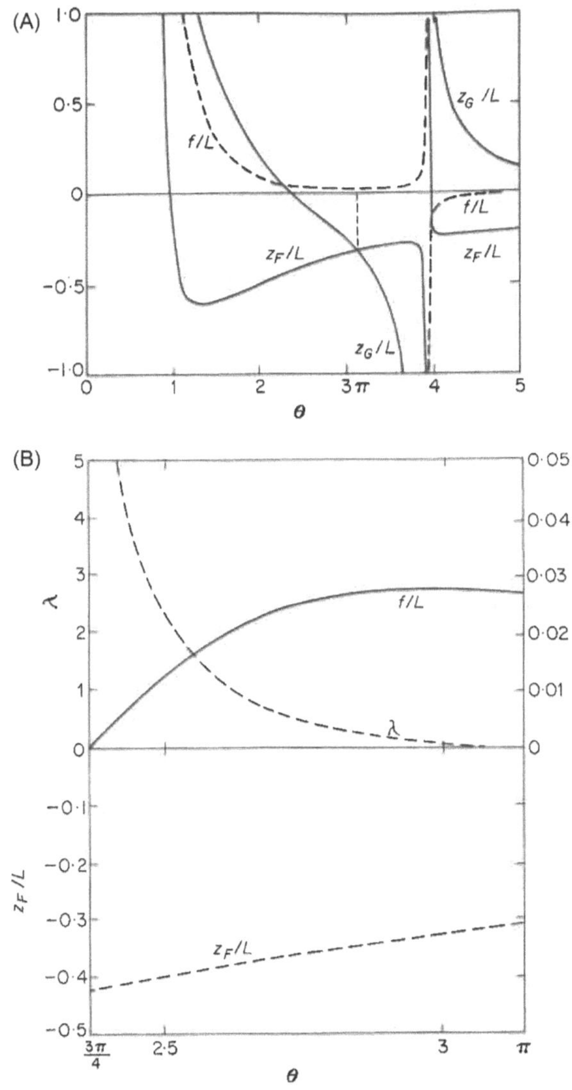

Fig. 15 (A) The focal length and focal distance (measured from the end of the second lens) of an antisymmetric doublet (rectangular model) as a function of θ, lenses in contact. (B) The focal length, focal distance and lens separation of a regular antisymmetric doublet (rectangular model) as a function of θ. (After Dhuicq, 1961.)

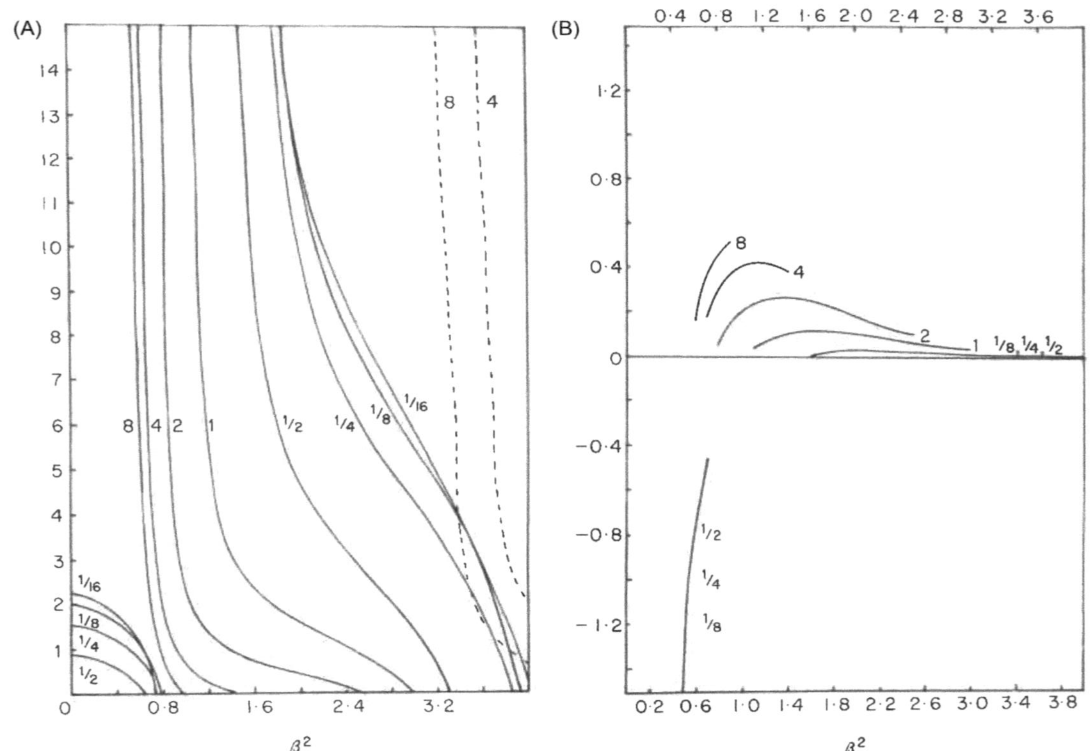

Fig. 16 (A) The lens separation in an antisymmetric doublet (symmetric and asymmetric bell-shaped model). (B) and (C) The corresponding cardinal elements.

(Continued)

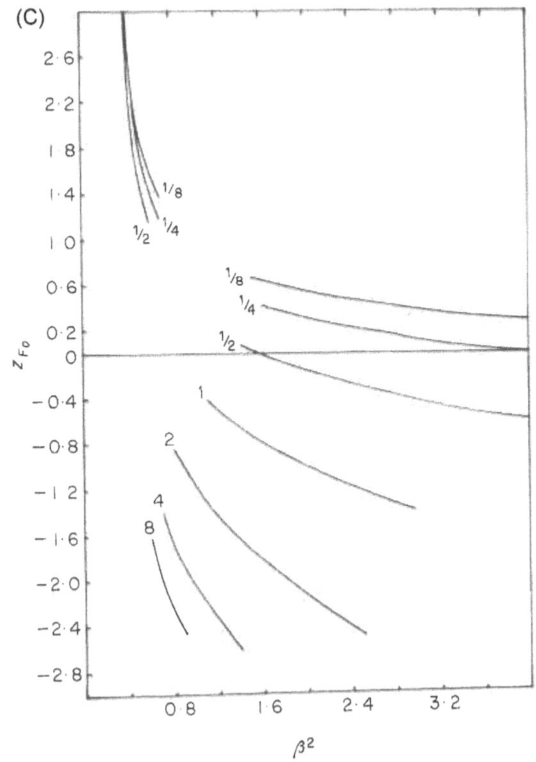

Fig. 16—Cont'd

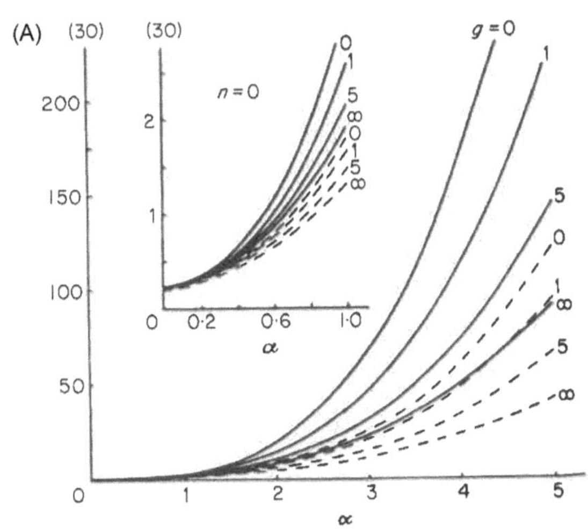

Fig. 17 The aperture aberration coefficients of a stigmatic doublet of magnetic quadrupoles as a function of the object distance α (measured from the end-face of the first quadrupole). g denotes the image distance (measured from the end-face of the second quadrupole); full curves correspond to the case of zero gap between the lenses (which are of equal length, $L = 1$) and dotted lines to the case in which the gap is equal to L. (A): (30); (B) (12) = (21); (C) (03). (After Fishkova, Shpak, et al., 1968.)

(Continued)

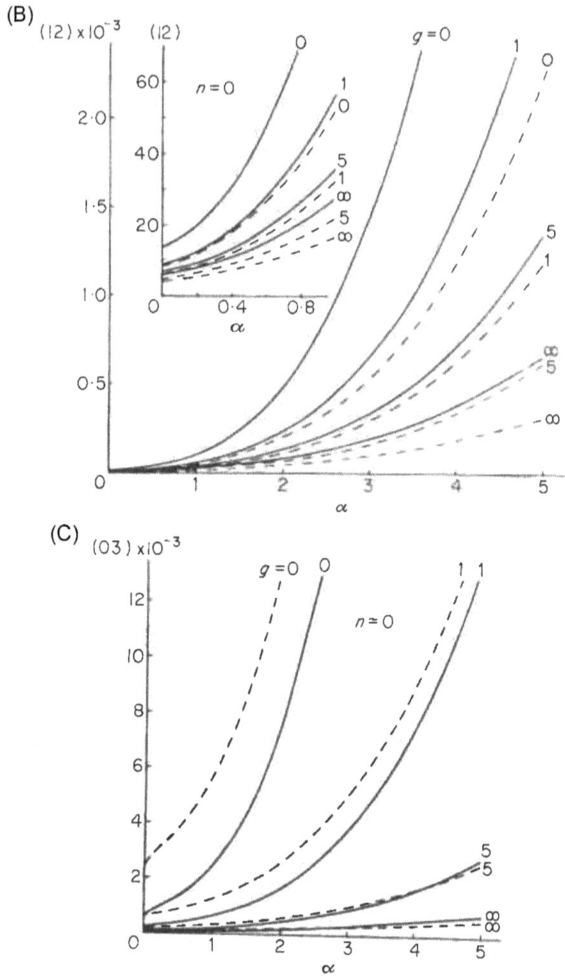

Fig. 17—Cont'd

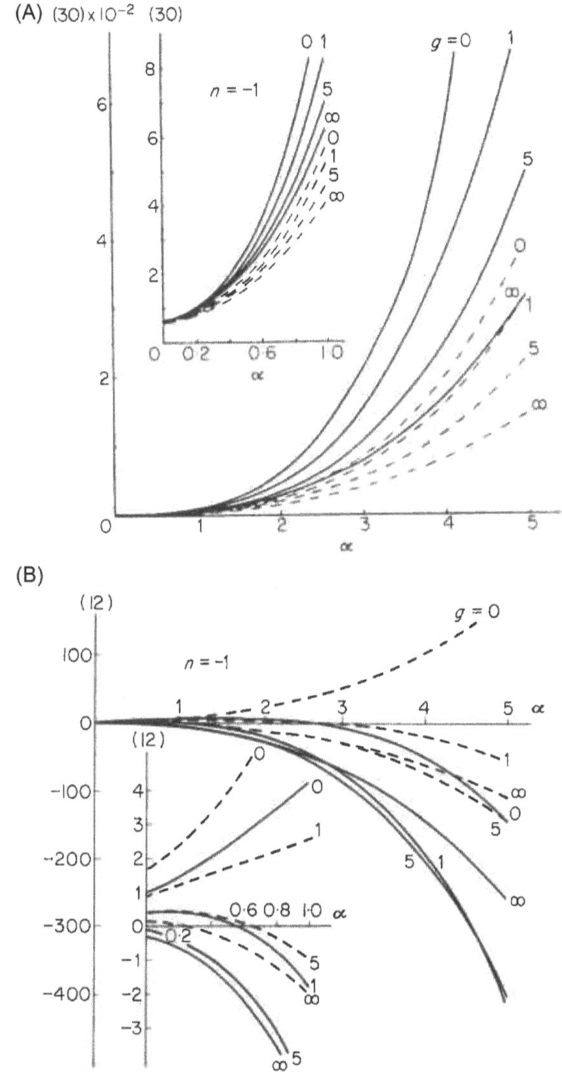

Fig. 18 The parameters of Fig. 17 in the case of electrostatic quadrupoles. (After Fishkova, Shpak, et al., 1968.)

(Continued)

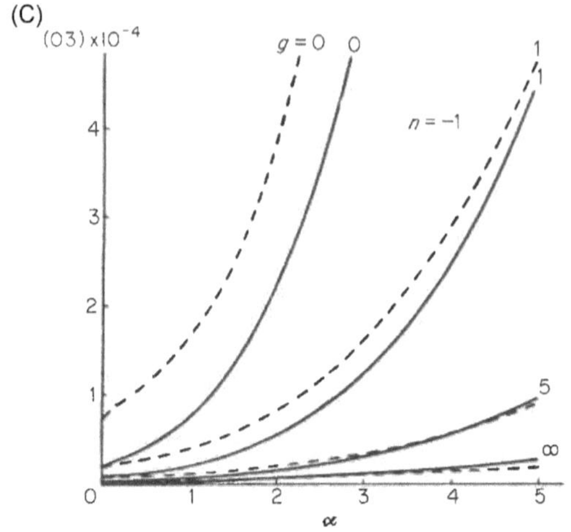

Fig. 18—Cont'd

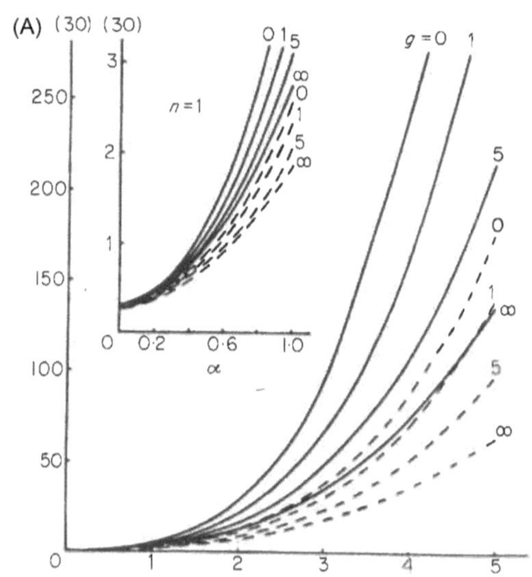

Fig. 19 The parameters of Fig. 17 in the case of achromatic quadrupoles. (After Fishkova, Shpak, et al., 1968.)

(Continued)

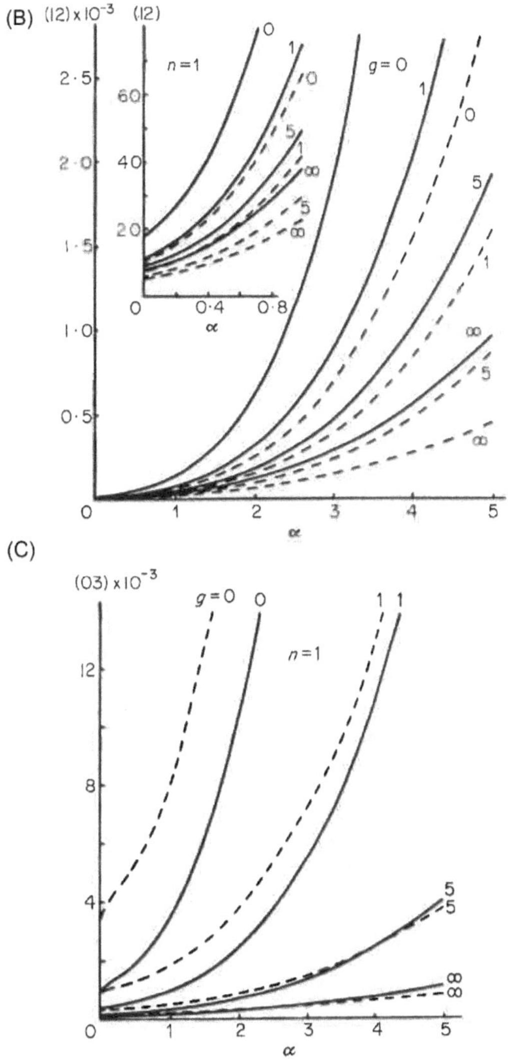

Fig. 19—Cont'd

3. The quadrupole triplet

Although the triplet follows the doublet in order of increasing complexity, it is not an inviting combination from the optical point of view. Since the number of components is odd, it cannot be antisymmetric, so that in general regularity will be achieved by adjusting the three excitations and the two separations, once the design of the individual components has been decided (in particular, their effective lengths). The symmetric triplet does, however, offer a convenient means of converting a virtual image into a real one with unit magnification.

The unsymmetric triplet is more difficult to analyze. Deltrap (1964a) and King (1963) have used the thin lens approximation to give a theoretical indication of the behavior of such a multiplet, and Antony-Spies and Junior (1967) have studied triplets in which the first two lenses create a stigmatic image and the final, very strong quadrupole converts this virtual stigmatic object into a real stigmatic image.

We have considered how a series of triplets, consisting of three geometrically identical quadrupoles, with the excitations and spacings as variables, can be made regular. Using the rectangular model, we find that the conditions for regularity imply that if two excitations are set (β_1 and β_2, say) then $\beta_3 = f(\beta_2, \beta_1)$ and $s = s(\beta_3)$, $t = t(\beta_3)$. The third excitation, β_3, and the spacings s and t have therefore been evaluated for a range of values of β_1 and $\beta_2 (0 \cdot 2 < \beta_1 < 2 \cdot 0$ and $0 \cdot 2 < \beta_2 < 2 \cdot 0)$, together with the corresponding cardinal elements. (L is the same for all three quadrupoles, and we have taken it as a unit of length, so that $\beta_i \equiv \theta_i$). Several classes of regular systems have emerged, details of which will be published separately.

Blewett (1958) has tabulated a few values of the parameters of symmetric triplets for which the imagery is stigmatic. Most of the published information is in graphical form, however. Enge (1961) has plotted families of curves illustrating the behavior of a symmetric triplet producing pseudo-stigmatic imagery; the length of the central quadrupole is l (equal to our L) and the outer components are each $l/2$ in length. Enge's curves are reproduced as Fig. 20A–E. Fig. 21A,B show the result of Dhuicq's unsuccessful search for a regular symmetric triplet (Dhuicq, 1961). King (1963) has used the thin lens model to study triplets, and his working curves are reproduced as Fig. 22A,B. An extensive set of curves illustrating triplet behavior have been prepared by Brown and Hindmarsh (1965); a selection of their results is reproduced as Figs. 23A,B, 24, 25A–C, 26A–D. Fig. 27 (Bauer, 1965–66)

(A)

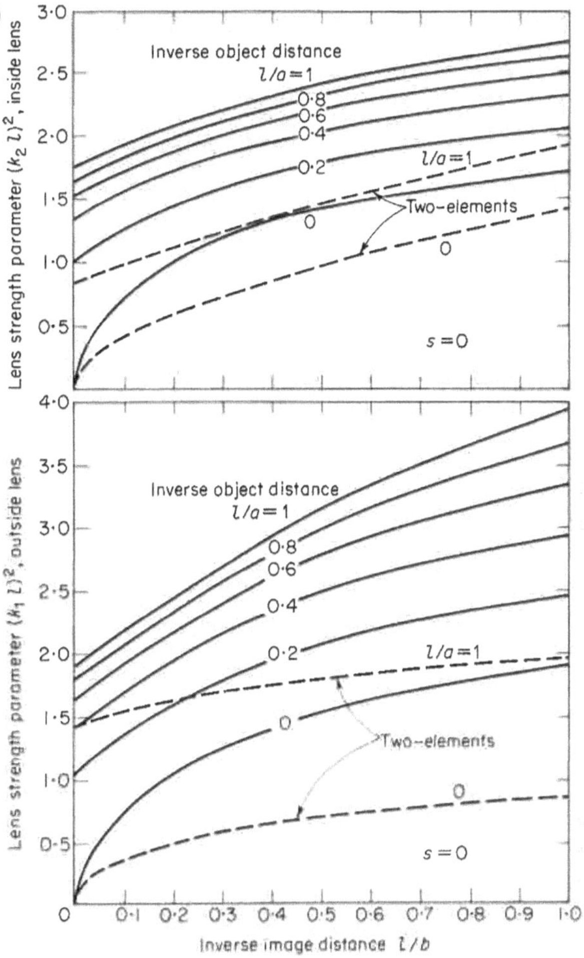

Fig. 20 The symmetric triplet (rectangular model). Centre lens-length, *l*; outer lens-length, *l*/2. Object distance, *a*; image distance, *b*. Gap s. (A) Lens strengths $(kl)^2(=\theta^2)$ against *l*/*b* for a range of object distances, $s=0$. Dotted curves correspond to the doublet. (Enge, 1959.) (B) The same parameters, $s=l/2$. (C) Magnifications against *b*/*l* for various object distances, $s=0$. (D) Magnifications against *b*/*l* for various object distances, $s=l/2$. (E) A comparison between the distortions, $M_D/M_C(=M_y/M_x)$, of a doublet and a triplet; object distance $=4l$. (After Enge, 1961. With the permission of AIP Publishing.)

(Continued)

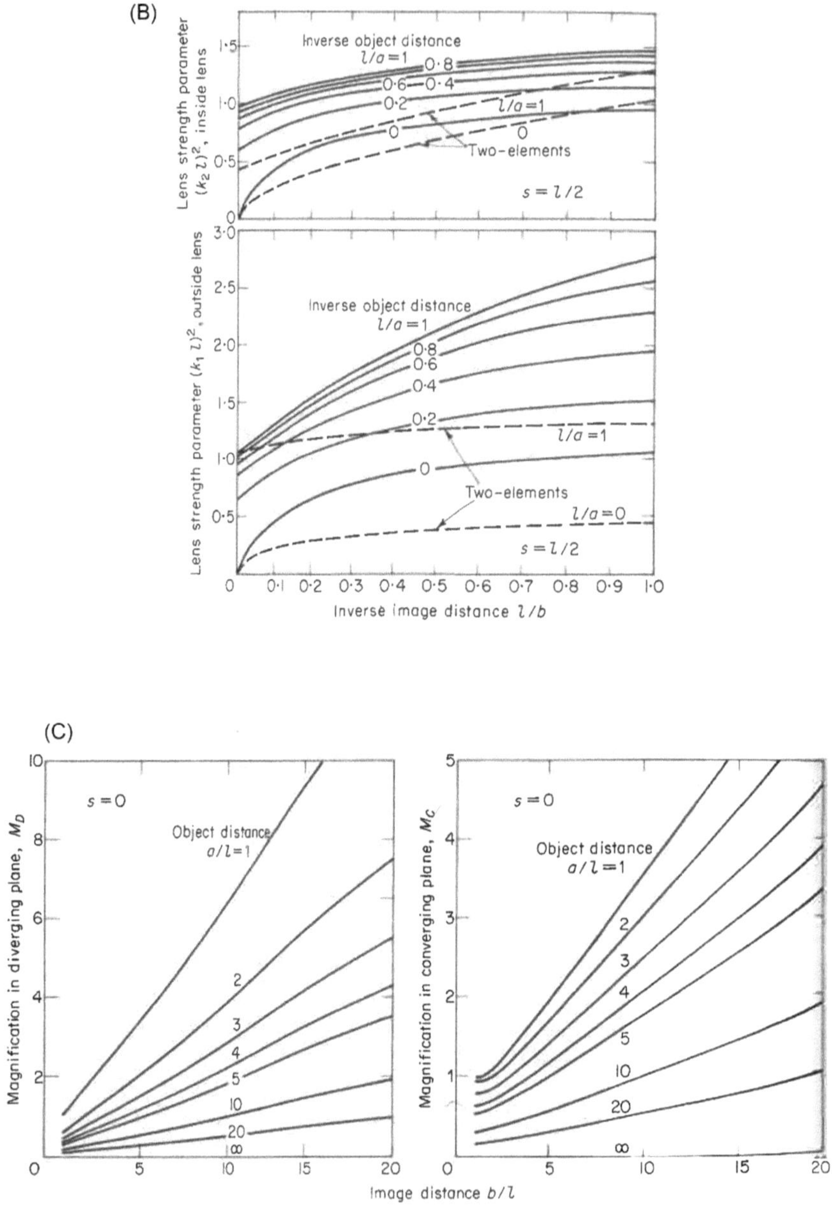
(B)
Lens strength parameter $(k_2 l)^2$, inside lens
Inverse object distance
$l/a=1$
0·8
0·6
0·4
0·2
0
$l/a=1$
0
Two-elements
$s=l/2$
Lens strength parameter $(k_1 l)^2$, outside lens
Inverse object distance
$l/a=1$
0·8
0·6
0·4
0·2
0
$l/a=1$
Two-elements
$l/a=0$
$s=l/2$
Inverse image distance l/b
(C)
Magnification in diverging plane, M_D
$s=0$
Object distance
$a/l=1$
2
3
4
5
10
20
∞
Magnification in converging plane, M_C
$s=0$
Object distance
$a/l=1$
2
3
4
5
10
20
∞
Image distance b/l

Fig. 20—Cont'd

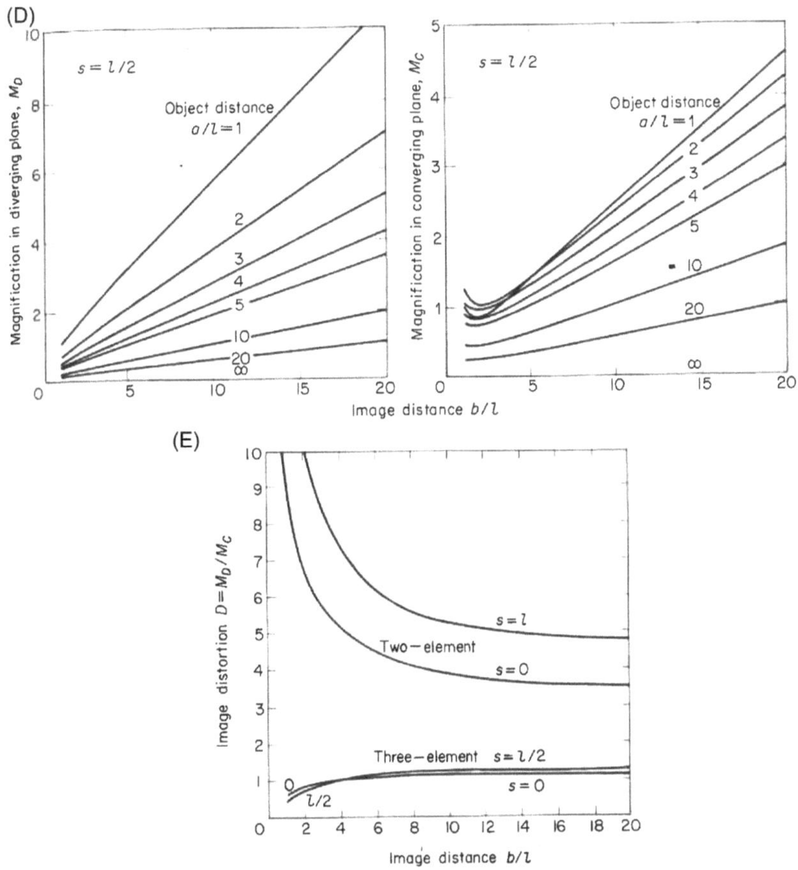

Fig. 20—Cont'd

shows how the inner excitation and object-distance are related to the outer excitation for a triplet of three quadrupoles of equal length, producing a pseudo-stigmatic image at unit magnification. Fig. 28 (Antony-Spies & Junior, 1967) shows how the magnification of a triplet objective varies with distance between object and image (the final quadrupole is very strong, as it is operating stigmatically). Regenstreif (1967b, 1967e, 1967f, 1967g) has examined whether it is possible to achieve turnover with a quadrupole triplet. We refer to the original papers for details of his findings.

Finally, we return to the work of Foster (1968) on the symmetric triplet as a lens spectrometer operating on the focal separation principle

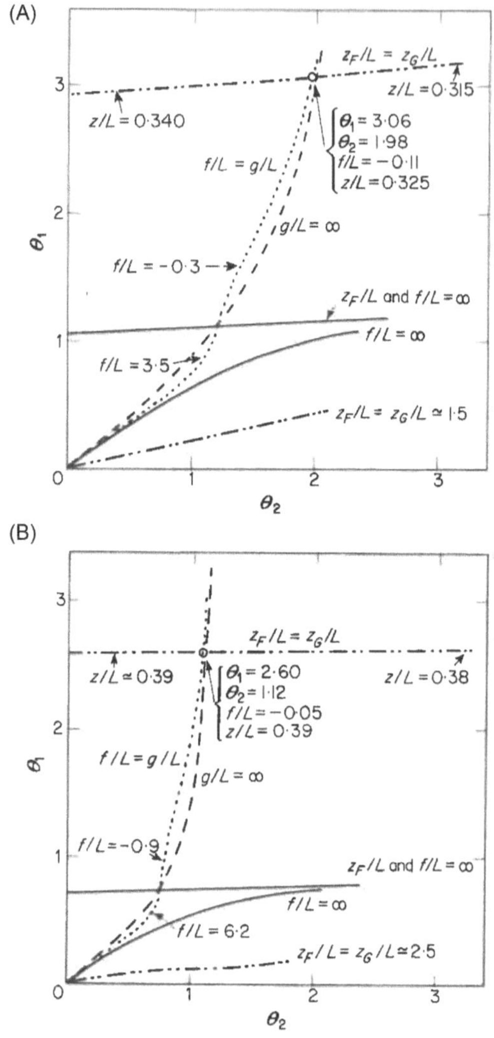

Fig. 21 The symmetric triplet (rectangular model). Contours of infinite focal length, infinite focal distance, equal focal lengths and equal focal distances in the $\theta_1 - \theta_2$ plane, where θ_1 is the excitation of the outer quadrupoles and θ_2 that of the central one. (A) Lenses of equal length, in contact. (B) Lenses of equal length, L, separated by L. (After Dhuicq, 1961.)

(see Fig. 13 in Definitions). Her computed results are set out in Table 32. The third-order quantities tabulated are the coefficients that appear in $\Delta x_i = A_3 x_i'^3 + B_3 x_i' y_i'^2$, $\Delta y_i = C_3 y_i'^3 + D_3 y_i' x_i'^2$ and $(\text{rosette})_3 = (A_3 - 2B_3 + C_3)/8$, $(\text{star})_3 = (A_3 - C_3)/8$ and $(\text{spherical})_3 = (3A_3 + 2B_3 + 3C_3)/8$; the fifth-order

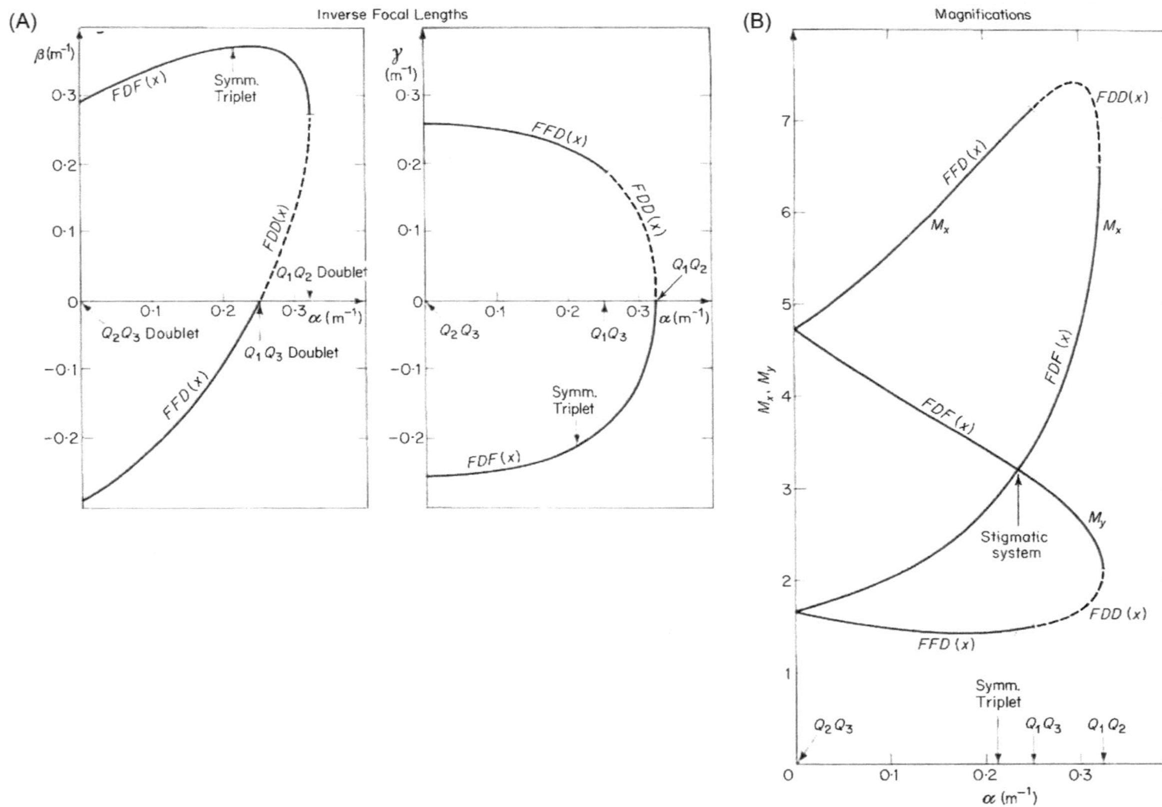

Fig. 22 A thin-lens triplet study. Object distance: 7·5 m, first drift space = second drift space =2·0 m, image distance =28·36 m. α, β and γ denote the reciprocals of the focal lengths of the three components. (A) β and γ as a function of α (*F* and *D* denote focusing, or convergent and defocusing, or divergent, respectively). (B) Magnifications. (After King, 1963.)

(A)

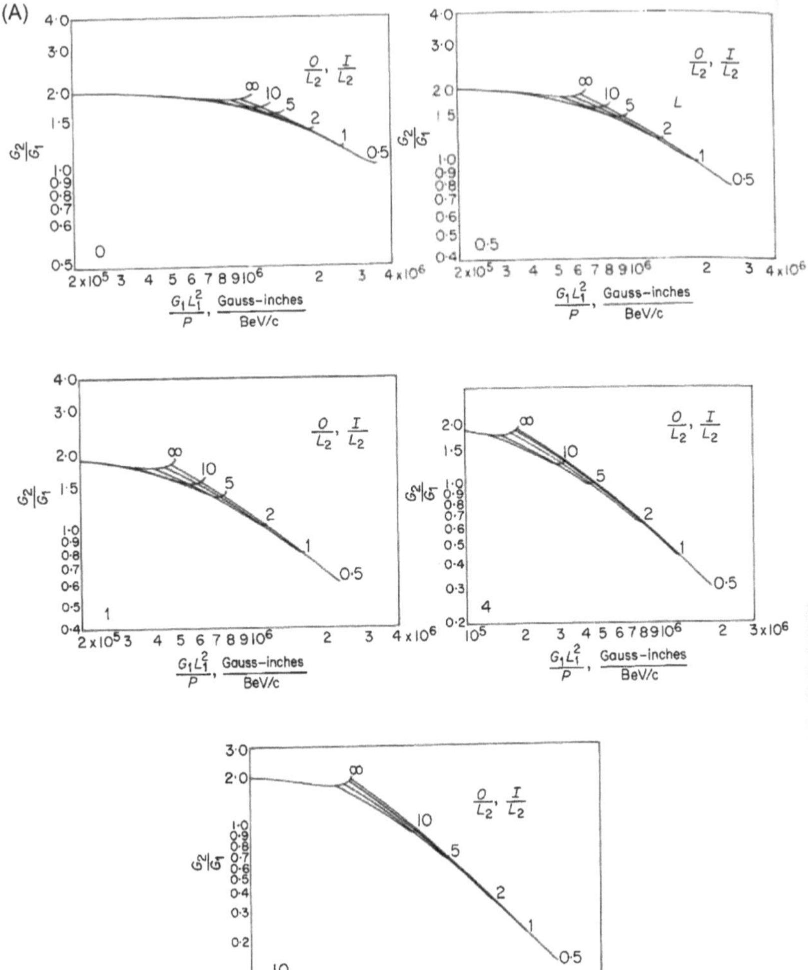

Fig. 23 (A) Symmetric triplet, $L_2 = L_1$; the ratio of the field gradients as a function of the outer lens strength ($G_1/G_2 = \theta_1^2/\theta_2^2$; $G_1 L_1^2/P \propto \theta_1^2$; $\beta^2 = 7\cdot62 \times 10^{-7} G$ [gauss/in.]/P[Be V/c]). Five values of the ratio s/L are shown: 0, $\frac{1}{2}$, 1, 4 and 10. (B) Symmetric triplet, $L_2 = 2L_1$; the ratio of the field gradients as a function of the outer lens strength. Five values of the ratio s/L are shown: 0, $\frac{1}{2}$, 2, 5 and 15. (After Brown & Hindmarsh, 1965. Courtesy SLAC. https://SLAC.Stanford.edu/pubs/slacreports/reports04/slac-r-047/pdf.)

(Continued)

(B)

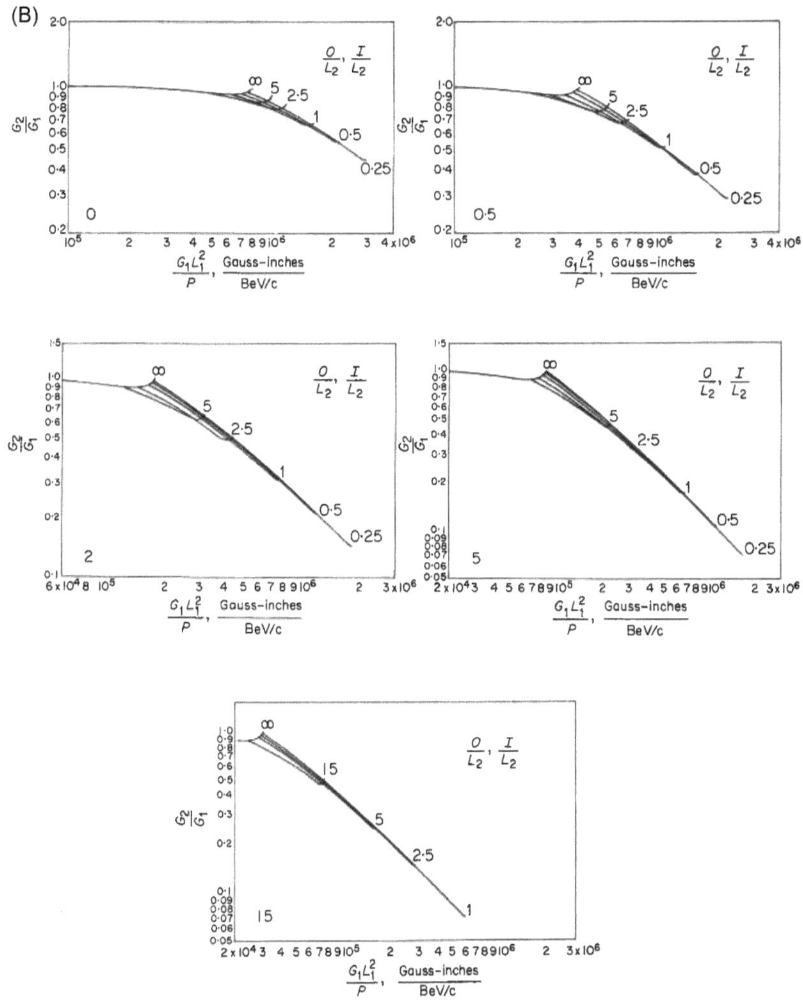

Fig. 23—Cont'd

coefficients are the four[a] independent quantities that appear in $\Delta x_i =$ third order terms $+ a_x\alpha^5 + 2b\alpha^3\gamma^2 + c\alpha\gamma^4$, $\Delta y_i =$ third order terms $+ a_y\gamma^5 + 2c\alpha^2\gamma^3 + b\alpha^4\gamma$ with $a_x = A_5 + 6B_5 + 3C_5 + D_5$, $b = -5A_5 - 10B_5 + C_5 + D_5$, $c = 5A_5 - 10B_5 - C_5 + D_5$, $a_y = -A_5 + 6B_5 - 3C_5 + D_5$. The experimental work was hampered by parasitic aberrations; Foster concludes that "The experiments

[a] See Section 5 of Definitions for further comment on this point.

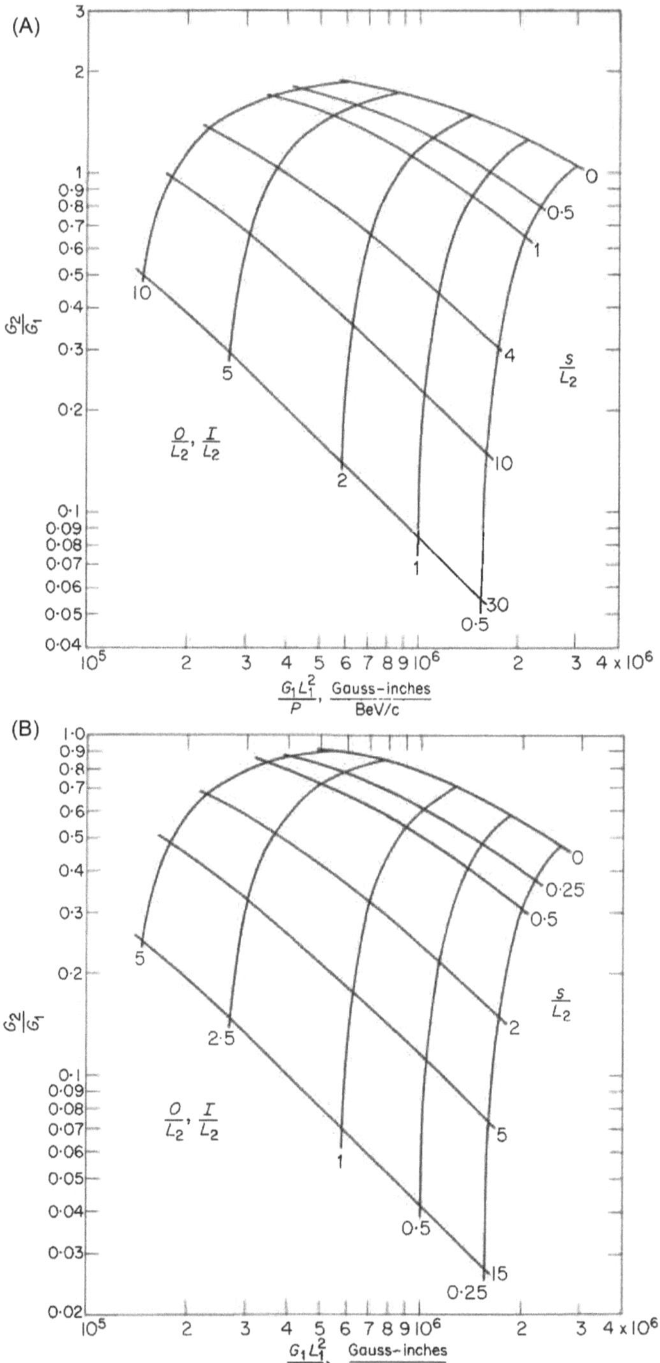

Fig. 24 The symmetric triplet, equal object and image distances. (A) $L_2 = L_1$; (B) $L_2 = 2L_1$. (After Brown & Hindmarsh, 1965.)

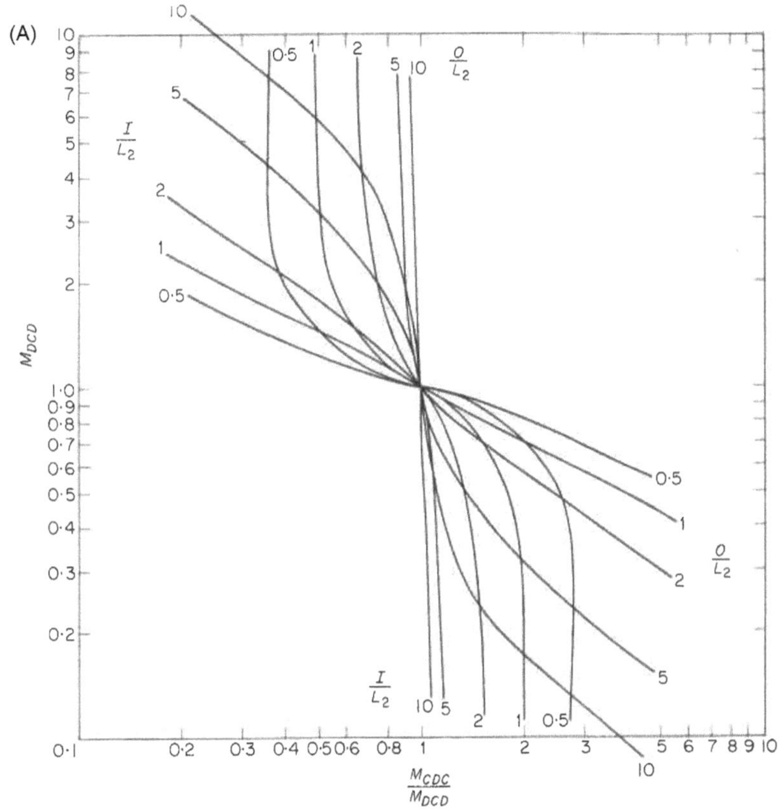

Fig. 25 Magnifications in the symmetric triplet. (*CDC* and *DCD* correspond to the converging–diverging–converging and diverging–converging–diverging principal sections.) $L_2 = L_1$. (A) $s/L_2 = 0$; (B) $s/L_2 = 1$; (C) $s/L_2 = 4$. (After Brown & Hindmarsh, 1965.)
(Continued)

have shown that the paraxial properties of the quadrupoles are in agreement with the theoretical investigations. The excitations agree closely with those found for the finite solenoids.

The star and rosette[b] aberrations of the quadrupole system agree closely with those found using the point pole model. The excitation of the central quadrupole required to produce a threefold circle [with an annular beam]

[b] See Hawkes (1967e), where the relations between these coefficients introduced by Burfoot (1954, 1956) and the (*pqrs*) are set out.

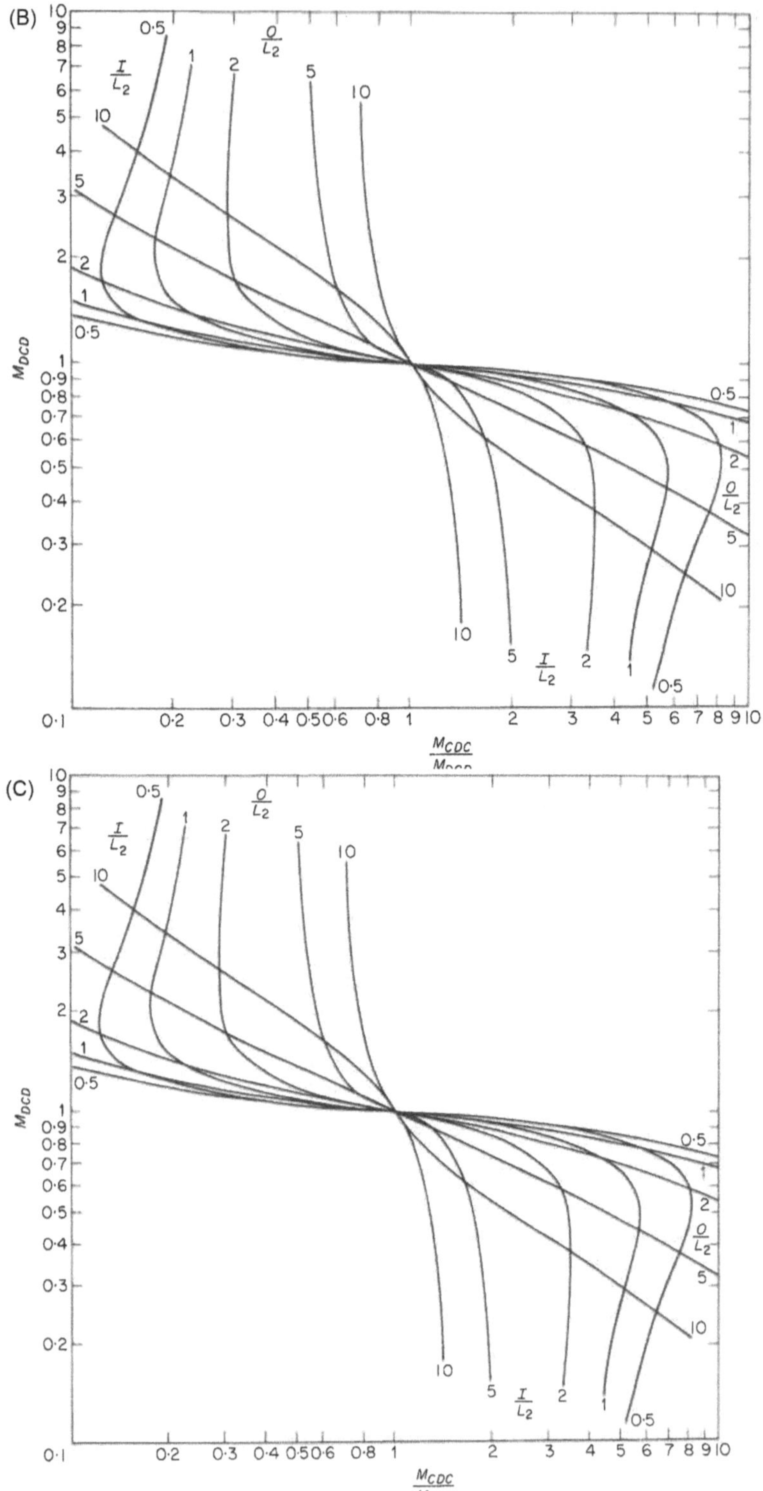
(B)
M_{DCD}
$\frac{M_{CDC}}{M_{DCD}}$
$\frac{I}{L_2}$
$\frac{O}{L_2}$
(C)
M_{DCD}
$\frac{M_{CDC}}{M_{DCD}}$
$\frac{I}{L_2}$
$\frac{O}{L_2}$

Fig. 25—Cont'd

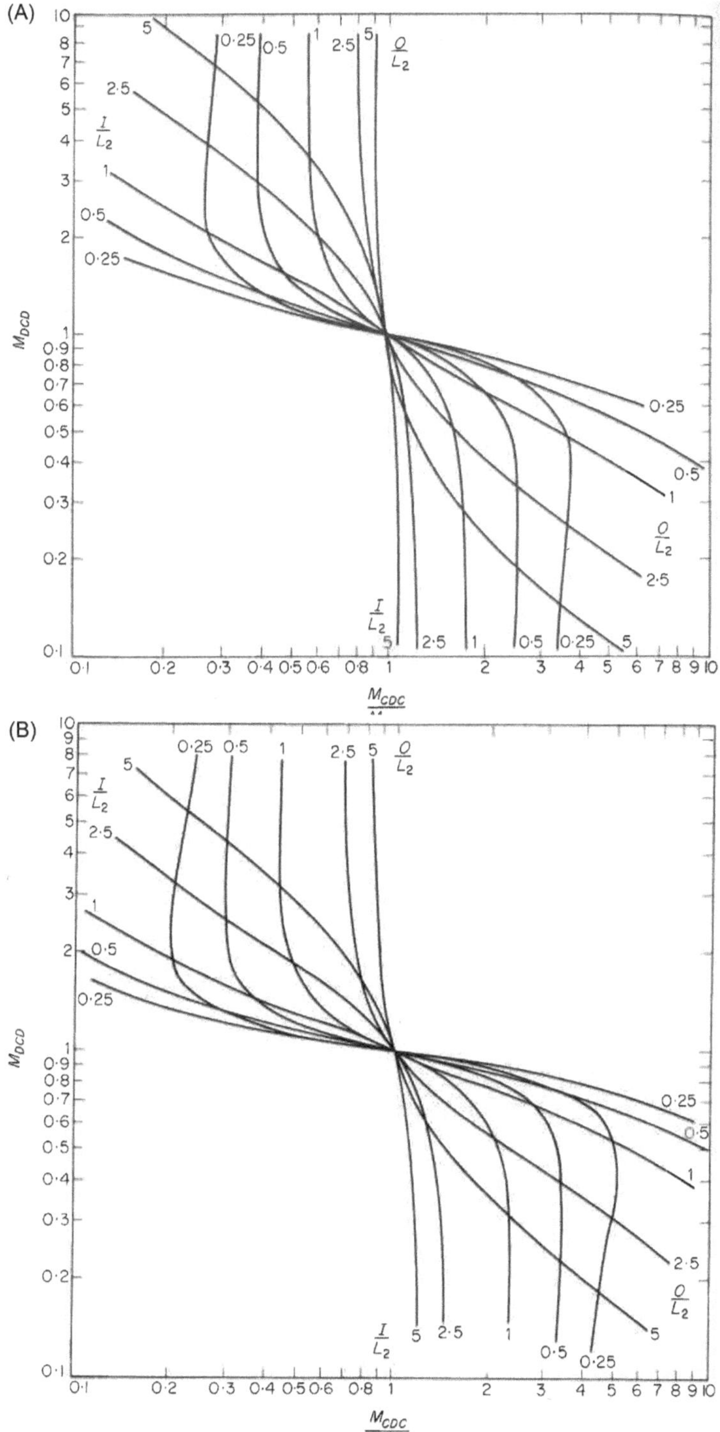

Fig. 26 Magnifications in the symmetric triplet. $L_2 = 2L_1$. (A) $s/L_2 = 0$; (B) $s/L_2 = \frac{1}{2}$; (C) $s/L_2 = 2$; (D) $s/L_2 = 5$. (After Brown & Hindmarsh, 1965.)

(Continued)

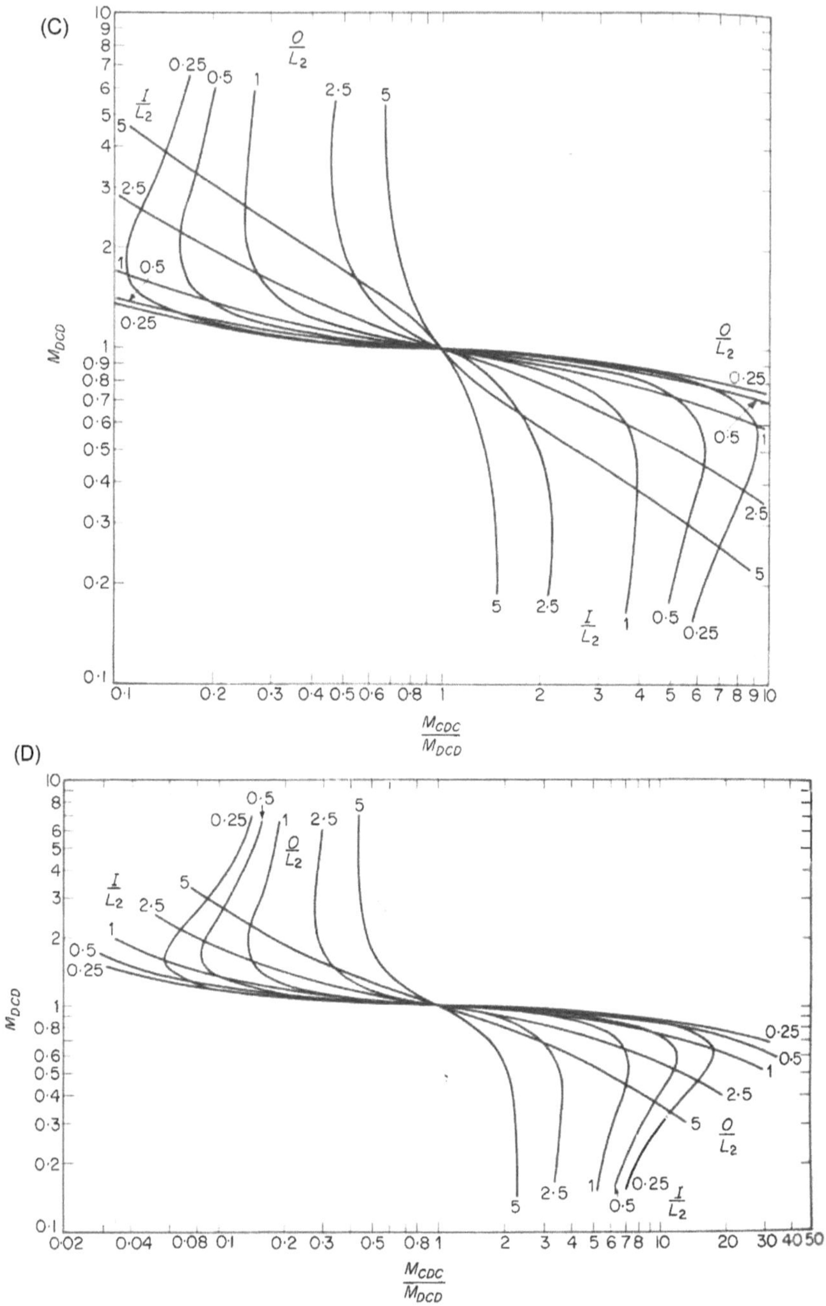
(C)
(D)
$\frac{O}{L_2}$
$\frac{I}{L_2}$
M_{DCD}
$\frac{M_{CDC}}{M_{DCD}}$
0·25
0·5
1
2·5
5
0·1
0·2
0·3
0·4 0·5 0·6
0·8
2
3
4 5 6 7 8 9 10
0·02
0·04
0·08
20
30 40 50

Fig. 26—Cont'd

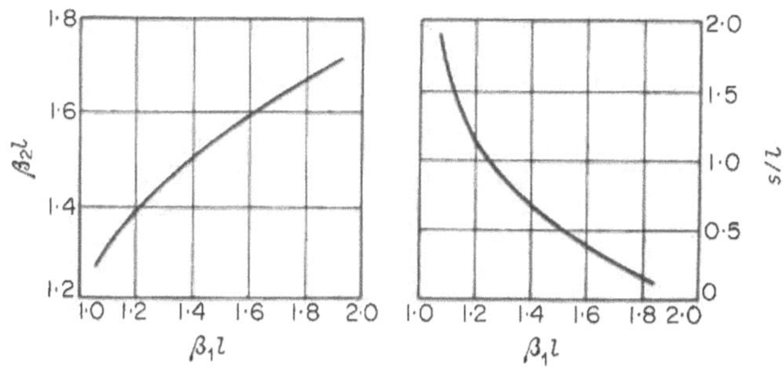

Fig. 27 The symmetric triplet. The relation between $\theta_2(\beta_2 l)$ and $\theta_1(\beta_1 l)$ and between object or image distance (s) and θ_1 for stigmatic imagery. The quadrupoles are of equal length (l) and in contact. (After Bauer, 1965–66.)

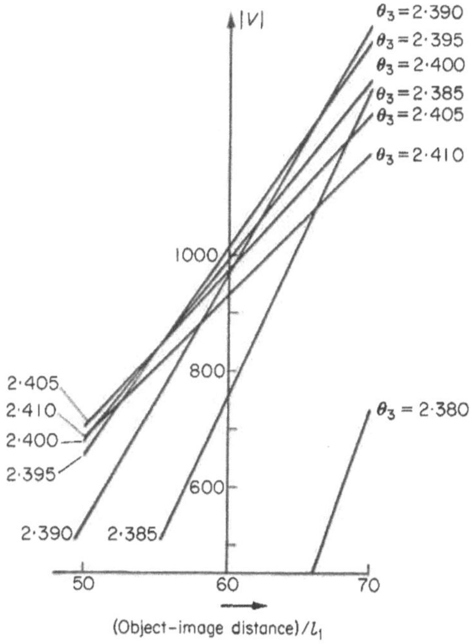

Fig. 28 The triplet objective of Antony-Spies and Junior (1967). Optimization of the lens with respect to magnification (V) for object-image distance $= 60l_1$, $l_2 = l_1$, $l_3 = 2l_1$. (l denotes lens length and $kl = \theta$.).

agrees with that calculated, and is not far discrepant for the octopole in the mid-position. It could not be determined for the outer octopole as it would have been excessively large; the condition is not, however, of practical importance.

Table 32 The triplet of air-cored quadrupoles (Foster, 1968); aperture aberration coefficients.

System data: Triplet of air-cored solenoid quadrupoles with various octopoles superimposed.

Bore-radius: 6·5 cm	solenoid windings: 100 turns/cm
Lens separation: 30 cm	solenoid length: 6 cm
Source-screen: 128 cm	solenoid radius: 1·7 cm
Beam momentum: 300 gauss-cm (8 kV)	

Systems considered:

A: basic system, three four-pole quadrupoles (no intermediate image, Fig. 13A in Definitions).
B: system A, together with three octopoles, one at the center, one at an outer quadrupole and one midway between them.
C: basic system, three eight-pole quadrupoles.
D: as system A (forming an intermediate image, Fig. 13B).
E: system D with octopoles.
F: as system A, calculations based on finite solenoid field instead of point-pole approximation.
G: three 12-pole elements, each representing a 4-pole quadrupole and an 8-pole quadrupole-plus-octopole.

Third- and fifth-order aberration coefficients (cm)

System	C_3	A_3	$(C+D)_3$	$(A+B)_3$	(Rosette)$_3$	(Star)$_3$	(Spherical)$_3$
A	147	11,408	1947	13,208	994	1408	4783
B	0	0	2	1	—	—	—
C	147	11,407	1946	13,210	994	1408	4783
D	316	35,189	3439	38,311	3658	4359	14,095
E	−1	−10	3	− 1	—	—	—
F	165	11,661	2087	13,575	999	1436	4914
G	146	11,407	1947	13,210	993	1408	4783

	$A_5 \times 10^{-6}$	$B_5 \times 10^{-6}$	$C_5 \times 10^{-6}$	$D_5 \times 10^{-6}$
A	2.4	1.9	8.0	14.9
B	2.4	1.9	8.0	14.9
C	−2.5	−2.0	−8.3	−15.5
D	6.5	6.6	32.8	65.6
E	6.4	6.6	32.8	65.9
F	2.0	1.7	6.8	12.8
G	0	0	−0.2	−0.3

The excitations required to cancel star and rosette simultaneously could not be determined experimentally. The spherical aberration could likewise not be determined."

4. The quadrupole quadruplet

Only quadrupole quadruplets with some degree of symmetry have so far been studied in any detail, and it does not seem likely that combinations as complex as quadruplets could be used in practice without some simplifying symmetry. Two classes of quadruplets have been extensively studied: the symmetric and the antisymmetric configurations. It is convenient to subdivide the latter into two further groups: antisymmetric quadruplets in which the quadrupole excitations in a principal section alternate in sign and those in which the first two excitations have the same sign and the second two also have the same sign, opposite to that of the first pair (Figs. 29A–D). It is convenient to refer to these as the *CDCD* configuration and the *CCDD* configurations, where the letters *C* and *D* indicate the order in which electrons in a principal section encounter converging and diverging elements. The symmetric quadruplet will of course be *CDDC*.

Some other quadruplets have been studied, notably by Regenstreif (1967c) and Meads (see Hawkes, 1966b). Meads has considered the CDDC quadruplet in which the excitations follow the pattern $(\beta_1^2, -\beta_2^2, -\beta_1^2, \beta_2^2)$ and the spacings are those shown in Fig. 29E; this will automatically be orthomorphic, as Dymnikov et al. (1963) had already observed, because if, in the x–z plane, the magnifications of the first and second doublets are $M^{(1)}$ and $M^{(2)}$, those on the y–z plane will be $M^{(2)}$ and $M^{(1)}$. Regenstreif (1967c) has considered a very wide range of combinations of four quadrupoles in an elegant attempt to obtain general conditions for "turn-over". This is directed in particular at accelerator engineers but becomes immediately relevant when we realize that the transfer matrices of Archard–Deltrap correctors are identical in form with those of turn-over matrices: both describe telescopic systems.

Meads has also considered quadruplets that are electrically but not geometrically antisymmetric but these were not required to be regular.

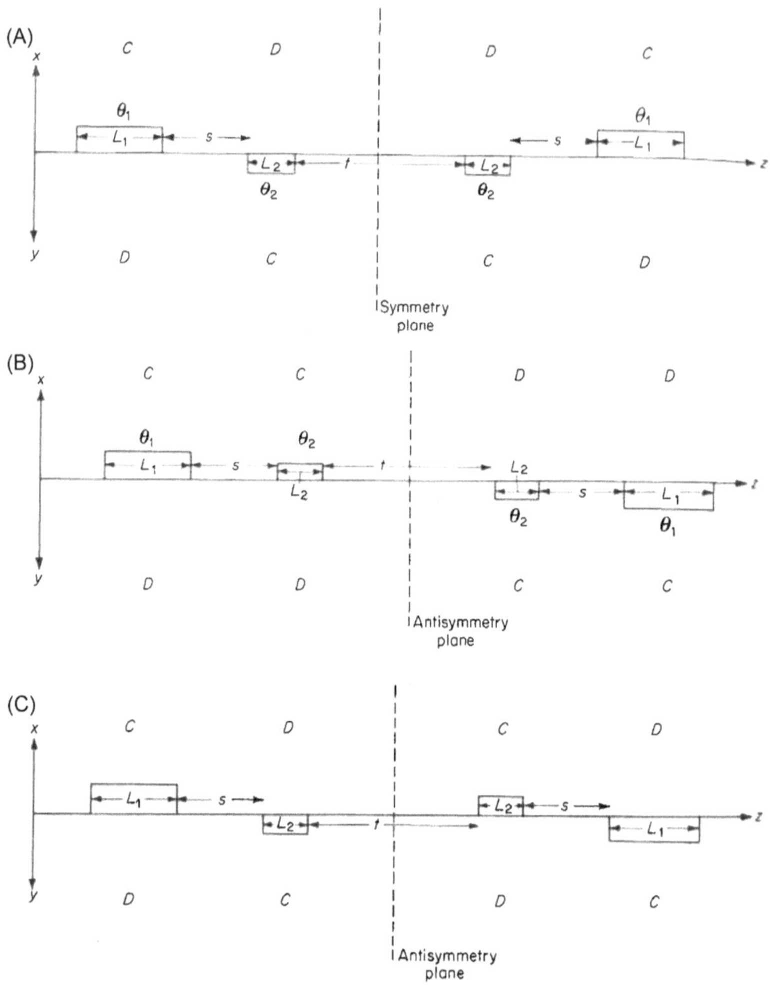

Fig. 29 The notation employed in connection with quadruplets. (A) The symmetric quadruplet (rectangular model). (B) The antisymmetric quadruplet, *CCDD* configuration (rectangular model). (C) The antisymmetric quadruplet, *CDCD* configuration (rectangular model). (D) The antisymmetric quadruplet, *CDCD* configuration (bell-shaped model). (E) Meads' doublet pair.

(Continued)

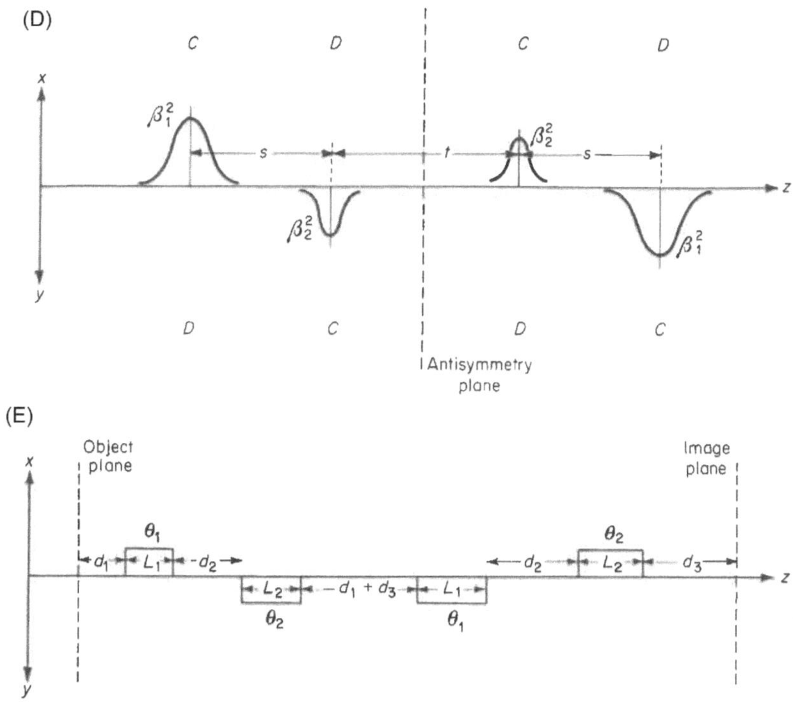

Fig. 29—Cont'd

5. The symmetric quadruplet

Dhuicq (1961) gives formulae for the cardinal elements of a symmetric quadruplet, using the rectangular model, and the conditions for regularity; these are highly complex and Dhuicq therefore considers a number of special cases: (i) all quadrupoles equal in length, $\beta_2^2 = -\beta_1^2$; (ii) all quadrupoles equal in length, outer lenses in contact; (iii) outer lenses in contact, inner quadrupoles different in length from outer quadrupoles. A particular system is then studied experimentally and theoretically. The results of this work are summarized in *QO* and briefly in Dhuicq (1968a) and we shall therefore not reproduce them here.

6. The antisymmetric quadruplet

This has been studied using both the rectangular and the bell-shaped models. We first consider the *CDCD* and the *CCDD* configurations using the rectangular model, and then consider both configurations with the bell. Finally, telescopic systems are studied using the rectangular model.

6.1 Rectangular model, CDCD configuration

The Gaussian properties of the antisymmetric quadruplet were first studied by the Leningrad group, and are conveniently collected together in Dymnikov, Fishkova, and Yavor (1965a); many of their results are reproduced in QO, but because of the importance of this combination we repeat some of these here as well. The notation employed is illustrated in Fig. 29C. In any antisymmetric system, the focal lengths f and g are automatically equal and if $z_{Fi} = z_{Gi}$ then $z_{Fo} = z_{Go}$; in consequence, only one condition need be satisfied for regularity. For a given geometry, therefore, the system will be regular for a range of pairs of values of θ_1 and θ_2; the curve $\theta_1 = f(\theta_2)$ is called the *load characteristic*. Dymnikov et al. (1965a) plot the load characteristic for two geometries: $L_1 = L_2 = L$; $s/L = 0$; $t/L = 1$ and 5 (Fig. 30A). They have investigated a wider range than this, however, and their subsequent figures cover the range $\frac{1}{4} \leq L_2/L_1 \leq 4; 0 \leq s/L_i \leq 1; 0 \leq t/L_i \leq 10$ where L_i is the shorter of the lens-lengths. Fig. 30B, C illustrate the variation in position of the focus and the focal length for different geometries and various branches of the load characteristic. We refer to the original paper for a full account of this work. Correction of the aperture aberrations with the aid of octopoles is described in Fishkova et al. (1968b) and in Yavor, Fishkova, and Baranova (1968).

Dhuicq (1968a) has also studied this system, in the range $s/L_1 = 0, \frac{1}{2}$ and 1, $L_2/L_1 = \frac{1}{4}, \frac{1}{2}, 1, 2$ and 4 and sought combinations for which $0 \leq t/L_1 \leq 5, 0 \leq z_F/L_1 \leq 3, 0 \leq f/L_1 \leq 3$ where z_F is the position of the image focus measured from the end of the last lens. Fig. 31A–C show "contours" of $t/L_1 = \text{const}$, $f/L_1 = \text{const}$ and $z_F/L_1 = \text{const}$ for $s = 0, L_2/L_2 = \frac{1}{4}$ and $s = 0$, $L_2/L_1 = 2$.

Dhuicq then studies the aberrations of a particular quadruplet, for which $s = 0$, $L_2 = 2L_1$, $\theta_1 = 1 \cdot 95$, $\theta_2 = 4 \cdot 5$, $t/L_1 = 0 \cdot 155$ and hence $f/L_1 = 0 \cdot 0408$ and $z_F/L_1 = 0 \cdot 267$. The aperture aberrations are (30) = 39f, (03) = 141,100f. (12) = 73f and (21) = 70f (electrostatic case) and (30) = 12f, (03) =48,920f, (12) = 558f and (21) = 544f; the slight inequality between (12) and (21) is the result of residual astigmatism. These aberrations are corrected by placing octopoles in the third and fourth lenses (Fig. 32); using Dhuicq's notation, we write $\Omega = 2(L_1/a)^4 V/\Phi_0$ or $\Omega = -4\eta(L_1/a)^4 \mu_0 NI/\Phi_0^{1/2}$ in which V and NI are the octopole voltages or currents (amp-turns). Denoting the octopoles in the third quadrupole

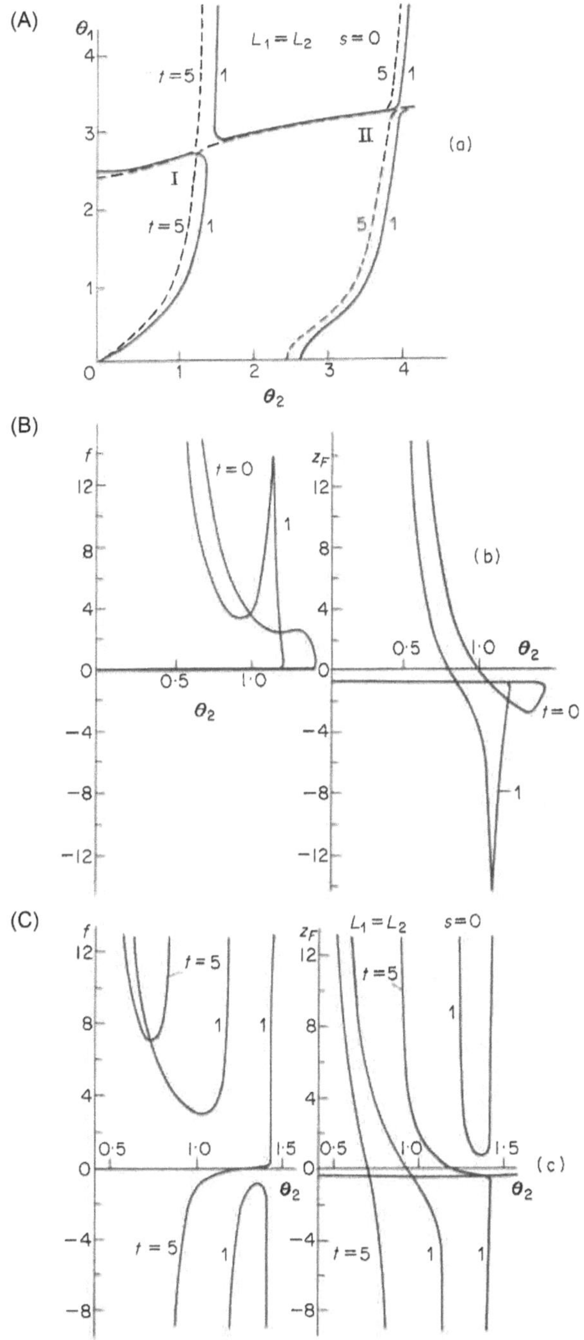

Fig. 30 (A) The load-characteristic for the antisymmetric quadruplet, *CDCD* configuration, rectangular model. (B) Focal length and focal distance, $L_2 = L_1/2$, $s = 0$. (C) Focal length and focal distance, $L_2 = L_1$, $s = 0$. (After Dymnikov et al., 1965a.)

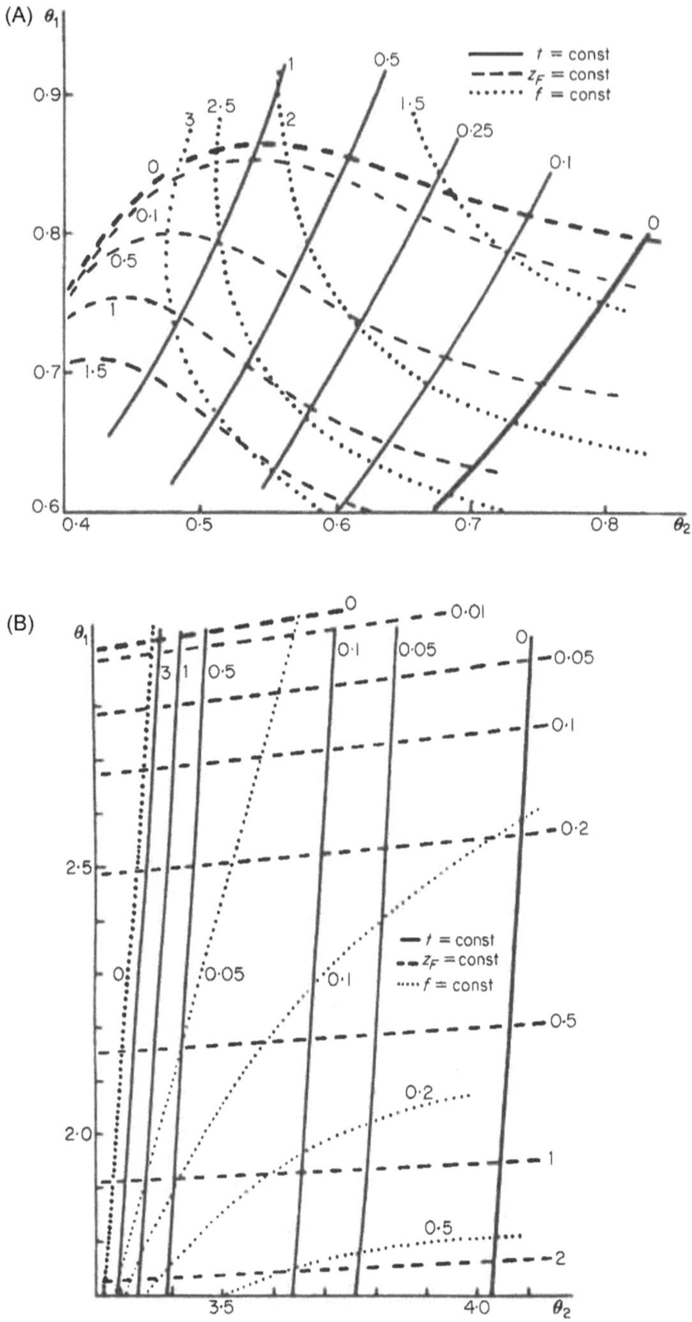

Fig. 31 The antisymmetric quadruplet, rectangular model, *CDCD* configuration. Contours of equal t, equal focal distance and equal focal length in the $\theta_1 - \theta_2$ plane; $L_1 = 4L_2$. (A) Weak lenses, $L_1 = 4L_2$; (B) Strong lenses, $L_1 = 4L_2$; (C) $L_2 = 2L_1$. (After Dhuicq, 1968a.)

(Continued)

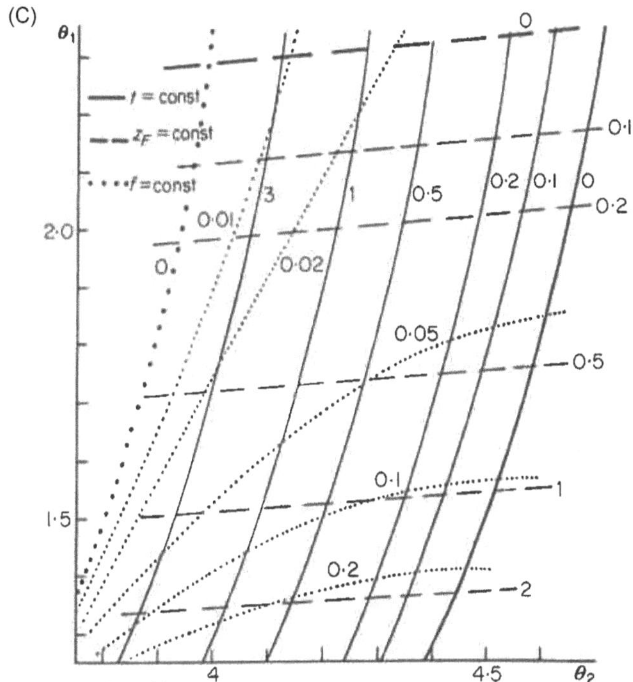

Fig. 31—Cont'd

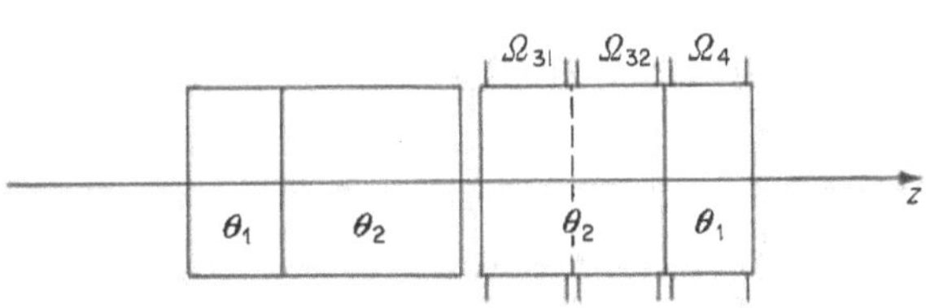

Fig. 32 The positions of the octopoles in Dhuicq's system.

by Ω_{31} and Ω_{32} and the octopole in the fourth lens by Ω_4, Dhuicq finds that the aperture aberrations are corrected when $\Omega_{31} = 98 \cdot 18$, $\Omega_{32} = -33 \cdot 92$, $\Omega_4 = 18 \cdot 14$ in the electrostatic case, and $\Omega_{31} = 55 \cdot 04$, $\Omega_{32} = -32 \cdot 44$, $\Omega_4 = 6 \cdot 35$ in the magnetic case (see also Dhuicq & Septier, 1968). Fig. 33 shows how these are arrived at.

6.2 Rectangular model, CCDD configuration

This has been studied by Vosburgh (1967) and the results are summarized in Kawakatsu, Vosburgh, and Siegel (1968a). Vosburgh observed that the focal length remains small and varies little over the branch of the load characteristic nearest to the origin (Fig. 34A,B) and both the distortion and chromatic

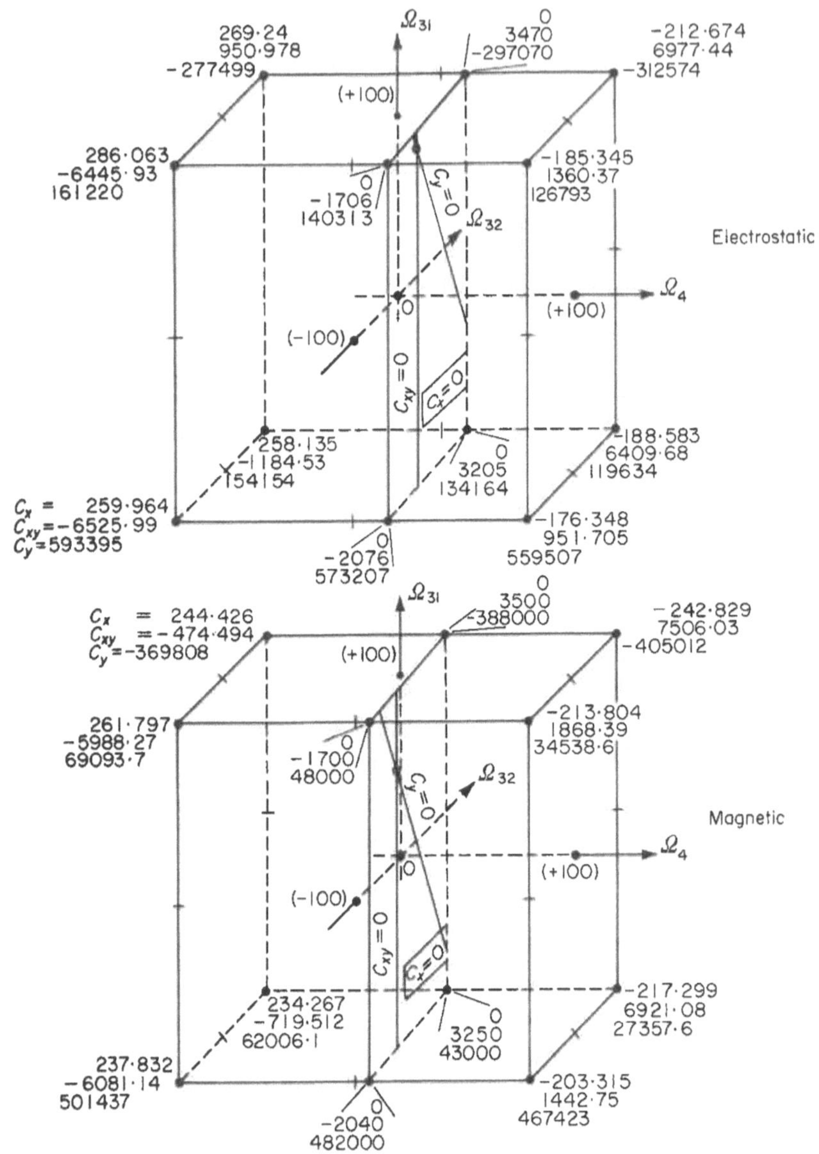

Fig. 33 The octopole strengths that cancel the various aperture aberration coefficients in Dhuicq's system. (Dhuicq, 1968a.)

aberration coefficients are also acceptable (Fig. 35A–D). These curves all correspond to quadrupoles of equal length, equally spaced at 4·7 cm center-to-center; the poles (in the form of cylindrical segments, of radius 1·15*a*) are 3·0 cm long and the bore-radius (*a*) is 1·5 cm. If the rectangular model is used, the quadrupoles will therefore be in contact.

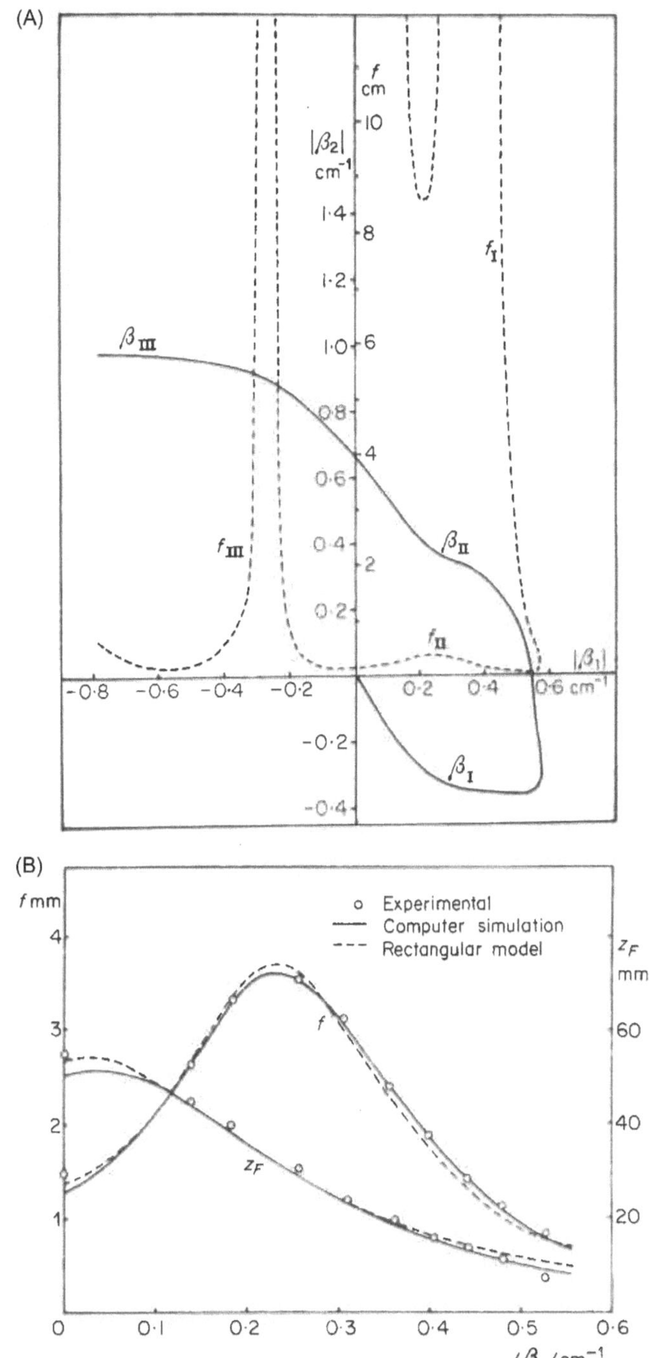

Fig. 34 The antisymmetric quadruplet, *CDCD* and *CCDD* configurations, rectangular model. (A) The load characteristic in the $\beta_1 - \beta_2$ plane together with the corresponding focal length. (Unscaled, $L = 4{\cdot}7$ for all four quadrupoles.) (B) CCDD configuration. Experimental and computed focal length and focal distance from the terminal end-face; "computer-simulation" corresponds to the sech field of Fig. 15 in Definitions. (After Vosburgh, 1967 and Kawakatsu et al., 1968a. With the permission of AIP Publishing.)

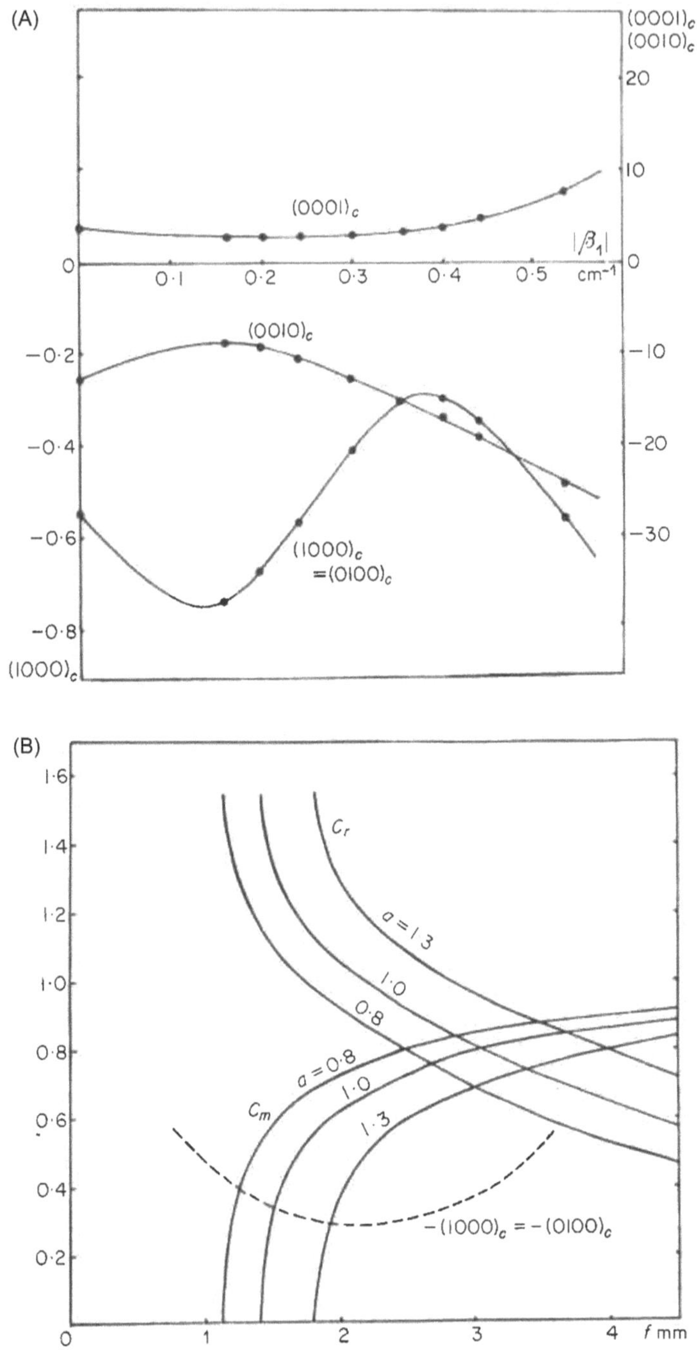

Fig. 35 (A) Chromatic aberration coefficients of Vosburgh's quadruplet (*CCDD*) as a function of β_1. The coefficients are defined by $z_F - z_G = f\{(0010)_c - (0001)_c\}\Delta\Phi_0/\Phi_0$ and $\Delta M/M = (1000)_c\Delta\Phi/\Phi$. (B) Comparison of the results of (A) with round lens data. a = half-width of Glaser's bell-shaped field, C_m = chromatic aberration of magnification and C_r = chromatic aberration of image rotation. (C) Distortion of Vosburgh's quadruplet projector as a function of β_1. (D) Comparison of the results of Fig. 35C with data obtained by Liebmann (1952). (C_d = radial distortion and C_{SD} = anisotropic or spiral distortion.) (After Vosburgh, 1967 and Kawakatsu et al., 1968a.)

(Continued)

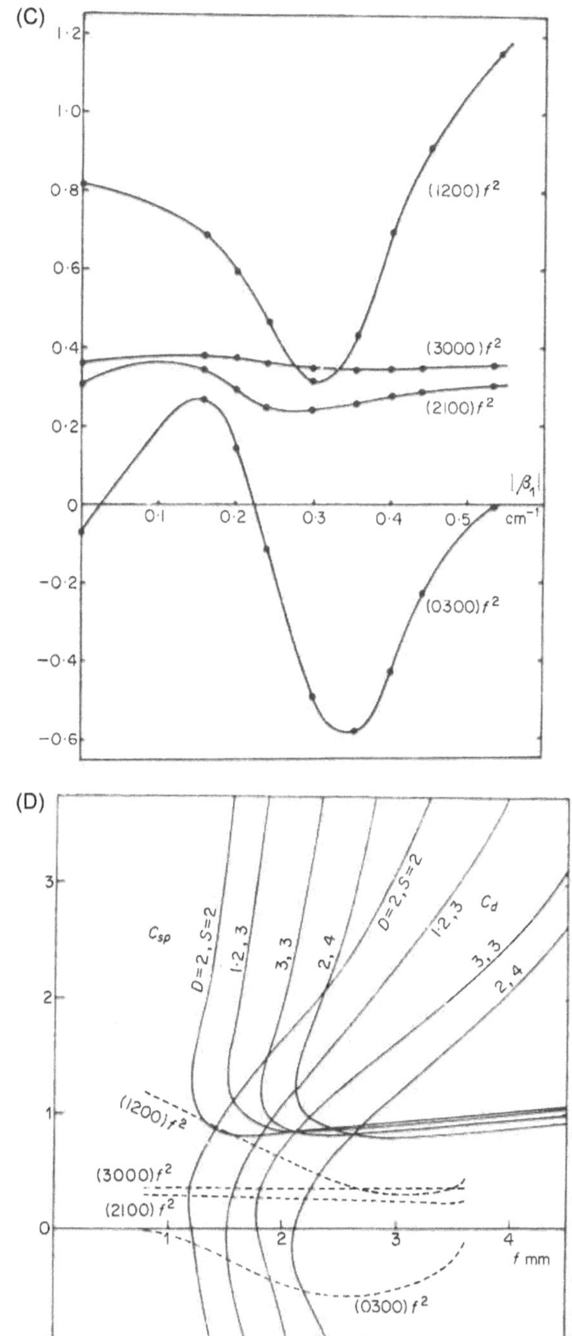
(C)
1·2
1·0
0·8
0·6
0·4
0·2
0
−0·2
−0·4
−0·6
0·1
0·2
0·3
0·4
0·5
cm⁻¹
|β₁|
(1200) f²
(3000) f²
(2100) f²
(0300) f²
(D)
3
2
1
0
−1
1
2
3
4
f mm
C_sp
D=2, S=2
1·2, 3
3, 3
2, 4
D=2, S=2
1·2, 3
C_d
3, 3
2, 4
(1200) f²
(3000) f²
(2100) f²
(0300) f²

Fig. 35—Cont'd

6.3 The bell-shaped model

We have made a computer study of the quadrupole quadruplet, based on the bell-shaped model and the asymptotic approximation; by this, we mean that the rays are assumed to reach their asymptotes after leaving one quadrupole and before entering the next. Although this cannot of course be strictly true, and is scarcely realistic when the lenses are close together, the fact that Kawakatsu et al. (1968a) found that the rectangular model gave such an accurate indication of the behavior of their quadruplets suggests that the approximation is a reasonable one. Nevertheless, we have not pursued our calculations in great detail where the behavior of the quadruplet is very sensitive to small changes in the parameters, since the predictions will be least reliable here.

Fig. 36A–D show the load characteristic for four geometries. We also indicate (dotted lines) the telescopic characteristic, for which the quadruplet

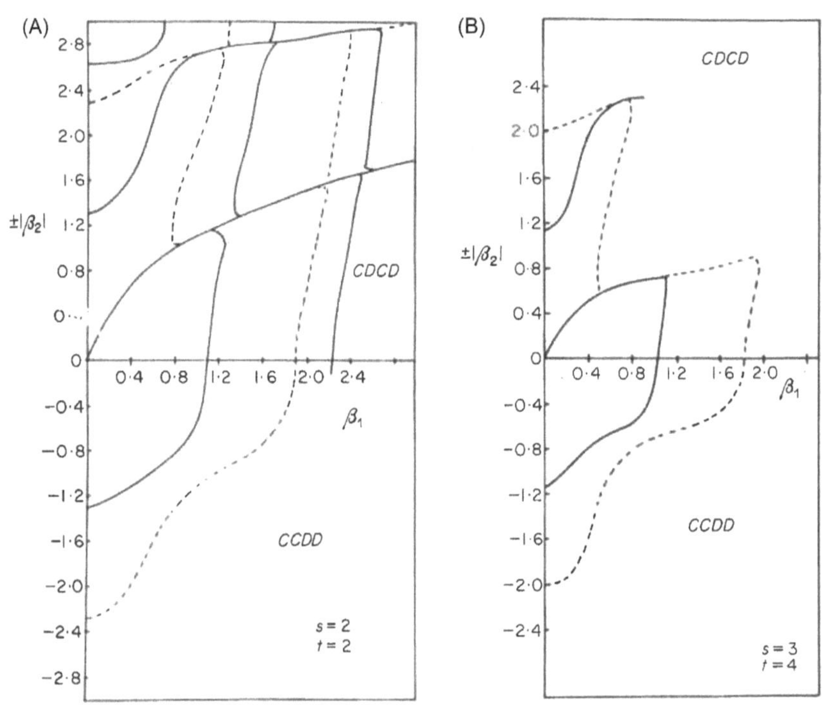

Fig. 36 The antisymmetric quadruplet: bell-shaped model. The load characteristic for four geometries. Full lines correspond to regular quadruplets, broken lines to the telescopic case for which *f* and *g* are infinite. Where the two curves meet, the quadruplet behaves like an Archard–Deltrap corrector.

(Continued)

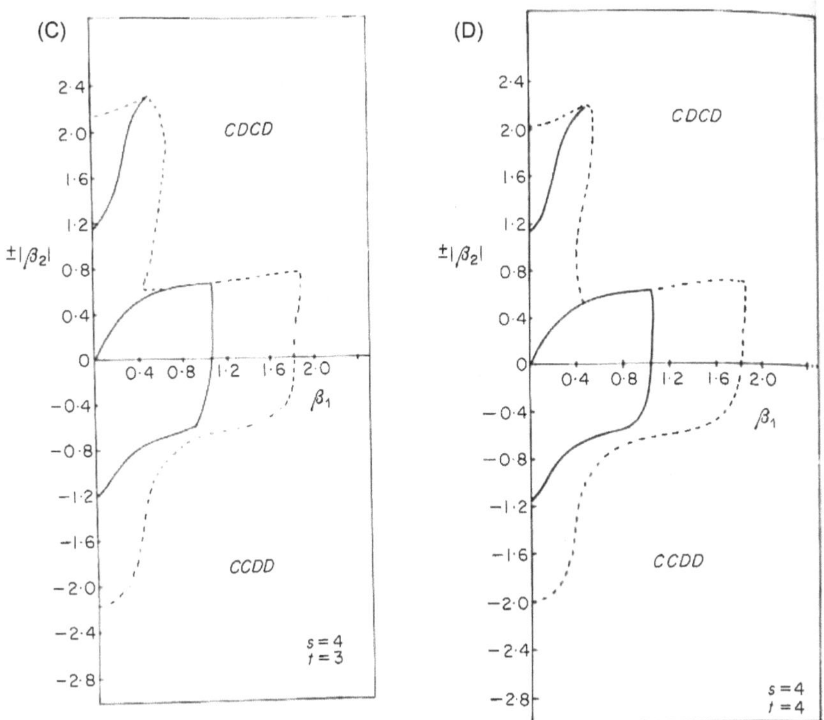

Fig. 36—Cont'd

is afocal; where this intersects the load characteristic, the system will be suitable for use as an Archard–Deltrap corrector, in principle at least–rays have then to be traced to discover whether there are acceptable octopole positions. The cardinal elements and aberrations are listed in Tables 33–36. We have also derived the coefficients of the aberration polynomials, but these are not listed here.

6.4 The rectangular model: telescopic systems

Möller, Dhuicq, and Septier (1966a, 1966b) give contours in the $\theta_1 - \theta_2$ plane of $s/L_1 = \text{const}$, $t/L_1 = \text{const}$ and $b/2 = \text{const}$ such that the transfer matrix has the form $\begin{pmatrix} 1 & -b \\ 0 & 1 \end{pmatrix}$ for $L_2/L_1 = 2$ and they state that the curves have the same general form for $L_2/L_1 = \frac{1}{4}, \frac{1}{2}, 1$ and 4 (Fig. 37A). They also give similar contours for the matrix $\begin{pmatrix} -1 & b \\ 0 & -1 \end{pmatrix}$ for $L_2/L_1 = 4$ (Fig. 37B).

Table 33 The regular antisymmetric quadruplet: The bell-shaped model, $s = 2$, $t = 2$. The aperture aberrations are listed for high magnification.

s = 2, *t* = 2, *CDCD*				Electrostatic		
β_1	β_2	f	z_F	(30)	(03)	(12) = (21)
1.000⋆-1	1.732⋆-1	1.656⋆3	1.653⋆3	1.721⋆10	1.721⋆10	−7.369⋆9
2.000⋆-1	3.450⋆-1	1.058⋆2	1.028⋆2	4.551⋆6	4.474⋆6	−1.902⋆6
3.000⋆-1	5.095⋆-1	2.287⋆1	1.977⋆1	5.093⋆4	4.265⋆4	−1.853⋆4
4.000⋆-1	6.552⋆-1	8.994	5.686	4.371⋆3	1.895⋆3	−1.043⋆3
5.000⋆-1	7.731⋆-1	5.349	1.611	1.797⋆3	1.548⋆2	−1.899⋆2
6.000⋆-1	8.642⋆-1	4.304	−2.110⋆-1	2.405⋆3	2.781⋆1	−6.569⋆1
7.000⋆-1	9.358⋆-1	4.332	−1.556	7.758⋆3	9.879⋆1	−5.933⋆1
8.000⋆-1	9.953⋆-1	5.278	−3.213	5.334⋆4	6.969⋆3	−1.583⋆3
9.000⋆-1	1.048	8.058	−6.429	9.015⋆5	4.032⋆5	−6.334⋆4
1.100	1.142	−2.020⋆2	2.031⋆2	3.451⋆12	3.492⋆12	−4.560⋆11
1.200	1.187	3.668⋆1	−3.571⋆1	1.174⋆10	1.141⋆10	−1.615⋆9
1.100	2.094⋆-1	6.249⋆-2	5.211⋆-1	1.871	−1.586⋆-1	−5.003⋆-1
1.150	5.412⋆-1	5.121⋆-2	5.997⋆-1	2.314	−1.680⋆-1	−4.988⋆-1
1.200	7.873⋆-1	4.434⋆-2	6.739⋆-1	2.834	−1.750⋆-1	−5.101⋆-1
1.250	1.020	4.933⋆-2	7.371⋆-1	3.432	−1.631⋆-1	−5.866⋆-1
1.000⋆-1	1.334	1.027⋆-1	−1.140	−2.192⋆-1	4.051	−3.911⋆-1
2.000⋆-1	1.384	8.577⋆-2	−1.171	−2.198⋆-1	3.082	−3.983⋆-1
3.000⋆-1	1.472	6.009⋆-2	−1.195	−2.184⋆-1	1.873	−4.361⋆-1
4.000⋆-1	1.629	3.325⋆-2	−1.169	−2.162⋆-1	9.403⋆-1	−5.568⋆-1
5.000⋆-1	1.841	1.425⋆-2	−1.056	−1.898⋆-1	5.697⋆-1	−8.145⋆-1
6.000⋆-1	2.114	5.225⋆-3	−8.681⋆-1	−4.480⋆-2	6.992⋆-1	−1.199
7.000⋆-1	2.422	2.565⋆-3	−7.093⋆-1	−1.784	2.863	−1.888
8.000⋆-1	2.609	3.941⋆-3	−8.238⋆-1	−3.298⋆1	−5.120⋆1	−5.455
9.000⋆-1	2.673	9.537⋆-3	−1.237	1.321⋆2	2.160⋆4	−2.595⋆1
1.100	2.729	8.604⋆-2	−3.647	−9.177⋆3	−1.341⋆7	−1.867⋆3
1.200	2.749	7.481⋆-1	−1.057⋆1	−7.537⋆5	−7.510⋆9	−1.534⋆5

s = 2, *t* = 2, *CCDD*				Electrostatic		
β_1	β_2	f	z_F	(30)	(03)	(12) = (21)
1.000⋆-6	1.318	1.085⋆-1	−1.127	4.412	−2.186⋆-1	−3.907⋆-1
1.000⋆-1	1.302	1.143⋆-1	−1.113	4.791	−2.176⋆-1	−3.910⋆-1
2.000⋆-1	1.259	1.312⋆-1	−1.066	6.012	−2.130⋆-1	−3.962⋆-1
3.000⋆-1	1.197	1.569⋆-1	−9.738⋆-1	8.207	−1.997⋆-1	−4.129⋆-1
4.000⋆-1	1.126	1.839⋆-1	−8.303⋆-1	1.109⋆1	−1.740⋆-1	−4.430⋆-1
5.000⋆-1	1.055	2.017⋆-1	−6.455⋆-1	1.344⋆1	−1.418⋆-1	−4.805⋆-1
6.000⋆-1	9.846⋆-1	2.033⋆-1	−4.387⋆-1	1.352⋆1	−1.154⋆-1	−5.157⋆-1
7.000⋆-1	9.121⋆-1	1.881⋆-1	−2.272⋆-1	1.070⋆1	−1.033⋆-1	−5.376⋆-1
8.000⋆-1	8.293⋆-1	1.606⋆-1	−2.182⋆-2	6.378	−1.060⋆-1	−5.413⋆-1
9.000⋆-1	7.207⋆-1	1.267⋆-1	1.723⋆-1	2.902	−1.194⋆-1	−5.310⋆-1
9.500⋆-1	6.462⋆-1	1.093⋆-1	2.645⋆-1	1.956	−1.284⋆-1	−5.230⋆-1
1.050	3.832⋆-1	7.648⋆-2	4.389⋆-1	1.574	−1.485⋆-1	−5.065⋆-1

Magnetic			Achromatic		
(30)	(03)	(12) = (21)	(30)	(03)	(12) = (21)
4.918⋆9	4.917⋆9	1.473⋆10	7.377⋆9	7.375⋆9	2.209⋆10
1.299⋆6	1.277⋆6	3.756⋆6	1.949⋆6	1.917⋆6	5.642⋆6
1.450⋆4	1.214⋆4	3.487⋆4	2.179⋆4	1.824⋆4	5.268⋆4
1.243⋆3	5.383⋆2	1.777⋆3	1.869⋆3	8.097⋆2	2.717⋆3
5.125⋆2	4.415⋆1	2.874⋆2	7.695⋆2	6.628⋆1	4.464⋆2
6.889⋆2	7.698	1.053⋆2	1.032⋆3	1.172⋆1	1.623⋆2
2.230⋆3	2.999⋆1	1.578⋆2	3.335⋆3	4.375⋆1	2.301⋆2
1.537⋆4	2.034⋆3	1.854⋆3	2.297⋆4	3.021⋆3	3.000⋆3
2.606⋆5	1.168⋆5	4.677⋆4	3.887⋆5	1.741⋆5	8.347⋆4
1.007⋆12	1.019⋆12	1.802⋆11	1.496⋆12	1.514⋆12	3.923⋆11
3.452⋆9	3.353⋆9	4.693⋆8	5.109⋆9	4.964⋆9	1.164⋆9
8.365⋆-1	−2.702⋆-1	−2.366⋆-1	1.043	−2.479⋆-1	−1.486⋆-1
9.637⋆-1	−2.770⋆-1	−1.919⋆-1	1.234	−2.552⋆-1	−8.960⋆-2
1.116	−2.837⋆-1	−1.471⋆-1	1.460	−2.620⋆-1	−2.607⋆-2
1.295	−2.872⋆-1	−9.895⋆-2	1.722	−2.624⋆-1	6.361⋆-2
−3.076⋆-1	1.479	−7.714⋆-2	−2.899⋆-1	1.994	2.751⋆-2
−3.288⋆-1	1.124	−1.117⋆-1	−3.070⋆-1	1.516	−1.618⋆-2
−3.734⋆-1	6.500⋆-1	−1.658⋆-1	−3.424⋆-1	8.946⋆-1	−7.565⋆-2
−4.408⋆-1	2.314⋆-1	−2.469⋆1	−3.959⋆-1	3.732⋆-1	−1.436⋆-1
−4.682⋆-1	5.591⋆-3	−4.222⋆-1	−4.132⋆-1	1.184⋆-1	−2.914⋆-1
−3.761⋆-1	−6.143⋆-3	−7.579⋆-1	−2.465⋆-1	1.350⋆-1	−6.110⋆-1
−1.904	8.578⋆-1	−1.310	1.081	1.034	−1.117
−5.624⋆1	−2.359⋆1	−3.238	−2.660⋆1	−3.531⋆1	−2.498
1.100⋆2	2.662⋆4	−1.112⋆1	−2.757⋆2	−1.380⋆4	−6.179
−7.414⋆3	−1.548⋆7	−2.423⋆2	−1.450⋆4	−4.585⋆7	2.993⋆2
−1.132⋆6	−8.698⋆9	9.674⋆4	−1.103⋆6	−1.265⋆10	1.801⋆5

Magnetic			Achromatic		
(30)	(03)	(12) = (21)	(30)	(03)	(12) = (21)
1.608	−3.014⋆-1	−6.553⋆-2	2.169	−2.849⋆-1	4.287⋆-2
1.741	−2.958⋆-1	−5.403⋆-2	2.351	−2.801⋆-1	5.831⋆-2
2.159	−2.813⋆-1	−2.137⋆-2	2.930	−2.677⋆-1	1.036⋆-1
2.882	−2.639⋆-1	2.039⋆-2	3.947	−2.510⋆-1	1.648⋆-1
3.790	−2.486⋆-1	3.886⋆-2	5.251	−2.337⋆-1	1.995⋆-1
4.490	−2.377⋆-1	−2.180⋆-3	6.280	−2.186⋆-1	1.573⋆-1
4.469	−2.323⋆-1	−1.038⋆-1	6.278	−2.089⋆-1	3.348⋆-2
3.569	−2.324⋆-1	−2.258⋆-1	4.994	−2.066⋆-1	−1.219⋆-1
2.235	−2.375⋆-1	−3.176⋆-1	3.064	−2.112⋆-1	−2.431⋆-1
1.173	−2.462⋆-1	−3.474⋆-1	1.518	−2.208⋆-1	−2.862⋆-1
9.822⋆-1	−2.516⋆-1	−3.372⋆-1	1.097	−2.270⋆-1	−2.753⋆-1
7.543⋆-1	−2.637⋆-1	−2.782⋆-1	9.182⋆-1	−2.407⋆-1	−2.021⋆-1

Table 34 The regular antisymmetric quadruplet: the bell-shaped model, $s = 3$, $t = 4$. The aperture aberrations are listed for high magnification.

$s = 3, t = 4$, *CDCD*				Electrostatic		
β_1	β_2	f	z_F	(30)	(03)	(12) = (21)
1.000⋆-1	1.579⋆-1	5.058⋆2	5.008⋆2	1.079⋆8	1.076⋆8	−4.578⋆7
2.000⋆-1	3.099⋆-1	3.538⋆1	3.020⋆1	3.958⋆4	3.350⋆4	−1.376⋆4
3.000⋆-1	4.359⋆-1	1.041⋆1	4.550	1.661⋆3	4.342⋆2	−2.486⋆2
4.000⋆-1	5.225⋆-1	6.921	−8.490⋆-1	1.407⋆3	5.426⋆1	−4.688⋆1
5.000⋆-1	5.784⋆-1	8.414	−5.065	1.157⋆4	8.785⋆2	3.556⋆2
6.000⋆-1	6.168⋆-1	3.523⋆1	−3.303⋆1	1.242⋆7	8.951⋆6	1.006⋆6
7.000⋆-1	6.458⋆-1	−1.045⋆1	1.201⋆1	3.208⋆5	6.248⋆5	3.250⋆4
8.000⋆-1	6.999⋆-1	−6.464⋆-1	7.849⋆-1	2.933⋆1	7.119⋆3	5.666⋆1
9.000⋆-1	6.913⋆-1	−2.914	3.852	1.988⋆4	5.175⋆4	1.317⋆3
1.100	7.293⋆-1	−5.245	5.882	1.967⋆6	2.134⋆6	3.060⋆4
1.120	7.304⋆-1	3.122	−2.503	3.055⋆5	2.851⋆5	1.457⋆3
1.120	6.041⋆-1	5.897⋆-2	5.192⋆-1	2.030	−1.187⋆-1	−6.143⋆-1
1.100	4.574⋆-1	4.115⋆-2	5.062⋆-1	1.867	−1.448⋆-1	−5.476⋆-1
1.000⋆-1	1.183	7.724⋆-2	−2.440	−1.552⋆-1	1.883	−4.599⋆-1
2.000⋆-1	1.332	5.051⋆-2	−2.494	−6.276⋆-2	1.406	−5.206⋆-1
3.000⋆-1	1.589	2.788⋆-2	−2.495	1.273⋆-1	1.913	−8.701⋆-1
4.000⋆-1	1.884	1.896⋆-2	−2.461	3.163⋆-1	8.366	−1.951
5.000⋆-1	2.102	2.419⋆-2	−2.646	2.752⋆-1	1.338⋆2	−6.081
6.000⋆-1	2.209	5.683⋆-2	−3.463	5.128	4.747⋆3	−3.625⋆1
7.000⋆-1	2.261	2.197⋆-1	−5.886	8.045⋆1	2.724⋆5	−5.758⋆2

$s = 2, t = 2$, *CCDD*				Electrostatic		
β_1	β_2	f	z_F	(30)	(03)	(12) = (21)
1.000⋆-2	1.137	8.896⋆-2	−2.409	2.192	−1.743⋆-1	−4.629⋆-1
1.000⋆-1	1.094	1.016⋆-1	−2.373	2.577	−1.859⋆-1	−4.757⋆-1
2.000⋆-1	9.865⋆-1	1.445⋆-1	−2.229	4.269	−1.799⋆-1	−5.568⋆-1
3.000⋆-1	8.682⋆-1	2.114⋆-1	−1.926	8.642	−9.081⋆-2	−7.385⋆-1
4.000⋆-1	7.798⋆-1	2.607⋆-1	−1.512	1.566⋆1	2.622⋆-2	−9.210⋆-1
5.000⋆-1	7.200⋆-1	2.642⋆-1	−1.086	2.129⋆1	7.869⋆-2	−9.999⋆-1
6.000⋆-1	6.751⋆-1	2.316⋆-1	−7.051⋆-1	2.105⋆1	6.336⋆-2	−9.636⋆-1
7.000⋆-1	6.344⋆-1	1.829⋆-1	−3.792⋆-1	1.501⋆1	1.536⋆-2	−8.576⋆-1
8.000⋆-1	5.886⋆-1	1.327⋆-1	−1.028⋆-1	7.470	−3.844⋆-2	−7.368⋆-1
9.000⋆-1	5.222⋆-1	8.851⋆-2	1.332⋆-1	2.623	−8.621⋆-2	−6.339⋆-1

Magnetic			Achromatic		
(30)	(03)	(12) = (21)	(30)	(03)	(12) = (21)
3.083⋆7	3.074⋆7	9.171⋆7	4.624⋆7	4.611⋆7	1.375⋆8
1.135⋆4	9.593⋆3	2.832⋆4	1.700⋆4	1.437⋆4	4.234⋆4
4.829⋆2	1.265⋆2	5.628⋆2	7.187⋆2	1.880⋆2	8.333⋆2
4.145⋆2	1.622⋆1	1.481⋆2	6.130⋆2	2.383⋆1	2.131⋆2
3.432⋆3	2.731⋆2	1.044⋆3	5.059⋆3	3.942⋆2	1.273⋆3
3.692⋆6	2.665⋆6	2.326⋆6	5.438⋆6	3.922⋆6	2.766⋆6
9.525⋆4	1.853⋆5	8.875⋆4	1.404⋆5	2.732⋆5	1.075⋆5
9.037	2.161⋆3	1.245⋆2	1.310⋆1	3.152⋆3	1.471⋆2
5.878⋆3	1.531⋆4	5.720⋆3	8.679⋆3	2.259⋆4	7.188⋆3
5.751⋆5	6.241⋆5	3.472⋆5	8.535⋆5	9.261⋆5	4.527⋆5
8.908⋆4	8.311⋆4	5.099⋆4	1.324⋆5	1.235⋆5	6.750⋆4
8.877⋆-1	−2.670⋆-1	−2.178⋆-1	1.116	−2.373⋆-1	−8.560⋆-2
8.374⋆-1	−2.692⋆-1	−2.449⋆-1	1.043	−2.443⋆-1	−1.440⋆-1
−2.701⋆-1	7.930⋆-1	−2.105⋆-1	−2.471⋆-1	1.011	−1.274⋆-1
−2.645⋆-1	5.366⋆-1	−1.897⋆-1	−2.242⋆-1	7.104⋆-1	−7.944⋆-2
−3.461⋆-1	5.983⋆-1	3.297⋆-2	−2.514⋆-1	8.613⋆-1	3.340⋆-1
−6.988⋆-1	3.038	6.709⋆-1	−4.948⋆-1	4.103	1.545
−1.656	5.368⋆1	3.659	−1.510	6.964⋆1	6.905
−3.655	1.911⋆3	3.178⋆1	−7.606⋆-1	2.485⋆3	5.445⋆1
1.355⋆1	1.087⋆5	5.629⋆2	−7.243⋆1	1.396⋆5	9.425⋆2

Magnetic			Achromatic		
(30)	(03)	(12) = (21)	(30)	(03)	(12) = (21)
9.297⋆-1	−2.755⋆-1	−1.979⋆-1	1.182	−2.553⋆-1	−1.095⋆-1
1.090	−2.810⋆-1	−1.780⋆-1	1.387	−2.620⋆-1	−7.872⋆-2
1.737	−2.917⋆-1	−9.470⋆-2	2.244	−2.693⋆-1	5.933⋆-2
3.252	−2.831⋆-1	−1.279⋆-3	4.330	−2.446⋆-1	2.445⋆-1
5.511	−2.528⋆-1	−6.939⋆-2	7.541	−1.970⋆-1	2.145⋆-1
7.212	−2.283⋆-1	−2.758⋆-1	1.003⋆1	−1.669⋆-1	−3.448⋆-2
7.020	−2.200⋆-1	−4.601⋆-1	9.825	−1.633⋆-1	−2.923⋆-1
5.042	−2.234⋆-1	−5.409⋆-1	7.035	−1.756⋆-1	−4.354⋆-1
2.642	−2.325⋆-1	−5.231⋆-1	3.608	−1.937⋆-1	−4.518⋆-1
1.112	−2.441⋆-1	−4.440⋆-1	1.414	−2.125⋆-1	−3.807⋆-1

Table 35 The regular antisymmetric quadruplet: the bell-shaped model: $s = 4$, $t = 3$. The aperture aberrations are listed for high magnification.

$s = 4, t = 3$, *CDCD*				Electrostatic		
β_1	β_2	f	z_F	(30)	(03)	(12) = (21)
1.000⋆-1	1.909⋆-1	2.653⋆2	2.598⋆2	1.618⋆7	1.606⋆7	−6.798⋆6
2.000⋆-1	3.653⋆-1	2.088⋆1	1.508⋆1	9.343⋆3	6.441⋆3	−2.705⋆3
3.000⋆-1	4.863⋆-1	8.019	1.153	1.171⋆3	1.022⋆2	−9.268⋆1
4.000⋆-1	5.544⋆-1	6.868	−2.751	2.787⋆3	5.212⋆1	1.404⋆1
5.000⋆-1	5.938⋆-1	1.010⋆1	−7.684	5.126⋆4	9.361⋆3	2.066⋆3
6.000⋆-1	6.194⋆-1	4.984⋆1	−4.821⋆1	1.119⋆8	9.221⋆7	5.272⋆6
7.000⋆-1	6.382⋆-1	−1.811⋆1	1.930⋆1	6.863⋆6	9.599⋆6	2.205⋆5
8.000⋆-1	6.532⋆-1	−1.118⋆1	1.210⋆1	3.377⋆6	4.793⋆6	3.614⋆4
9.000⋆-1	6.658⋆-1	−6.928⋆1	6.998⋆1	1.632⋆10	1.692⋆10	−1.070⋆8
1.060	2.081⋆-1	3.729⋆-2	4.434⋆-1	1.581	−1.400⋆-1	−5.416⋆-1
1.080	4.557⋆-1	4.217⋆-2	4.715⋆-1	1.723	−1.388⋆-1	−5.555⋆-1
1.000⋆-1	1.277	8.075⋆-2	−3.373	−1.398⋆-1	2.229	−4.385⋆-1
2.000⋆-1	1.552	3.977⋆-2	−3.470	1.665⋆-1	2.277	−6.407⋆-1
3.000⋆-1	1.935	2.273⋆-2	−3.620	1.158	1.370⋆1	−1.711
4.000⋆-1	2.189	3.568⋆-2	−4.381	6.828	5.894⋆2	−7.574
5.000⋆-1	2.289	1.447⋆-1	−7.313	1.119⋆2	6.653⋆4	−9.702⋆1
$s = 4, t = 3$, *CCDD*				Electrostatic		
β_1	β_2	f	z_F	(30)	(03)	(12) = (21)
1.000⋆-1	1.116	1.333⋆-1	−3.247	3.552	−2.126⋆-1	−4.503⋆-1
2.000⋆-1	9.472⋆-1	2.393⋆-1	−2.958	8.297	−1.497⋆-1	−5.994⋆-1
3.000⋆-1	8.110⋆-1	3.808⋆-1	−2.407	2.312⋆1	7.730⋆-2	−9.036⋆-1
4.000⋆-1	7.350⋆-1	4.394⋆-1	−1.787	4.631⋆1	2.376⋆-1	−1.198
5.000⋆-1	6.906⋆-1	4.051⋆-1	−1.240	6.206⋆1	2.457⋆-1	−1.324
6.000⋆-1	6.590⋆-1	3.258⋆-1	−7.902⋆-1	5.812⋆1	1.689⋆-1	−1.236
7.000⋆-1	6.304⋆-1	2.380⋆-1	−4.249⋆-1	3.842⋆1	7.287⋆-2	−1.029
8.000⋆-1	5.971⋆-1	1.597⋆-1	−1.261⋆-1	1.701⋆1	−1.111⋆-2	−8.218⋆-1
9.000⋆-1	5.451⋆-1	9.717⋆-2	1.227⋆-1	4.553	−7.519⋆-2	−6.679⋆-1
1.050	1.073	−7.389⋆-3	4.428⋆-1	2.951	1.168⋆8	−3.538⋆1

Magnetic			Achromatic		
(30)	(30)	(12) = (21)	(30)	(30)	(12) = (21)
4.623⋆6	4.590⋆6	1.365⋆7	6.934⋆6	6.884⋆6	2.047⋆7
2.693⋆3	1.849⋆3	5.758⋆3	4.023⋆3	2.767⋆3	8.579⋆3
3.441⋆2	3.019⋆1	2.244⋆2	5.095⋆2	4.459⋆1	3.301⋆2
8.282⋆2	1.627⋆1	1.369⋆2	1.220⋆3	2.344⋆1	1.779⋆2
1.527⋆4	2.841⋆3	7.168⋆3	2.247⋆4	4.145⋆3	3.868⋆3
3.329⋆7	2.745⋆7	3.040⋆7	4.902⋆7	4.041⋆7	3.877⋆7
2.039⋆6	2.850⋆6	2.340⋆6	3.004⋆6	4.199⋆6	3.047⋆6
1.002⋆6	1.422⋆6	1.134⋆6	1.477⋆6	2.096⋆6	1.500⋆6
4.836⋆9	5.013⋆9	4.622⋆9	7.134⋆9	7.395⋆9	6.199⋆9
7.557⋆-1	−2.643⋆-1	−2.834⋆-1	9.207⋆-1	−2.395⋆-1	−1.974⋆-1
7.968⋆-1	−2.664⋆-1	−2.662⋆-1	9.820⋆-1	−2.409⋆-1	−1.698⋆-1
−2.575⋆-1	8.759⋆-1	−1.735⋆-1	−2.340⋆-1	1.147	−8.518⋆-2
−1.842⋆-1	7.982⋆-1	2.512⋆-2	−1.141⋆-1	1.094	2.471⋆-1
−1.677⋆-1	5.218	1.687	8.584⋆-2	6.914	2.820
−7.917⋆1	2.367⋆2	1.854⋆1	1.219	3.065⋆2	2.724⋆1
1.293⋆1	2.697⋆4	4.703⋆2	7.435	3.470⋆4	6.595⋆2

Magnetic			Achromatic		
(30)	(30)	(12) = (21)	(30)	(30)	(12) = (21)
1.399	−3.042⋆-1	−9.743⋆-2	1.830	−2.859⋆-1	2.019⋆-2
3.041	−3.277⋆-1	1.456⋆-1	4.092	−2.921⋆-1	3.940⋆-1
7.756	−2.878⋆-1	3.269⋆-1	1.083⋆1	−2.148⋆-1	7.370⋆-1
1.482⋆1	−2.315⋆-1	1.187⋆-1	2.112⋆1	−1.377⋆-1	5.577⋆-1
1.945⋆1	−2.049⋆-1	−2.125⋆-1	2.797⋆1	−1.148⋆-1	1.580⋆-1
1.808⋆1	−2.040⋆-1	−4.372⋆-1	2.609⋆1	−1.294⋆-1	−1.712⋆-1
1.200⋆1	−2.149⋆-1	−5.375⋆-1	1.728⋆1	−1.573⋆-1	−3.737⋆-1
5.472	−2.289⋆-1	−5.345⋆-1	7.779	−1.853⋆-1	−4.387⋆1
1.686	−2.429⋆-1	−4.563⋆-1	2.259	−2.093⋆-1	−3.857⋆-1
1.125	4.455⋆7	1.991⋆2	1.491	5.901⋆7	2.773⋆2

Table 36 The regular antisymmetric quadruplet: the bell-shaped model. $s = 4$, $t = 4$. The aperture aberrations are listed for high magnification.

$s = 4, t = 4$, *CDCD*				Electrostatic		
β_1	β_2	f	z_F	(30)	(03)	(12) = (21)
1.000⋆-1	1.727⋆-1	2.853⋆2	2.793⋆2	1.497⋆7	1.485⋆7	−6.277⋆6
2.000⋆-1	3.311⋆-1	2.247⋆1	1.609⋆1	8.789⋆3	5.810⋆3	−2.437⋆3
3.000⋆-1	4.423⋆-1	8.785	1.001	1.244⋆3	9.617⋆1	−8.892⋆1
4.000⋆-1	5.060⋆-1	8.171	−3.793	4.258⋆3	1.244⋆2	4.900⋆1
5.000⋆-1	5.430⋆-1	1.733⋆1	−1.478⋆1	3.448⋆5	1.351⋆5	2.244⋆4
6.000⋆-1	5.673⋆-1	−2.202⋆1	2.373⋆1	3.340⋆8	5.076⋆6	2.996⋆5
7.000⋆-1	5.852⋆-1	−6.596	7.839	9.479⋆4	2.290⋆5	7.628⋆3
8.000⋆-1	5.997⋆-1	−4.353	5.299	6.107⋆4	1.451⋆5	3.334⋆3
9.000⋆-1	6.121⋆-1	−4.331	5.080	1.971⋆5	3.351⋆5	5.548⋆3
1.050	7.047⋆-2	3.495⋆-2	4.253⋆-1	1.517	−1.367⋆-1	−5.468⋆-1
1.000⋆-1	1.223	6.882⋆-2	−3.465	−1.179⋆-1	1.824	−4.636⋆-1
2.000⋆-1	1.498	3.583⋆-2	−3.575	2.374⋆-1	2.269	−6.714⋆-1
3.000⋆-1	1.859	2.570⋆-2	−3.817	1.600	2.050⋆1	−1.898
4.000⋆-1	2.076	5.205⋆-2	−4.890	1.183⋆1	1.171⋆3	−9.948

$s = 4, t = 3$, *CCDD*				Electrostatic		
β_1	β_2	f	z_F	(30)	(03)	(12) = (21)
1.000⋆-1	1.060	1.141⋆-1	−3.339	2.826	−1.989⋆-1	−4.912⋆-1
2.000⋆-1	8.890⋆-1	2.102⋆-1	−3.047	6.668	−1.266⋆-1	−6.856⋆-1
3.000⋆-1	7.531⋆-1	3.432⋆-1	−2.477	1.883⋆1	1.234⋆-1	−1.043
4.000⋆-1	6.786⋆-1	3.992⋆-1	−1.830	3.768⋆1	2.843⋆-1	−1.313
5.000⋆-1	6.356⋆-1	3.676⋆-1	−1.262	4.998⋆1	2.751⋆-1	−1.369
6.000⋆-1	6.052⋆-1	2.945⋆-1	−8.005⋆-1	4.622⋆1	1.824⋆-1	−1.232
7.000⋆-1	5.778⋆-1	2.143⋆-1	−4.295⋆-1	3.018⋆1	7.801⋆-2	−1.016
8.000⋆-1	5.460⋆-1	1.432⋆-1	−1.282⋆-1	1.325⋆1	−8.988⋆-3	−8.158⋆-1
9.000⋆-1	4.962⋆-1	8.697⋆-2	1.215⋆-1	3.629	−7.403⋆-2	−6.681⋆-1
1.020	3.055⋆-1	4.105⋆-2	3.711⋆-1	1.375	−1.278⋆-1	−5.598⋆-1
1.040	1.848⋆-1	3.650⋆-2	4.078⋆-1	1.459	−1.341⋆-1	−5.499⋆-1

Magnetic			Achromatic		
(30)	(03)	(12) = (21)	(30)	(03)	(12) = (21)
4.279⋆6	4.244⋆6	1.261⋆7	6.417⋆6	6.365⋆6	1.890⋆7
2.534⋆3	1.669⋆3	5.197⋆3	3.785⋆3	2.497⋆3	7.741⋆3
3.654⋆2	2.858⋆1	2.205⋆2	5.411⋆2	4.210⋆1	3.236⋆2
1.266⋆3	3.949⋆1	2.844⋆2	1.864⋆3	5.647⋆1	3.628⋆2
1.028⋆5	4.058⋆4	5.884⋆2	1.512⋆5	5.947⋆4	7.098⋆4
9.951⋆5	1.511⋆6	1.007⋆6	1.464⋆6	2.224⋆6	1.243⋆6
2.824⋆4	6.817⋆4	3.430⋆4	4.155⋆4	1.003⋆5	4.319⋆4
1.819⋆4	4.322⋆4	2.122⋆4	2.676⋆4	6.360⋆4	2.719⋆4
5.866⋆4	9.975⋆4	5.675⋆4	8.635⋆4	1.468⋆5	7.381⋆4
7.381⋆-1	−2.630⋆-1	−2.944⋆-1	8.939⋆-1	−2.377⋆-1	−2.102⋆-1
−2.442⋆-1	7.604⋆-1	−2.082⋆-1	−2.189⋆-1	9.730⋆-1	−1.231⋆-1
−1.425⋆-1	8.360⋆-1	6.621⋆-2	−6.650⋆-2	1.123	3.121⋆-1
1.066⋆-2	8.220	2.411	3.239⋆-1	1.067⋆1	3.847
1.650	4.816⋆2	3.167⋆1	3.854	6.195⋆2	4.554⋆1

Magnetic			Achromatic		
(30)	(03)	(12) = (21)	(30)	(03)	(12) = (21)
1.189	−3.008⋆-1	−1.500⋆-1	1.516	−2.804-1	−3.622⋆-2
2.581	−3.292⋆-1	9.325⋆-2	3.398	−2.887⋆-1	3.529⋆-1
6.551	−2.847⋆-1	2.806⋆-1	9.006	−2.031⋆-1	7.216⋆-1
1.236⋆1	−2.250⋆-1	5.403⋆-2	1.743⋆1	−1.231⋆-1	5.098⋆-1
1.598⋆1	−2.002⋆-1	−3.021⋆-1	2.278⋆1	−1.052⋆-1	5.351⋆-2
1.464⋆1	−2.020⋆-1	−5.314⋆-1	2.095⋆1	−1.251⋆-1	−2.978⋆-1
9.603	−2.142⋆-1	−6.052⋆-1	1.372⋆1	−1.558⋆1	−4.683⋆-1
4.372	−2.287⋆-1	−5.663⋆-1	6.147	−1.847⋆-1	−4.832⋆-1
1.413	−2.428⋆-1	−4.656⋆-1	1.856	−2.090⋆-1	−3.981⋆-1
7.011⋆1	−2.590⋆-1	−3.258⋆-1	8.358⋆-1	−2.328⋆-1	−2.478⋆-1
7.226⋆-1	−2.617⋆-1	−3.045⋆-1	8.700⋆-1	−2.362⋆-1	−2.227⋆-1

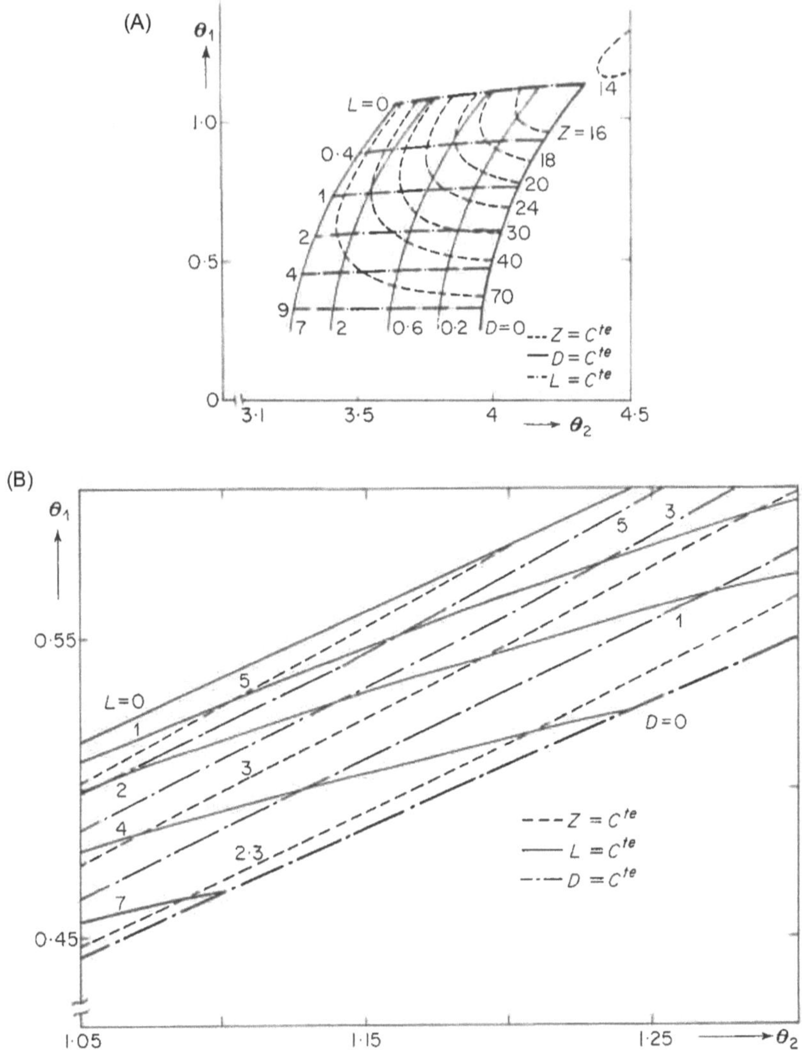

Fig. 37 The antisymmetric quadruplet (*CDCD* configuration, rectangular model) with transfer matrix $\begin{pmatrix} 1 & b \\ 0 & 1 \end{pmatrix}$ or $\begin{pmatrix} -1 & b \\ 0 & -1 \end{pmatrix}$. (A) Positive case. Contours in the $\theta_1 - \theta_2$ plane of equal $L = s/L_1$, $D = t/L_1$ and $Z = -b/2L_1$ for the case $L_2/L_1 = 2$. (The curves are similar for $L_2/L_1 = \frac{1}{4}, \frac{1}{2}$, 1 and 4.) (B) Negative case. Contours in the $\theta_1 - \theta_2$ plane of equal L, D and $Z = b/2L_1$ for the case $L_2/L_1 = 4$. (Similar results are obtained for other values of L_2/L_1 but there is no solution for $L_1 = L_2$.) (After Möller et al., 1966b. Courtesy Académie des Sciences de Paris.)

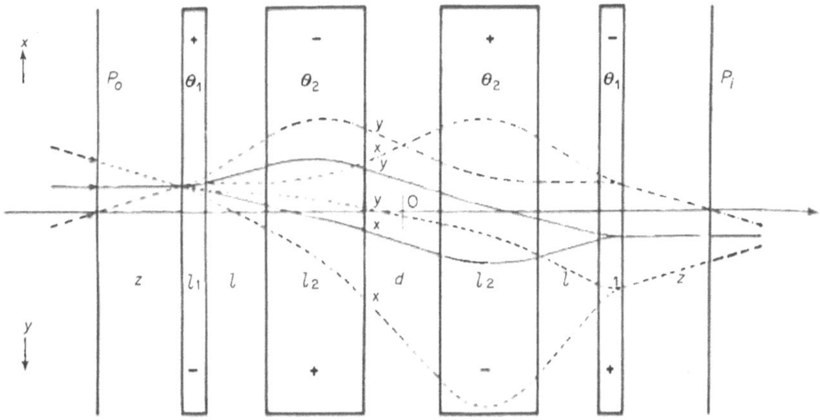

Fig. 38 Trajectories in an antisymmetric quadruplet (*CDCD* configuration) with transfer matrix $\begin{pmatrix} -1 & b \\ 0 & -1 \end{pmatrix}$. Between the planes P_o and P_i, the transfer matrices have the form $\begin{pmatrix} -1 & 0 \\ 0 & -1 \end{pmatrix}$. In this example, $\theta_1 = 0.51$, $\theta_2 = 1 \cdot 10$ and $L_2/L_1 = 4$. (After Möller et al., 1966b.)

Here, they find no comparable set for $L_2 = L_1$. Fig. 38 shows the form of the trajectories in the case $L_2/L_1 = 4$, $b = 7L_1$, $s = 2 \cdot 5L_1$, $t = 3L_1$, $\theta_1 = 0 \cdot 51$ and $\theta_2 = 1 \cdot 1$.

Regenstreif (1967c, 1968) has made a systematic study of the antisymmetric quadruplet, seeking a transfer matrix of the form $\begin{pmatrix} -1 & b \\ 0 & -1 \end{pmatrix}$ such that $b > 0$. The following summary of the results follows Regenstreif's text closely. When $\theta_1 > \theta_2$, the drift spaces s and t and the parameter b are simultaneously positive provided $D_1 S_2 / F_{12} < \kappa < H$ and when $\theta_1 < \theta_2$, they are simultaneously positive for $H < \kappa < F_{21}/S_1 D_2$, where

$$
\begin{aligned}
\kappa &= \beta_2/\beta_1 \\
H(\theta_1, \theta_2) &= \frac{\sinh\theta_1 \cos\theta_2 + \sin\theta_1 \cosh\theta_2}{\sinh\theta_2 \cos\theta_1 + \sin\theta_2 \cosh\theta_1} \\
F_{12} &= 2 - \frac{1}{\sin\theta_1 \sinh\theta_1} \left(\frac{\sinh\theta_2}{\sin\theta_2} - \frac{\sin\theta_2}{\sinh\theta_2} \right) \\
F_{21} &= 2 - \frac{1}{\sin\theta_2 \sinh\theta_2} \left(\frac{\sinh\theta_1}{\sin\theta_1} - \frac{\sin\theta_1}{\sinh\theta_1} \right) \\
S(\theta) &= \coth\theta + \cot\theta \\
D(\theta) &= \coth\theta - \cot\theta.
\end{aligned}
\tag{58}
$$

Regenstreif tabulates the upper and lower limits of κ for a range of values of θ_1 and θ_2.

Regenstreif then considers a number of degenerate cases: $\beta_1 = \beta_2$; $\theta_1 = \theta_2$ but $L_2 \neq L_1$; and $\theta_1 = \theta_2$, $L_1 = L_2$. In the first case, no solution is found unless $L_2 = b = 0$ and in the second, no solution is found at all. In the third case, the normal turn-over quadruplet, only one condition remains:

$$(s + D/\beta)(t + D/\beta) = \left(\operatorname{cosec}^2\theta + \operatorname{cosech}^2\theta\right)/\beta^2. \tag{59}$$

We now have $b = t$. If in addition we set $s = t = \sigma$, say, the condition gives

$$\beta\sigma = \cot\theta - \coth\theta + \left(\operatorname{cosech}^2\theta + \operatorname{cosec}^2\theta\right)^{1/2}. \tag{60}$$

The solution is illustrated in Fig. 39. The total length of the system is an important parameter for accelerator engineers and is not unimportant in

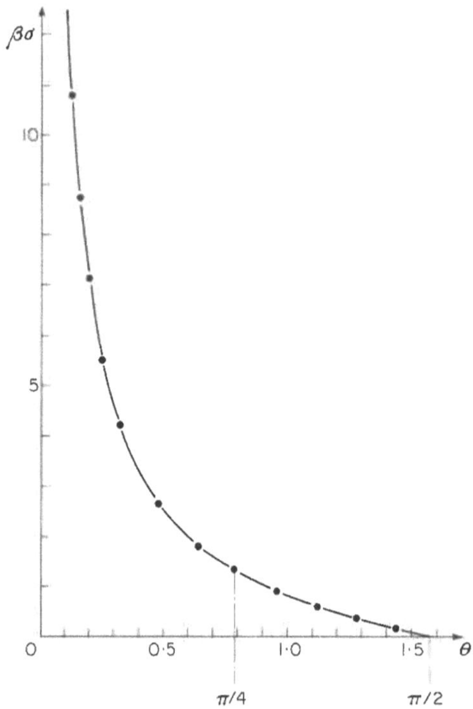

Fig. 39 The normal antisymmetric "turn-over" quadruplet. Lens separation as a function of lens excitation.

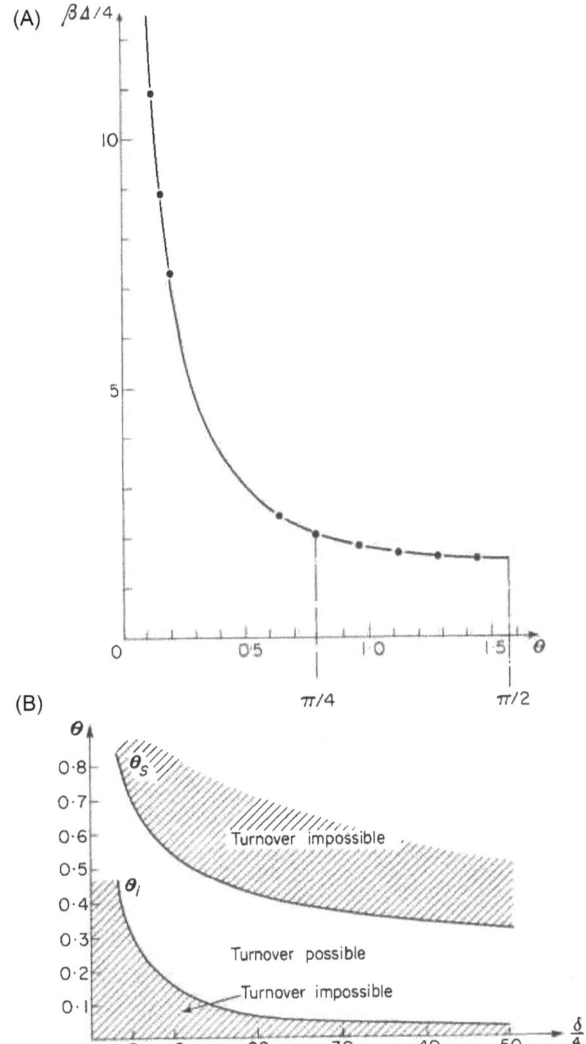

Fig. 40 The normal antisymmetric "turn-over" quadruplet. (A) System length as a function of lens separation. (B) The regions in the $\theta - \delta(\delta = \beta\Delta)$ plane in which turn-over is possible and impossible. (C) The values of $l_o(=\beta s)$ and $l_i(=\beta t)$ as a function of θ for a range of values of $\delta(=\beta\Delta)$. (After Regenstreif, 1967c, 1968. Courtesy Académie des Sciences de Paris.)

(Continued)

corrector design, and Regenstreif discusses the case $s \neq t$ in terms of this total length. The length parameter selected is Δ, the distance between the planes P_1 and P_2, distant p and q from the ends of the quadruplet, between which the transfer matrix has the form $\begin{pmatrix} -1 & 0 \\ 0 & -1 \end{pmatrix}$. For the normal turnover

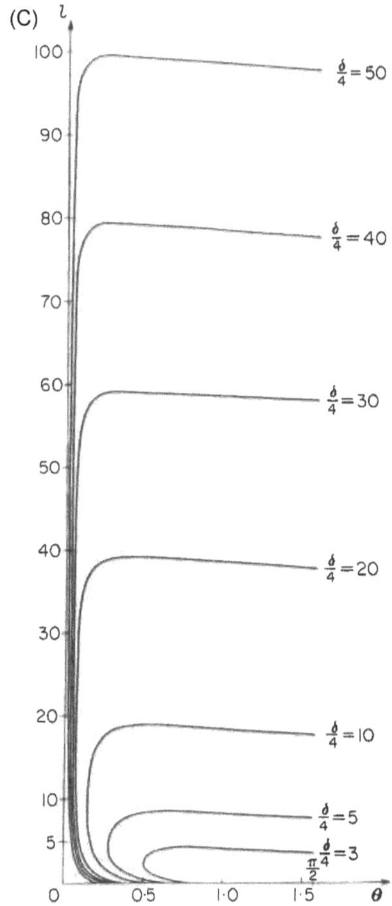

Fig. 40—Cont'd

quadruplet, $p + q = t$, so that Δ exceeds the physical length of the quadruplet by an amount t. When $s = t$, we have

$$\beta\Delta/4 = \theta + \cot\theta - \coth\theta + \left(\operatorname{cosec}^2\theta + \operatorname{cosech}^2\theta\right)^{1/2} \tag{61}$$

(Fig. 40A). When $s \neq t$ and Δ is given, we have

$$\begin{aligned} \beta s &= \beta\Delta/4 - \theta \pm \left\{(\beta\Delta/4 - \theta + D)^2 - \operatorname{cosec}^2\theta - \operatorname{cosech}^2\theta\right\}^{1/2} \\ \beta t &= \beta\Delta/4 - \theta \mp \left\{(\beta\Delta/4 - \theta + D)^2 - \operatorname{cosec}^2\theta - \operatorname{cosech}^2\theta\right\}^{1/2} \end{aligned} \tag{62}$$

(Fig. 40C). Fig. 40B shows the upper and lower limits imposed on θ as Δ varies.

7. Other systems

7.1 General comments

The simplicity of the antisymmetric quadrupole doublet makes it a particularly attractive system, but it is never capable of forming a real image of a real object. This fact led Bauer (1964, 1965–66) to associate a triplet with a doublet, thus forming a quintuplet objective, the role of the triplet being to convert the immersed image into a real image with unit magnification. This system has been pursued by Dhuicq (1968a) and, briefly, by Meads (see Hawkes, 1966b). We have already mentioned many features of Bauer's system, and we merely refer to the schematic diagram showing the principal parameters (Fig. 22A in Quadrupole systems: Their suitability for specific purposes).

Fig. 22b shows the corresponding quantities for Dhuicq's system. Since two of the three octopoles have the same effect on the beam, Dhuicq gives them the same excitation, so that only two of the three aperture aberrations

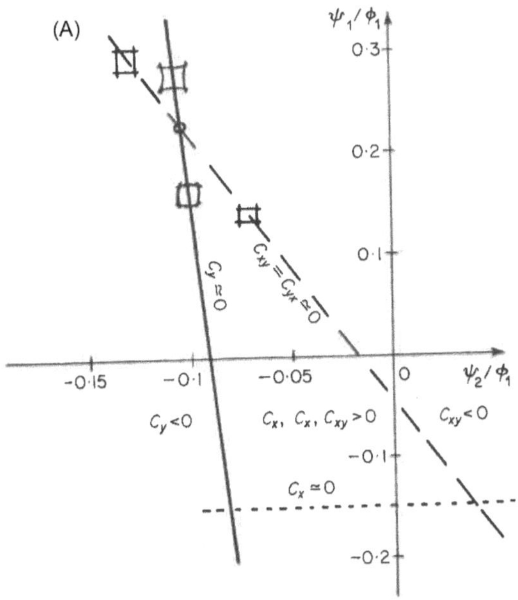

Fig. 41 (A) Correction of the various aperture aberration coefficients of a quintuplet lens. C_x, C_y and C_{xy} correspond to (30), (03) and (12) = (21), respectively. ψ_1 and ψ_2 are the octopole potentials applied in the triplet. (B) Contours showing equal values of the aperture aberration coefficients in the $\overline{\psi}_1 - \overline{\psi}_2$ plane ($\overline{\psi} = 100\psi/\Phi_0$). The numbers represent $10^{-4}C$ (C in mm). (After Dhuicq, 1968a.)

(Continued)

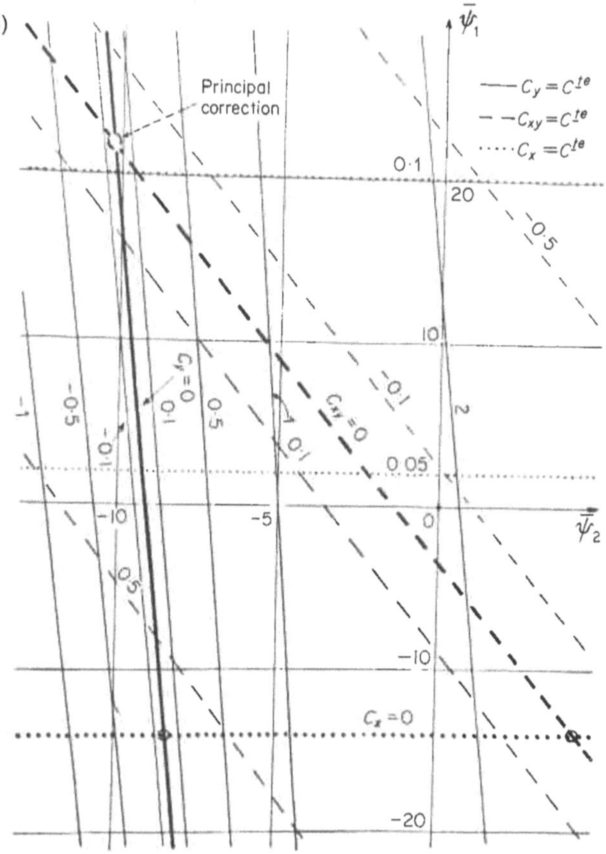

Fig. 41—Cont'd

can be corrected. Fig. 41A shows what octopole excitations correct the various coefficients; the points of intersection of the curves indicate that two coefficients are corrected simultaneously. The corresponding contours are shown in Fig. 42B.

The only other combinations about which information is available are the quintuplet and sextuplet studied by Regenstreif (1967b). We shall not consider these here, but we mention in conclusion that the problem of designing a system with a particular transfer matrix is simplified by the introduction of Regenstreif's formalism: if the excitations of the members of a multiplet are specified, it is a simple matter to calculate the spacing. The problem of calculating the excitations when the spacings are fixed invariably leads to transcendental equations, which can only be solved by the rather elaborate machinery of inverse interpolation.

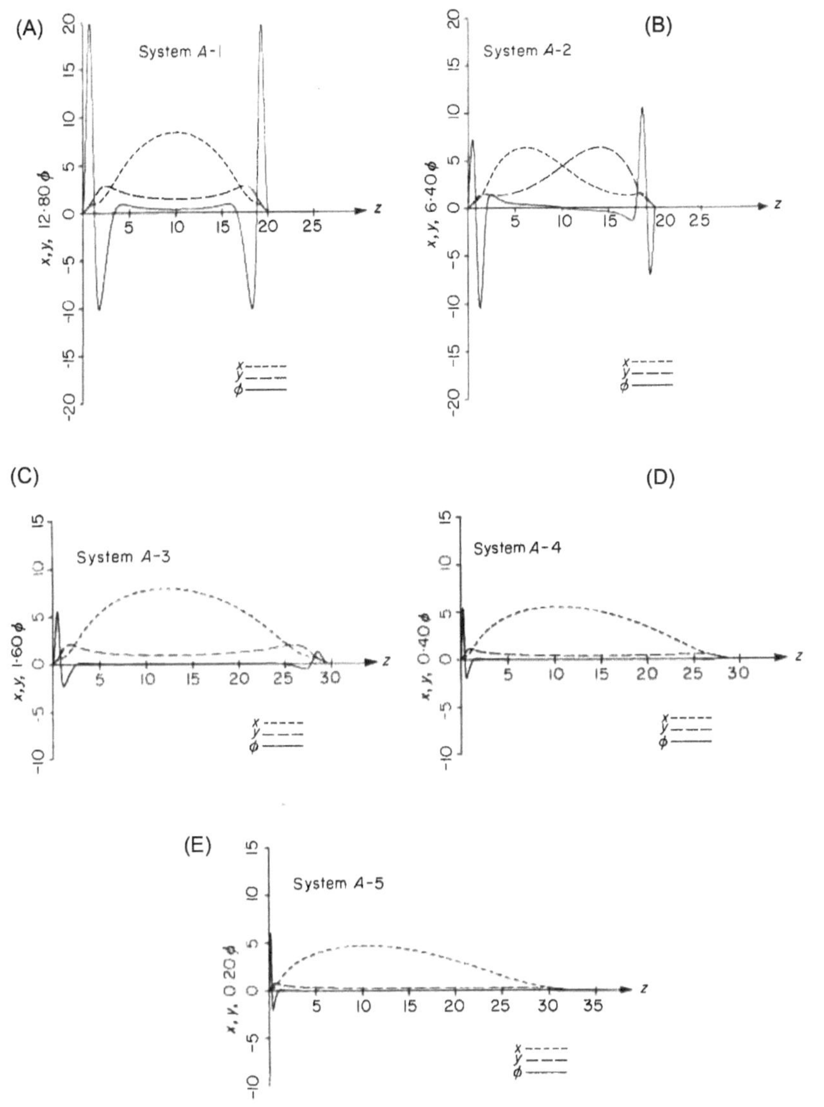

Fig. 42 Lens systems (sextuplets) producing a stigmatic, orthomorphic image. (A) and (B), $M = -1$; in (B), which is inferior to (A) so far as the aberrations are concerned, the trial function is antisymmetric. (C) $M = -2$; (D) $M = -5$; (E) $M = -10$. (After Moses, 1968. With the permission of AIP Publishing.) Corresponding results are tabulated in Table 37.

7.2 Variational methods of minimizing aperture aberrations (Moses, 1968)

In the preceding paragraphs, we have provided a variety of answers to the question: given a system, what are its aberrations? Of more interest to the lens designer is the complementary question: what systems will have aberrations below some specified level and produce a stigmatic undistorted image with a particular magnification? This latter problem has been investigated by the methods of variational calculus by Tretner (1959), in the case of round lenses, and recently by Moses (1968), in a study of quadrupole lens systems. It is worth pointing out that their approach differs considerably from that of Glaser (1940), Kas'yankov (1952, 1967), Dutova and Kas'yankov (1963), Taganov (1966) and Taganov and Kas'yankov (1964, 1965, 1967). Glaser obtained a differential equation for the field distribution in a round lens free of spherical aberration by setting the integrand in the formula for this aberration equal to zero. Kas'yankov and his colleagues have followed a similar procedure, concentrating on the other aberrations, in which the various integrands are set equal to some test function, the integral of which, from the object to the image plane, vanishes; this again yields a differential equation for the field or potential distribution. Although the resulting distribution will be free of the aberration in question, there is no guarantee that it can be created in practice and the lens strength cannot be specified beforehand. In the work of Tretner and Moses, however, the first-order desiderata and the practical limitations on field strengths are inserted as constraints at the outset, and the aberration coefficients are not set equal to zero but minimized.

Moses considers systems of magnetic quadrupoles in the standard orientation; a figure of merit is selected for minimization, subject to constraints that prevent the field gradient from becoming impractically high and determine the magnifications in the principal sections. First, however, Moses points out that the system can be conveniently scaled with respect to the distance from the object (at $z = 0$) to the image (at $z = z_{i1}$); for, consider a second system in which the image is formed at $z = z_{i2}$. In the x–z plane, we have

$$\frac{d^2 h_{x1}(z)}{dz^2} + \phi_1(z) h_{x1}(z) = 0 \tag{63a}$$

and

$$\frac{d^2 h_{x2}(z)}{dz^2} + \phi_2(z) h_{x2}(z) = 0 \tag{63b}$$

in which we follow Moses in writing $\phi(z) = \beta^2 k(z)$; $h_{x1}(0) = h_{x1}(z_{i1}) = 0$ and $h_{x2}(0) = h_{x2}(z_{i2}) = 0$. We set

$$z_{i2}/z_{i1} = L \tag{64}$$

and it is clear that the left-hand side of Eq. (63b) takes the form

$$\frac{1}{L}\left\{\frac{d^2 h_{x1}(z/L)}{d(z/L)^2} + \phi_1(z/L)h_{x1}(z/L)\right\} \tag{65}$$

if we have

$$h_{x2}(z) = Lh_{x1}(z/L) \tag{66}$$

and

$$\phi_2(z) = \phi_1(z/L)/L^2. \tag{67}$$

Furthermore, the magnifications of the two systems are the same if Eq. (67) is satisfied. The aberration coefficients are proportional to L, and Eq. (67) shows that for practical reasons, they cannot be reduced indefinitely by shortening the system.

The aperture aberration coefficients are combined into a single coefficient by Moses; he writes

$$\begin{aligned}\Delta x(z_i) &= M_x\left(C_1\alpha^3 + C_2\alpha\gamma^2\right)\\ \Delta y(z_i) &= M_y\left(D_1\gamma^3 + D_2\alpha^2\gamma\right)\end{aligned} \tag{68}$$

so that $C_1 = (30)$, $C_2 = (12) = D_2 = (21)$ and $D_1 = (03)$–we retain his notation which is more convenient in this context. Setting $\alpha = \rho\cos\theta$ and $\gamma = \rho\sin\theta$, the "radial aperture aberration coefficient," $C_s(\theta)$, is defined by

$$\begin{aligned}C_s^2(\theta) = {}& \cos^2\theta\left(C_1\cos^2\theta + C_2\sin^2\theta\right)^2\\ &+ \sin^2\theta\left(D_1\sin^2\theta + D_2\cos^2\theta\right)^2\end{aligned} \tag{69}$$

and Moses minimizes the root-mean-square average of $C_s(\theta)$:

$$\begin{aligned}\tilde{C}(C_1, C_2, D_2) &= \left\{\frac{1}{2\pi}\int_0^{2\pi} C_s^2(\theta)d\theta\right\}^{1/2}\\ &= \frac{1}{4}\left\{(C_1 + C_2)^2 + (D_1 + D_2)^2 + 4\left(C_1^2 + D_1^2\right)\right\}^{1/2}\end{aligned} \tag{70}$$

$\tilde{C}$ has the dimensions of length, and the dimensionless figure of merit is obtained by combining $\tilde{C}$ with a "gradient strength factor," S. Moses selects the latter in two ways. If ϕ_m is the maximum gradient, so that $\phi_m > |\phi(z)|$ in the region between object and image (or between object and arbitrary current plane when the emergent beam is parallel to the axis), then $S = \phi_m^{1/2}$ will have the dimensions of (length)$^{-1}$. For a system of given length, therefore, $SL = L\phi_m^{1/2}$ will be a measure of the maximum field; if this is treated as a constraint subject to which $\tilde{C}S$ is varied, a minimum value of $\tilde{C}$ will be obtained for given maximum field gradient. An alternative expression for S is $S = I^{1/3}$, with

$$I = \int_{z_0}^{z_c} \phi^2(z)dz. \tag{71}$$

The problem, then, is to minimize $\tilde{C}S$, given LS and the paraxial properties of the system. This involves introducing a number of Lagrange multipliers, and we now describe Moses' procedure very briefly; there are numerous points of interest in Moses' analysis that we cannot reproduce here, however. In order to maintain S constant, we introduce a weighting function W and the corresponding integral, I_W:

$$I_W = \int_{z_o}^{z_c} W(\phi^2, z)dz. \tag{72}$$

W is selected in such a way that its first-order variation vanishes when $\delta S = 0$; thus when $S = I^{1/3}$, we may set

$$I_W = I = \int_{z_o}^{z_c} \phi^2 dz \tag{73a}$$

and when $S = \phi_m^{1/2}$,

$$I_W = \int_{z_o}^{z_c} W_m(\phi^2)dz \tag{73b}$$

with $W_m = 0$ for $\phi^2 < \phi_m^2$ and $W_m \to \infty$ for $\phi^2 > \phi_m^2$. The sum

$$C_W = \tilde{C} + \Gamma_W I_W \tag{74}$$

is then formed, in which Γ_W is a Lagrange multiplier.

The fact that the paraxial equations of motion must be satisfied introduces two further multipliers, $\eta(z)$ and $\zeta(z)$; the expression

$$\frac{\partial \tilde{C}}{\partial C_2}\int_{z_o}^{z_c}\left\{\eta\left(h_x''+\phi h_x\right)+\zeta\left(h_y''-\phi h_y\right)\right\}dz \tag{75}$$

is added to C_W and after some manipulation, the following equations for the extremal system emerge:

$$\begin{aligned}\eta''+\phi\eta &= -3\phi\left\{h_x\left(h_y'^2+4\lambda h_x'^2\right)-2h_x'h_yh_y'\right\}-2\phi^2h_x\left(h_y^2+4\lambda h_x^2\right)\\ &\equiv F\left(h_x,h_y,h_x',h_y',\phi\right)\end{aligned} \tag{76a}$$

$$\begin{aligned}\zeta''-\phi\zeta &= 3\phi\left\{h_y\left(h_x'^2+4\mu h_y'^2\right)-2h_x'h_xh_y'\right\}-2\phi^2h_y\left(h_x^2+4\mu h_y^2\right)\\ &\equiv K\left(h_x,h_y,h_x',h_y',\phi\right)\end{aligned} \tag{76b}$$

$$h_x''+\phi h_x=0 \tag{76c}$$

$$h_y''-\phi h_y=0 \tag{76d}$$

$$\phi=\frac{-\eta h_x+\zeta h_y}{2\left(h_x^2h_y^2+2\lambda h_x^4+2\mu h_y^4+\Gamma W_\phi\right)} \tag{76e}$$

in which

$$\begin{aligned}\lambda &= \frac{\partial\tilde{C}/\partial C_1}{6\partial\tilde{C}/\partial C_2}=\frac{5C_1+C_2}{6(C_1+D_1+2C_2)}\\ \mu &= \frac{\partial\tilde{C}/\partial D_1}{6\partial\tilde{C}/\partial C_2}=\frac{5D_1+C_2}{6(C_1+D_1+2C_2)}\\ \Gamma &= \frac{\Gamma_W}{\partial\tilde{C}/\partial C_2}\\ W_\phi &= \frac{\partial W}{\partial\phi^2}.\end{aligned} \tag{77}$$

If, furthermore, δh_x, δh_y and the corresponding variations of slope do not all vanish at the boundaries, we must also have

$$\left[\left(4\lambda h_x'^3+3h_x'h_y'^2-\eta'\right)\delta h_x+\eta\delta h_x'+\left(4\mu h_y'^3+3h_x'^2h_y'-\zeta'\right)\delta h_y+\zeta\delta h_y'\right]_{z_0}^{z_c}=0. \tag{78}$$

Moses solves these equations for three classes of systems: systems forming a stigmatic and orthomorphic image of an object with a pre-set magnification; "objective-type systems" that form an image at infinity and have the same focal length in the two principal sections, $f = g$; and similar systems with different focal lengths in the two principal sections, $f \neq g$. We conclude this section with some account of his results; this is of necessity highly selective, and in the original work, each of about 50 systems is fully specified numerically and illustrated graphically.

The first case considered by Moses is the orthomorphic system, with four values of the magnification: $M = -1, -2, -5$ and -10; I is held constant. Table 37 and Figs. 42A–E summarize the results of the minimization procedure. Moses then investigates orthomorphic and anamorphotic objective lenses, using both forms of S. His results are reproduced as Fig. 43A,B; Figs. 44 and 45 illustrate some of the systems considered and Tables 38 and 39 contain the corresponding numerical results.

The remainder of Moses' work is concerned with the effect of choosing the weighting function, $W(\phi^2, z)$, differently and with comparisons of his results with published information. The choice of weighting function is important, because this function can be used not only to confine the field gradient within physically acceptable bounds but also to ensure that there are no discontinuities or unduly abrupt changes in $\phi(z)$.

From this succinct account of Moses' work, it will be apparent that he has developed powerful techniques for designing quadrupole systems of high quality with specific first-order properties; it must, however, be stressed that (a) the constraints that he applies are imposed on the field gradient, for which there is a good practical reason, and on the h-rays. This is convenient, mathematically, but it has the practical drawback that the ensuing systems are not regular: they are stigmatic and orthomorphic only for the operating conditions specified and such inflexibility is a handicap in systems that are never

Table 37 Stigmatic orthomorphic lens systems (Moses, 1968).

System	$M_x = M_y$	L	$LI^{1/3}$	$L\phi_M^{1/2}$	C_1	D_1	C_2	$\tilde{C}$	$\tilde{C}I^{1/3}$	$\tilde{C}\phi_M^{1/2}$
A1	−1	200.0	30.61	24.90	201.6	180.3	241.7	204.2	31.25	25.43
A2	−1	200.0	31.63	25.71	208.6	208.6	229.9	214.0	33.84	27.51
A3	−2	295.3	53.86	55.45	66.60	65.88	79.98	69.76	12.72	13.10
A4	−5	576.8	98.40	105.6	56.66	59.36	67.84	60.53	10.32	11.08
A5	−10	956.8	160.1	174.6	54.62	58.24	65.72	58.81	9.84	10.73

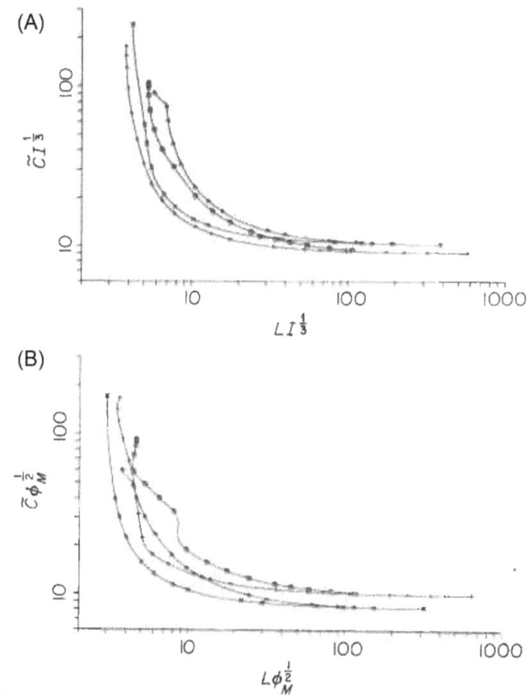

Fig. 43 (A) Comparison of a range of multiplet objectives, using the gradient strength factor $S = I^{1/3}$. Notice that systems B and C have minimum values of $\tilde{C}S$ for each value of LS. +B: anamorphotic systems; B-1 is a doublet, but for higher values of LS, triplets arise. OC: orthomorphic systems. ×D: anamorphotic systems calculated using $S = \phi_M^{1/2}$ (see Fig. 43B). ΔE: orthomorphic systems calculated using $S = \phi_M^{1/2}$ (see Fig. 43B). (B) Comparison of a range of multiplet objectives, using the gradient strength factor $S = \phi_M^{1/2}$. The labels +B, OC, ×D and ΔE are those of Fig. 43A. (After Moses, 1968.)

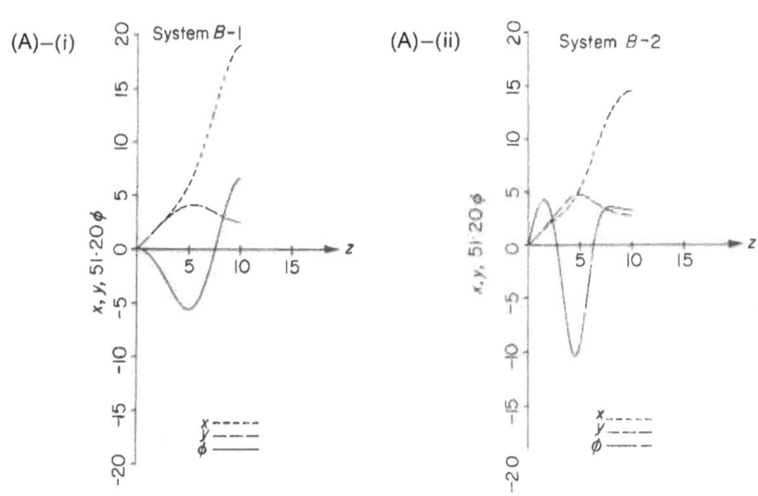

Fig. 44 (A) Systems B-1, B-2, B-3, B-4, and B-6. (B) Systems C-1, C-3, C-4, C-6, and C-7. (See Table 38 for details of the systems in Panel A and B.) (After Moses, 1968.)

(Continued)

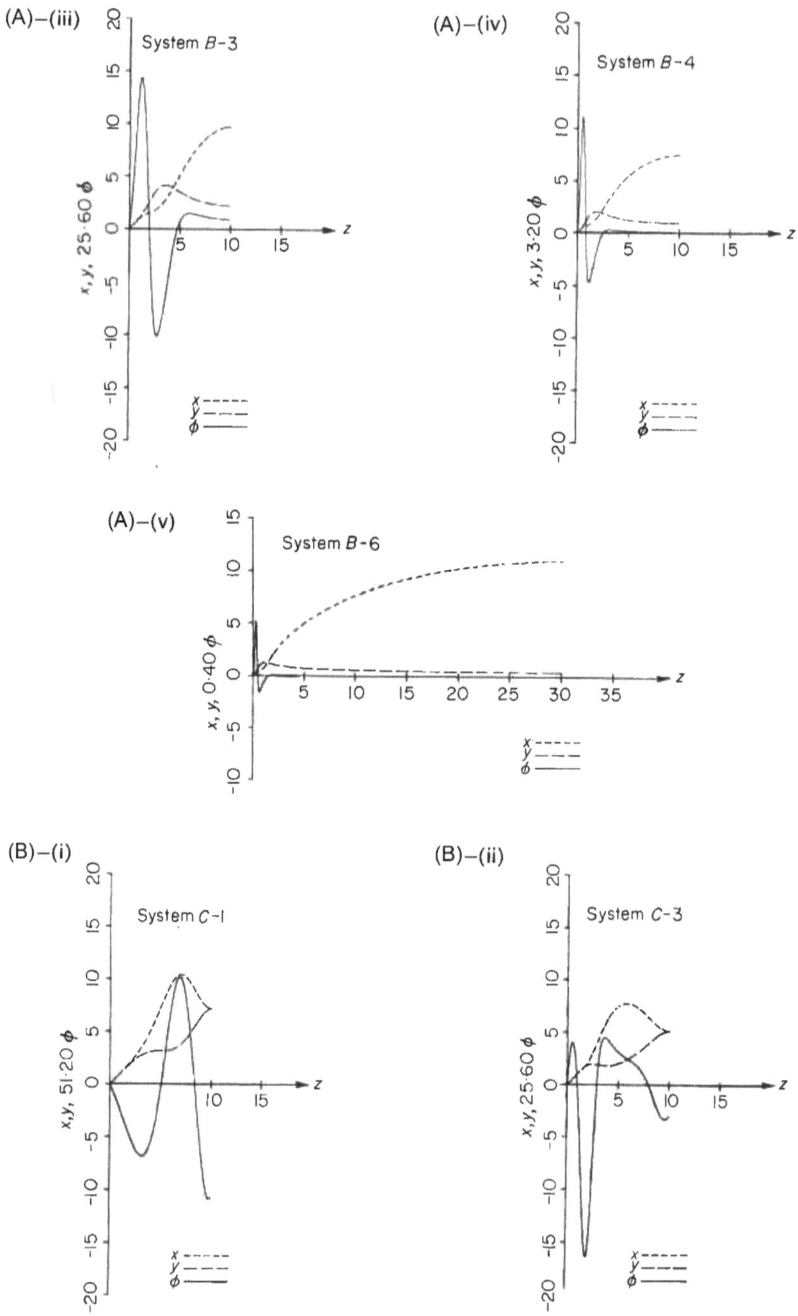
(A)–(iii)
System B-3
x, y, 25·60 ϕ
z
(A)–(iv)
System B-4
x, y, 3·20 ϕ
z
(A)–(v)
System B-6
x, y, 0·40 ϕ
z
(B)–(i)
System C-1
x,y, 51·20 ϕ
z
(B)–(ii)
System C-3
x,y, 25·60 ϕ
z
x
y
ϕ

Fig. 44—Cont'd

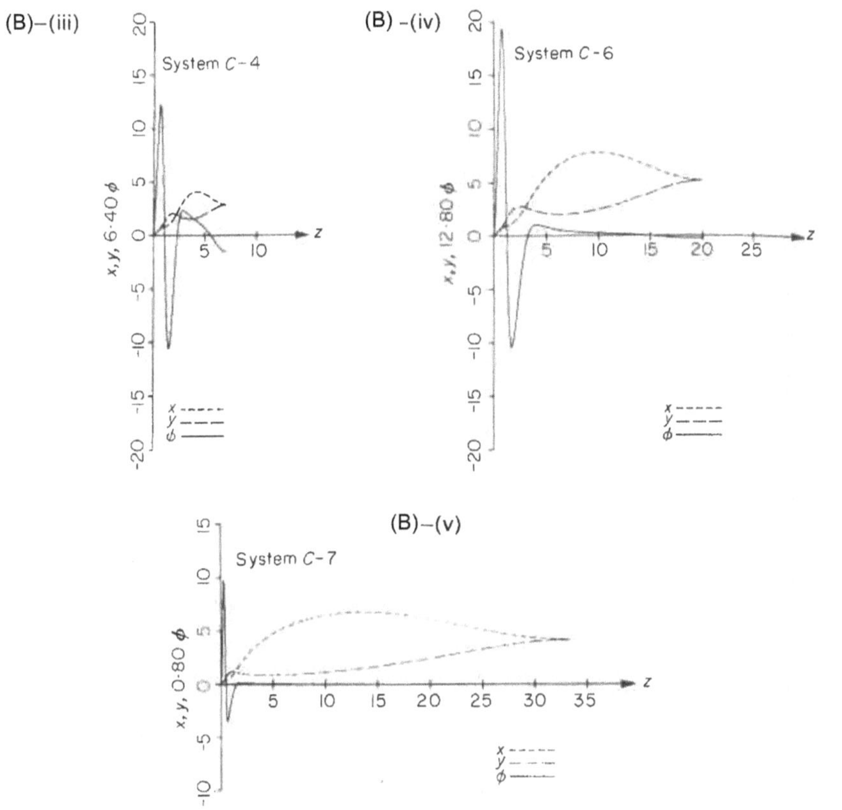

Fig. 44—Cont'd

easy to use in practice. Nonetheless, Moses has "found that the optimum quadrupole systems have aperture aberrations considerably smaller than those obtained in previous studies. The essential difference appears to be that optimum systems usually have two strong thin lenses near the object followed by one or two weak, long lenses, as opposed to systems with three or four lenses having similar dimensions and strengths.... the aperture aberrations of quadrupoles appear to be about 14 times larger than those of round lenses when both the fields and linear dimensions are comparable. When expressed in terms of resolutions..., these round lenses would be superior in resolution to the quadrupoles by a factor of 2."

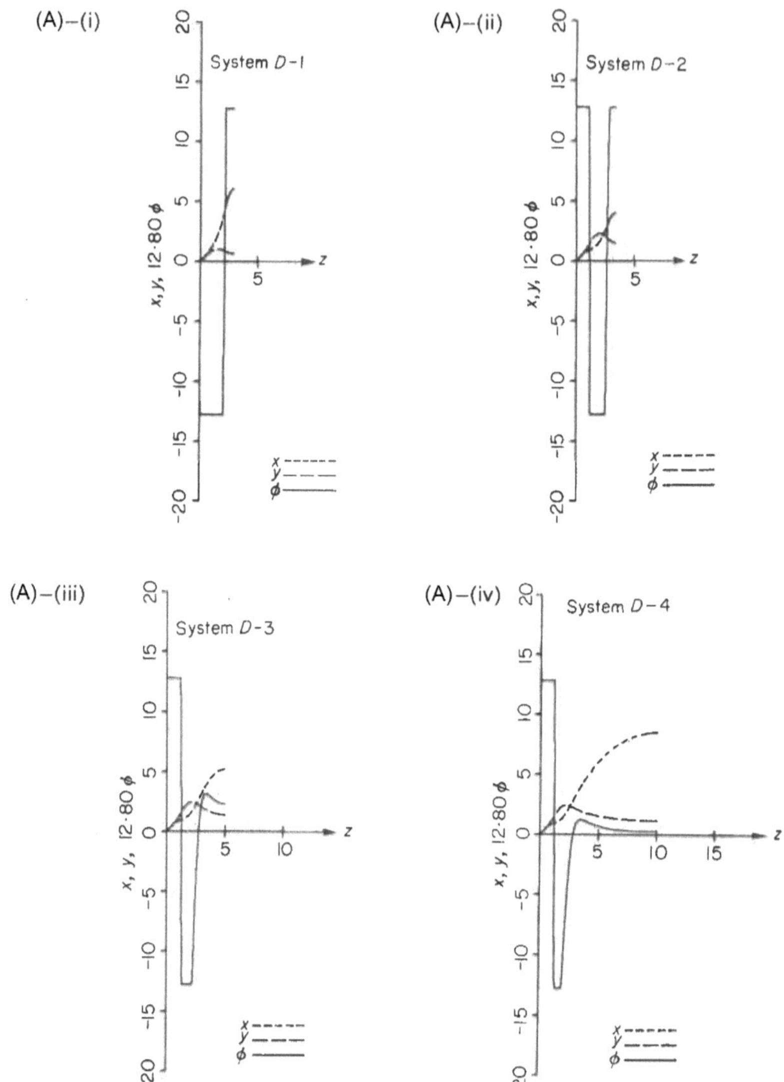

Fig. 45 (A) Systems D-1, D-2, D-3, D-4, and D-5. (B) Systems E-1, E-2, E-3, E-4, and E-5. (See Table 39 for details of the systems in Panel A and B.) (After Moses, 1968.)

(Continued)

(A)–(v)

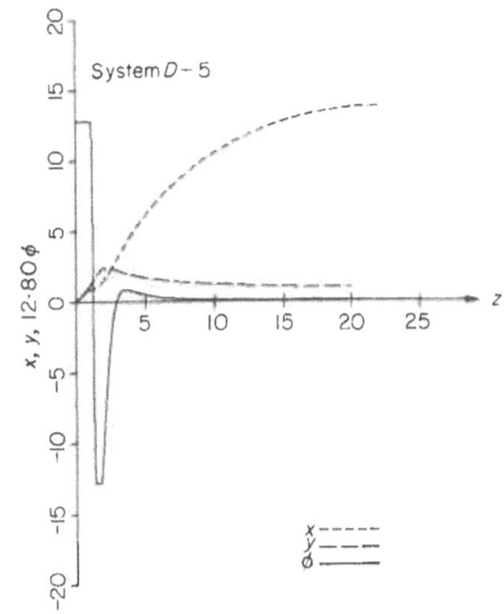

(B)–(i)

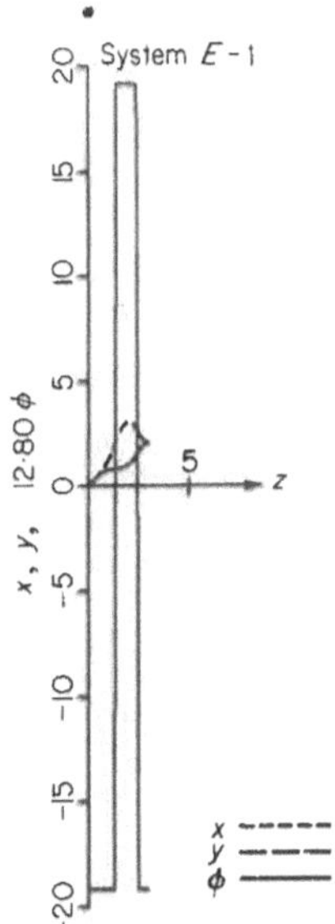

(B)–(ii)

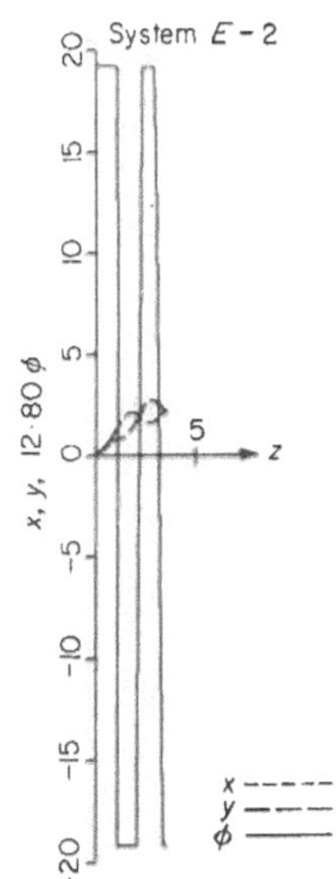

Fig. 45—Cont'd

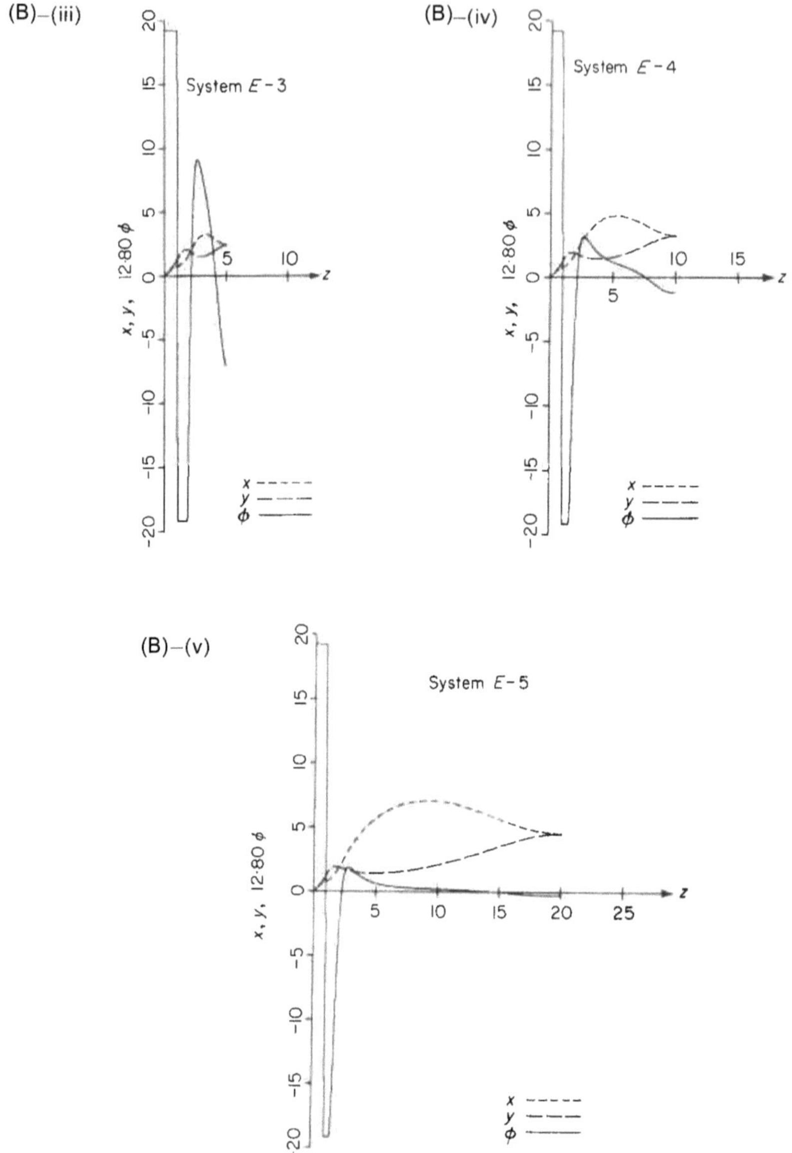
(B)–(iii)
System E-3
x, y, 12·80 φ
z
x
y
φ
(B)–(iv)
System E-4
x, y, 12·80 φ
z
x
y
φ
(B)–(v)
System E-5
x, y, 12·80 φ
z
x
y
φ

Fig. 45—Cont'd

Table 38 Objective systems B (anamorphotic) and C (orthomorphic) obtained with $S = I^{1/3}$ (Moses, 1968).

System	x_c	y_c	L	$LI^{1/3}$	$L\phi_M^{1/2}$	C_1	D_1	C_2	$\tilde{C}$	$\tilde{C}I^{1/3}$	$\tilde{C}\phi_M^{1/2}$
B1	190	25.4	100	3.805	3.585	8101	34.16	839.3	4631	176.2	166.0
B2	144	27.4	100	4.496	4.514	1675	106.9	660.3	1040	46.75	49.94
B3	97.1	22.0	100	7.778	7.492	216.1	158.4	254.2	205.9	16.02	15.43
B4	74.2	9.28	100	17.69	18.75	60.39	58.00	72.06	62.50	15.06	11.72
B6	221	8.16	600	97.61	107.8	52.68	58.05	63.09	57.37	9.33	10.31
C1	71.47		100	5.28	4.62	2806	373.7	2200	1996	105.3	92.26
C3	51.08		100	7.63	8.02	618.1	61.80	369.5	411.2	31.35	32.98
C4	28.56		70	10.56	9.68	140.1	89.44	193.0	137.2	20.70	18.98
C6	51.92		200	23.96	24.61	100.7	85.18	131.3	103.1	12.36	12.69
C7	124.9		1000	106.7	116.4	83.71	85.68	101.9	89.11	9.50	10.37

Table 39 Objective systems D (anamorphotic) and E (orthomorphic) obtained with $S = \phi_M^{1/2}$ (Moses, 1968).

System	x_c	y_c	L	$LI^{1/3}$	$L\phi_M^{1/2}$	C_1	D_1	C_2	$\tilde{C}$	$\tilde{C}I^{1/3}$	$\tilde{C}\phi_M^{1/2}$
D1	61.1	6.24	29.5	4.208	2.950	3005	5.13	221.5	1706	243.3	170.6
D2	40.2	14.6	34	5.003	3.400	520.4	113.4	463.4	390.1	57.41	39.01
D3	51.7	13.6	50	6.655	5.000	171.2	109.5	208.1	160.1	21.30	16.01
D4	85.4	11.3	100	12.75	10.00	111.6	82.19	130.7	106.3	13.56	10.63
D5	139	9.57	220	27.56	22.00	96.68	72.43	110.6	91.79	11.50	9.179
E1	20.6		30.5	5.764	3.735	696.5	76.16	518.7	487.0	92.02	59.64
E2	21.9		36	6.889	4.409	347.8	223.1	650.1	390.6	74.74	47.84
E3	24.2		50	8.510	6.124	178.1	128.0	294.5	192.7	32.80	23.60
E4	32.6		100	15.84	12.25	106.7	72.31	142.0	104.3	16.52	12.78
E5	45.5		200	30.93	24.49	86.04	58.73	104.0	81.40	12.59	9.97

8. Octopoles

Some at least of the aperture aberrations of quadrupoles, like those of round lenses, cannot change sign and complete correction of these aberrations requires octopoles. These may be separate elements with eight electrodes or poles or the quadrupoles may be designed in such a way that they produce octopole fields as well as quadrupolar effects. We consider these two situations separately. Octopoles may also be used to correct the distortion of cathode-ray tube deflectors (Kartashev & Utochkin, 1967).

8.1 Individual octopoles

These have been studied in detail by Dhuicq (1968a, 1968b) and Fishkova and Yavor et al. (1968). The latter evaluate the octopole aberration coefficients for several situations: the thin octopole separated from and superimposed upon a rectangular model quadrupole; they then consider in detail correction of a line image and a stigmatic doublet.

Dhuicq (1968a, 1968b) derives a systematic procedure for calculating the aberrations of a stigmatic quadrupole multiplet and the octopole strengths necessary to correct these. We have described his numerical results in the section on quadruplets (§ 4). If we consider the nonrelativistic form of Eq. (151) in Definitions, and set the four aberration coefficients equal to zero, we obtain

$$\begin{aligned}
&\int \left(\frac{3n^2 - 2n + 3}{6}\beta^4 k^2 + \frac{1}{24}\beta^2 k''\right) H_x^4\, dz = \int \Xi H_x^4 dz \\
&\int \left(\frac{3n^2 - 2n + 3}{6}\beta^4 k^2 - \frac{1}{24}\beta^2 k''\right) H_y^4\, dz = \int \Xi H_y^4 dz \\
&\int \left[\frac{n^2 - 2n + 1}{6}\beta^4 k^2 H_x^2 H_y^2 + \frac{1}{4}\beta^2 k' \left\{ H_x H_y \left(H_x H_y' - H_x' H_y \right) \right\} \right] dz \\
&\quad = \int \Xi H_x^2 H_y^2 dz.
\end{aligned} \tag{79}$$

These equations are substantially the same as Eq. (9) in Chapter 2 of Dhuicq (1968b). The integrals involving $\Xi(z)$ may be divided into the contributions from each octopole, Ξ_i, and if $\Xi_i = \xi_i t_i(z)$ where $t(z)$, like $k(z)$, reaches a maximum value of unity, then we have

$$
\begin{aligned}
\sum_i \xi_i \int t_i(z) H_x^4 dz &= I_x \\
\sum_i \xi_i \int t_i(z) H_y^4 dz &= I_y \\
\sum_i \xi_i \int t_i(z) H_x^2 H_y^2 dz &= I_{xy}
\end{aligned}
\tag{80}
$$

where I_x, I_y, I_{xy} denote the left-hand sides of Eq. (79). In a given system, that has been selected for its regularity and convenient cardinal elements, for instance, the requisite octopole strengths for correction of the aperture aberrations can be calculated from the three simultaneous equations (80).

Let us now turn to the paper by Fishkova and Yavor et al. (1968). A convenient notation analogous to that of Eq. (11) in Definitions, also proposed by the Leningrad group, is introduced: if R_M is the bore-radius of any octopole pole-pieces and R_E that of electrodes, and if V_E is the electrode potential and V_M the magnetostatic potential, then[c]

$$
\tau_E = -\frac{2V_E\sigma}{R_E^4 V} T_E, \quad \tau_M = -\frac{4\eta V_M}{R_M^4 V^{1/2}} T_M \tag{81}
$$

where T_E and T_M are constants, dependent upon the shape of the electrodes and pole-pieces. We have

$$
D_1(z) = \frac{V_E T_E}{R_E^4} t_E(z), \quad Q_1(z) = \frac{V_M T_M}{R_M^4} t_M(z) \tag{82}
$$

so that

$$
\begin{aligned}
\Xi(z) &= \frac{2\sigma D_1 - 4\eta Q_1 V^{1/2}}{V} \\
&= \tau_M t_M(z) - \tau_E t_E(z).
\end{aligned}
\tag{83}
$$

If the functions $t_E(z)$ and $t_M(z)$ are identical, we write

$$
t_E(z) = t_M(z) = t(z) \tag{84}
$$

and

$$
\tau = \tau_M - \tau_E \tag{85}
$$

[c] We have modified the definition in Fishkova and Yavor et al. (1968) by including the relativistic factor σ.

giving

$$\Xi(z) = \tau t(z). \tag{86}$$

The contributions to (0030), (0003) and (0012) or (0021) are then of the form

$$\begin{aligned}
(30)_O &: -\tau\int_{-\infty}^{\infty} t(z)H_x^4 dz \\
(03)_O &: -\tau\int_{-\infty}^{\infty} t(z)H_y^4 dz \\
(21)_O \text{ or } (12)_O &: 3\tau\int_{-\infty}^{\infty} t(z)H_x^2 H_y^2 dz
\end{aligned} \tag{87}$$

in which we have used the short form (*rs*) instead of (00*rs*) and the subscript *O* indicates that the contribution arises from an octopole. If the octopole lies in field-free space, a distance λ beyond the center-plane of a quadrupole, then the functions H_x, H_y that appear in Eq. (87) are the asymptotes to the rays $H_x(z)$, $H_y(z)$ emerging from the quadrupole:

$$H_x = m_x(z - z_i) \quad H_y = m_y(z - z_j). \tag{88}$$

If the octopole is thin, we obtain

$$\begin{aligned}
(30)_O &= -\tau L_o m_x^4 (\lambda - z_i)^4 \\
(03)_O &= -\tau L_o m_y^4 (\lambda - z_j)^4 \\
(12)_O &= (21)_O \\
&= 3\tau L_o m_x^2 m_y^2 (\lambda - z_i)^2 (\lambda - z_j)^2 (\text{thin-lens approximation})
\end{aligned} \tag{89}$$

in which L_o is the effective length of the octopole, defined by

$$L_o = \int_{-\infty}^{\infty} t(z)dz. \tag{90}$$

If the octopole is not short, and the rectangular model is used, then $t(z) = 1$ inside the octopole and $t(z) = 0$ elsewhere. Writing

$$\lambda_i = \lambda - z_i \quad \lambda_j = \lambda - z_j \tag{91}$$

we find

$$
\begin{aligned}
(30)_O &= -\frac{1}{5}\tau m_x^4\{(L_o+\lambda_i)^5-\lambda_i^5\}\\
(03)_O &= -\frac{1}{5}\tau m_y^4\left\{\left(L_o+\lambda_j\right)^5-\lambda_j^5\right\}\\
(12)_O &= (21)_O = 3\tau m_x^2 m_y^2\Big\{L_o^5/5+L_o^4(\lambda_i+\lambda_j)/2+L_o^3\left(\lambda_i^2+\lambda_j^2+4\lambda_i\lambda_j\right)/3\\
&\qquad +L_o^2\lambda_i\lambda_j(\lambda_i+\lambda_j)+L_o\lambda_i^2\lambda_j^2\Big\}.
\end{aligned}
\tag{92}
$$

Fishkova and Yavor also give formulae for the octopole contributions to (rs) when the octopole field is superimposed on a quadrupole field; the rectangular model is used and the effective lengths of quadrupole and octopole are both set equal to L. They find

$$
\begin{aligned}
\frac{(30)_O}{L} &= -\frac{\tau L^4}{8\theta^4}\Big\{3(1+\theta^2\zeta_x^2)^2-4\zeta_x(1+\theta^2\zeta_x^2)(1-\cos 2\theta)-4(1-\theta^4\zeta_x^4)\frac{\sin 2\theta}{2\theta}\\
&\quad +\zeta_x(1-\theta^2\zeta_x^2)(1-\cos 4\theta)+(1-6\theta^2\zeta_x^2+\theta^4\zeta_x^4)\frac{\sin 4\theta}{4\theta}\Big\}\\
\frac{(03)_O}{L} &= -\frac{\tau L^4}{8\theta^4}\Big\{3\left(1-\theta^2\zeta_y^2\right)^2-4\zeta_y\left(1-\theta^2\zeta_y^2\right)(1-\cosh 2\theta)-4\left(1-\theta^4\zeta_y^4\right)\frac{\sinh 2\theta}{2\theta}\\
&\quad +\zeta_y\left(1+\theta^2\zeta_y^2\right)(1-\cosh 4\theta)+\left(1+6\theta^2\zeta_y^2+\theta^4\zeta_y^4\right)\frac{\sinh 4\theta}{4\theta}\Big\}\\
\frac{(21)_O}{L} &= \frac{(12)_O}{L} = \frac{3\tau L^4}{8\,\theta^4}\Big[-2(1+\theta^2\zeta_x^2)\left(1-\theta^2\zeta_y^2\right)+\zeta_x\left(1-3\theta^2\zeta_y^2\right)\\
&\quad +\zeta_y(1+3\theta^2\zeta_x^2)+2\left(1-\theta^2\zeta_y^2\right)\Big\{(1-\theta^2\zeta_x^2)\frac{\sin 2\theta}{2\theta}-\zeta_x\cos 2\theta\Big\}\\
&\quad +2(1+\theta^2\zeta_x^2)\Big\{\left(1+\theta^2\zeta_y^2\right)\frac{\sinh 2\theta}{2\theta}-\zeta_y\cosh 2\theta\Big\}\\
&\quad -\Big\{\zeta_x\left(1+\theta^2\zeta_y^2\right)-\zeta_y(1-\theta^2\zeta_x^2)\Big\}\sin 2\theta\sinh 2\theta\\
&\quad +\Big\{\zeta_x\left(1+\theta^2\zeta_y^2\right)+\zeta_y(1-\theta^2\zeta_x^2)\Big\}\cos 2\theta\cosh 2\theta\\
&\quad -\Big\{(1-\theta^2\zeta_x^2)\left(1+\theta^2\zeta_y^2\right)-4\theta^2\zeta_x\zeta_y\Big\}\cosh 2\theta\frac{\sin 2\theta}{2\theta}\\
&\quad -\Big\{(1-\theta^2\zeta_x^2)\left(1+\theta^2\zeta_y^2\right)+4\theta^2\zeta_x\zeta_y\Big\}\cos 2\theta\frac{\sinh 2\theta}{2\theta}\Big]
\end{aligned}
\tag{93}
$$

in which

$$
\theta=\beta L,\quad \zeta_x=\frac{z_{ox}}{L}+\frac{1}{2},\quad \zeta_y=\frac{z_{oy}}{L}+\frac{1}{2}
\tag{94}
$$

(so that ζ_x and ζ_y denote the object distances when the origin is situated at the beginning of the rectangle, measured in units of L: in the notation of Fishkova and Yavor, $a_C = -L\zeta_x$, $a_D = -L\zeta_y$). We may rearrange these terms into the polynomial form discussed in § IID. We find

$$\begin{aligned}
\langle 30\rangle_4 &= -\tau L^4(3 + 2\sin 2\theta/\theta + \sin 4\theta/4\theta)/8 \\
\langle 30\rangle_3 &= \tau L^4(5 - 4\cos 2\theta - \cos 4\theta)/8\theta^2 \\
\langle 30\rangle_2 &= -3\tau L^4(1 - \sin 4\theta/4\theta)/4\theta^2 \\
\langle 30\rangle_1 &= \tau L^4(3 - 4\cos 2\theta + \cos 4\theta)/8\theta^4 \\
\langle 30\rangle_0 &= -\tau L^4(3 - 2\sin 2\theta/\theta + \sin 4\theta/4\theta)/8\theta^4
\end{aligned} \tag{95}$$

and

$$\begin{aligned}
\langle 12\rangle_{22} &= 3\tau L^4(2 + \sin 2\theta/\theta + \sinh 2\theta/\theta + \cosh 2\theta\sin 2\theta/2\theta + \cos 2\theta\sinh 2\theta/2\theta)/8 \\
\langle 12\rangle_{21} &= 3\tau L^4(3 - 2\cosh 2\theta - \sin 2\theta\sinh 2\theta - \cos 2\theta\cosh 2\theta)/8\theta^2 \\
\langle 12\rangle_{12} &= 3\tau L^4(-3 + 2\cos 2\theta - \sin 2\theta\sinh 2\theta + \cos 2\theta\cosh 2\theta)/8\theta^2 \\
\langle 12\rangle_{20} &= 3\tau L^4(-2 - \sin 2\theta/\theta + \sinh 2\theta/\theta + \cosh 2\theta\sin 2\theta/2\theta + \cos 2\theta\sinh 2\theta/2\theta)/8\theta^2 \\
\langle 12\rangle_{11} &= 3\tau L^4(\cosh 2\theta\sin 2\theta/\theta - \cos 2\theta\sinh 2\theta/\theta)/4\theta^2 \\
\langle 12\rangle_{02} &= 3\tau L^4(2 - \sin 2\theta/\theta + \sinh 2\theta/\theta - \cosh 2\theta\sin 2\theta/2\theta - \cos 2\theta\sinh 2\theta/2\theta)/8\theta^2 \\
\langle 12\rangle_{10} &= 3\tau L^4(1 - 2\cos 2\theta - \sin 2\theta\sinh 2\theta + \cos 2\theta\cosh 2\theta)/8\theta^4 \\
\langle 12\rangle_{01} &= 3\tau L^4(1 - 2\cosh 2\theta + \sin 2\theta\sinh 2\theta + \cos 2\theta\cosh 2\theta)/8\theta^4 \\
\langle 12\rangle_{00} &= 3\tau L^4(-2 + \sin 2\theta/\theta + \sinh 2\theta/\theta - \cosh 2\theta\sin 2\theta/2\theta - \cos 2\theta\sinh 2\theta/2\theta)/8\theta^4
\end{aligned} \tag{96}$$

in which

$$\frac{(30)_O}{L} = \sum \langle 30\rangle_i \zeta_x^i, \quad \frac{(12)_O}{L} = \sum \langle 12\rangle_{ij}{}^i\zeta_x^i\zeta_y^j. \tag{97}$$

Introducing the magnification, we find

$$\frac{(30)_O}{L} = \sum (30)_i m_x^i, \quad \frac{(12)_O}{L} = \sum (12)_{ij} m_x^i m_y^j \tag{98}$$

with

$$
\begin{aligned}
(30)_4 &= (30)_0 = -\tau L^4(3 + 2\sin 2\theta/\theta + \sin 4\theta/4\theta)/8\theta^4 \sin^4\theta \\
(30)_3 &= (30)_1 = \tau L^4(12\cos\theta + 11\sin\theta/\theta + 3\sin 3\theta/\theta)/8\theta^4 \sin^4\theta \\
(30)_2 &= -3\tau L^4(4 + 5\sin 2\theta/\theta + 2\cos 2\theta)/8\theta^4 \sin^4\theta
\end{aligned}
\tag{99}
$$

and

$$
\begin{aligned}
(12)_{22} = (12)_{00} = {}& 3\tau L^4(2 + \sin 2\theta/\theta + \sinh 2\theta/\theta + \cosh 2\theta \sinh 2\theta/2\theta \\
& + \cos 2\theta \sinh 2\theta/2\theta)/8\theta^4 \sin^2\theta \sinh^2\theta \\
(12)_{21} = (12)_{01} = {}& -3\tau L^4(4\cosh\theta + 5\sinh\theta/\theta + \cos 2\theta \sinh\theta/\theta \\
& + 3\cosh\theta \sin 2\theta/\theta)/8\theta^4 \sin^2\theta \sinh^2\theta \\
(12)_{12} = (12)_{10} = {}& -3\tau L^4(4\cos\theta + 5\sin\theta/\theta + \cosh 2\theta \sin\theta/\theta \\
& + 3\cos\theta \sinh 2\theta/\theta)/8\theta^4 \sin^2\theta \sinh^2\theta \\
(12)_{20} = (12)_{02} = {}& -3\tau L^4(4 + 3\sin 2\theta/\theta + 3\sinh 2\theta/\theta)/16\theta^4 \sin^2\theta \sinh^2\theta \\
(12)_{11} = {}& 3\tau L^4(2\cosh\theta\cos\theta + 3\cos\theta\sinh\theta/\theta + 3\cosh\theta\sin\theta/\theta)/2\theta^4 \sin^2\theta \sinh^2\theta.
\end{aligned}
\tag{100}
$$

Fishkova and Yavor finally discuss the correction of the aperture aberrations of the real line image of a single quadrupole and of a stigmatic doublet. They show that two (thin) octopoles are necessary to correct the line image in the converging plane of a (thin) quadrupole and give formulae for the strengths of the two octopoles and the position of the second when the first coincides with the quadrupole. Two octopoles can also correct the two aperture aberrations in one plane of a pseudo-stigmatic doublet (the x–z *or* the y–z plane) while three are necessary to correct all three aberrations at the point image. In the latter situation, the octopole strengths are as follows for complete correction, when two of the octopoles (suffices 1,2) coincide with quadrupoles and the third (suffix 3) is situated a distance λL_1 beyond the first quadrupole (thin-lens approximation):

$$
\begin{aligned}
\tau_1 L_o^{(1)} L_1^2 &= \frac{1}{6}\left[\left\{2\frac{\kappa_1}{L_1}n_1(n_1-1)-\frac{\kappa_2}{L_2}\left(n_2^2+2\right)\right\}\theta_1^4\right.\\
&\left.+\frac{\left\{\frac{\kappa_1}{L_1}\left(n_1^2+2\right)+\frac{\kappa_2}{L_2}\left(n_2^2+2\right)\right\}\left\{\theta_1^4\zeta^2 s^2+(s-\zeta)^2\right\}\left\{\theta_1^4\zeta^2\lambda^2+(\lambda-\zeta)^2\right\}}{4\zeta^2 s\lambda(s-\zeta)(\lambda-\zeta)}\right]\\
\tau_2 L_o^{(2)} L_1^2 &= \frac{1}{6}\zeta\left[2\frac{\kappa_2}{L_2}\left(3n_2^2-2n_2+2\right)\frac{\theta_1^4\zeta^4}{\left\{0_1^4\zeta^2 s^2-(s-\zeta)^2\right\}^2}\right.\\
&\left.+\frac{\left\{\frac{\kappa_1}{L_1}\left(n_1^2+2\right)+\frac{\kappa_2}{L_2}\left(n_2^2+2\right)\right\}\left\{\theta_1^4\zeta^2\lambda^2+(\lambda-\zeta)^2\right\}}{4s(s-\zeta)(s-\lambda)\left\{\theta_1^4\zeta^2 s\lambda-(s-\zeta)(\lambda-\zeta)\right\}}\right]\\
\tau_3 L_o^{(3)} L_1^2 &= -\frac{\zeta}{24}\frac{\left\{\frac{\kappa_1}{L_1}\left(n_1^2+2\right)+\frac{\kappa_2}{L_2}\left(n_2^2+2\right)\right\}\left\{\theta_1^4\zeta^2 s^2+(s-\zeta)^2\right\}}{\lambda(\lambda-\zeta)(s-\lambda)\left\{\theta_1^2\zeta^2 s\lambda-(s-\zeta)(\lambda-\zeta)\right\}}
\end{aligned}
\tag{101}
$$

in which the object is situated at $z=\zeta L_1$, with the origin at the center of the first quadrupole; $\theta_1=\beta_1 L_1$ and the second quadrupole is centered on $z=sL_1$ (Fig. 46). The numbers κ_1 and κ_2 are the values of κ (Eq. 152b, Section 4 of Definitions) for the two lenses. Fig. 47 shows how the octopole strengths required to correct a doublet of high magnification vary with object distance when the quadrupoles have the same length, $L_1=L_2=L$, $S=5$ and $\lambda=s/2=2\cdot5$; the ordinates correspond to $T_1=\tau_1 L_o^{(1)} L^3/\kappa$ (curve 1), $T_2=\tau_2 L_o^{(2)} L^3/\kappa$ (curve 2) and $T_3=\tau_3 L_o^{(3)} L^3/\kappa$ (curve 3).

8.2 Combined quadrupoles and octopoles

The idea of perturbing the potential distribution within a quadrupole in order to create a beneficial octopole component is an old one. Septier

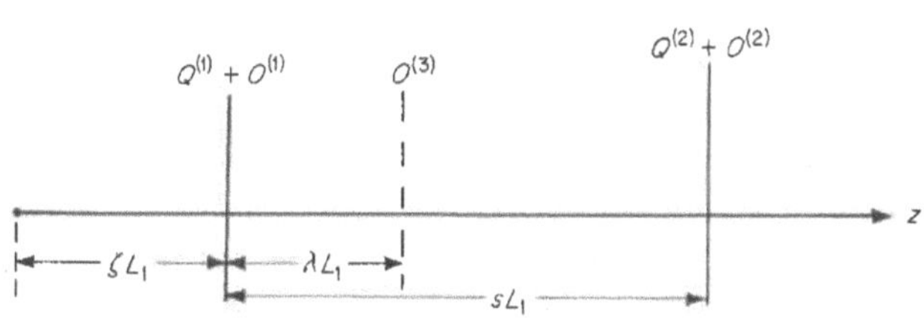

Fig. 46 The position of the octopole in the quadruplet doublet discussed by Fishkova and Yavor et al. (1968); Q: quadrupole, O: octopole.

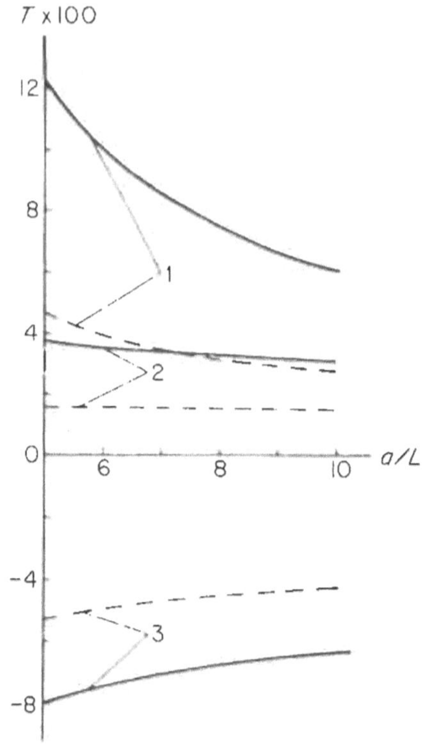

Fig. 47 The octopole strengths required to correct the aperture aberration of a high magnification doublet as a function of object distance. (After Fishkova & Yavor, 1968.)

(1957) improved the line image of a single magnetic quadrupole by inserting soft iron rods as did Reisman (1957), and Septier and van Acker (1960a, 1960b, 1961) corrected a single electrostatic quadrupole and a quadrupole doublet by exciting the electrodes asymmetrically. In a mainly theoretical paper, Markovich and Tsukkerman (1960) explained how magnetic quadrupoles could be modified (Fig. 2C in Definitions) so that they would produce an octopole as well as a quadrupole component and verified experimentally that this could be used to vary the aberrations at a real line focus. Strashkevich (1963) gave the theory of asymmetrically excited quadrupoles (Fig. 1C in Definitions) in the thin-lens approximation and the "two-dimensional" lens approximation (in which the electrodes are assumed to be of unlimited length parallel to the optic axis); Dymnikov et al. (1964a, 1964b) subsequently investigated this two-dimensional case in greater detail. In practice, it has been found convenient to combine octopole and quadrupole effects; Deltrap (1964a, 1964b) wound his magnetic eight-pole elements in such a way that both components were created while Dhuicq (1968a) inserted plates between the electrodes of his electrostatic quadrupoles.

One of the main mechanical problems of quadrupole systems arises from the large number of pole pieces or electrodes that must be aligned; it is clear that if all the octopole effects could be obtained by exciting the electrostatic lenses asymmetrically or by building the magnetic quadrupoles with the poles inclined at an oblique angle to one another, this problem would be greatly simplified. A systematic study of octopole effects created in this way is now being made by Preston (1968).

Nomenclature and Notation

1. Nomenclature

The reader who is not familiar with the literature of quadrupoles may find the following glossary helpful; a similar German-language list has been compiled by Scherzer (1965).

ANAMORPHOTIC producing nonuniform or anisotropic magnification; an anamorphotic system suffers from first order distortion.

ANASTIGMATIC free of, or corrected for, third-order astigmatism (cf. STIGMATIC).

ANTISYMMETRIC term describing multiplets consisting of an even number of components and possessing mirror symmetry about the center plane with respect to geometry and excitation except that the polarities in the second half are the reverse of those in the first (Σ_c in Dhuicq's notation; geometrically symmetric and electrically antisymmetric in Regenstreif's terminology).

APERTURE ABERRATIONS the counterpart of the spherical aberration of a round lens; the aberrations that do not vanish when the object is a point on the axis.

DISTORTION-FREE orthomorphic; free of first-order distortion; magnifying uniformly.

ORTHOGONAL term commonly used to describe systems in which there exists a coordinate system (x_1, x_2, z) in which the Gaussian equations of motion separate into an equation for $x_1(z)$ and another for $x_2(z)$; often used loosely of systems for which the surfaces $x_1 - z$ and $x_2 - z$ are planes.

ORTHOMORPHIC converse of ANAMORPHOTIC; see DISTORTION-FREE.

PRINCIPAL SECTION (OR DIRECTION) the plane (or more generally, the surface) in which electrons experience no force tending to make them leave it; the $x_1 - z$, $x_2 - z$ planes of ORTHOGONAL systems. The coordinate axes x_1, x_2 may be referred to as PRINCIPAL DIRECTIONS (German: Hauptschnitt).

PSEUDO-STIGMATIC term describing the imagery of a system producing a point-to-point image of an object for particular location(s) of the object plane only; not REGULAR.

REGULAR producing a point-to-point undistorted image of an object for all positions of the object plane; equivalent to a round lens so far as Gaussian imagery is concerned.

ROSETTE the APERTURE ABERRATIONS were classified by Burfoot (1954) into ROSETTE, STAR and spherical aberration; see Hawkes (1967e) for the relation between these coefficients and (00*rs*).

STANDARD ORIENTATION the PRINCIPAL SECTIONS of a quadrupole system will be planes, the $x-z$ and $y-z$ planes, if these planes intersect all electrodes and pass mid-way between all pole-pieces; the quadrupoles are then in the standard orientation Used by extension for octopoles.

STAR See ROSETTE.

STIGMATIC producing point-to-point imagery of an object plane in some image plane; free of first order astigmatism (cf. ANASTIGMATIC).

SYMMETRIC term describing multiplets possessing mirror symmetry about the center plane with respect to geometry and excitation (Σ_p in Dhuicq's notation).

TRANSFER MATRIX matrix summarizing the first-order (Gaussian) properties of an optical system (see Section 2 of Definitions, Notation and Methods of Analysis).

TURN-OVER when the TRANSFER MATRIX (IN A PRINCIPAL SECTION) between two planes, z_1 and z_2, is of the form $\begin{pmatrix} -1 & 0 \\ 0 & -1 \end{pmatrix}$, the transfer matrix between a second pair of planes, $z_1{}'$ and $z_2{}'$, both of which lie between z_1 and z_2, will be of the form $\mathbf{T}_{-b} = \begin{pmatrix} -1 & b \\ 0 & -1 \end{pmatrix}$. If an optical system is placed between $z_1{}'$ and $z_2{}'$, it is said to produce turn-over if its transfer matrix is of the form $\mathbf{T}_{-b}$.

2. Notation

The following symbols are used consistently throughout

a	bore-radius of quadrupoles
$B(z)$	magnetic field on the axis of a round magnetic lens

d	parameter occurring in symmetric bell-shaped model
e	charge on the electron is $-e$; $e \simeq 1 \cdot 6 \times 10^{-19}$C
f	asymptotic focal length of a round lens and of a quadrupole in the $x-z$ plane
g	asymptotic focal length of a quadrupole lens in the $y-z$ plane
$k(z)$	function defining shape of quadrupole potential, normalized to rise to a maximum value of unity
L	effective length, $L=\int_{-\infty}^{\infty} k(z)dz$
m, M	M denotes the magnification of a round lens system; $m=1/M$
m_0	rest mass of the electron; $m_0 \simeq 9 \times 10^{-31}$ kg
m_x, m_y, M_x, M_y	M_x and M_y denote the magnifications in the principal sections of a quadrupole lens system; $m_x=1/M_x$ and $m_y=1/M_y$
n	parameter characterizing type of quadrupole : $n=-1$, electrostatic; $n=0$, magnetic; $n=1$, achromatic (nonrelativistically)
p, q	parameters occurring in unsymmetric bell-shaped model
V	relativistically modified potential, $V=\Phi(1+\varepsilon\Phi)$
V_0	relativistically modified accelerating voltage, $V_0=\Phi_0(1+\varepsilon\Phi_0)$
z_{Fo}, z_{Fi}	position of the asymptotic object (o) and image (i) foci of a round lens, and of a quadrupole (system) in the $x-z$ plane
z_{Go}, z_{Gi}	position of the asymptotic object (o) and image (i) foci of a quadrupole (system) in the $y-z$ plane
z_{P_o}, z_{P_i}, z_{Q_o}, z_{Q_i}	position of the asymptotic principal planes in the $x-z$ plane (P_o, P_i) and in the $y-z$ plane (Q_o, Q_i)
α	angular aperture at the object plane or x-component at the object in the $x-z$ plane, for quadrupoles
β^2	$\beta_M^2-\beta_E^2$
β_E^2	measure of electrostatic quadrupole strength; $\beta_E^2 k(z)=\sigma D(z)/4V_0$
β_M^2	measure of magnetic quadrupole strength; $\beta_M^2 k(z)=\eta Q(z)/V_0^{1/2}$
γ	y_component of angular aperture at the object in the $y-z$ plane, for quadrupoles

ε	relativistic modification factor: $\varepsilon = e/2m_0c^2 \simeq 10^{-6}\ \mathrm{V}^{-1}$ or $1\ \mathrm{MV}^{-1}$
ζ	commonly used for scaled values of the coordinate z
η	reserved for the constant $(e/2m_0)^{1/2} \simeq 3 \times 10^5 \mathrm{C}^{1/2}\ \mathrm{kg}^{-1/2}$; also used for scaled values of the coordinate y
θ	the dimensionless quantity βL in rectangular model formulae
κ	quadrupole form factor, defined by $\kappa = \int_{-\infty}^{\infty} k^2(z)dz/L$
$\mu, \bar{\mu}$	the "projection" of the refractive index onto the z-axis; $\bar{\mu} = \{\phi(1+\varepsilon\phi)\}^{1/2} - \eta(\boldsymbol{A}_x x' + \boldsymbol{A}_y y' + \boldsymbol{A}_z)$ where $\boldsymbol{A}$ denotes vector potential; $\mu = \bar{\mu}/V_0^{1/2}$
ξ	used for scaled values of the coordinate x
Ξ	the octopole potential function, $\Xi(z) = 2(\sigma D_1 - 2\eta Q_1 V^{1/2})/V$
σ	$1 + 2\varepsilon\Phi \simeq 1 + 2\Phi$, Φ in MV
$\phi(x, y, z)$	electrostatic potential
$\Phi(z)$	electrostatic potential on the axis
Φ_0	accelerating voltage
$\psi(x, y, z)$	magnetostatic potential
ω_x^2, ω_y^2	dimensionless parameters occurring in bell-shaped model formulae; $\omega_x^2 = 1 + \beta^2 d^2$, $\omega_y^2 = 1 - \beta^2 d^2$
$\Omega(z)$	conveniently scaled magnetic field: $\Omega(z) = \eta B(z)/2V_0^{1/2}$

APPENDIX II

The shadow projection method of measuring aberration coefficients

This is at present the standard method of measuring aperture aberrations. This brief account of the essentials follows Deltrap (1964a), who first applied the method to the measurement of quadrupole aberrations, and Hardy (1967) closely.

We consider a system that produces separated line images of a (stigmatic or astigmatic) object on the optic axis at $z = z_i$ and $z = z_j$, and we consider the points of intersection of rays that pass through these line images with two planes: the grid plane, $z = z_G$ and the plane of the fluorescent screen, $z = z_S$ (Fig. A1). Let us consider a ray making angles α, γ (in image space) with the optic axis in the principal sections. It will intersect a general plane at a point with coordinates x, y, which may be expanded in the form

$$\begin{aligned} x &= p_{10}\alpha + p_{30}\alpha^3 + p_{12}\alpha\gamma^2 + p_{50}\alpha^5 + p_{32}\alpha^3\gamma^2 + p_{14}\alpha\gamma^4 \\ y &= p_{01}\gamma + p_{03}\gamma^3 + p_{21}\alpha^2\gamma + p_{05}\gamma^5 + p_{23}\alpha^2\gamma^3 + p_{41}\alpha^4\gamma. \end{aligned} \tag{A.1}$$

If $z_G \simeq z_i$, we have

$$x_G = z_{iG}\alpha + (30)\alpha^3 + (12)\alpha\gamma^2 + (50)\alpha^5 + (32)\alpha^3\gamma^2 + (14)\alpha\gamma^4 \tag{A.2}$$

and if the line images are widely separated,

$$y_G \simeq z_{jG}\gamma \tag{A.3}$$

where $z_{iG} = z_i - z_G$ and $z_{jG} = z_j - z_G$. If the line images are coincident ($z_i = z_j$) or close to each other, the expression for y_G resembles that for x_G. If $z_G \simeq z_j$, and the line images are widely separated, then

$$\begin{aligned} x_G &\simeq z_{iG}\alpha \\ y_G &= z_{jG}\gamma + (03)\gamma^3 + (21)\alpha\gamma^2 + (05)\gamma^5 + (23)\alpha^2\gamma^3 + (41)\alpha^4\gamma. \end{aligned} \tag{A.4}$$

(The coefficients (rs) are the various aperture aberration coefficients ($00rs$), defined in terms of angles in image space: the definitions appropriate to probe-forming lenses.)

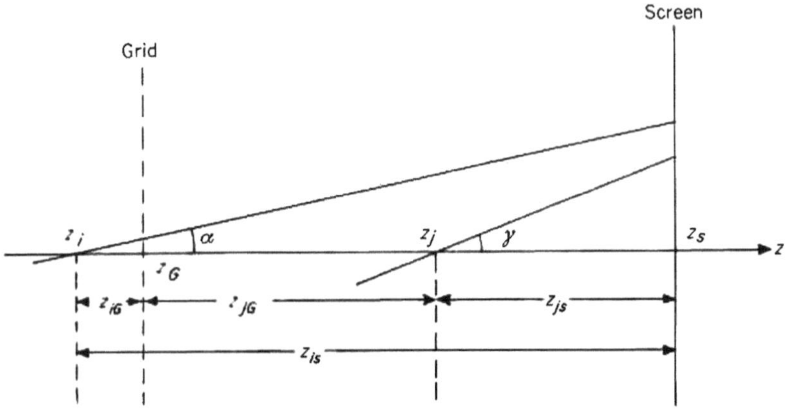

Fig. A1 Notation employed in the discussion of the shadow projection method of measuring aberration coefficients.

We now consider the position of the ray in the plane z_s; the coefficients p_{10} and p_{01} are now overwhelmingly large, provided that neither of the line images is near to the fluorescent screen, and hence

$$\begin{aligned} x_s &= z_{is}\alpha \\ y_s &= z_{js}\gamma. \end{aligned} \tag{A.5}$$

Let us assume that we have set $z_G \simeq z_i$ and that the line images are well separated. We eliminate the angles α and γ between Eqns. (A.2), (A.3), and (A.5), giving

$$\begin{aligned} x_G &= \frac{z_{iG}}{z_{is}} x_s + \frac{(30)}{z_{is}^3} x_s^3 + \frac{(12)}{z_{is} z_{js}^2} x_s y_s^2 + \text{fifth order terms} \\ y_G &= \frac{z_{jG}}{z_{js}} y_s. \end{aligned} \tag{A.6}$$

Temporarily neglecting the fifth-order terms, we find

$$x_s^3 + 3Hx_s + G = 0, \quad y_s = \frac{z_{j\,s}}{z_{jG}} y_G \tag{A.7}$$

where

$$3H = \left\{ z_{iG} + \frac{(12)}{z_{jG}^2} y_G^2 \right\} \frac{z_{is}^2}{(30)}, \quad G = -\frac{z_{is}^3}{(30)} x_G. \tag{A.8}$$

If therefore a square or rectangular mesh is placed in the grid plane, z_G, each grid-bar will produce a curve (x_s, y_s) in the plane of the fluorescent screen. The cubic equation for x_s has one real solution when $G^2 + 4H^3 > 0$ and three when $G^2 + 4H^3 \leqslant 0$ (two of which coincide at $G^2 + 4H^3 = 0$). The condition for three real roots implies

$$z_{iG}^3 \leqslant -\frac{27}{4} x_G^2 (30) \tag{A.9}$$

when $y_G = 0$ and

$$\left\{ z_{iG} + \frac{(12)}{z_{jG}^2} y_G^2 \right\}^3 \leqslant -\frac{27}{4} x_G^2 (30) \tag{A.10}$$

in general. When this condition is satisfied, the shadow of the grid-bar will consist of a closed loop and a line; when it is not, no closed loops will be seen. A typical set of patterns, showing both situations, is shown in Fig. A2.

Returning to the aberration coefficients and their measurement, let us consider how we can obtain the unknown quantities in Eqns. (A.2)–(A.4). Close to the center of the shadow pattern, the paraxial properties will dominate and in neighboring regions, each of the various aberration coefficients will be predominant. Hardy gives a convenient chart showing where each coefficient is important (Fig. A3). Measurements in the central region give z_{iG}, z_{jG} and since the distance between grid and screen is known, z_{is} and z_{js} are automatically given as well. Measurements of the positions of the various grid bars in more distant regions yield all the third- and fifth-order aperture aberration coefficients with reasonable accuracy; the precision depends to a large extent upon the centering of the grid and upon parasitic aberrations.

Deltrap obtains the aberrations by plotting x_G/x_s against x_s^2; Eqn. (A.6) shows that, neglecting fifth-order terms,

$$\frac{x_G}{x_s} = \frac{z_{iG}}{z_{is}} + \frac{(30)}{z_{is}^3} x_s^2 + \frac{(12)}{z_{is} x_{js}^2} y_s^2. \tag{A.11}$$

The slope of the curve yields (30) and the intercept where $x_s = 0$ gives $z_{iG}/z_{is} + [(12)/z_{is} z_{js}^2] y_s^2$; plotting the latter against y_s^2 for several values of y_s yields (12). Deltrap points out that this is equivalent to plotting longitudinal aberration (in terms of x_G/x_s) against angles (in terms of x_G, y_G). Mesh-lengths of the order of 30 μm (750 mesh per inch) or 50 μm are

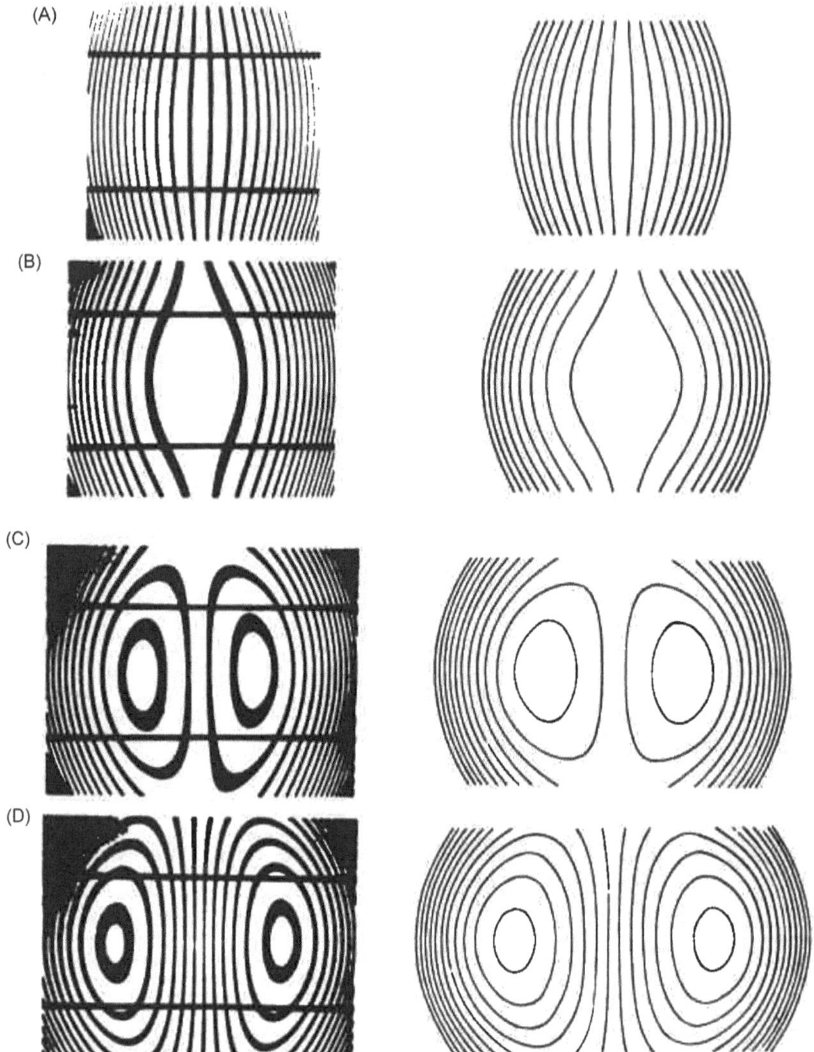

Fig. A2 Experimental and calculated shadow patterns; in (A) and (B), the cubic has only one real solution, in (C) and (D) there is a region in which it has three. Each loop is associated with a line in the other half of the pattern. (After Deltrap, 1964a.)

convenient in practice; it is usually necessary to insert a few bars of much wider spacing to give the parameters corresponding to a distant virtual line image.

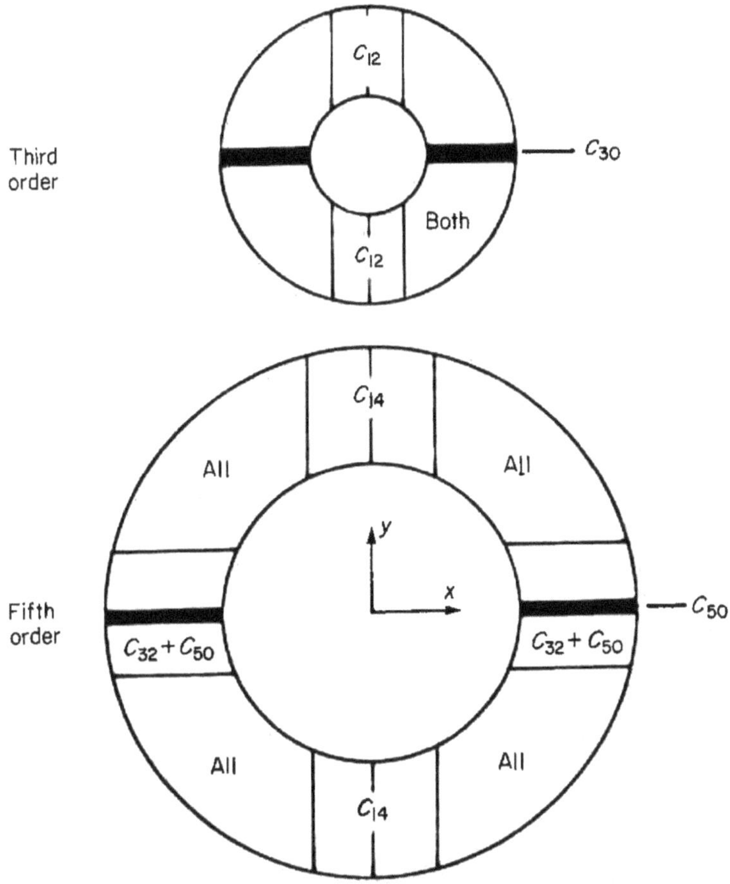

Fig. A3 Charts, showing which regions of the shadow pattern are influenced by specific third and fifth order aberrations. (After Hardy, 1967.)

Fishkova and Baranova (1968) have described a method using two grids, with 1.0 and 0.5 mm square meshes.

The most detailed, though by no means the earliest, published account of the shadow projection method of measuring aberration coefficients is to be found in the work of Kanaya, where round lenses are considered (Kanaya et al., 1959; Kanaya & Kawakatsu, 1960).

References†

Ahlberg, J. H., Nilson, E. N., & Walsh, J. L. (1967). *The theory of splines and their applications.* New York and London: Academic Press.

Antony-Spies, P., & Junior, P. (1967). *Zeitschrift für Angewandte Physik, 22,* 490.

Archard, G. D. (1954). *British Journal of Applied Physics, 5,* 294.

Archard, G. D. (1955). *Proceedings of the Physical Society of London, B68,* 156.

Archard, G. D. (1958a). *Proceedings of the Physical Society of London, 72,* 135.

Archard, G. D. (1958b). *The Review of Scientific Instruments, 29,* 1049.

Archard, G. D. (1960a). *Proc. Internat. Symp. X-ray Microscopy, X-ray Microanalysis, 2nd, Stockholm, 1959,* p. 337. Amsterdam: Elsevier.

Archard, G. D. (1960b). *British Journal of Applied Physics, 11,* 521.

Archard, G. D., Mulvey, T., & Petrie, D. P. R. (1960). *Proc. European Regional Conf. Electron Microscopy, Delft. Vol. 1,* p. 51. Delft: De Nederlandse Vereniging voor Electronenmicroscopie.

Baker, W. F. (1963). *Properties of quadrupole magnet doublets. BNL-795.*

Banford, A. P. (1966). *The transport of charged particle beams.* London: Spon.

*Baranova, L. A., Fishkova, T. Ya., & Yavor, S. Ya. (1968). Some types of electrostatic quadrupole-octopole lenses. *Radiotekhnika i Elektronika, 13,* 2108 (*Radio Eng. Electron Phys. (USSR)* (English Transl.) 13, 1854).

Bauer, H.-D. (1964). *Naturwissenschaften, 51,* 632.

Bauer, H.-D. (1965–66). *Optik, 23,* 596.

*Bauer, H.-D. (1968). Ein elektrostatisches Quadrupolsystem mit veränderlichen Öffnungsfehler. In L. Pátý (Ed.), *Fourth Czechoslovak conference on electronics and vacuum physics, Prague, 1968; texts of the contributed papers,* p. 490.

Belonozhko, O. B., Gaidukova, I. S., & Der-Shvarts, G. V. (1967). *Pribory i Tekhnika Eksperimenta No. 1, 78. Instr. Exp. Tech. (USSR)* (English Transl.) p. 76.

Bernard, M.-Y. (1953). La "focalisation forte" dans les accélérateurs linéaires d'ions. Thèse, Paris; (1954). *Annales de Physique (Paris), 9,* 633.

Bernard, M.-Y. (1955). *Comptes Rendus, 240,* 1612.

*Bezić, N., & Malecki, A. M. (1968). *Focal lengths and principal-plane positions of quadrupole doublet and triplet lenses. AECL-3160.*

Blewett, J. P. (1958). *On the design of quadrupole focusing systems. BNL:JPB-11.*

Blewett, J. P. (1959). *The focal properties of certain quadrupole lenses. BNL: JPB-13.*

Bok, A. B. (1968). *A mirror electron microscope.* Delft: Proefschrift.

Bok, A. B., Kramer, J., & Le Poole, J. B. (1964). *Proc. European Regional Conf. Electron Microscopy, 3rd, Prague, 1964 appendix,* p. 9. Prague: Publ. House Czech. Acad. Sci.

Bonshtedt, B. E., & Markovich, M. G. (1967). *The focusing and deflection of rays in electron-beam devices* (in Russian). Moscow: Sovetskoe Radio.

*Brown, K. L. (1968). A first- and second-order matrix theory for the design of beam transport systems and charged particle spectrometers. *Advances in Particle Physics, 1,* 71.

Brown, K. L., & Hindmarsh, A. C. (1965). *Properties of symmetric quadrupole magnet triplets. SLAC-47.*

Bruck, H. (1957). *Optique corpusculaire.* Paris: C.D.U. and S.E.D.E.S.

† Papers that appeared before August-September, 1968 are all discussed in the text; articles that became available between then and Summer 1969 and are especially relevant are indicated with an asterisk and their titles are included to show their scope.

Burfoot, J. C. (1952). *British Journal of Applied Physics, 3*, 22.

Burfoot, J. C. (1954). *Proceedings of the Physical Society of London, B67*, 523.

Burfoot, J. C. (1956). *Proc. Internat. Conf. Electron Microscopy, 3rd, London, 1954*, p. 105. London: Royal Microscopical Society.

Carathéodory, C. (1937). Geometrische Optik. *Ergebnisse der Mathematik und ihrer Grenzgebiete, 4*(5).

Carré, B. A. (1961). *The Computer Journal, 4*, 73.

Carré, B. A., & Wreathall, W. M. (1964). *Radio & Electronic Engineer, 27*, 446.

Castaing, R. (1951). *Application des sondes électroniques à une methode d'analyse ponctuelle chimique et cristallographique. Thèse, Paris (published by Editions de l'O.N.E.R.A., Paris, 1952).*

Cohen, D. (1966). *Proc. AMU-ANL Workshop High-voltage Electron Microscopy*, p. 25. Argonne Natl. Lab. ANL-7275.

Crewe, A. V. (1965). Some effects of quadrupole misalignments in a microscope. In *AMU-ANL High Voltage Electron Microscopy Newsletter No. 6.*

Crewe, A. V., Cohen, D., & Meads, P. F. (1968). *Proc. European Regional Conf. Electron Microscopy, 4th, Rome. Vol. 1*, p. 183. Rome: Tipografia Poliglotta Vaticana.

Crewe, A. V., Eggenberger, D. N., Welter, L. M., & Wall, J. (1967). *Journal of Applied Physics, 38*, 4257.

Deltrap, J. H. M. (1964a). *Correction of spherical aberration of electron lenses.* Dissertation, Cambridge University.

Deltrap, J. H. M. (1964b). *Proc. European Regional Conf. Electron Microscopy, 3rd, Prague. Vol. A*, p. 45. Prague: Publ. House Czech. Acad. Sci.

*Der-Shvarts, G. V., & Arkhipova, N. V. (1969). Estimate of the efficacy of full and partial correction of the third order spherical aberration of probe systems. *Radiotekhnika i Elektronika, 14*, 740 (*Radio Eng. Electron Phys. (USSR)* (English Transl.) *14*, 639).

*Der-Shvarts, G. V., Arkhipova, N. V., & Krupnova, E. A. (1969). Fifth order spherical aberration of axially symmetric round magnetic lenses. *Radiotekhnika i Elektronika, 14*, 738 (*Radio Eng. Electron Phys.* (*USSR*) (English Transl.) *14*).

Der-Shvarts, G. V., & Makarova, I. S. (1967). *Radiotekhnika i Elektronika, 12*, 168 (*Radio Eng. Electron Phys. (USSR)* (English Transl.) *12*, 161).

Der-Shvarts, G. V., & Netrebenko, K. A. (1959). *Izvestiya Akademii Nauk SSR, Seriya Fizicheskaya, 23*, 506 (*Bull. Acad. Sci. USSR (Phys. Ser.)* (English Transl.) *23*, 488).

Dhuicq, D. (1961). *Recherche de systèmes de lentilles quadrupolaires stigmatiques au ler ordre, Thèse 3ᵉ Cycle, Paris.*

Dhuicq, D. (1968a). *Contribution à la correction des aberrations d'ouverture du 3ème ordre dans les systèmes de lentilles quadrupolaires stigmatiques au ler ordre.* Thèse, Paris-Orsay.

Dhuicq, D. (1968b). *Comptes Rendus, B266*, 906.

Dhuicq, D. (1968c). *Journal de Microscopie, 7*, 297.

Dhuicq, D., & Septier, A. (1966a). *Comptes Rendus, B263*, 280.

Dhuicq, D., & Septier, A. (1966b). *Comptes Rendus, B263*, 364.

Dhuicq, D., & Septier, A. (1968). *Proc. European Regional Conf. Electron Microscopy, 4th, Rome. Vol. 1*, p. 177. Rome: Tipografia Poliglotta Vaticana.

*Dhuicq, D., & Septier, A. (1969). Premiers essais d'un objectif constitué par un ensemble de lentilles quadrupolaires. Colloque annuel, SFME, Lausanne. *Journal de Microscopie, 8*, 10a.

Doinikov, N. I. (1966a). *Elektrofizicheskaya Apparatura, 4*, 84.

Doinikov, N. I. (1966b). *Proc. Internat. Conf. High Energy Accelerators, 5th, Frascati. 1965*, p. 721. Rome: CNEN.

Doinikov, N. I., & Samsonov, G. N. (1967). *Pribory i Tekhnika Eksperimenta*, No. 3, 22 (*Instr. Exp. Tech. (USSR)* (English Transl.) *10*, 492).

Doinikov, N. I., & Titov, V. A. (1966). *Elektrofizicheskaya Apparatura, 5*, 21.

Dupouy, G., Perrier, F., & Trinquier, J. (1968). *Comptes Rendus, B266*, 1102.

Durand, E. (1966). *Electrostatique* (Vols. II and III). Paris: Masson.

Dušek, H. (1959). *Optik, 16*, 419.

Dutova, K. P., & Kas'yankov, P. P. (1963). *Izvestiya Akademii Nauk SSR, Seriya Fizicheskaya, 27*, 1127 (*Bull. Acad. Sci. USSR (Phys. Ser.)* (English Transl.) 27, 1108).
Dymnikov, A. D. (1968). *Zhurnal Tekhnicheskoi Fiziki, 38*, 1120 (*Soviet Phys.–Tech. Phys.* (English Transl.) *13*, 929).
Dymnikov, A. D., Fishkova, T. Ya., & Yavor, S. Ya. (1964a). *Zhurnal Tekhnicheskoi Fiziki, 34*, 1711 (*Soviet Phys.–Tech. Phys.* (English Transl.) *9*, 1322).
Dymnikov, A. D., Fishkova, T. Ya., & Yavor, S. Ya. (1964b). *Doklady Akademii Nauk SSSR, 154*, 1321 (*Soviet Phys. Doklady* (English Transl.) *9*, 182).
Dymnikov, A. D., Fishkova, T. Ya., & Yavor, S. Ya. (1965a). *Zhurnal Tekhnicheskoi Fiziki, 35*, 431 (*Soviet Phys.–Tech. Phys.* (English Transl.) *10*, 340).
Dymnikov, A. D., Fishkova, T. Ya., & Yavor, S. Ya. (1965b). *Zhurnal Tekhnicheskoi Fiziki, 35*, 759 (*Soviet Phys.–Tech. Phys.* (English Transl.) *10*, 592).
Dymnikov, A. D., Fishkova, T. Ya., & Yavor, S. Ya. (1965c). *Zhurnal Tekhnicheskoi Fiziki, 35*, 1068 (*Soviet Phys.–Tech. Phys.* (English Transl.) *10*, 822).
Dymnikov, A. D., Fishkova, T. Ya., & Yavor, S. Ya. (1965d). *Doklady Akademii Nauk SSSR, 162*, 1265 (*Soviet Phys. Doklady* (English Transl.) *10*, 547).
Dymnikov, A. D., Fishkova, T. Ya., & Yavor, S. Ya. (1965e). *Nuclear Instruments & Methods, 37*, 268.
Dymnikov, A. D., Fishkova, T. Ya., & Yavor, S. Ya. (1965f). *Physics Letters, 15*, 132.
Dymnikov, A. D., Fishkova, T. Ya., & Yavor, S. Ya. (1966). *Izvestiya Akademii Nauk SSR, Seriya Fizicheskaya, 30*, 739 (*Bull. Acad. Sci. USSR (Phys. Ser.)* (English Transl.) *30*, 766).
Dymnikov, A. D., Fishkova, T. Ya., & Yavor, S. Ya. (1967). *Radiotekhnika i Elektronika, 12*, 662 (*Radio Eng. Electron. Phys. (USSR)* (English Transl.) *12*, 613).
Dymnikov, A. D., Ovsyannikova, L. P., & Yavor, S. Ya. (1963). *Zhurnal Tekhnicheskoi Fiziki, 33*, 393 (*Soviet Phys.–Tech. Phys.* (English Transl.) *8*, 293).
Enge, H. (1959). *The Review of Scientific Instruments, 30*, 248.
Enge, H. (1961). *The Review of Scientific Instruments, 32*, 662.
Fifer, S. (1961). *Analogue Computation (4 volumes)*. New York: McGraw-Hill.
Fishkova, T. Ya., & Baranova, L. A. (1968). *Radiotekhnika i Elektronika, 13*, 1525 (*Radio Eng. Electron. Phys. (USSR)* (English Transl.) *13*, 1331).
Fishkova, T. Ya., Baranova, L. A., & Yavor, S. Ya. (1968a). *Zhurnal Tekhnicheskoi Fiziki, 38*, 694 (*Soviet Phys.–Tech. Phys.* (English Transl.) *13*, 520).
Fishkova, T. Ya., Baranova, L. A., & Yavor, S. Ya. (1968b). *Izvestiya Akademii Nauk SSR, Seriya Fizicheskaya, 32*, 1026 (*Bull. Acad. Sci. USSR (Phys. Ser.)* (English Transl.) *32*, 951).
Fishkova, T. Ya., Ovsyannikova, L. P., & Yavor, S. Ya. (1966). *Proc. Internat. Cong. Electron Microscopy, 6th, Kyoto. Vol. I*, p. 209. Tokyo: Maruzen.
*Fishkova, T. Ya., & Shpak, E. V. (1969). Compensation of spherical aberration with a stigmatic triplet of short quadrupole-octopole lenses. *Zhurnal Tekhnicheskoi Fiziki, 39*, 1333 (*Soviet Phys.–Tech. Phys.* (English Transl.) *14*).
*Fishkova, T. Ya., Shpak, E. V., Szilágyi, M., & Yavor, S. Ya. (1968). A two-dimensional field investigation of electrostatic quadrupole lenses composed of cylindrical surfaces. *Periodica Polytechnica Electrical Engineering, 12*, 301.
Fishkova, T. Ya., Siladi, M., & Yavor, S. Ya. (1967). *Radiotekhnika i Elektronika, 12*, 1311 (*Radio Eng. Electron. Phys. (USSR)* (English Transl.) *12*, 1222).
Fishkova, T. Ya., & Yavor, S. Ya. (1968). *Zhurnal Tekhniceskoj Fiziki, 38*, 686 (*Soviet Phys.–Tech. Phys.* (English Transl.) *13*, 514).
Forsythe, G. E., & Wasow, W. R. (1960). *Finite-difference methods for partial differential equations*. New York: Wiley.
Foster, A. (1968). *Correction of aperture aberrations in magnetic lens spectrometers*. Thesis, London.
Fox, L., & Goodwin, E. T. (1949). *Proceedings. Cambridge Philosophical Society, 45*, 373.
Francken, J. C. (1959). *Philips Technical Review, 21*, 10.
Francken, J. C. (1967). In A. Septier (Ed.), *Vol. I. Focusing of charged particles*, p. 101. New York: Academic Press.

*Freindl, L., & Karcz, W. (1968). Ion focusing properties of a double magnetic quadrupole lens pair. *Nukleonika, 13*, 859.

Germain, C. (1967). In A. Septier (Ed.), *Vol. I. Focusing of charged particles*, p. 163. New York: Academic Press.

Gershtein, G. M. (1958). *Izvestiya Vysshikh Uchebnykh Zavedenii, 1*(2), 13.

Gershtein, G. M. (1959a). *Izvestiya Vysshikh Uchebnykh Zavedenii, 2*, 602 (English transl. U.S. Rep. OTS 60-41,412 or JPRS-5610).

Gershtein, G. M. (1959b). *Radiotekhnika i Elektronika, 4*, 137 *Radio Eng. Electron Phys. (USSR)* (English Transl.) *4*, 226).

Gershtein, G. M. (1960). *Certain questions concerning the interaction of charged particles with an electric field* (in Russian). Saratov: Izdatel'stvo Saratovskogo Univ.

Gershtein, G. M. (Ed.). (1964). *Voprosy elektricheskogo modelirovaniya polei*, p. 1.

Gershtein, G. M. (Ed.). (1967). *Voprosy elektricheskogo modelirovaniya polei*, p. 2.

Glaser, W. (1940). *Zeitschrift für Physik, 116*, 19.

Glaser, W. (1952). *Grundlagen der Elektronenoptik*. Vienna: Springer.

Glaser, W. (1955). *Investigations of four-pole lenses; an electron microscope corrected with respect to spherical aberration*. Farrand Opt. Comp. Internal. Repts.

Glaser, W. (1956a). Elektronen- und lonenoptik. In S. Flügge (Ed.), *Handbuch der Physik, Vol. 33*, p. 123. Berlin: Springer.

Glaser, W. (1956b). *Electron lens*. U.S. Patent No. 2,919,381. (Application July 25, 1956, patented December *29*, 1959.).

Gouiran, R. (1967). Focalisation et aberrations chromatiques des faisceaux de particules dans des dispositifs périodiques. Thèse, Rennes; (1968). *Annales de Physique (Paris), 3*, 67.

Gregory, B. C., & Sander, K. F. (1962). *Journal of Electronics & Control, 13*, 123.

Grivet, P. (1965). *Electron optics*. Oxford: Pergamon.

Grivet, P., & Septier, A. (1958). *Les lentilles quadrupolaires magnétiques* (pp. 58-25). CERN.

Grivet, P., & Septier, A. (1960). *Nuclear Instruments & Methods, 6*, 126 & 243.

Grümm, H. (1956). *Annalen der Physik, 17*(6), 269. 275.

Gullstrand, A. (1915). Kgl. Svenska Vetenskapsakadamiens Handlingar. No. 55(1), 139 pp.

* Halbach, K. (1969). First order perturbation effects in iron-dominated two-dimensional symmetrical multipoles. UCRL-18841. *Nuclear Instruments & Methods, 74*, 147.

Hanszen, K. J., & Lauer, R. (1965–66). *Optik, 23*, 478.

Hardisty, B. M. M. (1964). *The computation of electron trajectories*. Dissertation, Cambridge University

Hardy, D. F. (1967). *Combined magnetic and electrostatic quadrupole electron lenses*. Dissertation, Cambridge University.

Hart, R. K. (1966). *Proc. AMU-ANL Workshop High-voltage Electron Microscopy*, p. 96. Argonne Natl. Lab. ANL-7275.

Hawkes, P. W. (1963). *The aberrations of electron optical systems in the absence of rotational symmetry*. Dissertation, Cambridge University.

Hawkes, P. W. (1965–66). *Optik, 23*, 145.

Hawkes, P. W. (1965a). *Philosophical Transactions. Royal Society of London, A257*, 479.

Hawkes, P. W. (1965b). *Philosophical Transactions. Royal Society of London, A257*, 523.

Hawkes, P. W. (1965c). *Optik, 22*, 543.

Hawkes, P. W. (1966–67). *Optik, 24*, 252. 275.

Hawkes, P. W. (1966a). *Quadrupole optics* (Springer Tracts Mod. Phys. *42*). Berlin: Springer.

Hawkes, P. W. (1966b). *Proc. AMU-ANL Workshop High-voltage Electron Microscopy*, p. 32. Argonne Natl. Lab. ANL-7275.

Hawkes, P. W. (1967a). *British Journal of Applied Physics, 18*, 545.

Hawkes, P. W. (1967b). *Journal de Microscopie, 6*, 917.

Hawkes, P. W. (1967c). *Optik, 25*, 439.

Hawkes, P. W. (1967d). *Optik, 26*, 497.

Hawkes, P. W. (1967e). In A. Septier (Ed.), *The focusing of charged particles, Vol. 1*, p. 411. New York: Academic Press.

Hawkes, P. W. (1968a). *British Journal of Applied Physics, 1*(2), 131.
Hawkes, P. W. (1968b). *Optik, 27,* 287.
Hawkes, P. W. (1968c). *Proc. European Regional Conf. Electron Microscopy, 4th, Rome. Vol. I,* p. 181. Rome: Tipografia Poliglotta Vaticana.
Hawkes, P. W. (1968d). *Journal of Physics D: Applied Physics 1,* 1549.
*Hawkes, P. W. (1969). Cardinal elements and aperture aberrations of asymmetric quadrupoles and quadrupole multiplets. I: The asymmetric quadrupole. II: The antisymmetric doublet of asymmetric quadrupoles. III: The antisymmetric quadruplet of symmetric quadrupoles. *Optik, 29,* 471. 485 & 507.
Hawkes, P. W., & Cham, E. L. (1967). *Optik, 26,* 507.
Hawkes, P. W., & Hardy, D. F. (1967). *Optik, 25,* 450.
*Hayashi, T., & Sakudo, N. (1968). *The Review of Scientific Instruments, 39,* 958.
*Hayashi, T., & Sakudo, N. (1969). Quadrupole field in circular concave electrodes with finite thickness. *The Review of Scientific Instruments, 40,* 923.
*Himmelbauer, E. E. (1968). *Design considerations for a wide-band oscilloscope tube with scan magnification.* Groningen: Proefschrift. (1969) Philips Res. Repts Suppl. 1.
Hutter, R. G. E. (1945). *Journal of Applied Physics, 16,* 670.
Jennings, J. C. E., & Pratt, R. G. (1955). *Proceedings of the Physical Society of London, B68,* 526.
Junior, P., & Antony-Spies, P. (1967). *Zeitschrift für Angewandte Physik, 22,* 420.
Kanaya, K., & Kawakatsu, H. (1960). Method of measuring spherical and astigmatic aberrations by shadow electron microscope images. *Journal of Electronmicroscopy, 8,* 1.
Kanaya, K., Kawakatsu, H., & Kato, A. (1959). Theoretical considerations for shadow electron microscope images. *Bulletin of the Electrotechnical Laboratory, 23,* 801.
*Kanaya, K., Kawakatsu, H., & Okayama, S. (1969). Spherical aberration of electrostatic symmetrical doublet. *Bulletin of the Electrotechnical Laboratory, 33.*
Kanaya, K., & Matsuura, N. (1967). Electron optical properties of quadrupole electron lenses. *Communicated to the St. Andrews Conf. Electron Optics, Instrumentation, and Quantitative Electron Microscopy.*
*Kanaya, K., & Okayama, S. (1969). Third order aberrations of an electrostatic quadrupole lens for a modified bell-shaped model. *Bulletin of the Electrotechnical Laboratory, 33.*
Kanaya, K., Okayama, S., & Matsuura, N. (1968a). *Bulletin of the Electrotechnical Laboratory, 32,* 565.
Kanaya, K., Okayama, S., & Mishina, H. (1968b). *Bulletin of the Electrotechnical Laboratory, 32,* 583.
Kanaya, K., Okayama, S., & Mishina, H. (1968c). *Bulletin of the Electrotechnical Laboratory, 32,* 665.
Karplus, W. J. (1958). *Analog simulation.* New York: McGraw-Hill.
Karplus, W. J., & Soroka, W. W. (1959). *Analog methods.* New York: McGraw-Hill.
Kartashev, V. P., & Kotov, V. I. (1966). *Zhurnal Tekhnicheskoi Fiziki, 36,* 1569 (*Soviet Phys.–Tech. Phys.* (English Transl.) *11,* 1173).
*Kartashev, V. P., Kotov, V. I., & Utochkin, B. A. (1968). Increasing the resolution of cathode ray tubes. *Radiotekhnika i Elektronika, 13,* 1854. (*Radio Eng. Electron Phys. (USSR)* (English Transl.) *13,* 1621). This paper contains triplet curves.
Kartashev, V. P., & Utochkin, B. A. (1967). *Radiotekhnika i Elektronika, 12,* 1132 (Radio Eng. Electron Phys. (USSR) (English Transl.) *12,* 1055).
Kas'yankov, P. P. (1952). *Zhurnal Tekhnicheskoi Fiziki, 22,* 80.
Kas'yankov, P. P. (1967). In G. I. Marchuk (Ed.), *Vychislitel'nye Sistemy: Chislennye Metody Rascheta Elektronno-Opticheskikh Sistem,* p. 4. Novosibirsk: Nauka.
Kawakatsu, H., Vosburgh, K. G., & Siegel, B. M. (1968a). *Journal of Applied Physics, 39,* 245.
Kawakatsu, H., Vosburgh, K. G., & Siegel, B. M. (1968b). *Journal of Applied Physics, 39,* 255.
Keck, C., Liebich, W., & Pose, H. (1967). *Kernenergie, 10,* 247. 273, 349.
*Kel'man, V. M. (1968). *Electron optics* (in Russian). Moscow: Nauka.
Kel'man, V. M., & Yavor, S. Ya. (1961). *Zhurnal Tekhniceskoj Fiziki, 31,* 1439 (*Soviet Phys.–Tech. Phys.* (English Transl.) *6,* 1052).

Kel'man, V. M., & Yavor, S. Ya. (1963a). *Zhurnal Tekhniceskoj Fiziki*, *33*, 368 (*Soviet Phys.–Tech. Phys.* (English Transl.) *8*, 271).

Kel'man, V. M., & Yavor, S. Ya. (1963b). *Electron optics* (in Russian) (2nd ed.). Moscow-Leningrad: Izd. Akad. Nauk USSR.

Kel'man, V. M., & Yavor, S. Ya. (1968). *Electron optics* (in Russian) (3rd ed.). Leningrad: Nauka.

Kel'man, V. M., Yavor, S. Ya., Dymnikov, A. D., & Ovsyannikova, L. P. (1963). *Izvestiya Akademii Nauk SSSR, Seriya Fizicheskaya*, *27*, 1135 (*Bull. Acad. Sci. USSR (Phys. Ser.)* (English Transl.) *27*, 1116).

King, N. M. (1962). *Some focusing properties of quadrupole doublets. CERN MPS/EP/22.*

King, N. M. (1963). *Focusing with quadrupole triplets: Thin lens approximation. CERN MPS/EP/63-6.*

Koltay, E., & Szabó, Gy. (1965). *Nuclear Instruments & Methods*, *35*, 88.

Le Poole, J. B. (1964). *Proc. European Conf. Electron Microscopy, 3rd, Prague, 1964 Appendix*, p. 8. Prague: Publ. House Czech. Acad. Sci.

*Lee-Whiting, G. E., & Bezić, N. (1969). Beam matching with quadrupole lenses. *Nuclear Instruments & Methods*, *71*, 61.

Lenormand, J.-M. (1968). *Comptes Rendus*, *B266*, 1313.

Lenz, F. (1958). *Optik*, *15*, 393.

Liebmann, G. (1950). *British Journal of Applied Physics*, *1*, 92.

Liebmann, G. (1952). *Proceedings of the Physical Society of London*, *B65*, 94.

Markovich, M. G., & Tsukkerman, I. I. (1960). *Zhurnal Tekhnicheskoi Fiziki*, *30*, 1362 (*Soviet Phys.–Tech. Phys.* (English Transl.) *5*, 1292).

McKeown, D. L., & Ferrari, R. L. (1967). The impedance of a resonance probe. *International Journal of Electronics*, *23*, 39.

Meads, P. F. (1963). *The theory of aberrations of quadrupole focusing arrays. Thesis, Part I.* University of California. issued as UCRL-10807.

Meads, P. F. (1967). *The Review of Scientific Instruments*, *38*, 1752.

Melkich, A. (1944). *Ausgezeichnete astigmatische Systeme der Elektronenoptik. Dissertation, Berlin (1947). Sitzungsberichte der Akademie der Wissenschaften in Wien*, Abt. IIA, 155, 393, 440.

Meyer, W. E. (1961). *Optik*, *18*, 69.

Möllenstedt, G. (1956a). *Proc. Internat. Conf. Electron Microscopy, 3rd, London, 1954*, p. 694. London: Royal Microscopical Society.

Möllenstedt, G. (1956b). *Optik*, *13*, 209.

*Möller, P.-A. (1968). Calcul de l'acceptance dans le plan de phases d'un quadruplet antisymétrique formé de lentilles quadrupolaires. Application au cas d'une matrice de transfert unité. *Comptes Rendus*, *B267*, 768.

Möller, P.-A., Dhuicq, D., & Septier, A. (1966a). *Comptes Rendus*, *B263*, 534.

Möller, P.-A., Dhuicq, D., & Septier, A. (1966b). *Comptes Rendus*, *B263*, 955.

Moses, R. W. (1966). *The Review of Scientific Instruments*, *37*, 1370.

Moses, R. W. (1968). *Minimum aperture aberrations of quadrupole lens systems*. Thesis, Univ. Wisconsin, Madison. *Review of Scientific Instruments 41* (1970) 729 and *42* (1971) 828 & 832.

My Chzhen, U. (1957). *Scientia Sinica*, *6*, 833.

*Norman, B. A., & Moore, W. H. (1967). *Theory and design of beam transport systems. BNL-12138.*

Ovsyannikova, L. P. (1968). *Radiotekhnika i Elektronika*, *13*, 1489 (*Radio Eng. Electron. Phys. (USSR)* (English Transl.) 13, 1292).

Ovsyannikova, L. P., Chechulin, V. N., & Yavor, S. Ya. (1968a). *Izvestiya Akademii Nauk SSR, Seriya Fizicheskaya*, *32*, 1031 (*Bull. Acad. Sci. USSR (Phys. Ser.)* (English Transl.) *32*, 956).

*Ovsyannikova, L. P., Chechulin, V. N., & Yavor, S. Ya. (1968b). Spherical aberration of a short quadrupole lens taking the shape of the poles and electrodes into account. *Zhurnal Tekhnicheskoi Fiziki*, *38*, 1953 (*Soviet Phys.–Tech. Phys.* (English Transl.) *13*, 1566).
Ovsyannikova, L. P., & Yavor, S. Ya. (1965). *Zhurnal Tekhnicheskoi Fiziki*, *35*, 940 (*Soviet Phys.–Tech. Phys.* (English Transl.) *10*, 723).
Ovsyannikova, L. P., & Yavor, S. Ya. (1967a). *Radiotekhnika i Elektronika*, *12*, 489 (*Radio Eng. Electron. Phys. (USSR)* (English Transl.) *12*, 449).
Ovsyannikova, L. P., & Yavor, S. Ya. (1967b). *Doklady Akademii Nauk SSSR*, *173*, 76 (*Soviet Phys. Doklady* (English Transl.) *12*, 261).
*Ovsyannikova, L. P., & Yavor, S. Ya. (1968). Aberrations of an electrostatic quadrupole lens, associated with departures from the field symmetry. *Zhurnal Tekhnicheskoi Fiziki*, *38*, 1810 (*Soviet Phys.–Tech. Phys.* (English Transl.) *13*, 1459).
*Ovsyannikova, L. P., & Yavor, S. Ya. (1969). Weak lens theory in electron optics. *Zhurnal Tekhnicheskoi Fiziki*, *39*, 412 [*Soviet Phys.–Tech. Phys.* (English Transl.) *14*].
Petrie, D. P. R. (1962). *Proc. Internat. Cong. Electron Microscopy, 5th, Philadelphia. Vol. I*, p. KK-2. New York: Academic Press.
Picht, J. (1963). *Einführung in die Theorie der Elektronenoptik* (3rd ed.). Leipzig: Barth.
Preikszas, D., & Rose, H. (1995). *Optik, 100*, 179.
Preston, H. E. R. (1968). *Proc. European Regional Conf. Electron Microscopy, 4th, Rome. Vol. I*, p. 179. Rome: Tipografia Poliglotta Vaticana.
*Preston, H. E. R. (1969). An analytic solution for the potential distribution within an asymmetrically excited electrostatic quadrupole lens. *Proceedings of the Royal Society A*, *313*, 217.
Regenstreif, E. (1964). *Chromatic aberrations in quadrupole multiplets*, CERN 64-2.
Regenstreif, E. (1965). *Focusing and chromatic aberrations of a periodic AG beam transport system*, CERN 65-42.
Regenstreif, E. (1966a). *Basic properties of turn-over matrices*. UCRL AS/Theoretical/02.
Regenstreif, E. (1966b). *General conditions for achieving phase shifts of π and 2π by means of quadrupole multiplets*. BN-10813.
Regenstreif, E. (1966c). *Comptes Rendus*, *B263*, 1205.
Regenstreif, E. (1966d). *Comptes Rendus*, *B263*, 1297.
Regenstreif, E. (1967a). *Phase-space transformations by means of quadrupole multiplets*, CERN 67-6.
Regenstreif, E. (1967b). *Possible and impossible phase-space transformations by means of alternating-gradient doublets and triplets*, CERN 67-8.
Regenstreif, E. (1967c). *Sur la theorie du quadruplet de retournement*, CERN 67-22.
Regenstreif, E. (1967d). *Comptes Rendus*, *B264*, 596.
Regenstreif, E. (1967e). *Comptes Rendus*, *B264*, 684.
Regenstreif, E. (1967f). *Comptes Rendus*, *B265*, 234.
Regenstreif, E. (1967g). *Comptes Rendus*, *B265*, 298.
Regenstreif, E. (1967h). In A. Septier (Ed.), *The focusing of charged particles. Vol. 1*, p. 353. New York: Academic Press.
Regenstreif, E. (1968). *Comptes Rendus*, *B266*, 383.
*Reichenbach, M., & Rose, H. (1968–69). Entwurf eines korrigierten magnetischen Objektivs. *Optik*, *28*, 475.
Reisman, E. (1957). *The imaging properties of crossed magnetic quadripoles*. Thesis, Ithaca: Cornell University.
Reisman, E., & Siegel, B. M. (1957). *Bulletin of the American Physical Society*, *2*(2), 174.
Rose, H. (1966a). *Optik*, *24*, 36.
Rose, H. (1966b). *Optik*, *24*, 108.
Rose, H. (1967a). *Optik*, *25*, 587.
Rose, H. (1967b). *Optik*, *26*, 289.

*Rose, H. (1968a). Über die Berechnung der Bildfehler elektronenoptischer Systeme mit gerader Achse. *Optik*, *27*, 466–497.
*Rose, H. (1968b). Der Zusammenhang der Bildfehler-Koeffizienten mit den Entwicklungs-Koeffizienten des Eikonals. *Optik*, *28*, 462.
Scherzer, O. (1936). *Zeitschrift für Physik*, *101*, 593.
Scherzer, O. (1947). *Optik*, *2*, 114.
Scherzer, O. (1949). *Journal of Applied Physics*, *20*, 20.
Scherzer, O. (1965). *Optik*, *22*, 314.
Scherzer, O., & Typke, D. (1967). *Optik*, *26*, 564.
Schiske, P. (1966–67). *Optik*, *24*, 521.
Schiske, P. (1967). Über die Korrektion des Farbfehlers durch gekreuzte magnetische und elektrische Quadrupollinsen. In *Communicated to the Marburg Meeting, Deutsche Gesellschaft für Elektronenmikroskopie*.
Schleich, F. (1960). *Proc. European Regional Conf. Electron Microscopy, Delft, 1960. Vol. 1*, p. 48. Delft: De Nederlanse Vereniging voor Electronenmicroscopie.
Schleich, F., Hoffmeister, D., Koops, H., & Lenz, F. (1968). *Optik*, *27*, 305.
Seeliger, R. (1948–49). *Optik*, *4*, 258.
Seeliger, R. (1949). *Optik*, *5*, 490.
Seeliger, R. (1951). *Optik*, *8*, 311.
Seeliger, R. (1953). *Optik*, *10*, 29.
Seman, O. I. (1955). *Trudy Instituta Fiziki i Astronomii Akademii Nauk Estonskoi SSR [Eesti NSV Teaduste Akadeemia Füüsika ja Astronoomia Instituudi Uurimused]*, *2*, 3 and 30.
Seman, O. I. (1958). *Uchenye Zapiski Rostovskogo Gosudarstvennogo Universiteta, Seriya Fizicheskaya*, *68*(8), 63.
Septier, A. (1957). *Comptes Rendus*, *245*, 2036.
Septier, A. (1958a). *Comptes Rendus*, *246*, 1835.
Septier, A. (1958b). *Comptes Rendus*, *246*, 1983.
Septier, A. (1961). Strong-focusing lenses. *Advances in Electronics and Electron Physics*, *14*, 85.
Septier, A. (1963). *Comptes Rendus*, *256*, 2325.
Septier, A. (1966). The struggle to overcome spherical aberration in electron optics. *Advances in Optical and Electron Microscopy*, *1*, 204.
Septier, A. (Ed.). (1967). *The focusing of charged particles*. New York: Academic Press.
Septier, A., & van Acker, J. (1960a). *Comptes Rendus*, *251*, 346.
Septier, A., & van Acker, J. (1960b). *Comptes Rendus*, *251*, 1750.
Septier, A., & van Acker, J. (1961). *Nuclear Instruments & Methods*, *13*, 335.
*Shpak, E. V., & Petrov, I. A. (1969). A doublet of quadrupole-octopole lenses with compensation of the spherical aberration of the line image. *Zhurnal Tekhnicheskoi Fiziki*, *39*, 934 (*Soviet Phys.–Tech. Phys.* (English Transl.) *14*, 699).
Shpak, E. V., & Yavor, S. Ya. (1964). *Zhurnal Tekhnicheskoi Fiziki*, *34*, 2003 (*Soviet Phys.–Tech. Phys.* (English Transl.) *9*, 1540).
Shpak, E. V., & Yavor, S. Ya. (1965). *Zhurnal Tekhnicheskoi Fiziki*, *35*, 947 (*Soviet Phys.–Tech. Phys.* (English Transl.) *10*, 727).
*Shpak, E. V., Yavor, S. Ya., Lyubchik, Y., & G. (1968). Optical diagrams of prism spectrometers with quadrupole lenses. *Nuclear Instruments & Methods*, *64*, 97.
Shukeilo, I. A. (1959). *Zhurnal Tekhnicheskoi Fiziki*, *29*, 1225 *Soviet Phys.–Tech. Phys.* (English Transl.) *4*, 1123).
**Siladi, M., Shpak, E. V., & Yavor, S. Ya. (1968). *Radiotekhnika i Elektronika*, *13*, 185 (*Radio Eng. Electron. Phys. (USSR)* (English Transl.) *13*, 159).
*Skarek, P. (1969). Beam optics and particle transport. *Fortschritte der Physik*, *17*, 277.
Steffen, K. G. (1965). *High energy beam optics*. New York: Wiley (Interscience).
Strashkevich, A. M. (1959). *The electron optics of electrostatic fields not possessing axial symmetry* (in Russian). Moscow: Fizmatgiz.

**Szilágyi, M.: the English transliteration of the Russian spelling of this name is Siladi or Silad'i.

Strashkevich, A. M. (1963). *Zhurnal Tekhnicheskoi Fiziki, 33*, 512 (*Soviet Phys.–Tech. Phys.* (English Transl.) *8*, 380).
Strashkevich, A. M. (1964). *Zhurnal Tekhnicheskoi Fiziki, 34*, 1401 (*Soviet Phys.–Tech. Phys.* (English Transl.) *9*, 1082).
Strashkevich, A. M. (1966). *The electron optics of electrostatic systems* (in Russian). Moscow -Leningrad: Energiya.
Sturrock, P. A. (1955). *Static and dynamic electron optics*. London and New York: Cambridge University Press.
Tadano, B. (1966). *Proc. AMU-ANL Workshop on High-voltage Electron Microscopy. Argonne Natl. Lab.*, p. 144. ANL-7275.
Taganov, I. N. (1966). *Radiotekhnika i Elektronika, 11*, 1329 (*Radio Eng. Electron. Phys. (USSR)* (English Transl.) *11*, 1160).
Taganov, I. N., & Kas'yankov, P. P. (1964). *Optiko-Mekhanicheskaya Promyshlennost', 31*(1), 14.
Taganov, I. N., & Kas'yankov, P. P. (1965). *Optiko-Mekhanicheskaya Promyshlennost', 32*(12), 21.
Taganov, I. N., & Kas'yankov, P. P. (1967). In G. I. Marchuk (Ed.), *Vychislitel'nye Sistemy: Chislennye Metody Rascheta Elektronno-Opticheskikh Sistem*, p. 11. Novosibirsk: Nauka.
Tanguy, P. (1967). *Comptes Rendus, B265*, 463.
Tanguy, P., & Joly, S. (1967). *Comptes Rendus, B265*, 532.
Thomson, M. G. R. (1966). *Proc. Internat. Cong. Electron Microscopy, 6th, Kyoto. Vol. 1*, p. 207. Tokyo: Maruzen.
Thomson, M. G. R. (1967). *The aberrations of quadrupole electron lenses*. Dissertation, Cambridge University.
Tretner, W. (1959). *Optik, 16*, 155.
Tschopp, P. (1961). Analogieverfahren zur Bestimmung von magnetischen Feldern in nichtlinearen nichtisotropen Medien. *Promotionsarbeit Nr 3180*. Zürich: ETH.
*Utochkin, B. A., Rodnov, Y. V., & Glupova, O. A. (1967). A sensitive high-speed oscillograph. *Pribory i Tekhnika Eksperimenta*, No. 3, 92. [*Instr. Exp. Tech. (USSR)* (English Transl.) *10*, 565]. The device includes a quadrupole triplet.
Varga, R. S. (1962). *Matrix iterative analysis*. Englewood Cliffs, New Jersey: Prentice-Hall.
Verster, J. L. (1963). *On the use of gauzes in electron optics. Thesis*. Delft: Technological Univ. published as (1963) Philips Res. Rept. *18*, 465.
Vosburgh, K. G. (1967). *The aberrations of a quadrupole quadruplet projector*. Thesis, Ithaca: Cornell University, Ithaca *(Rept. No. 646, Materials Sci. Center, Cornell University)*.
Weber, C. (1967a). *Analogue and digital methods for investigating electron-optical systems*. Eindhoven: Proefschrift. Philips Res. Rept. Suppl. 6.
Weber, C. (1967b). In A. Septier (Ed.), *Vol. I. Focusing of charged particles*, p. 45. New York: Academic Press.
Welton, T. A., & Allison, J. L. (1966). *Aberrations in a high-resolution electron optical system. ORNL-4082*.
Welton, T. A., & Allison, J. L. (1967). *Optimization of aberrations in a high-resolution electron optical system. ORNL-4230*.
*Worsham, R. E. (1969). Development of superconducting lenses for a 500 kV electron-microscope column. *Journal of Applied Physics, 40*, 2160.
*Worster, J. (1969). Integration of the paraxial ray equation for magnetic electron lenses, using cubic splines. *Journal of Physics D: Applied Physics, 2*, 1620.
Yavor, S. Ya. (1968). *The focusing of charged particles by means of quadrupole lenses* (in Russian). Moscow: Atomizdat.
Yavor, S. Ya., Fishkova, T. Ya., & Baranova, L. A. (1968). *Proc. European Regional Conf. Electron Microscopy, 4th, Rome. Vol. 1*, p. 169. Rome: Tipografia Poliglotta Vaticana.
Yavor, S. Ya., Fishkova, T. Ya., & Shpak, E. V. (1964). *Proc. European Regional Conf. Electron Microscopy, 3rd, Prague. Vol. A*, p. 35. Prague: Publ. House Czech. Acad. Sci.

Author index

Note: Page numbers followed by "*f*" indicate figures, "*t*" indicate tables, and "*ge*" indicate glossary terms.

H

J

K

L

M

N

O

P

R

Subject index

Note: Page numbers followed by "*f*" indicate figures, "*t*" indicate tables, and "*np*" indicate footnotes.